Principles of Modern Chemistry

EIGHTH EDITION

David Oxtoby
Pomona College

H.P. Gillis
University of California, Los Angeles

Laurie Butler
University of Chicago

Prepared by

Wade A. Freeman
University of Illinois, Chicago

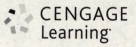
CENGAGE
Learning

Australia • Brazil • Mexico • Singapore • United Kingdom • United States

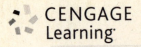ISBN: 978-1-305-09227-3

Cengage Learning
20 Channel Center Street
Boston, MA 02210
USA

Cengage Learning is a leading provider of customized learning solutions with office locations around the globe, including Singapore, the United Kingdom, Australia, Mexico, Brazil, and Japan. Locate your local office at: **www.cengage.com/global**.

Cengage Learning products are represented in Canada by Nelson Education, Ltd.

To learn more about Cengage Learning Solutions, visit **www.cengage.com**.

Purchase any of our products at your local college store or at our preferred online store **www.cengagebrain.com**.
Printed in the United States of America
Print Number: 01   Print Year: 2015

Contents

iv

FOREWORD

The eighth edition of *Principles of Modern Chemistry,* by David W. Oxtoby, H. P. Gillis, and Laura Butler presents a thorough introduction to university-level chemistry organized in six units:

I   Chapters 1 and 2 are introductory. They cover the classification of matter, evidence for the existence of atoms, the classical laws of chemical combination, the nuclear atom, the mole concept, empirical and molecular formulas, the writing of chemical equations and mass relationships in chemical reactions;

II   Chapters 3 through 8 give the classical description of chemical bonding and go on to discuss current quantum-mechanics-based understanding of ionic and molecular bonding and molecular structure. The unit then applies these ideas to bonding in organic and in inorganic compounds.

III   Chapters 9 through 11 cover kinetic-molecular theory as it explains the different states of matter and the nature and characterization of solutions;

IV   Chapters 12 through 17 cover thermodynamics and chemical equilibrium including acid-base equilibria, dissolution and precipitation equilibria, and electrochemistry;

V   Chapters 18 through 20 concern the rates of chemical and physical processes, nuclear chemistry, and molecular spectroscopy and photochemistry;

VI   Chapters 21 through 23 treat structure and bonding in the solid state and inorganic and organic polymeric materials.

All 23 chapters conclude with extensive problem sets. The problems are of different types. First come paired problems. Members of a pair treat the same concepts. Then come unpaired problems in two categories: *Additional Problems* and *Cumulative Problems*. *Additional Problems* provide further applications of the principles developed in the chapter. *Cumulative Problems* integrate material from the chapter with topics that appeared earlier in the book.

This Manual gives solutions for every odd-numbered problem in the text. Solutions for the even-numbered problems appear in the "Instructor's Manual," which is available to instructors from the publisher.

● **How to Study Using this Manual.** Success in a serious chemistry course requires the solving of problems, which are universally used to illustrate concepts and to test understanding. Obtain and learn to use a scientific calculator (one that accepts scientific notation, and computes trigonometric functions, logarithms, powers, and roots). Read the chapter. Then try some of the odd-numbered problems, devoting five to twenty minutes to an earnest attempt on each selected problem. The paired problems in each chapter of the text are organized according to the section headings in that chapter. Go back to the indicated section in the text or to the list of *Key Equations* at the end of the chapter and *write down* equations and definitions in a notebook as you grapple with the problem. Write your notes toward a solution in the same notebook. If you obtain an answer, check it against text Appendix G, which gives very brief answers to most of the odd-numbered, paired problems. If your answer is wrong or you cannot arrive at an answer, turn to the detailed solutions in this Manual for help. Just reading the solution in the Manual is not enough. Study the solution and try again to work the problem independently.

After completing several odd-numbered problems, take a rest. Later, go back and try the related even-numbered problems. The idea is to check and confirm your grasp of the material.

The next step is to move on to the *Additional Problems* and *Cumulative Problems*. These problems combine the concepts covered in the chapter in novel and sometimes challenging ways. If the paired problems are like a quiz, then the additional problems are like an examination. Select several problems in these groups. Try to solve them in writing in your notebook, referring freely to your notes and to the text. Use the lists of *Concepts and Skills* provided at the end of each chapter explicitly to identify exactly the ideas and techniques that are required to deal with the problem. Later, check your results on the odd-numbered problem against the solutions in this Manual.

The large number of problems in the text makes it unlikely that any one student will independently figure out solutions to them all. Do not however ignore unassigned or apparently duplicative problems. Instead read and study all of the solutions given in this Manual to confirm your understanding.

● **About Detailed Solutions.** Nearly all of the solutions in this Manual contain much more than just the answers. They include analysis of the chemical issues raised by the problem and references to tables, figures and equations in the text that furnish required data or that are needed in figuring out the answer. They also include step-by-step numerical details, suggest alternative methods of attack, and point out common pitfalls.

Tip. This heading indicates commentary on the problem or the method of soution or related information of interest and possible assistance.

The actual answers to questions appear in $\boxed{\text{boxes}}$ whenever appropriate to make them easier to spot.

Queries. Send queries about solutions to the problems and report difficulties in using this Manual directly to the author via Internet. The address is Wfreeman@uic.edu.

Wade A. Freeman

April, 2015

Chapter 1

The Atom in Modern Chemistry

Elements: The Building Blocks of Matter

1.1 Table salt consists of sodium chloride plus additives; the additives make it a heterogeneous mixture. Sodium chloride however is a substance (a compound, NaCl). Wood is a heterogeneous mixture; air (absent dust, pollen or fog) is a homogeneous mixture of several gases. Mercury is a substance (in fact, it is an elemental substance), and water is a substance (a compound, H_2O), but seawater is a homogeneous mixture of many compounds. Mayonnaise is a heterogeneous mixture of water with egg and oil. The latter two are themselves mixtures).

1.3 The chemist is writing about $\boxed{\text{substances.}}$ Mixtures of substances can be separated (resolved) into the individual components (elements and/or compounds) by physical means.

Indirect Evidence for the Existence of Atoms: Laws of Chemical Combination

1.5 According to the law of definite proportions (text page 1), a compound such as ascorbic acid has the same chemical composition regardless of source (as long as it is pure). Therefore, the ratio of carbon to oxygen in the sample isolated from lemons must equal the ratio in the sample synthesized in the laboratory. The laboratory sample contains 40.00 g of O for every 30.00 g of C. The mass of oxygen in the sample isolated from lemons is accordingly

$$m_{\text{oxygen}} = \left(\frac{40.00 \text{ g O}}{30.00 \text{ g C}}\right) \times 12.7 \text{ g C} = \boxed{16.9 \text{ g O}}$$

Tip. Notice the cancellation of the unit "g C".

1.7 **a)** 100.00 g of compound 1 contains 66.72 g of Si and 33.28 g of N. The desired quantities are just the ratio of these two masses

$$\text{compound 1} \quad \frac{66.72 \text{ g Si}}{33.28 \text{ g N}} = \boxed{\frac{2.005 \text{ g Si}}{1.000 \text{ g N}}} \qquad \text{compound 2} \quad \frac{60.06 \text{ g Si}}{39.94 \text{ g N}} = \boxed{\frac{1.504 \text{ g Si}}{1.000 \text{ g N}}}$$

b) To test the law of multiple proportions, compare the masses of Si associated with 1.000 g of N in the two compounds. The way to compare these two quantities is to form their ratio

$$\frac{2.005 \text{ g Si}/1.000 \text{ g N}}{1.504 \text{ g Si}/1.000 \text{ g N}} = 1.333$$

According to the law of multiple proportions, this ratio should equal a ratio of small whole numbers. Recognizing that $1.333 = 4/3$ (to four significant figures) confirms that the law of multiple proportions applies in this case.

1

Compound 1 has more Si per gram of N than compound 2; it is richer in Si by the factor 4/3. To obtain the formula of compound 1, take the formula of compound 2 (given as Si_3N_4) and multiply the subscript on the Si by this "richness factor." The result is Si_4N_4. When rewritten using the smallest possible whole-number subscripts, Si_4N_4 becomes "Si_1N_1". Subscripts equal to 1 are customarily omitted in chemical formulas, so the answer is $\boxed{SiN}$. Integral multiples (such as Si_2N_2, Si_3N_3) are also correct.

Tip. Memorize the decimal equivalents of small whole-number ratios such as 2/3, 3/4, 4/5, 5/8.

1.9   The problem asks for the *relative* number of atoms of oxygen combined with a given mass of vanadium in four compounds. "Relative" means "take a ratio," that is, divide. The first compound contains 23.90 g of O for every 76.10 g of V. Take a ratio of these two masses

$$\frac{23.90 \text{ g O}}{76.10 \text{ g V}} = \frac{0.3141 \text{ g O}}{1 \text{ g V}} \quad \text{for cmpd 1}$$

Compute similar ratios for the second, third and fourth compounds in the table

$$\frac{0.4710 \text{ g O}}{1 \text{ g V}} \quad \text{for cmpd 2} \qquad \frac{0.6281 \text{ g O}}{1 \text{ g V}} \quad \text{for cmpd 3} \qquad \frac{0.7851 \text{ g O}}{1 \text{ g V}} \quad \text{for cmpd 4}$$

The increasing size of the ratios indicates an increasing proportion of oxygen moving down the series of compounds from 1 to 4. Next, compare the ratios. For example, divide the second by the first

$$\frac{0.4710 \text{ g O/ g V}}{0.3141 \text{ g O/ g V}} = 1.500$$

This means that the second compound is 1.500 times richer in oxygen than the first. Compare the third and fourth compounds to the first in the same way

$$\frac{0.6281 \text{ g O/ g V}}{0.3141 \text{ g O/ g V}} = 2.000 \qquad \frac{0.7851 \text{ g O/ g V}}{0.3141 \text{ g O/ g V}} = 2.500$$

The preceding ratios hold for any mass of vanadium (note that the units cancelled out). The relative numbers of atoms of oxygen for any given mass of vanadium in these four compounds are therefore 1 to $1\frac{1}{2}$ to 2 to $2\frac{1}{2}$. This is the same as $\boxed{2 \text{ to } 3 \text{ to } 4 \text{ to } 5}$.

1.11   **a)** The law of combining volumes states that at constant temperature and pressure the volumes of gases combining to form a substance are in the ratio of small whole numbers. The problem describes the reverse of combination, namely the breakdown of a compound into two gases. Still, a law of *un*-combining volumes clearly must apply. Also, it is reasonable to assume that conditions of temperature and pressure are the same at the two electrodes. Therefore, the ratio of the number of particles of gaseous hydrogen to the number of particles of gaseous oxygen equals the ratio of the volume of gaseous hydrogen to the volume of gaseous oxygen. This ratio equals 14.4 mL/14.4 mL or 1.00 : 1.00. The particles of gaseous hydrogen (H_2 molecules) contain the same number of hydrogen atoms as the particles of gaseous oxygen (O_2 molecules) contain oxygen atoms. Therefore, the simplest chemical formula is H_1O_1, or $\boxed{HO}$. The experimental facts rule out the answer H_2O.

b) All formulas in which H and O have equal subscripts are also correct because the experiment reveals only the relative numbers of atoms of the two elements. Complete decomposition of *any* compound having a formula of the type H_nO_n gives gaseous H_2 and gaseous O_2 in the same 1 : 1 ratio of volumes (as long as the volumes are measured at the same temperature and pressure).

1.13   From the mention of "pure nitrogen dioxide", it is clear that the reaction between the dinitrogen oxide (N_2O) and oxygen (O_2) generates nitrogen dioxide (NO_2) exclusively and also goes to completion (does not stop as long as both reactants are available). According to the law of combining volumes, the volumes of gases taking part in this reaction are in the ratio of small whole numbers. These small whole numbers are just the ratios obtained by balancing the chemical equation that represents the reaction

$$2\,N_2O + 3\,O_2 \longrightarrow 4\,NO_2$$

Thus $\boxed{2.0 \text{ L of } N_2O}$ and $\boxed{3.0 \text{ L of } O_2}$ react to form 4.0 L of NO_2.

Tip. In this reaction, 2.0 L of one gas combines with 3.0 L of a second gas to give 4.0 L of a third gas. Clearly, no "principle of conservation of volume" exists.

The Physical Structure of Atoms

1.15   An electron speeds into the gap between two parallel plates. The plates are electrically charged and maintain an electric field of 200 volts per meter V m^{-1} across the gap. The statement of the problem points out that one V m^{-1} equals one J C^{-1} m^{-1} (joule per coulomb per meter). This unit also equals one N C^{-1}, because a joule (J) is defined as a newton-meter (N·m).

The fast-moving electron spends only a brief period of time t between the plates, which are 0.1 m long

$$t = \frac{0.10 \text{ m}}{8.0 \times 10^6 \text{ m s}^{-1}} = 1.25 \times 10^{-8} \text{ s}$$

During this time the electron experiences a constant unbalanced electric force that accelerates it toward the positively charged plate. This accelerating force is the product of the magnitude of the electric field E and the charge on the electron, which is symbolized e. Inserting this expression into Newton's second law of motion $F = ma$ and solving for a gives

$$a = \frac{Ee}{m_e} = \frac{(200 \text{ N C}^{-1})(1.602 \times 10^{-19} \text{ C})}{9.109 \times 10^{-31} \text{ kg}} = 3.5174 \times 10^{13} \text{ m s}^{-2}$$

The mass and charge of the electron come from the table on the inside back cover of the text. The cancellation of units uses the fact that a newton equals a kilogram meter per second squared.

The constant electric force pushes the electron downward, as shown in text Figure 1.11. The amount of the deflection depends on the duration of the passage of the electron through the space between the plates. The deflection is

$$\ell = \frac{1}{2}\,at^2 = \frac{1}{2}\,(3.5178 \times 10^{13} \text{ m s}^{-2})(1.25 \times 10^{-8} \text{ s})^2 = 2.748 \times 10^{-3} \text{ m}$$

(The preceding formula also gives the distance that an object falls during time t under the influence of the constant gravitational force exerted by the earth.) The small deflection ℓ from the original path grows larger after the electron exits the gap, as shown in text Figure 1.11. The magnification ratio, defined as S/D in the caption to this figure, is

$$\frac{S}{D} = \frac{2L}{\ell}$$

where L is the distance from the centerline of the two plates to the plane of a detecting screen and S is the actual magnification. In this case

$$S = \frac{2(0.95) \text{ m}}{0.1 \text{ m}}(2.748 \times 10^{-3} \text{ m}) = \boxed{0.052 \text{ m}}$$

Mass Spectrometry, Isotopes, and the Measurement of Relative Mass

1.17   The relative atomic mass of naturally-occurring Si is the *weighted* mean (weighted average) of the relative atomic masses of the three isotopes listed. What does it mean to *weight* an average? The *un*-weighted mean of the relative masses of the three isotopes would be

$$\text{un-weighted mean} = \frac{1}{3}(27.97693) + \frac{1}{3}(28.97649) + \frac{1}{3}(29.97376)$$

Weighting corresponds to replacing the $\frac{1}{3}$'s in this expression with values telling each isotope's *true* contribution to the total. These values are the abundances. Fractional abundances (which add up to exactly 1.00) rather than percent abundances (which add up to 100.0) must be used

$$\text{weighted mean} = \frac{9221}{10000}(27.97693) + \frac{470}{10000}(28.97649) + \frac{309}{10000}(29.97376) = \boxed{28.086}$$

1.19 The relative atomic mass of natural boron is the weighted mean of the relative masses of the two isotopes

$$A_{\text{boron}} = A_{^{10}B}\,p_{^{10}B} + A_{^{11}B}\,p_{^{11}B}$$

where the A's represent relative atomic masses and the p's represent fractional abundances. With one exception, all of the quantities in this equation are known:

$$10.811 = (10.013)(0.1961) + A_{^{11}B}(0.8039) \quad \text{Solving gives} \quad A_{^{11}B} = \boxed{11.01}$$

1.21 **a)** The atomic number Z of Pu equals 94. Hence, an atom of Pu has 94 protons in its nucleus. An atom of ^{239}Pu has a mass number A of 239, that is, a total of 239 protons and neutrons in its nucleus. Since the neutron number N equals $A - Z$, the atom has 145 neutrons. The requested ratio is 145/94, which equals $\boxed{1.54}$.

b) The Pu atom is electrically neutral. This means that its extranuclear electrons contribute sufficient negative charge exactly to balance the positive charge of the 94 protons in its nucleus. The charge on the electron equals the charge on the proton in magnitude, so a Pu atom has $\boxed{94 \text{ electrons}}$.

1.23 The atomic number of americium is 95; americium has $\boxed{95 \text{ protons}}$ in its nucleus. In the neutral atom there are also exactly $\boxed{95 \text{ electrons}}$ because the negative charge of the electrons balances the positive charge of the protons. Of the 241 nucleons, those that are not protons are neutrons. There are accordingly $\boxed{146 \text{ neutrons}}$.

The Mole: Counting Molecules by Weighing

1.25 Use Avogadro's number (the Avogadro constant) as follows

$$m_{\text{I atom}} = 1 \text{ I atom} \times \left(\frac{126.90447 \text{ g}}{1 \text{ mol I}}\right)\left(\frac{1 \text{ mol I}}{6.0221418 \times 10^{23} \text{ atom I}}\right) = \boxed{2.1072979 \times 10^{-22} \text{ g}}$$

Tip. The preceding is equivalent to

$$m_{\text{I atom}} = 1 \text{ I atom} \times \frac{126.90447 \text{ g mol}^{-1}}{6.0221418 \times 10^{23} \text{ atom mol}^{-1}} = 2.1072979 \times 10^{-22} \text{ g}$$

Also, the known mass of a single atom of ^{12}C (computed in text Section 2.1) can be used together with the relative atomic masses of ^{12}C and I

$$m_{\text{I atom}} = m_{^{12}\text{C atom}}\left(\frac{126.90447}{12.000000}\right) = 1.9926465 \times 10^{-23} \text{ g}\left(\frac{126.90447}{12.000000}\right)$$
$$= 2.1072979 \times 10^{-22} \text{ g}$$

1.27 Use the relative atomic masses from the inside back cover of the text.
a) P_4O_{10}: $4(30.974) + 10(15.999) = \boxed{283.886}$.
b) BrCl: $79.904 + 35.453 = \boxed{115.357}$.
c) $Ca(NO_3)_2$: $40.08 + 2\left(14.01 + 3(16.00)\right) = \boxed{164.09}$.
d) $KMnO_4$: $39.098 + 54.938 + 4(15.999) = \boxed{158.032}$.
e) $(NH_4)_2SO_4$: $2\left(14.007 + 4(1.0079)\right) + 32.066 + 4(15.999) = \boxed{132.14}$.

1.29   Set up the computation as a string of unit-factors

$$m_{Au} = 80 \text{ yr} \times \frac{365 \text{ d}}{1 \text{ yr}} \left(\frac{24 \text{ h}}{1 \text{ d}}\right) \left(\frac{3600 \text{ s}}{1 \text{ h}}\right) \left(\frac{1 \text{ atom Au}}{1 \text{ s}}\right) \left(\frac{1 \text{ mol Au}}{6.022 \times 10^{23} \text{ atom Au}}\right) \left(\frac{197 \text{ g Au}}{1.00 \text{ mol Au}}\right)$$

$$= \boxed{8.3 \times 10^{-13} \text{ g Au}}$$

Advanced microbalances can detect as little as about 10^{-10} g. Even after a lifetime of counting, the mass of the counted atoms remains far too small to detect.

1.31   According to the formula, 51 atoms of all kinds are contained in a single molecule of vitamin A. Use this with a series of unit-factors to find out how many atoms there are in 1.000 mol of vitamin A

$$N_{atoms} = 1.000 \text{ mol vit A} \times \left(\frac{N_A \text{ molecules}}{1 \text{ mol vit A}}\right) \left(\frac{51 \text{ atoms}}{1 \text{ molecule}}\right) = 51.00 \, N_A \text{ atoms}$$

Now compute the amount (in moles) of vitamin A_2 that contains this number of atoms

$$n_{A_2} = 51.00 \, N_A \text{ atoms} \times \left(\frac{1 \text{ molecule } A_2}{49 \text{ atoms}}\right) \left(\frac{1 \text{ mol } A_2}{N_A \text{ molecules}}\right) = \boxed{1.041 \text{ mol } A_2}$$

Tip. The N_A's cancel out. The numerical value of Avogadro's number is not needed to complete the problem, just the concept that such a number exists.

1.33   The volume of a "flask" of mercury equals the volume per unit mass of mercury multiplied by the mass of mercury contained in a flask. The volume per unit mass is the reciprocal of the density (the density divided into one)

$$V_{flask} = 34.5 \times 10^3 \text{ g} \times \left(\frac{1 \text{ cm}^3 \text{ Hg}}{13.6 \text{ g Hg}}\right) \left(\frac{1 \text{ L}}{1000 \text{ cm}^3}\right) = \boxed{2.54 \text{ L}}$$

1.35   Use unit-factors to progress from volume of Al_2O_3 to the number of atoms of Al. The correct answer must be on the order of 10^{23} atoms because the amount of corundum is on the ordinary human scale

$$N_{Al} = 15.0 \text{ cm}^3 \text{ Al}_2O_3 \times \left(\frac{3.97 \text{ g Al}_2O_3}{1 \text{ cm}^3 \text{ Al}_2O_3}\right) \left(\frac{1 \text{ mol Al}_2O_3}{101.96 \text{ g Al}_2O_3}\right)$$

$$\times \left(\frac{6.022 \times 10^{23} \text{ Al}_2O_3 \text{ units}}{1 \text{ mol Al}_2O_3}\right) \left(\frac{2 \text{ atoms Al}}{1 \text{ Al}_2O_3 \text{ unit}}\right) = \boxed{7.03 \times 10^{23} \text{ atoms Al}}$$

Tip. Unit-factors can be "flipped over" (numerator and denominator exchanged) at will. To make progress with a chain of unit-factors, arrange each one so the desired unit is in the top (numerator) and the unit to be canceled away is in the bottom (denominator). Note the rather creative last unit-factor in the preceding equation.

ADDITIONAL PROBLEMS

1.37   **a)** Soft-wood chips: wood is a $\boxed{\text{mixture}}$ of many substances. Water: H_2O is a $\boxed{\text{compound}}$. Sodium hydroxide: NaOH is a $\boxed{\text{compound}}$.

b) Because the iron vessel was sealed, nothing was able to enter or escape, including gases. Therefore, exactly the original mass remains contained in the vessel—no more, no less. The total mass is $17.2 + 150.1 + 22.43 = \boxed{189.7 \text{ kg}}$.

1.39   The density of the nucleus of ^{127}I equals its mass divided by its volume. The problem gives the nuclear mass explicitly and provides a route to the nuclear volume. Start with the formula for the volume of a sphere in terms of its radius r and substitute with the formula for the radius of a nucleus in terms of the mass number A

$$V_{^{127}I} = \frac{4}{3}\pi r^3 = \frac{4}{3}\pi(kA^{\frac{1}{3}})^3 = \frac{4}{3}\pi k^3 A$$

$$= \frac{4}{3}\pi(1.3 \times 10^{-13} \text{ cm})^3(127) = 1.17 \times 10^{-36} \text{ cm}^3$$

The density of the iodine nucleus then is

$$\rho_{^{127}I} = \frac{m_{^{127}I}}{V_{^{127}I}} = \frac{2.1 \times 10^{-22} \text{ g}}{1.17 \times 10^{-36} \text{ cm}^3} = \boxed{1.8 \times 10^{14} \text{ g cm}^{-3}}$$

This is billions of times more dense than solid iodine!

1.41   Dalton's postulates were:

1. Matter consists of indivisible atoms. Chemists now know that atoms are not indivisible, but can lose one or more (or all) of their electrons to give species (ions) having chemical properties quite different from the properties of the neutral atoms. Moreover, some elements (such as uranium and radium) are radioactive: the nuclei of their atoms spontaneously emit or absorb subatomic particles, a process that results in new, chemically distinct, atoms.

2. All atoms of a given chemical element are identical in mass and in all other properties. The existence of isotopes is in direct contradiction to this postulate. Different atoms of the same chemical element *can* have different masses. In fact, the majority of the elements have two or more naturally occurring isotopes. Isotopes have virtually identical chemical properties, but isotope effects, such as changes in the rates of reactions, have been observed.

3. Different chemical elements have different kinds of atoms, and in particular, such atoms have different masses. This statement (so far) needs no modification or extension.

4. Atoms are indestructible and retain their identity in chemical reactions. Atoms are not indestructible under all circumstances. They can be split apart (or fused together) at the nuclear level to give new kinds of atoms in particle accelerators. No instances of atoms changing their identity in chemical reactions are however known.

5. The formation of a compound from its elements occurs through combining atoms of unlike elements in small whole-number ratios. Certain solid compounds have compositions that vary within a range. They are non-stoichiometric compounds.[1] The law of definite proportions is strictly true for gaseous and liquid compounds but not for solid compounds.

1.43   The sample contains equal amounts of boron and chlorine when compared in terms of their chemical amounts (number of moles):

$$n_B = \frac{0.664 \text{ g B}}{10.811 \text{ g mol}^{-1}} = 0.0614 \text{ mol B} \qquad n_{Cl} = \frac{(2.842 \text{ g} - 0.664 \text{ g}) \text{ Cl}}{35.453 \text{ g mol}^{-1}} = 0.0614 \text{ mol Cl}$$

This means that the empirical formula of the compound is BCl. To obtain the molecular formula, write

$$n_{Cl_2} = 0.0614 \text{ mol Cl} \times \left(\frac{1 \text{ mol Cl}_2}{2 \text{ mol Cl}}\right) = 0.0307 \text{ mol Cl}_2$$

From Avogadro's hypothesis, the chemical amount of the gaseous compound in the original sample was

$$n_{compound} = 0.0307 \text{ mol Cl}_2 \times \left(\frac{0.153 \text{ L compound}}{0.688 \text{ L Cl}_2}\right) = 0.00683 \text{ mol compound}$$

---

[1]See text Section 21.4.

This allows computation of the molar mass

$$\mathcal{M}_{\text{compound}} = \frac{2.842 \text{ g}}{0.00683 \text{ mol}} = 416.3 \text{ g mol}^{-1}$$

The molar mass corresponding to the empirical formula BCl is only 46.264 g mol^{-1}.   The ratio 416.3/46.264 equals 9.00. Each molecule contains 9 atoms of boron and 9 atoms of chlorine .

Chapter 2

Chemical Formulas, Chemical Equations, and Reaction Yields

Chemical Formula and Percentage Composition

2.1 The mass percentage of an element in any compound is the mass contributed by that element to some sample of the compound divided by the mass of the sample and then multiplied by 100%. Suppose that you have a sample of $ClF_2O_2PtF_6$ amounting to exactly 1 mol. This is 414.52 g of the compound, a value that is obtained by multiplying the molar masses of the various elements in the formula by their subscripts in the formula and adding the results (as in problem **1.27**). Now, set up and carry out the following computations

$$\text{for Cl:} \quad \frac{1 \text{ mol Cl}}{414.52 \text{ g compound}} \quad \frac{35.453 \text{ g Cl}}{1 \text{ mol Cl}} \times 100\% = \boxed{8.553\% \text{ Cl}}$$

$$\text{for F:} \quad \frac{8 \text{ mol F}}{414.52 \text{ g compound}} \quad \frac{18.998 \text{ g F}}{1 \text{ mol F}} \times 100\% = \boxed{36.67\% \text{ F}}$$

$$\text{for O:} \quad \frac{2 \text{ mol O}}{414.52 \text{ g compound}} \quad \frac{15.999 \text{ g 0}}{1 \text{ mol O}} \times 100\% = \boxed{7.720\% \text{ O}}$$

$$\text{for Pt:} \quad \frac{1 \text{ mol Pt}}{414.52 \text{ g compound}} \quad \frac{195.08 \text{ g Pt}}{1 \text{ mol Pt}} \times 100\% = \boxed{47.06\% \text{ Pt}}$$

Although Pt ties with Cl as the least prevalent element in the compound on the basis of number of atoms, it is by far the most prevalent on the basis of mass.

Tip. The repetition of F in the formula $ClF_2O_2PtF_6$ means that the compound contains F in different chemical settings (some bonded to Pt and the rest bonded to Cl, perhaps). This does not matter in obtaining mass percentages, so it was all right to replace the formula $ClF_2O_2PtF_6$ with ClF_8O_2Pt.

2.3 The task is to arrange four compounds in increasing order of their percentage of hydrogen by mass. The mass percentage of H in each compound can be calculated,[1] and the resulting numbers used to get the required order. The results are 11.19% for H_2O, 15.38% for $C_{12}H_{26}$, 9.742% for N_4H_6, and 12.68% for LiH, Therefore

$$\boxed{N_4H_6 \; < \; H_2O \; < \; LiH \; < \; C_{12}H_{26}}$$

Tip. A (slightly) easier method is to settle for estimates of the mass percentage of H. Get the estimates by adding up the masses of the non-H atoms and dividing by the number of H's. Exact arithmetic is not necessary. Thus, $C_{12}H_{26}$ has $144/26 \approx 6$ units of non-H mass per hydrogen atom,

---

[1] As in problem **2.13**.

8

but LiH has $7.9/1 \approx 7.9$, H_2O has $16/2 \approx 8$, and N_4H_6 has $56/6 \approx 9$. The compound that has the *least* amount of non-H mass per hydrogen atom is the richest in hydrogen.

2.5 Calculate the fraction (not percentage) by mass of hydrogen (H) in the compound C_4H_{10} (butane) by the method of problem **2.13** and multiply the result by 0.0130, the fraction of butane in "Q-gas". This fraction-of-a-fraction method works because helium, the other component of Q-gas; contains no hydrogen

$$f_H = \left(\frac{10 \times (1.008) \text{ g H}}{(4 \times 12.011) + (10 \times 1.008) \text{ g butane}} \right) \times \left(\frac{0.0130 \text{ g butane}}{1 \text{ g Q gas}} \right) = \frac{0.00225 \text{ g H}}{1 \text{ g Q-gas}}$$

Multiply by 100% to obtain the desired percentage: $\boxed{0.225\% \text{ H}}$ by mass.

2.7 The empirical formula of zinc phosphate is the smallest whole-number ratio of the number of atoms of the different kinds (or of moles of atoms of the different kinds) in the compound. Calculate the number of moles of the three elements from the given masses

$$n_O = 16.58 \times 10^{-3} \text{ g O} \times \left(\frac{1 \text{ mol O}}{15.999 \text{ g O}} \right) = 1.036 \times 10^{-3} \text{ mol O}$$

$$n_P = 8.02 \times 10^{-3} \text{ g P} \times \left(\frac{1 \text{ mol P}}{30.97 \text{ g O}} \right) = 2.59 \times 10^{-4} \text{ mol P}$$

$$n_{Zn} = 25.40 \times 10^{-3} \text{ g Zn} \times \left(\frac{1 \text{ mol Zn}}{65.409 \text{ g Zn}} \right) = 3.883 \times 10^{-4} \text{ mol Zn}$$

Divide through all of the results by the smallest, which is n_P. Doing this allows an easy comparison of the relative number of moles of each

$$\frac{n_O}{n_P} = \frac{1.036 \times 10^{-3} \text{ mol}}{2.59 \times 10^{-4} \text{ mol}} = 4.00 \quad \frac{n_{Zn}}{n_P} = \frac{3.883 \times 10^{-4} \text{ mol}}{2.59 \times 10^{-4} \text{ mol}} = 1.50 \quad \frac{n_P}{n_P} = \frac{2.59 \times 10^{-4} \text{ mol}}{2.59 \times 10^{-4} \text{ mol}} = 1.00$$

The three elements are present in the molar ratio $4 : 1.5 : 1$. This corresponds to the formula $O_4Zn_{1.5}P_1$. The preferred format of chemical formulas avoids fractional subscripts. To meet this preference, simply clear the fractions by multiplying all subscripts by 2. The result is $\boxed{O_8Zn_3P_2}$.

Tip. The formula of zinc phosphate is customarily written $Zn_3(PO_4)_2$ to suggest the way that the atoms are bonded.

2.9 The percentages of Fe and Si determined by analysis of the single crystalline grain in the fulgurite are correct for any amount of the compound. Compute the number of moles of Fe and Si in an arbitrary 100.0 g sample

$$n_{Fe} = 100.0 \text{ g compound} \times \left(\frac{46.01 \text{ g Fe}}{100 \text{ g compound}} \right) \left(\frac{1 \text{ mol Fe}}{55.845 \text{ g Fe}} \right) = 0.8239 \text{ mol Fe}$$

$$n_{Si} = 100.0 \text{ g compound} \times \left(\frac{53.99 \text{ g Si}}{100 \text{ g compound}} \right) \left(\frac{1 \text{ mol Si}}{28.086 \text{ g Si}} \right) = 1.922 \text{ mol Si}$$

Dividing 1.922 mol by 0.8239 mol gives the ratio of the number of moles of Si to the number of moles of Fe. It is $2.333 : 1$. This is expressed by the formula $FeSi_{2.333}$. Multiplying both subscripts by 3 to eliminate fractions gives the empirical formula $\boxed{Fe_3Si_7}$.

Tip. It is instructive to confirm that the answer is the same using some other arbitrary mass of compound.

Tip. The wrong answer $FeSi_2$ is fairly common. It comes from reckless rounding off: 2.333 differs a *lot* from 2.00.

2.11   Consider the two cases separately. 100.000 g of the first compound contains 90.745 g of Ba and, by subtraction, 9.255 g of N. The numbers of moles of the two elements are

$$n_{Ba} = 90.745 \text{ g Ba} \times \left(\frac{1 \text{ mol Ba}}{137.33 \text{ g Ba}} \right) = 0.66078 \text{ mol Ba}$$

$$n_{N} = 9.255 \text{ g N} \times \left(\frac{1 \text{ mol N}}{14.007 \text{ g N}} \right) = 0.6607 \text{ mol N}$$

The Ba and N are present in equal numbers of moles, that is, in a 1 : 1 molar ratio. The empirical formula is $\boxed{\text{BaN}}$.

For the second compound: 100.000 g of it contains 93.634 g of Ba and, by subtraction, 6.366 g of N. Do the same kind of calculation

$$n_{Ba} = 93.634 \text{ g Ba} \times \left(\frac{1 \text{ mol Ba}}{137.33 \text{ g Ba}} \right) = 0.68182 \text{ mol Ba}$$

$$n_{N} = 6.366 \text{ g N} \times \left(\frac{1 \text{ mol N}}{14.007 \text{ g N}} \right) = 0.4545 \text{ mol N}$$

Dividing both of these numbers of moles by the smaller establishes that the Ba and N are present in a 1.500 : 1 molar ratio. Accordingly, the empirical formula is $\boxed{\text{Ba}_3\text{N}_2}$.

2.13   **a)** Burning the compound in oxygen gives 0.692 g of H_2O and 3.381 g of CO_2. Determine the masses of elemental H and C in these amounts of H_2O and CO_2

$$m_{H} = 0.692 \text{ g } H_2O \times \frac{2.016 \text{ g H}}{18.015 \text{ g } H_2O} = \boxed{0.0774 \text{ g H}}$$

$$m_{C} = 3.381 \text{ g } CO_2 \times \frac{12.01 \text{ g C}}{44.01 \text{ g } CO_2} = \boxed{0.9226 \text{ g C}}$$

b) The mass of C in the CO_2 and the mass of H in the H_2O add up to 1.000 g. The compound therefore contains $\boxed{\text{no other elements}}$.

c) The compound is $\boxed{7.74\% \text{ H}}$ and $\boxed{92.26\% \text{ C}}$ by mass.

d) To determine the empirical formula of the compound, convert the masses of C and H in the sample to numbers of moles and determine their ratio

$$n_{H} = 0.0774 \text{ g H} \times \left(\frac{1 \text{ mol H}}{1.008 \text{ g H}} \right) = 0.0768 \text{ mol H}$$

$$n_{C} = 0.9226 \text{ g C} \times \left(\frac{1 \text{ mol C}}{12.0107 \text{ g C}} \right) = 0.07681 \text{ mol C}$$

The C and H are present in a 0.07681/0.0768 molar ratio, which is a 1.00/1.00 molar ratio. The empirical formula is therefore $\boxed{\text{CH}}$.

2.15   The 1-L sample of fluorocarbon has a mass of 8.93 g, but the 1-L sample of fluorine has a mass of only 1.70 g under the same conditions of temperature and pressure. It follows by Avogadro's principle (discussed in text Chapter 1) that the molecules of the fluorocarbon are 8.93/1.70 times more massive than those of fluorine (F_2). The relative molecular mass of F_2 equals 38.0. Therefore

$$\text{Relative molecular mass of fluorocarbon} = 38.0 \times \left(\frac{8.93}{1.70} \right) = 200.$$

A relative molecular mass of 200. requires four CF_2 units (each of which contributes a relative mass of $12 + 2 \cdot 19 = 50$). Hence the molecular formula of the fluorocarbon is $(CF_2)_4$, or $\boxed{C_4F_8}$.

2.17   **a)** Imagine 1-L samples of the unknown gaseous compound and of gaseous oxygen stored side-by-side at the same pressure and temperature. The mass of the unknown exceeds the mass of the O_2 by a factor of 1.94. The sample of unknown is confined at the same temperature and pressure as the O_2, so it contains the same number of molecules as the sample of the O_2 (Avogadro's principle). Therefore, the mass of each molecule in the unknown sample must be 1.94 times larger than the mass of an O_2 molecule. An O_2 molecule has a relative molecular mass of 32.0; the relative molecular mass of the unknown is $1.94 \times 32.0 = \boxed{62.1}$.

b) The unknown compound consists of hydrogen (H) and one other element. Burning 1.39 g of it in oxygen gives 1.21 g of water. This water contains all of the H that was present in the unknown before it was burned. Compute the number of moles of H in this water

$$n_H = 1.21 \text{ g } H_2O \times \left(\frac{1 \text{ mol } H_2O}{18.0153 \text{ g } H_2O} \right) \left(\frac{2 \text{ mol H}}{1 \text{ mol } H_2O} \right) = 0.1343 \text{ mol H}$$

Now figure out the number of moles of the unknown in the 1.39 g sample that was burned

$$n_{unknown} = 1.39 \text{ g unknown} \times \left(\frac{1 \text{ mol unknown}}{62.08 \text{ g unknown}} \right) = 0.0224 \text{ mol}$$

Compare this to the number of moles of H that was in the sample by dividing the number of moles of H by the number of moles of unknown[2]

$$\frac{n_H}{n_{unknown}} = \frac{0.1343 \text{ mol H}}{0.0224 \text{ mol unknown}} = \frac{6.00 \text{ mol H}}{1 \text{ mol unknown}}$$

There is 6.00 mol of H per mole of the unknown, and therefore there are $\boxed{6}$ atoms of H per molecule of unknown.

c) The unknown contains H and one other element, call it Z. The relative molecular mass of the unknown is 62.1. The maximum relative atomic mass of Z is $62.08 - 6(1.00794) = \boxed{56.0}$. This is the relative atomic mass of Z *if* one atom of Z is present per molecule of unknown, that is, if the molecular formula of the unknown is ZH_6.

d) The answer is $\boxed{\text{yes,}}$ other values of the relative atomic mass of Z are possible. The molecules of the unknown might contain more than one atom of Z. Two atoms of Z would imply a relative atomic mass of 28.0; three atoms of Z would imply a relative atomic mass of 18.7. As the subscript of Z gets larger, the relative atomic mass of Z gets smaller as the following table shows

Formula	Rel. Atomic Mass of Z	Identity of Z
Z_1H_6	56.0	Fe?
Z_2H_6	28.0	Si?
Z_3H_6	18.7	F?
Z_4H_6	14.0	N?
Z_5H_6	11.2	B?
Z_6H_6	9.33	Be?
Z_9H_6	6.23	Li?
$Z_{14}H_6$	4.00	He?
$Z_{56}H_6$	1.00	H?

To identify element Z, compare the relative atomic masses in the table to those of the known elements. If the molecular formula of the unknown compound is Z_2H_6, then the relative atomic mass of element Z is quite close to that of $\boxed{\text{Si}}$ (28.08); if the formula is Z_4H_6 , then it is quite close to that of $\boxed{\text{N}}$

---

[2]Once again, comparing two things means dividing one by the other.

(14.01). The other formulas in the table give relative atomic masses differing substantially from those of authentic elements. It is true that a subscript of 56 gives Z a relative atomic mass of 1.00, which is very close to the relative atomic mass of H, but then the unknown would be "H_{62}," which is not a binary compound. A subscript of 14 makes Z's atomic mass come out to equal 4.00, which is close to the atomic mass of He. . . but compounds of helium are unknown.

e) The compound is either $\boxed{Si_2H_6}$ (disilane, relative molecular mass 62.2196) or $\boxed{N_4H_6}$ (tetrazane, relative molecular mass 62.0756). Both exist, but Si_2H_6 is more stable.

Tip. The relative molecular mass of the unknown is 62.1. This value has three significant digits because it derives from the relative density 1.94, which is an experimental result that is reported to three significant digits in the problem. The last digit in experimental results such as these is uncertain by *at least* ± 1 and might be uncertain by as much as ± 3. The true relative molecular mass of the unknown is therefore most likely between 62.0 and 62.2 and might be as high as 62.4 or as low as 59.8. The data are just not precise enough to tell whether the compound is disilane (relative molecular mass 62.22) or tetrazane (relative molecular mass 62.08).

Writing Balanced Chemical Equations

2.19 Balance the equations by inspection. For example, in part **a)**, assign 1 as the coefficient of NH_3. This obliges a coefficient on $\frac{1}{2}$ for N_2 because it takes $\frac{1}{2}$ mol of N_2 to furnish the 1 mol of N that is signified in "1 NH_3." Similarly, "1 NH_3" obliges a coefficient of $\frac{3}{2}$ for the H_2 because $\frac{3}{2}$ mol of H_2 contains the same number of H atoms as 1 mol of NH_3. The answers that follow clear the fractions from all the sets of coefficients by multiplying the members of each set by a suitable integer.

a) $3\,H_2 + N_2 \longrightarrow 2\,NH_3$

b) $2\,K + O_2 \longrightarrow K_2O_2$

c) $PbO_2 + Pb + 2\,H_2SO_4 \longrightarrow 2\,PbSO_4 + 2\,H_2O$

d) $2\,BF_3 + 3\,H_2O \longrightarrow B_2O_3 + 6\,HF$

e) $2\,KClO_3 \longrightarrow 2\,KCl + 3\,O_2$

f) $CH_3COOH + 2\,O_2 \longrightarrow 2\,CO_2 + 2\,H_2O$

g) $2\,K_2O_2 + 2\,H_2O \longrightarrow 4\,KOH + O_2$

h) $3\,PCl_5 + 5\,AsF_3 \longrightarrow 3\,PF_5 + 5\,AsCl_3$

Mass Relationships in Chemical Reactions

2.21 **a)** According to the balanced equation $Mg + 2\,HCl \longrightarrow H_2 + MgCl_2$, the reaction produces 1 mol of H_2 for every 1 mol of Mg consumed. Diatomic hydrogen has a relative molecular mass of $2 \times 1.00794 = 2.01588$, and Mg has a relative atomic mass of 24.305. Therefore, "1 mol Mg $\longrightarrow$ 1 mol H_2" implies "24.305 g Mg $\longrightarrow$ 2.01594 g H_2". This fact provides a unit-factor to compute the mass of Mg that yields 1.000 g of H_2

$$m_{Mg} = 1.000 \text{ g } H_2 \times \left(\frac{24.305 \text{ g Mg}}{2.01594 \text{ g } H_2} \right) = \boxed{12.06 \text{ g Mg}}$$

b) The equation $2\,CuSO_4 + 4\,KI \longrightarrow 2\,CuI + I_2 + 2\,K_2SO_4$ states that 1 mol of I_2 comes from 2 mol of $CuSO_4$. Use this fact as a unit-factor in a string of conversions, starting from 1.000 g I_2

$$m_{CuSO_4} = 1.000 \text{ g } I_2 \times \left(\frac{1 \text{ mol } I_2}{253.809 \text{ g } I_2} \right)\left(\frac{2 \text{ mol } CuSO_4}{1 \text{ mol } I_2} \right)\left(\frac{159.608 \text{ g } CuSO_4}{1 \text{ mol } CuSO_4} \right) = \boxed{1.258 \text{ g } CuSO_4}$$

c) According to the balanced equation, 1 mol of $NaBH_4$ yields 4 mol of H_2. Some might argue that such a reaction is not possible, reasoning that no chemical reaction can transform the 4 H atoms of $NaBH_4$ into the 8 H atoms of 4 H_2. In fact, the extra H comes from the other reactant, water. Write a series of unit-factors

$$m_{NaBH_4} = 1.000 \text{ g } H_2 \times \left(\frac{1 \text{ mol } H_2}{2.0158 \text{ g } H_2} \right)\left(\frac{1 \text{ mol } NaBH_4}{4 \text{ mol } H_2} \right)\left(\frac{37.833 \text{ g } NaBH_4}{1 \text{ mol } NaBH_4} \right) = \boxed{4.692 \text{ g } NaBH_4}$$

2.23 An examination of the formula establishes that 1 mol of $K_2Zn_3[Fe(CN)_6]_2$ contains 12 mol of C. Since all of this carbon is captured in the form of K_2CO_3, 12 mol of K_2CO_3 forms per mole of $K_2Zn_3[Fe(CN)_6]_2$. This fact provides the second unit-factor in the following. The other unit-factors are routine

$$m = 18.6 \text{ g } K_2CO_3 \times \left(\frac{1 \text{ mol } K_2CO_3}{138.2 \text{ g } K_2CO_3} \right) \left(\frac{1 \text{ mol } K_2Zn_3[Fe(CN)_6]_2}{12 \text{ mol } K_2CO_3} \right) \left(\frac{698.3 \text{ g } K_2Zn_3[Fe(CN)_6]_2}{1 \text{ mol } K_2Zn_3[Fe(CN)_6]_2} \right)$$

$$= \boxed{7.83 \text{ g } K_2Zn_3[Fe(CN)_6]_2}$$

2.25 Write and balance a chemical equation to learn the relationship between the number of moles of Si_2H_6 consumed and the number of moles of SiO_2 formed. By inspection

$$2 \text{ Si}_2H_6 + 7 \text{ O}_2 \longrightarrow 4 \text{ SiO}_2 + 6 \text{ H}_2O$$

Next, use the density and volume to obtain the number of moles of Si_2H_6. Then obtain the number of moles of the SiO_2 product and finally the mass of the SiO_2. The following does this all in a single series of unit-factors

$$m_{SiO_2} = 25.0 \text{ cm}^3 \times \left(\frac{2.78 \times 10^{-3} \text{ g}}{1.00 \text{ cm}^3} \right) \left(\frac{1 \text{ mol } Si_2H_6}{62.2196 \text{ g } Si_2H_6} \right) \left(\frac{4 \text{ mol } SiO_2}{2 \text{ mol } Si_2H_6} \right) \left(\frac{60.085 \text{ g } SiO_2}{1 \text{ mol } SiO_2} \right)$$

$$= \boxed{0.134 \text{ g } SiO_2}$$

2.27 Use a series of unit-factors to pass from grams of Al_2O_3 to grams of cryolite

$$m_{Na_3AlF_6} = 287 \text{ g } Al_2O_3 \times \left(\frac{1 \text{ mol } Al_2O_3}{101.962 \text{ kg } Al_2O_3} \right) \left(\frac{2 \text{ mol } Na_3AlF_6}{1 \text{ mol } Al_2O_3} \right) \left(\frac{209.94 \text{ g } Na_3AlF_6}{1 \text{ mol } Na_3AlF_6} \right)$$

$$= \boxed{1.18 \times 10^3 \text{ g } Na_3AlF_6}$$

2.29 It does not matter whether the substance in question is a product or reactant. The form of the unit-factors is similar

$$m_{KCl} = 567 \text{ g } KNO_3 \times \left(\frac{1 \text{ mol } KNO_3}{101.103 \text{ g } KNO_3} \right) \left(\frac{1 \text{ mol } KCl}{1 \text{ mol } KNO_3} \right) \left(\frac{74.551 \text{ g } KCl}{1 \text{ mol } KCl} \right) = \boxed{418 \text{ g}}$$

For the mass of the by-product switch direction after the first unit-factor

$$m_{Cl_2} = 567 \text{ g } KNO_3 \times \left(\frac{1 \text{ mol } KNO_3}{101.103 \text{ g } KNO_3} \right) \left(\frac{2 \text{ mol } Cl_2}{4 \text{ mol } KNO_3} \right) \left(\frac{70.906 \text{ g } Cl_2}{1 \text{ mol } Cl_2} \right) = \boxed{199 \text{ g}}$$

2.31 **a)** The small whole-number ratios in chemical formulas and balanced chemical equations always refer to numbers of moles, never to mass. The balanced equation given in this problem assures that the numbers of moles of XCl_2 and XBr_2 are equal. To use this fact, convert the mass of XBr_2 to the number of moles of XBr_2. Also, convert the mass of XCl_2 to a number of moles of XCl_2. The conversions require the molar masses of the two compounds, which in turn require the molar mass of element X, a quantity that is unfortunately not known. Call it x. Then

$$\text{molar mass}_{XBr_2} = x + 2(79.904) \text{ g mol}^{-1} \qquad \text{molar mass}_{XCl_2} = x + 2(35.453) \text{ g mol}^{-1}$$

The numbers of moles of the two compounds are

$$n_{XBr_2} = 1.500 \text{ g } XBr_2 \times \left(\frac{1 \text{ mol } XBr_2}{(x + 159.808) \text{ g } XBr_2} \right) = \frac{1.500}{x + 159.808} \text{ mol } XBr_2$$

$$n_{XCl_2} = 0.890 \text{ g } XCl_2 \times \left(\frac{1 \text{ mol } XCl_2}{(x + 70.906) \text{ g } XCl_2} \right) = \frac{0.890}{x + 70.906} \text{ mol } XCl_2$$

But the numbers of moles of the XBr_2 and XCl_2 are equal. Hence:

$$\frac{1.500}{159.808 + x} = \frac{0.890}{70.906 + x}$$

This equation is easily solved for x, the molar mass of the unknown element. It equals 58.8 g mol^{-1}; the relative atomic mass of the element is $\boxed{58.8}$.

b) A check of the table of atomic masses printed on the inside back cover of the text reveals that the unknown element is $\boxed{\text{probably nickel}}$, which has a relative atomic mass of 58.6934. Cobalt (relative atomic mass 58.93320) is very close as well.

2.33   By the law of definite proportions, the compounds AgCl, NaCl, and KCl contain set fractions of their mass as chlorine. These fractions are readily computed from the formulas of the compounds and a table of relative atomic masses. Thus, in addition to the obvious relationship

$$1.0000 \text{ g} = m_{NaCl} + m_{KCl}$$

the following holds

$$m_{Cl} \text{ (in AgCl)} = m_{Cl} \text{ (in NaCl)} + m_{Cl} \text{ (in KCl)}$$

Figure out the relative formula masses of AgCl NaCl, and KCl and use them to obtain the fraction of each compound that is Cl. Then substitute in the preceding equation

$$2.1476 \text{ g} \left(\frac{35.4527}{35.4527 + 107.8682} \right) = m_{NaCl} \left(\frac{35.4527}{35.4527 + 22.9898} \right) + m_{KCl} \left(\frac{35.4527}{35.4527 + 39.0983} \right)$$

Dividing both sides of this equation by 35.4527 and completing the additions gives

$$2.1476 \text{ g} \left(\frac{1}{143.3209} \right) = m_{NaCl} \left(\frac{1}{58.4425} \right) + m_{KCl} \left(\frac{1}{74.5510} \right)$$

Next, substitute for m_{NaCl} in terms of m_{KCl} and simplify

$$2.1476 \text{ g} \left(\frac{1}{143.3209} \right) = (1.000 \text{ g} - m_{KCl}) \left(\frac{1}{58.4425} \right) + m_{KCl} \left(\frac{1}{74.5510} \right)$$

$$2.1476 \text{ g} \left(\frac{1}{143.3209} \right) = 1.000 \text{ g} \left(\frac{1}{58.4425} \right) + m_{KCl} \left(\frac{1}{74.5510} - \frac{1}{58.4425} \right)$$

$$\frac{2.1476 \text{ g}}{143.3209} - \frac{1.000 \text{ g}}{58.4425} = m_{KCl} \left(\frac{1}{74.5510} - \frac{1}{58.4425} \right)$$

Completing the arithmetic in the last equation, solving it, and then remembering that the masses of the NaCl and the KCl add up to 1.000 g gives first

$$m_{KCl} = 0.5751 \text{ g} \qquad \text{and then} \qquad m_{NaCl} = 0.4249 \text{ g}$$

The mass percentages of NaCl and KCl in the original mixture of NaCl and KCl are $\boxed{42.49\%}$ and $\boxed{57.51\%}$ respectively.

Limiting Reactant and Percentage Yield

2.35   Write the balanced chemical equation for the reaction

$$HCl(g) + NH_3(g) \longrightarrow NH_4Cl(s)$$

One mole of HCl gas weighs 36.46 g, and one mole of NH_3 weighs only 17.03 g. It takes fewer heavy molecules than light molecules to make up a specific mass. Therefore, equal masses of HCl and NH_3 contain more molecules of NH_3. When the two react in a 1-to-1 molar ratio, the HCl is used up first. This means that HCl is the limiting reactant. When the HCl is used up, the reaction stops, leaving excess NH_3. The mass of NH_4Cl that is produced is

$$m_{NH_4Cl} = 10.00 \text{ g HCl} \times \left(\frac{1 \text{ mol HCl}}{36.46 \text{ g HCl}}\right) \left(\frac{1 \text{ mol } NH_4Cl}{1 \text{ mol HCl}}\right) \left(\frac{53.49 \text{ g } NH_4Cl}{1 \text{ mol } NH_4Cl}\right) = \boxed{14.7 \text{ g } NH_4Cl}$$

Since 20.0 g of matter was present originally, the mass of left-over NH_3 is $(20.0 - 14.7) = \boxed{5.3 \text{ g } NH_3}$.

The preceding is a conceptual method of identifying the limiting reactant. A more mechanical (but frequently recommended) approach is to carry out multiple calculations of the sort just given. One selects a product and calculates its yield based on the amount of each reactant in turn and assuming unlimited amounts of the other reactants. In this case, one first computes that 10.0 g of NH_3 and unlimited HCl would give 31.4 g of NH_4Cl and next computes that 10.0 g of HCl and unlimited NH_3 would give only 14.7 g of NH_4Cl. The reactant giving *the lowest* yield of the selected product is the limiting reactant.

Tip. A quick way to identify the limiting reactant is to divide the number of moles of each reactant by the coefficient that the reactant has in the balanced chemical equation. The smallest answer identifies the limiting reactant.

2.37 Use the equation

$$Fe_2O_3 + 3\,CO \longrightarrow 2\,Fe + 3\,CO_2$$

to compute the maximum possible yield (the theoretical yield or T.Y.) of Fe based on Fe_2O_3 as the limiting reactant (the CO is "in excess"):

$$\text{T.Y.}_{Fe} = 433.2 \text{ g } Fe_2O_3 \times \left(\frac{1 \text{ mol } Fe_2O_3}{159.69 \text{ g } Fe_2O_3}\right) \left(\frac{2 \text{ mol Fe}}{1 \text{ mol } Fe_2O_3}\right) \frac{55.85 \text{ g Fe}}{1 \text{ mol Fe}} = \boxed{303.0 \text{ g Fe}}$$

The percent yield, a ratio, is

$$\text{Percent Yield} = \frac{\text{Actual Yield}}{\text{Theoretical Yield}} \times 100\% = \frac{254.3 \text{ g Fe actual}}{303.0 \text{ g Fe theoretical}} \times 100\% = \boxed{83.93\%}$$

Tip. A computation of theoretical yield assumes that the reaction proceeds cleanly (without side-reactions) and "to completion" to give the products appearing on the right-hand side of the equation. It also assumes that the product in question can be separated from other products, solvents, and impurities without loss. Needless to say, actual reactions rarely work that way.

ADDITIONAL PROBLEMS

2.39 The solution requires more significant figures than the computations in problem **2.13** but follows the same principles. The relative molecular mass of the human parathormone is 13 931.98. The mass percentages are

$$\boxed{\text{C, } 59.571\%; \quad \text{H, } 6.4968\%; \quad \text{N, } 12.5670\%; \quad \text{O, } 18.8336\%; \quad \text{S, } 2.5318\%}$$

2.41 **a)** A binary compound contains only two elements. In this case, each of the three compounds contains oxygen O and the metal M and nothing else. Compound 1 contains 13.38 g of O for every 86.62 g of M, which, by division, is $\boxed{0.15447 \text{ g}}$ of O per 1 g of M. Compound 2 and Compound 3 contain $\boxed{0.1029 \text{ g}}$ of O and $\boxed{0.07721 \text{ g}}$ of O per 1 g of M, respectively.

b) If Compound 1 is MO_2, then Compound 2 is $MO_{4/3}$ because Compound 2 has 2/3 as much oxygen per given quantity of M as Compound 2. The formula $MO_{4/3}$ (equivalent to $M_1O_{4/3}$) is improved by multiplying both subscripts by 3 to get rid of the fraction. The result is $\boxed{M_3O_4}$. Compound 3 is $\boxed{MO}$ if the Compound 1 is MO_2 because it has almost exactly 1/2 as much oxygen per quantity of M.

c) The assumption that Compound 1 is MO_2 still goes. Compute the amount of metal that combines with 2 mol of oxygen

$$m_M = (2 \times 15.9994 \text{ g O}) \left(\frac{1 \text{ g M}}{0.15447 \text{ g O}} \right) = 207.15 \text{ g M}$$

Because "MO_2" means there is 1 mol of M per 2 mol of O, the relative atomic mass of M equals 207.15. Consulting the periodic table reveals that M is $\boxed{\text{lead}}$.

Tip. If the formula of Compound 1 is not known, then figuring out the identity of element M is probably not possible. Thus, if Compound 1 is M_2O_2, then the relative atomic mass of M is half of 207.15 or 103.6; if Compound 1 is M_3O_2, then the relative atomic mass of M is 69.05; if Compound 1 is M_4O_2, then the relative atomic mass of M is only 51.79, etc. (see problem **2.17**).

2.43 Imagine 1.000 g of the first oxide. It contains 0.6960 g of Mn and 0.3040 g of O. Dividing these masses by the relative atomic masses of Mn and O gives the relative numbers of moles of the two elements. The smallest whole-number ratio of these numbers of moles gives the subscripts in the compound's empirical formula:

$$\text{Mn}_{\frac{0.6960}{54.94}} \text{O}_{\frac{0.3040}{16.00}} \Longrightarrow \text{Mn}_{0.01267}\text{O}_{0.01900} \Longrightarrow \text{Mn}_{1.000}\text{O}_{1.500} \Longrightarrow \boxed{\text{Mn}_2\text{O}_3}$$

Repeat the procedure for the second oxide:

$$\text{Mn}_{\frac{0.6319}{54.94}} \text{O}_{\frac{0.3681}{16.00}} \Longrightarrow \text{Mn}_{0.01150}\text{O}_{0.02301} \Longrightarrow \text{Mn}_{1.000}\text{O}_{2.000} \Longrightarrow \boxed{\text{MnO}_2}$$

Tip. Getting from the second to the third formula in calculations such as these is easiest if one divides each subscript by the smallest of the subscripts.

2.45 The balanced chemical equation reads

$$C_{14}H_{18}N_2O_5 + 2\,H_2O \longrightarrow C_4H_7NO_4 + CH_3OH + \text{phenylalanine}$$

The right-hand side of the equation is short 9 atom of C, which must be part of the phenylalanine molecule. Balancing amounts of H, N and O in the same way establishes the molecular formula of phenylalanine to be $C_9H_{11}NO_2$.

2.47 Assume that all the carbon is evolved as CH_4. Then 1 mol Al_4C_3 generates 3 mol CH_4

$$m_{CH_4} = \frac{63.2 \text{ g Al}_4\text{C}_3}{143.96 \text{ g mol}^{-1}} \times \frac{3 \text{ mol CH}_4}{1 \text{ mol Al}_4\text{C}_3} \times \frac{16.043 \text{ g CH}_4}{1 \text{ mol CH}_4} = 21.1 \text{ g}$$

2.49 Let x represent the mass of $SrCO_3$ in the sample. Then $0.800 - x$ is the mass of $BaCO_3$.

$$n_{SrCO_3} = \frac{x \text{ g}}{147.63 \text{ g mol}^{-1}} \quad \text{and} \quad n_{BaCO_3} = \frac{(0.800 - x) \text{ g}}{197.34 \text{ g mol}^{-1}}$$

$$\text{and} \quad n_{CO_2} = \frac{0.211 \text{ g CO}_2}{44.010 \text{ g mol}^{-1}} = 0.004794 \text{ mol}$$

Each mole of $SrCO_3$ and $BaCO_3$ generates one mole of CO_2. This means

$$\left(\frac{x}{147.63} + \frac{0.800 - x}{197.34} \right) \text{ mol} = 0.004794 \text{ mol} \quad \text{which gives} \quad x = 0.434 \text{ g}$$

The percentage of $SrCO_3$ in the sample was $0.434 \text{ g}/0.800 \text{ g} \times 100\% = 54.2\%$.

2.51   Assume that your car gets 20 miles per U.S. gallon. Then reducing driving by 20 miles per week would save you 1 gallon a week, or 52 gallons a year. In the metric system this is

$$V = 52 \text{ gallons} \times \left(\frac{3.785 \text{ L}}{1 \text{ U.S. gallon}} \right) = 200 \text{ L gasoline}$$

Gasoline floats on water, so its density is less than the density of water (which is 1.0 g cm^{-3}). Take the density of gasoline to be 0.9 g cm^{-3}, or 0.9 kg L^{-1}. Your annual savings is then about 180 kg of gasoline. Approximate the chemical formula of gasoline as C_8H_{18} (octane). Then

$$n_{\text{gasoline}} \approx 180 \times 10^3 \text{ g } C_8H_{18} \times \left(\frac{1 \text{ mol}}{114.2 \text{ g } C_8H_{18}} \right) = 1.6 \times 10^3 \text{ mol}$$

Combustion of octane release 8 mol of CO_2 per mole. The mass of CO_2 not released is

$$m_{CO_2} \approx 1.6 \times 10^3 \text{ mol } C_8H_{18} \times \left(\frac{8 \text{ mol } CO_2}{1 \text{ mol } C_8H_{18}} \right) \times \left(\frac{44 \text{ g } CO_2}{1 \text{ mol } CO_2} \right) = 5.7 \times 10^5 \text{ g} \approx 570 \text{ kg}$$

$$\approx 570 \text{ kg} \left(\frac{1 \text{ lb}}{0.4536 \text{ kg}} \right) = 1300 \text{ lb } CO_2$$

This is on the order of 1000 pounds, the amount that is mentioned in the problem. If your car gets 30 miles per gallon, then the same weekly 20 mile cut in its use would reduce your CO_2 emissions by 2/3 of 1300 pounds.

2.53   Compute the theoretical yield of $KClO_4$ for comparison to the observed yield of 3.00 g. Balance the chemical equation: $4 \text{ KClO}_3(s) \longrightarrow 3 \text{ KClO}_4(s) + \text{KCl}(s)$. Then

$$m_{KClO_4} = 4.00 \text{ g } KClO_3 \times \left(\frac{1 \text{ mol } KClO_3}{122.549 \text{ g } KClO_3} \right) \left(\frac{3 \text{ mol } KClO_4}{4 \text{ mol } KClO_3} \right) \left(\frac{138.549 \text{ g } KClO_4}{1 \text{ mol } KClO_4} \right)$$

$$= 3.392 \text{ g } KClO_4$$

The percentage yield is $(3.00/3.392) \times 100\% = 88.4\%$.

2.55   The yield of the product is less than the theoretical, and this fact must be reckoned with. A very good way is to use the percentage yield to construct an additional unit-factor (the second factor in the following)

$$m_{Si} = 125 \text{ g } Si_3N_4 \text{ isolated} \times \left(\frac{100 \text{ g } Si_3N_4 \text{ formed}}{95.0 \text{ g } Si_3N_4 \text{ isolated}} \right) \times \left(\frac{1 \text{ mol } Si_3N_4}{140.285 \text{ g } Si_3N_4 \text{ formed}} \right)$$

$$\times \left(\frac{3 \text{ mol } Si}{1 \text{ mol } Si_3N_4} \right) \times \left(\frac{28.086 \text{ g } Si}{1 \text{ mol } Si} \right) = 79.0 \text{ g } Si$$

Chapter 3

Chemical Bonding: The Classical Description

The Periodic Table

3.1 According to the periodic law, scandium ("eka-boron, Eb") should have properties intermediate between those of calcium and titanium. Simply average the numerical data:

Property	Predicted	Observed
Melting point	1250°C	1541°C
Boiling point	2386°C	2831°C
Density	3.02 g cm^{-3}	2.99 g cm^{-3}

The information in the "observed" column come from text Appendix F.

3.3 Elements in Groups numbered higher than IV in the periodic table tend to form hydrides having $(8 - n)$ hydrogen atoms, where n is the group number. Antimony is in Group V, and, by this rule, its hydride is $\boxed{SbH_3}$ (not "SbH_5"); bromine is in Group VII and its hydride is $\boxed{HBr}$; tin is in Group IV and its hydride is $\boxed{SnH_4}$; selenium is in Group VI and its hydride is $\boxed{H_2Se}$. All four of these compounds exist.

Tip. Using the $(8 - n)$ rule with Group VIII indicates that the hydrides of He, Ne, Ar, Kr, Xe, and Rn contain *zero* hydrogen atoms. In fact, these elements do not form binary compounds with hydrogen.

Forces and Potential Energy in Atoms

3.5 Coulomb's law gives the electrostatic force F between a pair of charged particles as a function of the charges q_1 and q_2 on the particles and the distance r between them. Text equation 3.4 states Coulomb's law in a form that explicitly indicates the direction of the force vector. It does this by including a unit vector, a vector of length 1, that points in the proper direction. Unit vectors have no physical units and are useful only when they appear as a set of components. For example, the vector $(\sqrt{2}/2, \sqrt{2}/2)$ points "northeast" between the $+x$ and $+y$ axes in a plane and has length 1 while the vector $(1/2, \sqrt{3}/2)$ points north-northeast and also has length 1. Multiplying such directional indicators by a magnitude (including units) generates complete vectors in component form. Vectors are reviewed in detail in text Appendix C.7.

a) Electron 1 is at the origin $(0,0)$ of a Cartesian coordinate system and electron 2 is at $(0, 2 \text{ Å})$, a point on the y-axis. Write text equation 3.4 for the force on electron 1. The charges on the two electrons are $-e$, the negative of the fundamental unit of electrical charge. The vectorial version of

18

Coulomb's law for this case is

$$\vec{F_1}(r) = \frac{-(-e)(-e)}{4\pi\epsilon_0 r^2}\,\hat{r}_{21}$$

where $\vec{F_1}(r)$ represents the force acting on electron 1. The fraction on the right-hand side of the equation gives the magnitude of the force vector and $\hat{r}_{21}$, the unit vector pointing from electron 2 to electron 1, gives its direction. This unit vector (distinguished by a "hat" on its symbol) equals the vector $\vec{r}_{21}$ divided by its length[1]

$$\hat{r}_{21} = \frac{\vec{r}_{21}}{r} = \frac{\vec{r}_{21}}{2\text{ Å}} = \frac{(0\text{ Å}, 2\text{ Å})}{2\text{ Å}} = (0,1)$$

Assume for the sake of concreteness in the calculation that the coordinates of the electrons are given to three significant figures. Completing the arithmetic then gives

$$\vec{F_1} = \frac{-(-e)^2}{4\pi\epsilon_0 r^2}\,\hat{r}_{21} = \frac{-(-1.602\times 10^{-19}\text{ C})^2}{4\pi\,(8.854\times 10^{-12}\text{ C}^2\text{ J}^{-1}\text{ m}^{-1})\,(2.00\times 10^{-10}\text{ m})^2}\cdot\left(0.00, 1.00\right)$$

$$= -5.77\times 10^{-9}\frac{\text{C}^2}{\text{C}^2\text{ J}^{-1}\text{ m}^{-1}\text{ m}^2}\cdot\left(0.00, 1.00\right)$$

$$= -5.77\times 10^{-9}\text{ J m}^{-1}\cdot\left(0.00, 1.00\right) = -5.77\times 10^{-9}\text{ N}\cdot\left(0.00, 1.00\right)$$

Numerical values for the magnitude of charge on the electron (e) and the electrical permittivity of the vacuum (ϵ_0) come from the table on the inside of the back cover of the textbook. The x-component of $\vec{F_1}$ equals -5.77×10^{-9} N multiplied by the x-component of the unit vector, which is 0.00: $\boxed{F_x = 0.00\text{ N}}$. The y-component of $\vec{F_1}$ equals -5.77×10^{-9} N multiplied by the y-component of the unit vector, which equals 1.00: $\boxed{F_y = -5.77\times 10^{-9}\text{ N}}$. The negative sign indicates that the force acts to push electron 1 in the negative y direction. Electron 2 repels electron 1.

Tip. Make sure to understand the way that the units work out. Use the definitions in text Appendix B (Table B.2).

b) Text equation 3.1 gives the electrostatic (Coulombic) potential energy of the interaction between two charged particles. Assume again that the coordinates of the two electrons are given to three significant figures

$$V = \frac{q_1 q_2}{4\pi\epsilon_0 r} = \frac{(-e)(-e)}{4\pi\epsilon_0 r} = \frac{(-1.602\times 10^{-19}\text{ C})^2}{4\pi\,(8.854\times 10^{-12}\text{ C}^2\text{ J}^{-1}\text{ m}^{-1}(2.00\times 10^{-10}\text{ m})}$$

$$= \boxed{+1.15\times 10^{-18}\text{ J}}$$

$$= 1.153\times 10^{-18}\text{ J}\times\left(\frac{1\text{ eV}}{1.6022\times 10^{-19}\text{ J}}\right) = \boxed{+7.20\text{ eV}}$$

The positive sign of the answer means that the two particles repel each other. A pair of electrons held close to each other is like a coiled spring, a reservoir of energy. This energy is potential energy.

Tip. The electrostatic force between two charged particles immersed in a medium is always less than the force between the same particles held the same distance apart in a vacuum. This is because the electrical permittivity (ϵ_0 in the preceding) of matter always exceeds that of a vacuum. For example, the permittivity of water is about 80 times larger than that of a vacuum.

3.7   **a)** Modify text equation 3.4, which gives the Coulombic force vector between two charged particle, by stripping out its directional aspect. The result is

$$|F(r)| = \left|\frac{q_1 q_2}{4\pi\epsilon_0 r^2}\right|$$

---

[1]Any vector can be converted to a unit vector by dividing its components, one by one, by its length.

where q_1 and q_2 are the charges on the particles and r is the distance between them. Marking both sides of the equation as absolute values emphasizes that the relationship concerns magnitudes only. The particles are a proton of charge $+e$ and an electron of charge $-e$. Convert the first inter-particle distance from Ångstroms to meters so that the units work out and use the preceding formula

$$|F| = \left| \frac{(-e)(+e)}{4\pi\epsilon_0 r^2} \right| = \left| \frac{(-1.602 \times 10^{-19}\ \text{C})(1.602 \times 10^{-19}\ \text{C})}{4\pi\,(8.854 \times 10^{-12}\ \text{C}^2\ \text{J}^{-1}\ \text{m}^{-1})\,(2.12 \times 10^{-10}\ \text{m})^2} \right| = \boxed{5.13 \times 10^{-9}\ \text{N}}$$

After the electron moves, its distance from the proton is $0.529/2.12 = 1/4$ of what it was originally. Coulombic forces varies inversely as to the square of the separation of the interacting charges. The magnitude of the new force is therefore larger by a factor of 16: $|F| = \boxed{82.4 \times 10^{-9}\ \text{N}}$.

b) The change in the potential energy of the proton-electron system equals the final potential energy minus the initial

$$V_2 - V_1 = \frac{q_{\text{p}}\,q_{\text{e}}}{4\pi\,\epsilon_0} \left[\frac{1}{r_2} - \frac{1}{r_1} \right]$$

$$= \frac{(+1.602 \times 10^{-19}\ \text{C})\,(-1.602 \times 10^{-19}\ \text{C})}{4\pi\,(8.854 \times 10^{-12}\ \text{C}^2\ \text{J}^{-1}\ \text{m}^{-1})} \left[\frac{1}{r_2} - \frac{1}{r_1} \right]$$

$$= (-2.307 \times 10^{-28}\ \text{J m}) \left[\frac{1}{(0.529 \times 10^{-10}\ \text{m})} - \frac{1}{(2.12 \times 10^{-10}\ \text{m})} \right] = \boxed{-3.27 \times 10^{-18}\ \text{J}}$$

$$= -3.27 \times 10^{-18}\ \text{J} \times \frac{1\ \text{eV}}{1.60217 \times 10^{-19}\ \text{J}} = \boxed{-20.4\ \text{eV}}$$

The two particles attract each other. The potential energy V of the interaction equals zero if the distance between the two is infinite. Otherwise it is negative. Decreasing the distance between the two particles makes the potential energy of the system more negative.

c) The change in the potential energy of the proton-electron system when the electron is moved in from a distance of 2.12 Å to a distance of 0.529 Å was found in the previous part to equal -3.27×10^{-18} J. Assume that this energy stays in the system and is not lost to the surroundings by some means. Then the decrease in the potential energy must be compensated for by an increase in the kinetic energy amounting to $+3.27 \times 10^{-18}$ J. This change, which is internal to the system, can be accounted for only by an increase in the velocity of one particle or the other. For simplicity, take the proton to have started out motionless and to remain motionless. The kinetic energy $\mathcal{T}$ of the electron is given by the formula $\frac{1}{2}m_{\text{e}}v^2$. Change in its kinetic energy implies change in its velocity:

$$\mathcal{T}_2 - \mathcal{T}_1 = \tfrac{1}{2}m_{\text{e}}(v_2 - v_1)^2$$

$$+3.27 \times 10^{-18}\ \text{J} = \tfrac{1}{2}m_{\text{e}}(v_2 - v_1)^2$$

$$(v_2 - v_1) = \sqrt{\frac{2 \times 3.27 \times 10^{-18}\ \text{J}}{9.109 \times 10^{-31}\ \text{kg}}} = \pm 2.68 \times 10^{6}\ \text{m s}^{-1}$$

The problem says to assume that the electron was originally motionless ($v_1 = 0$). Its final velocity v_2 is therefore $\pm 2.68 \times 10^{6}\ \text{m s}^{-1}$. The final speed of the electron is $|v_2| = \boxed{2.68 \times 10^{6}\ \text{m s}^{-1}}$.

Tip. The $\pm$ in the result for the velocity corresponds to the two possible senses (clockwise and counterclockwise) for the motion of the electron in an orbit. See text Figure 3.5.

Tip. The problem specifies that the electron is initially stationary at a distance of 2.12 Å. Under this condition the system is not really a hydrogen atom. In all real atoms the total kinetic energy of the electrons is related to the sum of the potential energies of all electron-electron repulsions and electron-nucleus attractions (as computed using text equation 3.8) by the equation $\bar{\mathcal{T}}_{\text{tot}} = -\frac{1}{2}\bar{V}_{\text{tot}}$. The bars over the $\mathcal{T}$ and V in this equation indicate time-averaged totals. Time-averages are used because the kinetic and potential energies of dynamic systems can fluctuate even as the total energy, which is the sum of $\mathcal{T}$ and V, remains constant. This happens for example if one particle moves in an elliptical orbit around another.

Ionization Energies, the Shell Model of the Atom, and Shielding

3.9   Use the periodic trends in ionization energy shown in text Figure 3.11.

a) $\boxed{\text{Sr}}$ has a higher IE_1 than Rb   c) $\boxed{\text{Xe}}$ has a higher IE_1 than Cs

b) $\boxed{\text{Rn}}$ has a higher IE_1 than Po   d) $\boxed{\text{Sr}}$ has a higher IE_1 than Ba

3.11   Text Table 3.1 gives the successive ionization energies of elements 1 through 21. Beryllium, element 4, has four ionization energies labelled IE_1 through IE_4. Compute $\log IE_n$ for n equal 1 through 4. To do this, first divide each entry by the unit eV as indicated in the label on the vertical axis in the following graph. This is necessary because it is impossible to take the logarithm of a quantity that has units. Next, graph the result versus n. The big increase between IE_2 and IE_3 indicates that electrons 3 and 4 are much more tightly held than are 1 and 2.

Logarithm of Successive Ionization Energies of Beryllium

Electron Affinity and Electronegativity: The Tendency of Atoms to Attract Electrons

3.13   Use the periodic trends in electron affinity exhibited in text Table 3.2.

a) $\boxed{\text{Cs}}$ has a larger EA than Xe   c) $\boxed{\text{K}}$ has a larger EA than Ca

b) $\boxed{\text{F}}$ has a larger EA than Pm   d) $\boxed{\text{At}}$ has a larger EA than Po

3.15 $\boxed{\text{K} < \text{Si} < \text{S} < \text{O} < \text{F}}$.   Electronegativity tends to increase moving up a Group (column) in the periodic table and from left to right across the Groups (rows) from I to VIII. Thus, electronegativity generally increases diagonally from lower left to upper right in the periodic table

Forces and Potential Energy in Molecules: Formation of Chemical Bonds

3.17   Obtain a factor to convert kJ mol^{-1} to eV, as suggested in the problem.

$$\frac{1 \text{ kJ}}{\text{mol}} \times \frac{1 \text{ mol}}{6.02214 \times 10^{23} \text{ molecules}} \times \frac{1000 \text{ J}}{\text{kJ}} \times \frac{1 \text{ eV}}{1.60217 \times 10^{-19} \text{ J}} = \frac{0.010364 \text{ eV}}{\text{molecule}}$$

Therefore, the H_2 molecule has a bond dissociation energy of 4.49 eV, and the F_2 molecule has a bond dissociation energy of 1.61 eV. A plot of the V_{eff} as a function of the interatomic distance in H_2 must have a minimum of -4.49 eV at 0.751 Å, and a similar plot for F_2 must have its minimum of -1.61 eV at 1.417 Å. The plots both have the shape shown in text Figure 3.20.

3.19   Using the same scale that was used for problem **3.17** helps in making comparisons.

The H—F bond is stronger than either the H—H bond or the F—F bond, as shown by the greater depth of its potential energy "well." The H—F bond is also shorter than one-half of the bond distance in H—H added to one-half of the bond distance in F—F.

Ionic Bonding

3.21   **a)** An atom of radon has 86 electrons. Of these, 78 are core electrons, and 8 are valence electrons.

The Lewis diagram is $:\overset{\cdot\cdot}{\underset{\cdot\cdot}{Rn}}:$

b) The monopositive strontium ion has a total of 37 electrons. Of these, 36 are core electrons and 1 is a valence electron. $Sr\cdot\ ^+$

c) The selenide(2−) ion has a total of 36 electrons. Of these 28 are core electrons and 8 are valence electrons. The Lewis diagram is $:\overset{\cdot\cdot}{\underset{\cdot\cdot}{Se}}:\ ^{2-}$

d) The antimonide(1−) ion has a total of 52 electrons. Of these 46 are core electrons and 6 are valence electrons. $\cdot\underset{\cdot\cdot}{Sb}\cdot\ ^-$

3.23   **a)** The two equations show the transfer of an electron between a K(g) atom and a Cl(g) atom in the two possible directions. The ΔE of the first reaction is the molar IE_1 (first ionization energy) of K(g) added to the negative of the molar EA (electron affinity) of Cl. The ΔE of the second reaction is the molar IE_1 of Cl added to the negative of the molar EA of K. Take data from text Appendix F

$$\Delta E \text{ (for } K^+Cl^-)(g) = 418.8 + (-349.0) = \boxed{69.8 \text{ kJ mol}^{-1}}$$

$$\Delta E \text{ (for } K^-Cl^+)(g) = 1251.1 + (-48.384) = \boxed{1202.7 \text{ kJ mol}^{-1}}$$

b) Combine the ionization energies and electron affinities of Na and Cl as in the preceding

$$\Delta E \text{ (for Na}^+\text{Cl}^-)(g) = 495.8 + (-349.0) = \boxed{146.8 \text{ kJ mol}^{-1}}$$

$$\Delta E \text{ (for Na}^-\text{Cl}^+)(g) = 1251.1 + (-52.867) = \boxed{1198.2 \text{ kJ mol}^{-1}}$$

Imagine that K^-Cl^+ (or Na^-Cl^+) were to form. A subsequent reaction transferring electrons to form K^+Cl^- (or Na^+Cl^-) would be strongly favored energetically, and no barrier would exist to stop it from occurring quickly.

3.25 The problem is to estimate the energy of dissociation of $KCl(g)$ given that the distance between the bonded K^+ and Cl^- ions equals 2.67 Å.[2] The energy of dissociation is the ΔE of the reaction

$$K\text{—}Cl(g) \longrightarrow K(g) + Cl(g)$$

The answer must be positive because breaking chemical bonds always requires energy. Approximate the bonding in the K—Cl molecule as a purely electrostatic (Coulombic) attraction between a K^+ ion and a Cl^- ion. The dissociation of such a molecule can be imagined to take place in two steps: 1) the separation of the two ions to an infinite distance from each other; 2) the transfer of an electron from the negative ion to the positive ion to give a pair of atoms

Step 1 $\qquad K\text{—}Cl(g) \longrightarrow K^+(g) + Cl^-(g) \qquad \Delta E_1 = -\Delta V_{\text{Coulomb}}$

Step 2 $\qquad K^+(g) + Cl^-(g) \longrightarrow K(g) + Cl(g) \qquad \Delta E_2 = -\Delta E_\infty$

The desired answer is the sum of the two ΔE's

$$\Delta E_d = \Delta E_1 + \Delta E_2 = -\Delta E_{\text{Coulomb}} - \Delta E_\infty$$

The Coulombic potential energy of a K^+ and Cl^- ion separated by 2.67 Å is

$$V_{\text{Coulomb}} = \frac{q_{K^+} q_{Cl^-}}{4\pi\epsilon_0 R_e} = \frac{(+1.602 \times 10^{-19} \text{ C})(-1.602 \times 10^{-19} \text{ C})}{4(3.1416)(8.854 \times 10^{-12} \text{ C}^2\text{J}^{-1}\text{m}^{-1})(2.67 \times 10^{-10} \text{ m})} = -8.64 \times 10^{-19} \text{ J}$$

This potential energy is for one K^+ to Cl^- interaction. For a *mole* of these pair-wise interactions, multiply by Avogadro's number

$$V_{\text{Coulomb}} = (-8.64 \times 10^{-19} \text{ J pair}^{-1}) \times (6.022 \times 10^{23} \text{ pair mol}^{-1}) = -520 \times 10^3 \text{ J mol}^{-1}$$

This result means that it requires +520 kJ to dissociate a mole of the KCl molecules to ions: to break up all the K—Cl pairs, move the Cl^- ions an infinite distance away from the K^+ ions to which they were bonded, and create a collection of non-interacting stationary $K^+(g)$ ions and a second collection of non-interacting stationary $Cl^-(g)$ ions.

Now for the electron transfer. The text defines the quantity ΔE_∞ in an ionic attraction as the electron affinity of the negative ion subtracted from the first ionization energy of the positive ion. In this case

$$\Delta E_\infty = IE_K - EA_{Cl}$$

Locate numbers for IE_K and EA_{Cl} in text Appendix F and insert them

$$\Delta E_\infty = 418.8 \text{ kJ mol}^{-1} - 349.0 \text{ kJ mol}^{-1}$$

Substituting in the equation for the energy of dissociation gives the answer

$$\Delta E_d = -V_{\text{Coulomb}} - \Delta E_\infty$$

$$= -(-520 \text{ kJ mol}^{-1}) - (418.8 \text{ kJ mol}^{-1} - 349.0 \text{ kJ mol}^{-1}) = \boxed{450 \text{ kJ mol}^{-1}}$$

---

[2]This problem resembles text Example 3.6.

Covalent and Polar Covalent Bonding

3.27 The As—H bond length should lie between the 1.42 Å of P—H and the 1.71 Å of Sb—H. A length of $\boxed{1.56 \text{ Å}}$, (the average) is a reasonable guess. The X—H bond will be weakest in $\boxed{SbH_3}$, which has the longest bonds.

Tip. The experimental As—H bond length is 1.519 Å.

3.29 The closeness of the bond length in H—I to the simple sum of the covalent radii of H and I indicates that the bond has little ionic character, that is, $\boxed{\text{it is mostly non-polar}}$. A bond distance *shorter* than the sum of the covalent radii of H and I would suggest the presence of an electrostatic attraction between the oppositely charged ends of the H—I dipole.

3.31 Take the electronegativities of C, N, O and P from text Figure 3.10 or from text Appendix F. Figure the absolute value of the difference in electronegativity between element A and element B in each bond listed in the problem.

Bond	$|\chi_A - \chi_B|$
N—P	0.85
C—N	0.49
N—O	0.40
N—N	0

The most polar bond is the one with the largest difference in electronegativity: $\boxed{N—P}$.

3.33 The two elements in binary ionic compounds have a large difference in electronegativity; the elements in binary covalent compounds have only a small difference. Higher vapor pressure in a compound is associated with relatively weak intermolecular attractions and so with molecular (covalent) compounds. For this reason, the compound with the higher vapor pressure is in each case the one with the smaller difference in electronegativity between its elements.

a) $\boxed{CI_4}$ **b)** $\boxed{OF_2}$ **c)** $\boxed{SiH_4}$

3.35 In diatomic molecules, the fractional ionic character δ is given by the formula

$$\delta = (0.2082 \text{ Å D}^{-1}) \left(\frac{\mu}{R}\right)$$

In this equation (obtained by rearranging text equation 3.24) the dipole moment μ must be in Debye (D) and the bond distance R in Ångstroms (Å) for the units to cancel out and give the required unit-less number as the fractional ionic character. The percent ionic character is 100 times the fractional ionic character. Substitute in the formula to obtain

Compound	Bond Length / Å	Dipole Moment / D	δ	% Ionic Character
ClO	1.573	1.239	0.16	16
KI	3.051	10.82	0.74	74
TlCl	2.488	4.543	0.38	38
InCl	2.404	3.79	0.33	33

3.37 The Δ in the expression $(16\,\Delta + 3.5\,\Delta^2)$ in this problem is *not* the Δ in the Pauling definition of electronegativity in text equation 3.11. Instead, it is $\chi_A - \chi_B$, the difference in electronegativity. Substitute into the formula to obtain

Compound	$|\chi_A - \chi_B|$	Calc. % Ionic	Dipole % Ionic
HF	1.78	40	41
HCl	0.96	19	18
HBr	0.76	14	12
HI	0.46	8	6
CsF	3.19	87	70

Tip. The results show that ionic character calculated from differences in electronegativity agrees fairly well with ionic character based on dipole moment. See also text Figure 3.23.

Electron Pair Bonds and Lewis Diagrams for Molecules

3.39   **a)** For the Lewis structure of SO_4^{2-} in which a central S atom is surrounded by four O atoms linked to the S by single bonds

> The four O atoms all have formal charges (f.c.'s) of -1; the central S has f.c. $+2$.

Tip. Oxygen atoms with one single bond *always* have f.c. -1.

b) For the Lewis structure of $S_2O_3^{2-}$ in which a central S atom is surrounded by three O atoms and a second S atom, with all four peripheral atoms linked to the central S atom by single bonds

> The central S has f.c. $+2$. The three O atoms and the S sulfur have f.c. -1.

c) For the Lewis structure of SbF_3 in which a central Sb atom is surrounded by three F atoms linked to it by single bonds   All atoms have f.c. zero.

d) For the Lewis structure of SCN^- ion in which a central C atom is triple-bonded to an N and single-bonded to an S atom   The S atom has f.c. -1; the C and N have f.c. zero.

Tip. The formal charges on atoms in a Lewis structure change if the locations of any electrons are shifted. For example, in part **d)**, the N has f.c. -1 and the S has f.c. zero if the central C is bonded to each by a double bond.

Tip. Whenever C has four covalent bonds and no lone pairs in a Lewis structure, its formal charge is zero; whenever N has three covalent bonds and one lone pair, its formal charge is zero; whenever O has two covalent bonds and two lone pairs, its formal charge is zero.

3.41   The structure H—N̈=Ö: has f.c. zero on all atoms. The isomeric structure H—Ö=N̈: has f.c.'s of zero on the H, $+1$ on the O and -1 on the N. The first of the isomers is favored because it minimizes formal charges on the atoms.

Tip. The structures for HNO and HON are *not* resonance structures. They show the atoms in different orders of linkage (different skeletons), a much more profound difference.

3.43   **a)** In this structure of "ZO_2" the oxygen atoms both possess a formal charge of 0. Since there is no net charge on the molecule, Z must also have a f.c. of 0. Therefore, it has 4 valence electrons and belongs to Group IV. The compound CO_2 (carbon dioxide) is an example.

b) In this Lewis structure of "Z_2O_7" each of the six peripheral O atoms has f.c. -1, but the bridging O atom has f.c. zero. Since the molecule has no net charge and is symmetrical, the f.c. on each Z is $+3$. Therefore, Z has 7 valence electrons and is in Group VII. An example is Cl_2O_7 (dichlorine heptaoxide).

c) In "ZO_2^-" one of the oxygen atoms has f.c. zero, but the other has f.c. -1. Since the species has a -1 net charge, Z must have f.c. zero. Therefore, it comes from Group V (5 valence electrons). An example is NO_2^- (nitrite ion).

d) In "$HOZO_3^-$" three of the O atoms have f.c. -1, and the other O atom and the H atom have f.c. zero. Since the ion has a -1 net charge, Z must have f.c. $+2$. Therefore Z has 6 valence electrons and comes from Group VI. An example is $HOSO_3^-$ (hydrogen sulfate ion).

3.45   The octet rule is satisfied for all atoms in these structures. Non-zero formal charges are indicated near the atom, and the non-zero overall charges of molecular ions are shown outside of brackets:

Tip. In Lewis structures, a line equals a pair of dots, but lines are rarely used for lone pairs.

3.47   Referring to text Table 3.6, the lengths of the bonds should be: N—H, 1.01×10^{-10} m; N—C, 1.47×10^{-10} m; C=O, 1.20×10^{-10} m. The following Lewis structure for urea has f.c. zero on all atoms

3.49   A Lewis structure for S_8

The molecule has 48 valence electrons. The octet rule is obeyed on all sulfur atoms, and all sulfur atoms have f.c. zero. Despite its appearance in the diagram the S_8 ring is *not* planar, but puckered.

3.51   **a)** The nitrogen and boron atoms would both get 4 valence electrons if all bonds were broken and the two electrons from each pair parcelled out evenly between the two atoms that share them. The nitrogen atom is supposed to have 5 valence electrons, so its formal charge is $\boxed{+1}$. The f.c. is $\boxed{-1}$ on the boron atom, and $\boxed{\text{zero}}$ on the rest of the atoms:

b) The single-bonded O atom in each resonance structure has $\boxed{\text{f.c. } -1}$. All other f.c.'s are $\boxed{\text{zero}}$. A double-headed arrow indicates resonance structures

c) The hydrogen carbonate ion has 24 valence electrons. The C atom contributes 4, the H atom contributes 1, and the 3 O atoms contribute 6 each; a final electron comes from elsewhere to make the overall charge -1. The resonance Lewis structures are

Formal charges on the atoms are ⌈zero⌉ except that the single bonded O's have f.c. -1, as indicated.

Tip. Resonance structures differ only in the positions of the electrons. A common error is to include a third structure in which the oxygen atom on the upper left shares two pairs of electrons with the central carbon atom and the H atom is moved onto the oxygen atom at the lower left. Such a structure is *not* a resonance structure, because an atom as well as electrons has moved. Resonance structures always use the same atomic skeleton.

3.53   The main resonance structures (others break the octet rule) for NO_2^- ion are

$$\left[:\ddot{O}\!-\!N\!=\!\ddot{O} \quad \longleftrightarrow \quad \ddot{O}\!=\!N\!-\!\ddot{O}: \right]^-$$

The two N-to-O bonds in this ion should be equal in length. The length should be intermediate in length between the lengths of a N-to-O single bond and a N-to-O double bond. Take these lengths from text Table 3.6: ⌈between 1.43 Å and 1.18 Å⌉.

3.55   Resonance structures of methyl isocyanate include

$$\left[H_3C\!-\!\overset{0}{N}\!=\!\overset{0}{C}\!=\!\overset{0}{\ddot{O}} \quad \longleftrightarrow \quad H_3C\!-\!\overset{-1}{\ddot{N}}\!-\!\overset{0}{C}\!\equiv\!\overset{+1}{O}: \quad \longleftrightarrow \quad H_3C\!-\!\overset{+1}{N}\!\equiv\!\overset{0}{C}\!-\!\overset{-1}{\ddot{O}}: \right]$$

The left structure has f.c. zero on all atoms; the center structure has f.c. -1 on the N, f.c. $+1$ on the O and f.c. zero on the other atoms; the right structure has f.c. $+1$ on the N, -1 on the O and zero on all of the other atoms. The predominant resonance contributor is the left structure, which sets f.c. zero on all atoms.

Tip. The bonding in the H_3C— (methyl) group is not shown explicitly. It is the same as in methane with one of the H's in methane replaced by the —NCO (isocyanate) group.

Tip. The "iso" in the name "methyl isocyanate" suggests that the compound is an isomer of something. Here are three resonance structures for the bonding in the compound methyl cyanate, which contains exactly the same atoms but bonded in a different order. These are *not* resonance structures of the preceding three

$$\left[H_3C\!-\!\overset{+1}{\ddot{O}}\!=\!\overset{0}{C}\!=\!\overset{-1}{\ddot{N}} \quad \longleftrightarrow \quad H_3C\!-\!\overset{0}{\ddot{O}}\!-\!\overset{0}{C}\!\equiv\!\overset{0}{N}: \quad \longleftrightarrow \quad H_3C\!-\!\overset{+2}{O}\!\equiv\!\overset{0}{C}\!-\!\overset{-2}{\ddot{N}}: \right]$$

3.57   **a)** In the structure for PF_5, all atoms have formal charge (f.c.'s) of zero. Bonding on the F atoms obeys the octet rule but the P atom possesses an expanded octet (10 electrons).

b) In the structure for SF_4, bonding on all F atoms obeys the octet rule. The S has an expanded octet (10 electrons). All f.c.'s equal 0.

c) In the structure of XeO_2F_2, the F atoms and the O atoms obey the octet rule. The Xe "sees" an expanded octet of 12 electrons. All atoms have f.c. zero.

The Shapes of Molecules: Valence Shell Electron-Pair Repulsion Theory

3.59   Construct the following table based on text Table 3.8 and Figure 3.37

SN of A	Molecular Type X is a bonded atom; E is a lone pair	Predicted Shape
2	AX_2	linear
3	AX_3	trigonal planar
	AX_2E	bent
4	AX_4	tetrahedral
	AX_3E	trigonal pyramidal
	AX_2E_2	bent
5	AX_5	trigonal bipyramidal
	AX_4E	distorted see-saw
	AX_3E_2	distorted T
	AX_2E_3	linear
6	AX_6	octahedral
	AX_5E	square pyramidal
	AX_4E_2	square planar

a)   b)   c)   d)   e)

tetrahedral   trigonal planar   octahedral   pyramidal   distorted T

a) In CBr_4, the central C atom has SN $\boxed{4}$. There are no lone pairs on the central carbon atom so this is an AX_4 case. As shown in the table, the molecule is $\boxed{\text{tetrahedral}}$.

b) In SO_3, the central S atom has SN $\boxed{3}$ and no lone pairs. The fact that one or more of the S-to-O bonds can be shown in a Lewis structure as a double bond does *not* affect the steric number. The sulfur trioxide molecule is $\boxed{\text{trigonal planar}}$.

c) In SeF_6, the central Se atom has SN $\boxed{6}$. There are no lone pairs on the central Se atom. The molecule is $\boxed{\text{octahedral}}$.

d) In $OSCl_2$, the central S has SN $\boxed{4}$ and one lone pair. It is an AX_3E case (see the preceding table). The disposition of the electron pairs about the S is approximately tetrahedral. The molecular geometry however is defined solely by the locations of atoms; the molecule is $\boxed{\text{pyramidal}}$.

e) In ICl_3, the central I atom has SN $\boxed{5}$. It is surrounded by 3 Cl atoms and 2 lone pairs. The disposition of the electron pairs is trigonal bipyramidal. The molecule itself has a $\boxed{\text{distorted T}}$ shape.

3.61   **a)** The molecular ion ICl_4^- is square planar. The central I atom has SN $\boxed{6}$, which means the geometry of the electron pairs about the central I is octahedral. The 2 lone pairs lie opposite each other on the octahedral pattern, minimizing lone-pair to lone-pair interactions. The 4 Cl atoms surround the central I atom in a $\boxed{\text{square-planar}}$ fashion.

b) In OF_2, the central O atom has SN $\boxed{4}$. The molecule is of the type AX_2E_2.[3] The molecule is $\boxed{\text{bent}}$ to accommodate the two lone pairs on the O atom. The F—O—F angle should be less than $109.5°$ (experimentally it is $103.7°$).

c) In BrO_3^-, the central Br atom has SN $\boxed{4}$ making the molecular type of the molecular ion AX_3E. The single lone pair on the central Br atom occupies one corner of a tetrahedron about the Br atom.

---

[3]Refer to the table in problem **3.59** above.

The resulting molecule is $\boxed{\text{pyramidal}}$. The presence of the lone pair is expected to force the O atoms together somewhat, so that the O—Br—O angles are less than 109.5° (experimentally, 104.1°).

d) In CS_2, the central C atom has SN $\boxed{2}$. Both of the C-to-S bonds are double bonds, but this plays no part in figuring the SN of the C atom. The molecule, which is of the type AX_2, is $\boxed{\text{linear}}$, corresponding to a bond angle of 180°.

3.63 The symbol "B" in the generic formulas in this problem is equivalent to the "X" in text Table 3.8 and in the solution to problem **3.59** in this manual. It is not to be confused with the symbol for the element boron.

a) Planar AB_3 BF_3, BH_3, SO_3 **b)** Pyramidal AB_3 NH_3, NF_3
c) Bent AB_2^- ClO_2^-, NO_2^- **d)** Planar AB_3^{2-} CO_3^{2-}

3.65 A molecule or molecular ion has a dipole moment when the center of its spatial distribution of positive charge does not coincide with the center of its distribution of negative charge. The *bonds* in the listed compounds are all polar. The symmetry of certain molecular shapes however causes the vector sum of the individual bond dipoles to equal zero. Thus, CBr_4 (tetrahedral), SO_3 (trigonal planar), and SeF_6 (octahedral) are $\boxed{\text{non-polar molecules}}$. The molecules of ICl_3 (distorted T) and of $SOCl_2$ (pyramidal) are less symmetrical, and the vector sums of their bond dipoles are not zero. They are $\boxed{\text{polar}}$.

3.67 The fact that the molecule is bent $\boxed{\text{does not help}}$ in deciding between the isomeric formulations

$$\overset{-1\ +1\ \ 0}{:\!N\!=\!O\!-\!\ddot{F}\!:}\quad\text{and}\quad\overset{0\ \ \ 0\ \ \ 0}{O\!=\!N\!-\!\ddot{F}\!:}$$

Both structures feature a central atom having SN 3 with 2 bonds and 1 lone pair. VSEPR theory predicts a bent molecule in both cases.

Tip. The preceding Lewis structures include formal charges. As in problem **3.55**, Lewis structures with a build-up of non-zero formal charges are *disfavored* energetically. This suggests that the structure on the right might exist but not the one on the left. In fact, the one on the right is a colorless gas of reasonable stability named nitrosyl fluoride. The one on the left is highly unstable but has been characterized spectroscopically during the brief period between its synthesis and decomposition.

3.69 **a)** The resonance structures

$$\left[:\!\ddot{N}\!=\!N\!=\!\ddot{O}\!:\quad\longleftrightarrow\quad:\!N\!\equiv\!N\!-\!\ddot{\ddot{O}}\!:\right]$$

can be written for the NNO molecule. When either is considered, the SN of the central nitrogen atom equals two. The predicted molecular geometry is $\boxed{\text{linear}}$.

b) The linear geometry in NNO would cause the N—O and N—N bond dipoles to add vectorially to zero if they were equal in magnitude. The observed net dipole moment means that the two bond dipoles differ in magnitude. The N—O bond should be more polar than the N—N bond because O is more electronegative than N. The $\boxed{\text{N end}}$ of the molecule is therefore expected to have the positive partial charge.

Oxidation Numbers

3.71 The oxidation numbers are determined by the standard rules:

$SrBr_2$	Sr +2	Br −1	$Zn(OH)_4^{2-}$	Zn +2	O −2	H +1
SiH_4	Si −4	H +1	$CaSiO_3$	Ca +2	Si +4	O −2
$Cr_2O_7^{2-}$	Cr +6	O −2	KO_2	K +1	O −1/2	
CsH	Cs +1	H −1	$Ca_5(PO_4)_3F$	Ca +2	P +5	O −2 F −1

Inorganic Nomenclature

3.73 The theme here is the way that simple rules on the transfer of valence electrons explain the formation of ionic compounds.

a) Cesium chloride (CsCl) is a compound between chlorine and cesium. The Lewis symbols for the elements before and after reaction are $\text{Cs}\cdot\ +\ :\!\overset{\cdot\cdot}{\underset{\cdot\cdot}{\text{Cl}}}\!\cdot\ \longrightarrow\ \text{Cs}^+\ :\!\overset{\cdot\cdot}{\underset{\cdot\cdot}{\text{Cl}}}\!:^-$

b) Calcium and astatine form $CaAt_2$, calcium astatide. Each At atom gains an electron and the Ca atom loses two $\cdot\text{Ca}\cdot\ +\ 2\ :\!\overset{\cdot\cdot}{\underset{\cdot\cdot}{\text{At}}}\!\cdot\ \longrightarrow\ :\!\overset{\cdot\cdot}{\underset{\cdot\cdot}{\text{At}}}\!:^-\ (\text{Ca})^{2+}\ :\!\overset{\cdot\cdot}{\underset{\cdot\cdot}{\text{At}}}\!:^-$

c) Aluminum and sulfur form Al_2S_3, aluminum sulfide. Each S atom gains two electrons and each Al atom loses three

$$2\ \cdot\overset{\cdot}{\text{Al}}\cdot\ +\ 3\ \cdot\overset{\cdot\cdot}{\underset{\cdot\cdot}{\text{S}}}\cdot\ \longrightarrow\ (\text{Al}^{3+})_2\left(:\!\overset{\cdot\cdot}{\underset{\cdot\cdot}{\text{S}}}\!:\right)_3^{2-}$$

d) Potassium and tellurium form K_2Te, potassium telluride. Each Te atom gains two electrons and each K atom loses one $2\ \text{K}\cdot\ +\ \cdot\overset{\cdot\cdot}{\underset{\cdot\cdot}{\text{Te}}}\cdot\ \longrightarrow\ (\text{K})^+\ (:\!\overset{\cdot\cdot}{\underset{\cdot\cdot}{\text{Te}}}\!:)^{2-}\ (\text{K})^+$

3.75 It is probably best just to memorize the patterns of the nomenclature of simple inorganic compounds.

a) Al_2O_3 aluminum oxide   **b)** Rb_2Se rubidium selenide
c) $(NH_4)_2S$ ammonium sulfide   **d)** $Ca(NO_3)_2$ calcium nitrate
e) Cs_2SO_4 cesium sulfate   **f)** $KHCO_3$ potassium hydrogen carbonate

Another name for the last item is potassium bicarbonate.

3.77 The formulas of the anions come from text Table 3.9.
a) Silver cyanide $AgCN$ **b)** Calcium hypochlorite $Ca(OCl)_2$
c) Potassium chromate K_2CrO_4 **d)** Gallium oxide Ga_2O_3
e) Potassium superoxide KO_2 **f)** Barium hydrogen carbonate $Ba(HCO_3)_2$

3.79 The phosphate ion has the formula PO_4^{3-} (text Table 3.9). Trisodium phosphate (TSP) must have the formula $\boxed{Na_3PO_4}$.

The correct systematic name for TSP is $\boxed{\text{sodium phosphate}}$. The prefix "tri" is superfluous because the phosphate ion always has a charge of -3, and the sodium ion always has a charge of $+1$. The requirement for charge balance then gives the subscripts in the formula.

3.81 **a)** SiO_2 **b)** $(NH_4)_2CO_3$ **c)** PbO_2 **d)** P_2O_5 **e)** CaI_2 **f)** $Fe(NO_3)_3$.

3.83 **a)** Copper(I) sulfide and copper(II) sulfide   **b)** Sodium sulfate
c) Tetraarsenic hexaoxide (or hexoxide)   **d)** Zirconium(IV) chloride
e) Dichlorine heptaoxide or chlorine(VII) oxide   **f)** Gallium(I) oxide

ADDITIONAL PROBLEMS

3.85 The difference in electronegativities of the atoms in HF is 1.78; the analogous difference in LiCl is 2.18. The two compounds $\boxed{\text{differ}}$ greatly in their bonding according to the evidence of their physical properties. Lithium chloride is an ionic compound (high boiling, high melting); hydrogen fluoride is a covalent compound (low melting, low boiling).

3.87 **a)** Refer to text Figure 3.21, which shows the energetics of ionic bonding for a specific case. The critical distance R_c for the purposes of this problem is the point in this figure at which the potential energy curve for the ionic interaction (the red line) crosses the horizontal axis, that is, the value of R at which $V(R) = 0$. It is *not* the distance labelled R_c.

The quantity ΔE_∞ in text Figure 3.21 equals the difference between the first ionization energy of the alkali metal atom $M(g)$ and the electron affinity of the halogen atom $X(g)$. It is the energy required to extract an electron from an isolated $M(g)$ atom and put it on an isolated $X(g)$ atom. Under the approximation established in the problem, the decrease in potential energy from the Coulombic (electrostatic) attraction between an $M^+(g)$ and $X^-(g)$ becomes equal to ΔE_∞ at R_c:

$$\Delta E_\infty = -V_{\text{Coulomb}} = -\frac{q_1 q_2}{4\pi\epsilon_0 R_c}$$

The constant ϵ_0 is 8.854×10^{-12} $C^2\,J^{-1}m^{-1}$, and, in alkali halides, q_1 and q_2 are $+1.60218 \times 10^{-19}$ C and -1.60218×10^{-19} C. Solve the preceding for R_c and substitute the various values:

$$R_c = -\frac{q_1 q_2}{4\pi\epsilon_0 \Delta E_\infty} = \frac{+(2.3071 \times 10^{-28} \text{ J m})}{(IE - EA) \text{ J}}$$

Text Appendix F and other sources give IE's and EA's on a per mole basis. Revise the preceding equation, which applies to a single pair of particles, to allow the use of IE's and EA's in joule per mole. Do this by multiplying the numerator and denominator on the right side by N_A

$$R_c = \frac{(6.022 \times 10^{23} \text{ mol}^{-1})(2.3071 \times 10^{-28} \text{ J m})}{(IE - EA) \text{ J mol}^{-1}} = \boxed{\frac{1.3894 \times 10^{-4} \text{ J m mol}^{-1}}{(IE - EA) \text{ J mol}^{-1}}}$$

b) From Appendix F for LiF, $(IE - EA)$ is $520.2 - 328.0 = 192.2 \times 10^3$ J mol^{-1}. Substitution in the preceding equation gives an R_c of $\boxed{7.23 \times 10^{-10} \text{ m}}$.

For KBr, $(IE - EA)$ equals 94.1×10^3 J mol^{-1} giving R_c equal to $\boxed{14.8 \times 10^{-10} \text{ m}}$.

For NaCl, $(IE - EA)$ equals 146.8×10^3 J mol^{-1}, making R_c equal to $\boxed{9.46 \times 10^{-10} \text{ m}}$.

3.89   The molecule C_2 has no single-, double- or triple-bonded Lewis structure that satisfies the octet rule. A double-bonded structure, unlike the others, at least puts zero formal charges on both atoms. On this basis, predict that the $\boxed{\text{bond order is 2.}}$

The bond length in double-bonded C_2 should be close to 1.34 Å, which is the bond length given in text Table 3.5 for the double bond in C_2H_4. The value 1.31 Å $\boxed{\text{is consistent}}$ with this prediction, being only slightly smaller.

3.91   **a)** A Lewis structure for OPCl in which the octet rule is obeyed for all atoms and all atoms have a formal charge of zero is $\ddot{\underset{\cdot\cdot}{O}}{=}\ddot{P}{-}\ddot{\underset{\cdot\cdot}{Cl}}{:}$

Tip. The oxidation numbers of the atoms in OPCl are oxygen -2, chlorine -1, phosphorus $+3$. The formal charges are zero on all three atoms. The oxidation numbers equal the charge that each atom would have if all of the bonds were broken and the electrons in each bond went with the more electronegative atom. The formal charges equal the charge that each atom would have if all of the bonds were broken and the electrons in the bonds were shared equally between the atoms that they connected before the break-up.

b) Three resonance Lewis structure for O_2PCl are given below. In the structure on the left, all atoms have f.c. zero, but the octet rule is violated on the P atom, which sees 10 electrons. In the two structures to the right, the octet rule is obeyed for every atom, but formal charges exist as shown. All three contribute to the "true" bonding:

3.93   Taken together, the following resonance structures of nitryl chloride imply equivalent N-to-O bonds:

Tip. Nitryl chloride is a reactive gas that decomposes readily to nitrogen dioxide (see problem **3.95**) and chlorine. Despite this, it is named as a salt: nitryl ion, a +1 ion, in combination with chloride ion, a −1 ion. The following Lewis structure takes up the suggestion of this name by showing ionic bonding to the Cl. Note that the two N-to-O bonds are still equivalent, but the pair of electrons previously between N and Cl now belongs entirely to the Cl. The octet rule is satisfied for all atoms:

3.95   **a)** The molecule of nitrogen dioxide has 17 valence electrons. At least one atom cannot achieve an octet. The following four resonance structures are the candidates for best single Lewis structure:

$$\left[\overset{0}{:}\overset{+1}{O}=\overset{-1}{N}-\overset{}{O}: \longleftrightarrow :\overset{-1}{O}-\overset{+1}{N}=\overset{0}{O}: \longleftrightarrow \cdot\overset{0}{O}-\overset{0}{N}=\overset{0}{O}: \longleftrightarrow \overset{0}{O}=\overset{0}{N}-\overset{0}{O}\cdot \right]$$

Other candidate structures break the octet rule on more than one atom or use octet expansion rather than octet deficiency. If the N atom is octet-deficient (left two structures), then formal charges build up as indicated just above the symbols for the atoms. If an O is octet-deficient (right two structures), all atoms have f.c. zero. The best single structure puts the odd electron on the ⌐O atoms⌐.

b) Four resonance structure can be drawn varying the relative positions of the double and single bonds:

3.97   The xenon atom in the XeF⁺ ion has an octet of electrons but bears a formal charge of +1. In the XeF₂ molecule the xenon atom has an expanded octet but a formal charge of zero.

a) $\left[:\overset{..}{\underset{..}{Xe}}-\overset{..}{\underset{..}{F}}: \right]^{+}$ **b)** $:\overset{..}{\underset{..}{F}} - \overset{..}{Xe} - \overset{..}{\underset{..}{F}}:$
$\quad\quad\quad\ _{+1}$

3.99   **a)** The Lewis structures are

$$\overset{..}{N}=\overset{..}{S}-\overset{..}{\underset{..}{F}}: \quad \text{and} \quad \overset{..}{\underset{..}{S}}=\overset{..}{N}-\overset{..}{\underset{..}{F}}: \quad \text{and} \quad \overset{..}{\underset{..}{N}}=\overset{..}{\underset{..}{F}}-\overset{..}{\underset{..}{S}}:$$

In the first, the formal charges are (from left to right) −1, +1, and 0. In the second, all three atoms have zero formal charge. In the third, the formal charges are −1, +2, and −1.

b) The structure having the least separation of formal charge has the central N. The observation of a central S atom is ⌐not consistent⌐ with the hopeful statement that appears in the problem.

c) The electronegativity of N exceeds that of S; that is, N has a greater tendency than S to accept electrons in a chemical bond. This ⌐does help⌐ explain why the observed structure (the one on the left in the preceding) corresponds to a formal build-up of negative charge on the N and formal build-up of positive charge on the S. Also, the two most electronegative atoms (N and F) are separated, reducing electron-electron repulsions.

3.101   **a)** In $SbCl_5^{2-}$ ion, the central Sb atom has *SN* 6. It is surrounded by five bonding pairs (single bonds to the five chlorine atoms) and one lone pair. As an AX_5E case,[4] the molecular ion is $\boxed{\text{square pyramidal}}$.

b The central Sb atom in $SbCl_6^{3-}$ ion has *SN* $\boxed{7}$. This steric number is rare. The required extension of VSEPR theory is inclusion of *SN* 7 and other higher *SN*'s. In fact, three geometries have been observed for *SN* 7: pentagonal bipyramidal, capped octahedral (in which the seventh atom occupies one face of an octahedron surrounding the central atom) and capped trigonal prism (in which the seventh atom occupies one rectangular face of a trigonal prism about the central atom).

3.103   The central S atom in F_4SO has *SN* 5 and falls into the class AX_5. The geometry of the molecule is therefore based on the $\boxed{\text{trigonal bipyramid}}$. The real question is whether the oxygen atom is equatorial or axial. Putting the double-bonded oxygen atom at an equatorial position minimizes 90° interactions with the four fluorine atoms and should be preferred, according to VSEPR theory. Also, the F—S—F angles will be slightly less than 90°, 120°, and 180°.

3.105   Set up a coordinate system and position the oxygen atom at its origin. Let the y axis bisect the angle θ defined by H—O—H:

The dipole moments of the two O—H bonds are symbolized μ_{OH}. They parallel the two bonds and are represented by vectors (the two arrows) in the diagram. The x components of the vectors oppose each other and cancel. The y components point in the same direction and add. The magnitude of the y components is $\mu_{OH}\cos(\theta/2)$, and the sum of the two y components equals the dipole moment of the molecule as a whole. Therefore

$$\mu(H_2O) = 2\mu_{OH}\cos\left(\frac{\theta}{2}\right)$$

Compute μ_{OH} in the case that $\mu(H_2O) = 1.86$ D by substitution in the preceding

$$1.86\ \text{D} = 2\mu_{OH}\cos\left(\frac{104.5°}{2}\right) \qquad \text{from which} \qquad \mu_{OH} = \boxed{1.52\ \text{D}}$$

3.107   From the formula Bi_5^{3+}, the oxidation number (or at least the average oxidation number) of bismuth is $\boxed{+3/5}$. Because the oxidation number of F is -1 by convention, the oxidation number of the As is $\boxed{+5}$. The elevated dot in the formula means that SO_2 is loosely associated with this salt in its solid state. The oxidation number of S in SO_2 is $\boxed{+4}$; the oxidation number of O is $\boxed{-2}$.

3.109   **a)** The oxidation number of lead in PbO is $+2$; in PbO_2, it is $+4$; in Pb_2O_3, $+3$; in Pb_3O_4, $+8/3$.

b) Observe that the formula "Pb_2O_3" is the sum of PbO and PbO_2. Perhaps in Pb_2O_3 half of the lead is Pb(II) and half is Pb(IV), as in "$(PbO)(PbO_2)$." Similarly, the lead in Pb_3O_4 may be 2/3 Pb(II) and 1/3 Pb(IV) as in "$(PbO)_2(PbO_2)$."

---

[4]See the table on page 28 of this Manual.

CUMULATIVE PROBLEMS

3.111 **a)** The element M loses two electrons in forming compounds, since it forms compounds MCl_2 and MO. It belongs to group II, the alkaline-earth metals.

b) The relative molecular mass of MCl_2 equals the sum of the relative masses of the three constituent atoms: $x + 2(35.453) = x + 70.906$ where x stands for the relative atomic mass of the element M. The fraction of mass that is chlorine is 0.447. So:

$$0.447 = \frac{70.906}{(x + 70.906)}$$

Solving gives $x = 87.7$. A look at a list of relative atomic masses identifies M as strontium.

3.113 **a)** The elemental analysis reveals that the compound contains F, Cl, O, and no other elements. Divide the respective mass percentages by the molar masses of the elements to obtain the relative number of moles of each. The ratios correspond to the empirical formula ClO_3F.

b) The central atom in the molecule ClO_3F is certainly the Cl. It is the least electronegative of the choices. One good Lewis structure is

Tip. This structure shows an expanded octet on the central Cl, but the formal charges equal zero on all atoms, which is desirable. Resonance structures can be written in which a pair of electron is shifted from one of the double bonds to reside completely on an O. These structures put a formal charge of −1 on the oxygen that becomes single-bonded to the Cl and a formal charge of +1 on the Cl. Doing this with all three oxygen gives a Lewis structure with a formal charge of +3 on the Cl and formal charges of −1 on the oxygens. Such a build-up of formal charge is too heavy a price to pay for attaining an octet on the Cl. A Lewis structure with all single bonds is a poor answer to this problem.

c) The central Cl has a steric number of four. VSEPR theory predicts a structure based on the tetrahedron. Because F is more electronegative than O, it tends to attract electrons away from the central Cl, reducing the electron-pair repulsion and causing the F—Cl—O angles to become smaller than tetrahedral while the O—Cl—O angles become larger than tetrahedral.

Chapter 4

Introduction to Quantum Mechanics

Preliminaries: Wave Motion and Light

4.1 The speed of propagation of a wave equals its frequency multiplied by its wavelength. A wave-crest hits the beach once every 3.2 s, which means that slightly less than one-third of a wave reaches the beach per second. More exactly, the frequency of these waves equals the reciprocal of 3.2 s, which is 0.3125 s^{-1}. Multiply this frequency by the wavelength, which is given, to obtain the speed of the waves

$$\text{speed} = \nu\lambda = \frac{1}{3.2\text{ s}}(2.1\text{ m}) = \boxed{0.66\text{ m s}^{-1}}$$

Tip. The problem quotes the duration between identical recurring points on a wave. This interval of time is the *period* of the wave. The period equals the reciprocal of the frequency.

4.3 The speed of propagation of electromagnetic radiation (light) through a vacuum is symbolized c. It equals 299 792 458 m s^{-1} exactly.[1] As with all other traveling waves, the speed of propagation of FM radio waves equals the product of its wavelength and frequency: $c = \lambda\nu$. Hence

$$\lambda = \frac{c}{\nu} = \frac{2.9979 \times 10^8\text{ m s}^{-1}}{9.86 \times 10^7\text{ s}^{-1}} = \boxed{3.04\text{ m}}$$

4.5 **a)** Solve $c = \lambda\nu$ for ν, the frequency and substitute

$$\nu = \frac{c}{\lambda} = \frac{2.9979 \times 10^8\text{ m s}^{-1}}{6.00 \times 10^2\text{ m}} = \boxed{5.00 \times 10^5\text{ s}^{-1}}$$

b) The time for a wave to travel a distance d equals the distance divided by the speed of the wave. These electromagnetic waves advance at the known speed c. Hence

$$t = \frac{d}{c} = \frac{8.0 \times 10^{10}\text{ m}}{3.00 \times 10^8\text{ m s}^{-1}} \times \left(\frac{1\text{ min}}{60\text{ s}}\right) = \boxed{4.4\text{ min}}$$

4.7 The wavelength of the sound waves can be determined from their frequency and speed of propagation

$$\lambda = \frac{\text{speed}}{\nu} = \frac{343.5\text{ m s}^{-1}}{261.6\text{ s}^{-1}} = \boxed{1.313\text{ m}}$$

The time required to travel 30.0 m is

$$t = \frac{d}{\text{speed}} = \frac{30.0\text{ m}}{343.5\text{ m s}^{-1}} = \boxed{0.0873\text{ s}}$$

---

[1]This value is *exact* because the meter is now defined as the distance travelled by light in vacuum during a time interval of 1/299 792 458 of a second.

Evidence for Energy Quantization in Atoms

4.9   Solve the equation given in the problem for the temperature T in terms of λ_{max} and then substitute. Note that k and k_B both serve to symbolize the Boltzmann constant.

$$T = \frac{0.20\, hc}{k_B\, \lambda_{max}} = \frac{0.20\,(6.626 \times 10^{-34}\ \text{J s})(2.9979 \times 10^8\ \text{m s}^{-1})}{(1.38066 \times 10^{-23}\ \text{J K}^{-1})(1.05 \times 10^{-3}\ \text{m})}$$

$$= 2.74 \frac{\text{J s m s}^{-1}}{\text{J K}^{-1}\ \text{m}} = \boxed{2.7\ \text{K}}$$

The existence of this radiation supports the "hot big bang" theory of the origin of the universe.

4.11   The strong emission from Li atoms is at a wavelength of 671 nm, which is 6.71×10^{-7} m. Text Figure 4.3 shows that light of this wavelength is $\boxed{\text{red}}$.

4.13   Compute the frequency corresponding to this transition energy in the barium atom using the Planck equation $\Delta E = h\nu$

$$\nu = \frac{\Delta E}{h} = \frac{3.6 \times 10^{-19}\ \text{J}}{6.626 \times 10^{-34}\ \text{J s}} = 5.43 \times 10^{14}\ \text{s}^{-1}$$

The wavelength equals the speed of propagation divided by the frequency

$$\lambda = \frac{c}{\nu} = \frac{2.9979 \times 10^8\ \text{m s}^{-1}}{5.43 \times 10^{14}\ \text{s}^{-1}} = \boxed{5.5 \times 10^{-7}\ \text{m}}$$

According to text Figure 4.3, this wavelength of light is $\boxed{\text{green}}$.

4.15   **a)** The energy change that an atom of sodium (or of anything else) experiences as it emits a quantum of radiation is inversely related to the wavelength λ of the radiation

$$-\Delta E_{Na} = \frac{hc}{\lambda}$$

The λ is 589.3 nm (or 5.893×10^{-7} m). Substitution of the values of λ, h, and c gives

$$\Delta E_{Na} = \frac{-(6.626 \times 10^{-34}\ \text{J s})(2.998 \times 10^8\ \text{m s}^{-1})}{5.893 \times 10^{-7}\ \text{m}} = \boxed{-3.371 \times 10^{-19}\ \text{J}}$$

The negative sign means that the final energy of the sodium atom is less than its original energy.

b) A mole of sodium atoms consists of Avogadro's number of sodium atoms. The energy change per mole is

$$\Delta E_{Na} = \left(\frac{-3.371 \times 10^{-19}\ \text{J}}{\text{atom}}\right)\left(\frac{6.022 \times 10^{23}\ \text{atom}}{\text{mol}}\right) = \boxed{-2.030 \times 10^5\ \text{J mol}^{-1}}$$

c) The sodium arc light is supposed to emit energy at the rate of 1000 W (watt), which is 1000 J s^{-1}, at the D-line. Then

$$n_{Na} = 1\ \text{s} \times \left(\frac{1000\ \text{J}}{1\ \text{s}}\right)\left(\frac{1\ \text{mol Na}}{+2.030 \times 10^5\ \text{J}}\right) = \boxed{4.926 \times 10^{-3}\ \text{mol Na}}$$

4.17   The observed voltage of the first excitation threshold in the Franck-Hertz experiment on Na atoms equals 2.103 V. This means that an electron accelerated across a potential difference of 2.103 V is just energetic enough to transfer a quantum of energy to a ground-state Na atom, which does not accept smaller quanta of energy. The change in the energy of the Na atom that accepts the quantum of energy equals the accelerating voltage multiplied by the charge on the electron

$$\Delta E_{Na} = (2.103\ \text{V})(1.602177 \times 10^{-19}\ \text{C}) = 3.3694 \times 10^{-19}\ \text{V C} = 3.3694 \times 10^{-19}\ \text{J}$$

Later, the excited Na atom spontaneously emits a quantum of light as it relaxes to its original state. Its ΔE during the relaxation equals the negative of its ΔE during the excitation. The wavelength of the light that it emits is

$$\lambda = \frac{hc}{-\Delta E_{\text{Na}}} = \frac{(6.626 \times 10^{-34} \text{ J s})(2.9979 \times 10^8 \text{ m s}^{-1})}{-(-3.3694 \times 10^{-19} \text{ J})} = \boxed{5.895 \times 10^{-7} \text{ m}} = 5895 \text{ Å}$$

Tip. The answer is quite close to 5893 Å (589.3 nm), the center of the close-spaced doublet mentioned in problem **4.15**. The wavelengths of the lines making up this doublet have been precisely measured. In air, they are 5890.00 Å and 5895.98 Å, which have an average of 5892.99 Å (589.299 nm). These wavelengths differ from the wavelengths measured in a vacuum because the speed of light is slightly slower in air than in a vacuum.

The Bohr Model: Predicting Discrete Energy Levels in Atoms

4.19 The B^{4+} ion is a hydrogen-like ion. Like H, it has only one electron. Unlike H, its atomic number Z is 5 (corresponding to a nuclear charge of $+5e$), not 1. It is also in the $n = 3$ excited state. This state is a higher energy state than the $n = 1$ state, which is the ground state. Substitute this Z and the $n = 3$ into text equations 4.12 and 4.14a, which arise from the Bohr model, and which are

$$r_n = \frac{n^2}{Z}(5.29 \times 10^{-11} \text{ m}) \quad \text{and} \quad E_n = -(2.18 \times 10^{-18} \text{ J})\frac{Z^2}{n^2}$$

The radius and energy of the B^{4+} ion in its $n = 3$ state equal

$$r_3 = \frac{3^2}{5}(5.29 \times 10^{-11} \text{ m}) = \boxed{9.52 \times 10^{-11} \text{ m}} \qquad E_3 = -\frac{5^2}{3^2}(2.18 \times 10^{-18} \text{ J}) = \boxed{-6.06 \times 10^{-18} \text{ J}}$$

The negative of the second answer is the input of energy needed to remove the electron from a single B^{4+} ion in its $n = 3$ state. Multiply by Avogadro's number to put this on the basis of a mole of B^{4+} ions

$$E = \frac{-(-6.06 \times 10^{-18} \text{ J})}{\text{atom}} \times \left(\frac{6.022 \times 10^{23} \text{ atom}}{\text{mol}}\right) = \boxed{3.65 \times 10^6 \text{ J mol}^{-1}}$$

The energy change in a B^{4+} ion undergoing the $3 \rightarrow 2$ transition equals the difference between the energies of the two states. The two energies are

$$E_3 = -\frac{5^2}{3^2}(2.18 \times 10^{-18} \text{ J}) = -\frac{25}{9}(2.18 \times 10^{-18} \text{ J})$$

$$E_2 = -\frac{5^2}{2^2}(2.18 \times 10^{-18} \text{ J}) = -\frac{25}{4}(2.18 \times 10^{-18} \text{ J})$$

The inviolable convention in getting a difference (a Δ) is to subtract the initial value from the final value. Do the subtraction

$$\Delta E_{3 \rightarrow 2} = E_2 - E_3 = \left(\frac{-25}{4} - \frac{-25}{9}\right)(2.18 \times 10^{-18} \text{ J}) = -7.57 \times 10^{-18} \text{ J}$$

This change in energy is negative because the B^{4+} ion loses energy. It relaxes from the higher energy state to the lower. The further transition $2 \rightarrow 1$ would cause the ion to lose more energy, relaxing into its ground state, from which it can relax no further.

The energy gained by the surroundings in the form of a photon in the $3 \rightarrow 2$ transition is $+7.57 \times 10^{-18}$ J. Dividing by h gives the frequency of the photon

$$\nu = \frac{-\Delta E}{h} = \frac{-(-7.57 \times 10^{-18} \text{ J})}{6.626 \times 10^{-34} \text{ J s}} = \boxed{1.14 \times 10^{16} \text{ s}^{-1}}$$

The wavelength of the photon is

$$\lambda = \frac{c}{\nu} = \frac{2.9979 \times 10^8 \text{ m s}^{-1}}{1.14 \times 10^{16} \text{ s}^{-1}} = \boxed{2.63 \times 10^{-8} \text{ m}}$$

4.21   Text equation 4.14a gives the energy of hydrogen-like ions in their different allowed quantum states

$$E_n = (-2.18 \times 10^{-18} \text{ J})\frac{Z^2}{n^2}$$

Use this equation to compute the difference between the energy of the Li^{2+} ion in its $n = 3$ state and in its $n = 2$ state. Set $Z = 3$ because $+3$ is the nuclear charge of lithium. The initial state is $n = 3$ and the final state is $n = 2$. As always, subtract the initial value from the final value

$$\Delta E = E_2 - E_3 = (-2.18 \times 10^{-18} \text{ J})\left(\frac{3^2}{2^2} - \frac{3^2}{3^2}\right) = -2.725 \times 10^{-18} \text{ J}$$

The negative answer means that the ion loses energy. This energy appears in the surroundings in the form of a photon. Figure the wavelength of the photon

$$\lambda = \frac{hc}{-\Delta E} = \frac{(6.626 \times 10^{-34} \text{ J s})(2.9979 \times 10^8 \text{ m s}^{-1})}{-(-2.725 \times 10^{-18} \text{ J})} = \boxed{72.90 \times 10^{-9} \text{ m}}$$

Light of this wavelength is in the $\boxed{\text{ultraviolet}}$ part of the spectrum.

Tip. The wavelength of the $3 \to 2$ emission in hydrogen (656.1 nm) was given but not used. Having it allows a quicker answer as follows: the $3 \to 2$ transition for the H atom ($Z = 1$) has its ΔE proportional to $1^2 \times (1/2^2 - 1/3^2)$ while the $3 \to 2$ transition for the Li^{2+} ion ($Z = 3$) has *its* ΔE proportional to $3^2 \times (1/2^2 - 1/3^2)$. The ΔE is obviously 9 times larger in the Li^{2+} case. Therefore the wavelength of the emitted light in the Li^{2+} ion's $3 \to 2$ transition is 9 times *shorter* than 656.1 nm. This is $(656.1/9)$ nm or 72.90 nm.

Evidence for Wave-Particle Duality

4.23   Blue light has a higher frequency than green light (see text Figure 4.3). Photons of blue light are therefore more energetic than photons of green light. Inasmuch as the work function of the surface of the potassium is the same for both colors of light, the $\boxed{\text{electrons ejected by blue light}}$ have higher average kinetic energy.

4.25   Combine the relationships $c = \lambda\nu$ and $E = h\nu$ to obtain an equation for the wavelength of light in terms of its energy. Then replace E with the work function of cesium, which is the minimum energy needed to eject electrons from a surface made of that metal

$$\lambda = \frac{c}{\nu} = \frac{hc}{E} = \frac{(6.626 \times 10^{-34} \text{ J s})(3.00 \times 10^8 \text{ m s}^{-1})}{3.43 \times 10^{-19} \text{ J}} = 5.80 \times 10^{-7} \text{ m}$$

This result is the *maximum* wavelength of light that can eject electrons from a cesium surface in the photoelectric experiment. Light of this wavelength is $\boxed{\text{yellow}}$ (see text Figure 4.3). Light of shorter wavelengths, such as $\boxed{\text{green, blue, indigo, violet}}$, also works.

For selenium, the work function is larger (more energy is required to eject electrons)

$$\lambda = \frac{hc}{E} = \frac{(6.626 \times 10^{-34} \text{ J s})(3.00 \times 10^8 \text{ m s}^{-1})}{9.5 \times 10^{-19} \text{ J}} = 2.1 \times 10^{-7} \text{ m}$$

This maximum wavelength is well into the ultraviolet. All visible light has longer wavelengths than this. $\boxed{\text{No visible light}}$ will eject electrons from selenium.

4.27 a) The photoelectric experiment is performed with a chromium surface and 250 nm radiation. Use text equation 4.18

$$E_{\max} = h\nu - \Phi = \frac{hc}{\lambda} - \Phi$$

$$= \frac{(6.626 \times 10^{-34} \text{ J s})(3.00 \times 10^8 \text{ m s}^{-1})}{2.50 \times 10^{-7} \text{ m}} - 7.21 \times 10^{-19} \text{ J} = \boxed{7.4 \times 10^{-20} \text{ J}}$$

b) The kinetic energy $\mathcal{T}$ of a particle depends on its speed and mass: $\mathcal{T} = \frac{1}{2}mv^2$. Solve this equation for v, substitute $E_{\max}$ from the previous part for $\mathcal{T}$, and insert the mass of the electron

$$v = \sqrt{\frac{2\mathcal{T}}{m}} = \sqrt{\frac{2 E_{\max}}{m_e}} = \sqrt{\frac{2(0.74 \times 10^{-19} \text{ J})}{9.109 \times 10^{-31} \text{ kg}}} = \boxed{4.0 \times 10^5 \text{ m s}^{-1}}$$

Tip. This speed is less than 0.2% of the speed of light, so relativistic effects (not mentioned in this chapter of the text) are safely ignored.

4.29 The wave in a guitar string is a standing wave. Its allowed wavelength satisfies the equation

$$\frac{n\lambda}{2} = L$$

in which L is the length of the string and n is an integer.

a) The first and third harmonics have $n = 1$ and $n = 3$ respectively. Substitution in the preceding equation gives

$$\lambda_1 = \frac{2L}{1} = \frac{2(50 \text{ cm})}{1} = \boxed{100 \text{ cm}} \qquad \lambda_3 = \frac{2L}{3} = \frac{2(50 \text{ cm})}{3} = \boxed{33 \text{ cm}}$$

b) The number of nodes in a standing wave in a vibrating string that is fixed at both ends is always one less than the number of the harmonic; the third harmonic has $\boxed{\text{two nodes}}$.

4.31 The deBroglie wavelength λ of an object is given by $\lambda = h/p$ where p is the momentum of the object. The momentum of an object equals its mass multiplied by its velocity.

a) For an electron moving at 1.00×10^3 m s^{-1}

$$\lambda_e = \frac{h}{p} = \frac{h}{m_e v} = \frac{6.626 \times 10^{-34} \text{ J s}}{(9.11 \times 10^{-31} \text{ kg})(1.00 \times 10^3 \text{ m s}^{-1})} = \boxed{7.27 \times 10^{-7} \text{ m}}$$

b) For a proton moving at the same speed

$$\lambda_p = \frac{h}{m_p v} = \frac{6.626 \times 10^{-34} \text{ J s}}{(1.673 \times 10^{-27} \text{ kg})(1.00 \times 10^3 \text{ m s}^{-1})} = \boxed{3.96 \times 10^{-10} \text{ m}}$$

c) A speed of 75 km h^{-1} is equivalent to 20.8 m s^{-1} (multiply by 1000 m km^{-1} and then divide by 3600 s h^{-1}). A 145 g baseball has a mass of 0.145 kg.

$$\lambda_{\text{ball}} = \frac{h}{(mv)_{\text{ball}}} = \frac{6.626 \times 10^{-34} \text{ J s}}{(0.145 \text{ kg})(20.8 \text{ m s}^{-1})} = \boxed{2.2 \times 10^{-34} \text{ m}}$$

4.33 a) A moving electron has a wavelength that depends inversely on its momentum (the DeBroglie relationship). The momentum in turn depends on the kinetic energy $\mathcal{T}$

$$\lambda_e = \frac{h}{p_e} = \frac{h}{m_e v} = \frac{h}{\sqrt{2 m_e \mathcal{T}}}$$

The wave property of electrons allows them to be diffracted. Use the preceding equation to figure out the wavelength of the electron when its kinetic energy is 45 eV

$$\lambda_e = \frac{6.626 \times 10^{-34} \text{ J s}}{\sqrt{2 \left(9.11 \times 10^{-31} \text{ kg}\right)\left(45 \text{ eV}\right)\left(1.602 \times 10^{-19} \text{ J eV}^{-1}\right)}} = 1.828 \times 10^{-10} \frac{\text{J s}}{\sqrt{\text{kg J}}}$$

$$= 1.828 \times 10^{-10} \frac{\left(\text{kg m}^2 \text{ s}^{-2}\right) \text{ s}}{\text{kg}^{1/2} \left(\text{kg}^{1/2} \text{ m s}^{-1}\right)} = 1.8 \times 10^{-10} \text{ m}$$

The LEED (low-energy electron diffraction) experiment is a modern version of the Davisson-Germer experiment. A beam of low-energy electrons hits the surface of a crystalline specimen at right angles to the surface. Diffracted electrons come back from the surface at various angles relative to the in-coming beam and are registered by a detector. The angles are related to the electron wavelength by the equation[2]

$$D \sin \phi = n \lambda_e$$

where n is the order of the diffraction (an integer) and D is the crystal spacing, the atom-to-atom distance along the lines of atoms on the surface that are causing the diffraction. Solve for D and substitute the given ϕ, which is 53°

$$D = n \frac{\lambda_e}{\sin \phi} = n \frac{1.828 \times 10^{-10} \text{ m}}{\sin 53°} = n \left(289 \times 10^{-10} \text{ m}\right)$$

Assume this to be a first-order diffracted beam. Then n is 1, and the spacing is $\boxed{2.3 \times 10^{-10} \text{ m}}$.

b) The wavelength of electrons having energies of 90 eV is

$$\lambda_e = \frac{6.626 \times 10^{-34} \text{ J s}}{\sqrt{2 \left(9.11 \times 10^{-31} \text{ kg}\right)\left(90 \text{ eV}\right)\left(1.602 \times 10^{-19} \text{ J eV}^{-1}\right)}} = 1.293 \times 10^{-10} \text{ m}$$

Solve $D \sin \phi = n \lambda_e$ for ϕ and substitute for D and λ_e

$$\phi = \arcsin \frac{n \lambda_e}{D} = \arcsin \left(\frac{(1)(1.293 \times 10^{-10} \text{ m})}{2.289 \times 10^{-10} \text{ m}}\right) = \boxed{34°}$$

Tip. The 90 eV electron in part **b)** has double the energy of the 45 eV electron in part **a)**. Notice that its wavelength equals the wavelength of the 45 eV electron divided by $\sqrt{2}$.

4.35   **a)** According to the Heisenberg uncertainty principle

$$(\Delta x)(\Delta p) \geq \frac{h}{4\pi} \quad \text{hence} \quad \Delta p_{\min} = \frac{h}{4\pi}\left(\frac{1}{\Delta x}\right)$$

The minimum uncertainty in the momentum of the electron is then

$$\Delta p_{\min} = \frac{6.626 \times 10^{-34} \text{ J s}}{4\pi(1.0 \times 10^{-9} \text{ m})} = \frac{6.626 \times 10^{-34} \text{ kg m}^2 \text{ s}^{-2} \text{ s}}{4\pi(1.0 \times 10^{-9} \text{ m})} = 5.27 \times 10^{-26} \text{ kg m s}^{-1}$$

Compute the minimum uncertainty in the velocity using the definition of linear momentum. It is the product of velocity and mass

$$\Delta v_{\min} = \frac{\Delta p_{\min}}{m_e} = \frac{(5.27 \times 10^{-26} \text{ kg m s}^{-1})}{9.11 \times 10^{-31} \text{ kg}} = \boxed{5.8 \times 10^4 \text{ m s}^{-1}}$$

b) The minimum uncertainty in the momentum of a He atom is the same as for an electron. Computing the minimum uncertainty of the velocity requires the mass of a helium atom, which can be

---

[2]Note that this equation differs from Bragg's law (text equation 4.24).

computed from the molar mass of He (by dividing it by Avogadro's number) or looked up. It is 6.647×10^{-27} kg. Substitution then gives

$$\Delta v_{\min} = \frac{\Delta p_{\min}}{m_{\mathrm{He}}} = \frac{5.27 \times 10^{-26} \text{ kg m s}^{-1}}{6.647 \times 10^{-27} \text{ kg}} = \boxed{7.9 \text{ m s}^{-1}}$$

Quantum Mechanics of Particle-in-a-Box Models

4.37   **a)** The wave equation for a particle contained in a one-dimensional box has the very familiar sine function as its solution. The crucial steps toward drawing the graphs are to set up the axes with the proper units, to locate the zeroes of the sine wave, to figure the locations and magnitudes of its maxima and minima.

Text equation 4.36 gives the form of the functions that solve the wave equation for a particle in a one-dimensional box

$$\psi_n(x) = \sqrt{\frac{2}{L}} \sin\left(\frac{n \pi x}{L}\right)$$

where the integer n indicates which wave-function is to be written and L gives the size of the box. The problem says that $n = 3$ and $L = 7$ Å. Therefore

$$\psi_3(x) = \sqrt{\frac{2}{7 \text{ Å}}} \sin\left(\frac{3 \pi x}{7 \text{ Å}}\right)$$

The argument of the sine function must be unit-less. This makes the Ångstrom the natural choice for the unit of L, which should be placed on the x-axis in both graphs. The factor $\sqrt{2/7 \text{ Å}}$ in front of the sine function then has the unit $\text{Å}^{-1/2}$. This means that ψ_3 comes out in units of $\text{Å}^{-1/2}$. This unit should be used for marking out the y-axis of the first graph.

The form of $\psi_3(x)$ is a sine curve with zeroes at $x = 0$ and $x = 7$ Å. These two zeroes occur because the wave-function must vanish at the boundaries of the box, which has walls that are infinitely high. Two more zeroes occur *inside* the box, at $x = (7/3)$ Å and $x = (14/3)$ Å. These values of x cause the argument of the sine function to equal π and 2π, and $\sin \pi = \sin 2\pi = 0$. Interior zeroes are *nodes* of the wave-function. The function $\psi_3(x)$ attains alternating maxima and minimum as the argument of the sine function equals $\pi/2$, $3\pi/2$, $5\pi/2$ and so forth. The values of x between 0 and 7 Å that generate these arguments are $(7/6)$ Å, $(21/6)$ Å, and $(35/6)$ Å. The corresponding values of $\psi_3(x)$ are $+0.535 \text{ Å}^{-1/2}$, $-0.535 \text{ Å}^{-1/2}$, and $+0.535 \text{ Å}^{-1/2}$.

The square of the $n = 3$ wave-function is

$$\left(\psi_3(x)\right)^2 = \frac{2}{7 \text{ Å}} \sin^2\left(\frac{3 \pi x}{7 \text{ Å}}\right) = \frac{2}{7 \text{ Å}} \left(\sin\left(\frac{3 \pi x}{7 \text{ Å}}\right)\right)^2$$

The unit of $\left(\psi_3(x)\right)^2$ is the Å^{-1}. The y-axis should be marked off with this unit in the graph. The squared wave-function goes to zero at the boundaries of the box just like the un-squared wave-function. It has interior zeroes at $x = (7/3)$ Å and $x = (14/3)$ Å, just like the un-squared wave-function. Unlike the un-squared wave-function, it has three maxima, at $x = (7/6)$ Å, $(21/6)$ Å, and $(35/6)$ Å. The values of $(\psi_3)^2$ at these maxima are all 0.286 Å^{-1}, the square of $\pm 0.535 \text{ Å}^{-1/2}$.

The preceding information allows production of good plots of $\psi_3(x)$ and $(\psi_3(x))^2$ versus x in this particular box.

The probability p of finding the particle between $x = 3$ and $x = 4$ Å equals the integral of the square of the wave-function evaluated between these limits. The antiderivative of $\sin^2 x$ is $x/2 - 1/4 \sin 2x$. Therefore

$$p = \frac{2}{7 \text{ Å}} \int_{3 \text{ Å}}^{4 \text{ Å}} \left(\sin\left(\frac{3 \pi x}{7 \text{ Å}}\right)\right)^2 dx = \frac{2}{7 \text{ Å}} \left[\frac{6\pi x - 7 \text{ Å} \sin(6\pi x/7 \text{ Å})}{12\pi}\right]_{3 \text{ Å}}^{4 \text{ Å}}$$

Evaluation requires figuring the value of the quantity in brackets at the lower limit of the integration, which is $x = 3$ Å, and subtracting it from the value at the upper limit of the integration, which is $x = 4$ Å. Doing this gives

$$p = \frac{2}{7\,\text{Å}} \left(\frac{6\pi\,[4\,\text{Å}] - 6\pi\,[3\,\text{Å}] - 7\,\text{Å}\,\sin(6\pi\,[4\,\text{Å}]/7\,\text{Å}) + 7\,\text{Å}\,\sin(6\pi\,[3\,\text{Å}]/7\,\text{Å})}{12\pi} \right)$$

All of the Å's cancel out, leading to

$$p = \frac{2}{7} \left(\frac{24\pi - 18\pi - 7\sin(24\pi/7) + 7\sin(18\pi/7)}{12\pi} \right) = \frac{2}{7}(0.86205) = \boxed{0.2463}$$

Tip. The presence of nodes inside the box has interesting implications. The particle has some probability of existing to the left of a node and some of existing to the right of a node, but it has zero probability of existing *at* a node. How does it get across?

Tip. Mathematics programs exist to evaluate definite integrals like the preceding. Some of them run as utilities on web-sites,

b) As explained on text page 159, the quantity $\mathcal{H}$ is neither a number nor a function. It is an operator, the Hamiltonian operator (named after the Irish mathematician William Hamilton). The statement $\mathcal{H}\psi(x)$ instructs the user to take the second derivative of the function $\psi(x)$ with respect to x and multiply it by a constant:

$$\mathcal{H}\psi(x) = \underbrace{\frac{-h^2}{8\pi^2 m}}_{\text{constant}} \cdot \underbrace{\frac{d^2}{dx^2}}_{\text{2nd deriv.}} \psi(x)$$

The derivative of the sine function is the cosine function, and the derivative of the cosine function is the negative of the sine function. Applying the Hamiltonian operator to $\psi_3(x)$ proceeds in these steps:

$$\phi_3(x) = \sqrt{\frac{2}{L}} \sin\left(\frac{3\pi x}{L}\right) \qquad \text{the function}$$

$$\frac{d\phi_3(x)}{dx} = \left(\frac{3\pi}{L}\right)\sqrt{\frac{2}{L}} \cos\left(\frac{3\pi x}{L}\right) \qquad \text{1st deriv.}$$

$$\frac{d^2\phi_3(x)}{dx^2} = -\left(\frac{3\pi}{L}\right)^2 \sqrt{\frac{2}{L}} \sin\left(\frac{3\pi x}{L}\right) \qquad \text{2nd deriv.}$$

$$\frac{-h^2}{8\pi^2 m}\frac{d^2\phi_3(x)}{dx^2} = \frac{h^2}{8\pi^2 m}\left(\frac{3\pi}{L}\right)^2 \cdot \sqrt{\frac{2}{L}} \sin\left(\frac{3\pi x}{L}\right) \qquad \mathcal{H}(\text{function})$$

The quantity following the dot on the right in the last line of the preceding is the original function $\psi_3(x)$.[3] This quantity preceding the dot equals the energy E_3 of the $n = 3$ state that is obtained by use of text equation 4.37

$$\frac{h^2}{8\pi^2 m}\left(\frac{3\pi}{L}\right)^2 = 3^2\left(\frac{h^2}{8mL^2}\right) = E_3$$

4.39 The allowed energies of a particle in a one-dimensional box are given by text equation 4.37

$$E_n = \frac{n^2 h^2}{8mL^2}$$

---

[3]Taking the derivatives to get to the last line requires the use of the chain rule. Review this rule on-line or in a calculus book, if necessary.

The particle in this case is an electron so m is known. The "box" is a bond that is 1.34 Å long. Convert all quantities to SI units, put them into the equation, and evaluate

$$E_n = n^2 \left(\frac{(6.626 \times 10^{-34} \text{ J s})^2}{8(9.109 \times 10^{-31} \text{ kg})(1.34 \times 10^{-10} \text{ m})^2} \right) = n^2 (3.36 \times 10^{-18} \text{ J})$$

Substitution of $n = 1, 2, 3$ gives

$$E_1 = \boxed{3.36 \times 10^{-18} \text{ J}} \qquad E_2 = \boxed{13.4 \times 10^{-18} \text{ J}} \qquad E_3 = \boxed{30.2 \times 10^{-18} \text{ J}}$$

To excite the electron from $n = 1$ (the ground state) to $n = 2$ (the first excited state) requires energy equal to the difference between E_1 and E_2. If this energy is supplied by one photon, then the wavelength of the photon must be

$$\lambda = \frac{hc}{E_2 - E_1} = \frac{(6{,}626 \times 10^{-34} \text{ J s})(2.9979 \times 10^8 \text{ m s}^{-1})}{(13.4 \times 10^{-18} - 3.36 \times 10^{-18}) \text{ J}} = \boxed{1.98 \times 10^{-8} \text{ m}}$$

This wavelength, 198 Å, occurs in the ultraviolet region of the spectrum.

Tip. In the formula for the allowed energies (text equation 4.37), a property of the *wave* (the quantum number n), appears along with a property of the *universe* (Planck's constant h), a property of the *particle* (its mass m), and a property of the *box* (its length L).

Wave Functions for Particles in Two- and Three-Dimensional Boxes

4.41 The Schrödinger wave equation for a particle moving in two-dimensional space is

$$\frac{-h^2}{8\pi^2 m} \left(\frac{\partial^2 \psi(x,y)}{\partial x^2} + \frac{\partial^2 \psi(x,y)}{\partial y^2} \right) + V(x,y)\psi = E\,\psi(x,y)$$

If the particle is confined inside a square two-dimensional box, this becomes

$$\frac{-h^2}{8\pi^2 m} \left(\frac{\partial^2 \psi(x,y)}{\partial x^2} + \frac{\partial^2 \psi(x,y)}{\partial y^2} \right) = E\,\psi(x,y)$$

because $V(x,y)$ is zero inside the box and infinite outside of the box. One solves this differential equation by the separation of variables, as explained in the text. Solutions to differential equations are functions (instead of numbers). The functions depend on x and y in this case and have the form

$$\psi(x,y) = \sqrt{\frac{4}{L^2}} \sin\left(\frac{n_x \pi x}{L} \right) \sin\left(\frac{n_y \pi y}{L} \right) \qquad n_x, n_y = 1, 2, 3 \dots$$

in which n_x and n_y are quantum numbers and L is the length of the side of the square box. This function appears in a slightly different form as text equation 4.41.

Tip. The sets of wave-functions describing a single particle confined along a line (a 1-d box), within a square (a 2-d box) and within a cube (a 3-d box) are respectively

$$\psi(x) = \sqrt{\frac{2}{L}} \sin\left(\frac{n_x \pi x}{L} \right) \qquad n_x = 1, 2, 3 \dots$$

$$\psi(x,y) = \sqrt{\frac{4}{L^2}} \sin\left(\frac{n_x \pi x}{L} \right) \sin\left(\frac{n_y \pi y}{L} \right) \qquad n_x, n_y = 1, 2, 3 \dots$$

$$\psi(x,y,z) = \sqrt{\frac{8}{L^3}} \sin\left(\frac{n_x \pi x}{L} \right) \sin\left(\frac{n_y \pi y}{L} \right) \sin\left(\frac{n_z \pi z}{L} \right) \qquad n_x, n_y, n_z = 1, 2, 3 \dots$$

Study the similarities and differences among these three sets of wave-functions.

a) The task is to compare $\psi(x,y)$ when $n_x = 2$ and $n_y = 1$ with $\psi(x,y)$ when $n_x = 1$ and $n_y = 2$ and confirm that the two states are degenerate (have exactly the same energy). Write the wave-functions

$$\psi_{21}(x,y) = \sqrt{\frac{4}{L^2}} \sin\left(\frac{2\pi x}{L}\right) \sin\left(\frac{1\pi y}{L}\right) \quad \text{and} \quad \psi_{12}(x,y) = \sqrt{\frac{4}{L^2}} \sin\left(\frac{2\pi y}{L}\right) \sin\left(\frac{1\pi x}{L}\right)$$

Inspection reveals that the two functions are identical except that the x and the y are switched on the right-hand side. Such an exchange corresponds to switching the labels on the edges of the square box. But the two edges are symmetrically equivalent. They are physically indistinguishable in all respects (such as their length L). Since the two wave-functions are the same except for their assigned labels, they have the same energy. They are degenerate.

Another way to confirm that these two wave-functions are degenerate is to use the formula for the energy of a particle confined in this kind of box

$$E_{n_x,n_y} = (n_x^2 + n_y^2)\frac{h^2}{8\,m\,L^2}$$

Clearly the energy is the same whether $n_x = 2$ and $n_y = 1$ or $n_x = 1$ and $n_y = 2$.

b) Exchanging x and y in ψ_{21} and ψ_{12} corresponds to a 90° rotation of graphs of the two functions. But making such exchanges converts ψ_{21} into ψ_{12} and vice versa.

c) Exchanging x and y as labels on our drawings or as subscripts in calculations cannot alter the energy of the confined particle, which is a physically observable quantity.

Tip. Think about variations of the problem. For example, what about a particle confined in a two-dimensional *rectangular* box? Such a box has $L_x \neq L_y$. What is the formula for the quantized energy of the confined particle then?

4.43　One must write down the $\psi_{222}(x,y,z)$ wave-function before one can consider it. Do this by substituting the three quantum numbers into the general form of the wave-function for a particle in a cubical box that appears as text equation 4.42

$$\psi_{222}(x,y,z) = \sqrt{\frac{8}{L^3}} \sin\frac{2\pi x}{L} \sin\frac{2\pi y}{L} \sin\frac{2\pi z}{L}$$

a) This wave-function depends on z according to $\sin(2\pi z/L)$. If z changes from $0.75L$ to $0.25L$ and everything else stays the same, the wave-function is multiplied through by -1. This is because

$$\sin\frac{2\pi(0.75L)}{L} = \sin 270° = -1 \quad \text{but} \quad \sin\frac{2\pi(0.25L)}{L} = \sin 90° = +1$$

b) Again consider $\sin(2\pi z/L)$, the z part of ψ_{222}, but now insert $z = 0.5L$. The result is $\sin\pi = 0$. This means that the ψ_{222} wave-function is zero everywhere in the $z = 0.5L$ plane. This plane is a node of ψ_{222}, which consequently $\boxed{\text{has no shape}}$ in the plane.

ADDITIONAL PROBLEMS

4.45　The wavelength is the speed of the wave divided by its frequency

$$\lambda = \frac{\text{speed}}{\nu} = \frac{343 \text{ m s}^{-1}}{440 \text{ s}^{-1}} = \boxed{0.780 \text{ m}}$$

Dividing the distance by the speed gives the time. The 10.0 m trip takes $\boxed{0.0292 \text{ s}}$.

4.47　As an object is heated, the wavelength at which it emits light with maximum intensity becomes shorter: $\lambda_{\max} = 0.20hc/k_B T$, see problem **4.49**. This would seem to predict a shift from red to orange to yellow to green to blue in the perceived color. However, as higher T brings the wavelength of maximum intensity into the yellow-green and green, which are at the center of the visible portion of the spectrum, emission becomes intense at all visible wavelengths. The intensities at green wavelengths are the greatest, but emissions at other wavelengths are enough to make the object glow white. Raising the temperature even more lowers intensities in the red while raising them in the blue. The result is white light with a blue cast: a blue-white star is hotter than a white star.

4.49   The problem gives a formula for the wavelength of the maximum intensity of blackbody radiation as a function of the absolute temperature. Solve the formula for T and substitute the given wavelength, Planck's constant, the speed of light, and the Boltzmann constant.[4] Take care with the units

$$T = \frac{0.20\, hc}{k_B\, \lambda_{max}} = \frac{0.20\,(6.626 \times 10^{-34}\text{ J s})(3.00 \times 10^{8}\text{ m s}^{-1})}{(1.38 \times 10^{-23}\text{ J K}^{-1})(465.0 \times 10^{-9}\text{ m})} = \boxed{6.2 \times 10^{3}\text{ K}}$$

4.51   The energy of the photon is sufficient to pry an electron loose from the polished nickel surface (overcome the work function Φ) and send it on its way with kinetic energy as large as 7.04×10^{-19} J

$$\frac{hc}{\lambda} = \Phi + \tfrac{1}{2}m_e v^2$$

$$\Phi = \left(\frac{(6.626 \times 10^{-34}\text{J s})(3.00 \times 10^{8}\text{ m s}^{-1})}{131 \times 10^{-9}\text{ m}} \right) - 7.04 \times 10^{-19}\text{ J} = \boxed{8.1 \times 10^{-19}\text{ J}}$$

4.53   The Lyman series is emitted as hydrogen atoms undergo transitions from electronic excited states $n = 2, 3, 4 \ldots$ to the ground state $n = 1$. The energies of the emitted photons are

$$E_n\,(\text{H}) = Z_H^2 \left(\frac{1}{n_f^2} - \frac{1}{n_i^2} \right)\text{ Ry} = (1^2)\left(\frac{1}{1} - \frac{1}{n_i^2} \right)\text{ Ry}$$

where 1 Ry equals 2.18×10^{-18} J, and n_i is the quantum number of the excited state. To be absorbed by a ground-state He^+ ion, the incoming photon must have exactly the energy required to excite the ion from its $n = 1$ state to its $n = 2, 3, 4 \ldots$ state. These energies are

$$E_n\,(\text{He}^+) = Z_{He+}^2 \left(\frac{1}{n_i^2} - \frac{1}{n_f^2} \right)\text{ Ry} = (2^2)\left(\frac{1}{1} - \frac{1}{n_f^2} \right)\text{ Ry}$$

Is there any combination of positive whole numbers n_i and n_f that makes these two energies equal? To find out, subtract the $E_n\,(\text{H})$ from $E_n\,(\text{He}^+)$ and set the difference equal to zero

$$E_n\,(\text{He}^+) - E_n\,(\text{H}) = 2^2\left(1 - \frac{1}{n_f^2} \right)\text{ Ry} - 1^2\left(1 - \frac{1}{n_i^2} \right)\text{ Ry}$$

$$0 = \left(4 - \frac{4}{n_f^2} \right) - \left(1 - \frac{1}{n_i^2} \right)$$

$$\frac{4}{n_f^2} = 3 + \frac{1}{n_i^2}$$

The right side of the last equation varies between $3\frac{1}{4}$ and 3 as n_i varies across its range of allowed values, which comprises the integers from $+2$ to infinity. The left side is always equal to or less than 1 as n_f takes on *its* allowed values, which are also the integers from $+2$ to infinity. Because no combination of allowed n's can make the equation valid, the answer to the question is $\boxed{\text{no}}$.

4.55   Bohr's quantization condition states that the angular momentum of an orbiting body is quantized in units of $h/2\pi$. The mass of the earth m, its velocity v in orbit, and the radius r of its orbit are all

---

[4]The Boltzmann constant is represented both by k and by k_B.

given in SI units. The orbital angular momentum of the earth is the product of these three quantities and is quantized in units of $h/2\pi$

$$n\frac{h}{2\pi} = mvr = (6.0 \times 10^{24} \text{ kg})(3.0 \times 10^4 \text{ m s}^{-1})(1.5 \times 10^{11} \text{ m}) = 2.7 \times 10^{40} \text{ kg m}^2 \text{ s}^{-1}$$

$$n = \frac{2\pi}{6.626 \times 10^{-34} \text{ J s}}(2.7 \times 10^{40} \text{ kg m}^2 \text{ s}^{-1}) = \boxed{2.6 \times 10^{74}}$$

Note that one J s is the same as one kg m^2 s^{-1}, so n is a pure number (that is, without units). Since n is truly huge, +1 in n has $\boxed{\text{no effect}}$ on the angular momentum of the earth in its orbit.

4.57 **a)** The defining equations for the kinetic energy $\mathcal{T}$ of a particle and its momentum p are

$$\mathcal{T} = \tfrac{1}{2}mv^2 \qquad \text{and} \qquad p = mv$$

where m and v stand for mass and velocity. Combining these two definitions gives

$$\mathcal{T} = \frac{m^2v^2}{2m} = \frac{p^2}{2m} \qquad \text{from which} \qquad p = \sqrt{2m\mathcal{T}}$$

The problem gives the indeterminacy in the kinetic energy of an electron in terms of a range. Use the preceding to compute the momentum of the electron at each end of this range

$$p_1 = \sqrt{2m\mathcal{T}_1} = \sqrt{2\,(9.109 \times 10^{-31} \text{ kg})(1.59 \times 10^{-19} \text{ J})} = 5.382 \times 10^{-25} \text{ kg m s}^{-1}$$

$$p_2 = \sqrt{2m\mathcal{T}_2} = \sqrt{2\,(9.109 \times 10^{-31} \text{ kg})(1.61 \times 10^{-19} \text{ J})} = 5.416 \times 10^{-25} \text{ kg m s}^{-1}$$

The difference between p_2 and p_1 is the indeterminacy in the momentum of this electron: $\Delta p = 0.034 \times 10^{-25}$ kg m s^{-1}. Now write the Heisenberg indeterminacy principle for position/momentum, insert this Δp and compute Δx, the indeterminacy in the position of this electron

$$\Delta x \geq \frac{h/4\pi}{\Delta p} = \frac{6.626 \times 10^{-34} \text{ J s}}{4\pi\,(0.034 \times 10^{-25} \text{ kg m s}^{-1})} = \boxed{1.6 \times 10^{-8} \text{ m} = 1.6 \text{ Å}}$$

This is the *minimum* indeterminacy in the position of this electron. The indeterminacy naturally might exceed this value (because of experimental difficulties in the electron's position), but even a flawless experiment cannot lower the indeterminacy.

b) The mass of a helium atom is 6.647×10^{-27} kg.[5] The indeterminacy in the momentum of a helium atom with the same $\Delta \mathcal{T}$ as the electron in the previous part is

$$\Delta p = \sqrt{2(6.647 \times 10^{-27} \text{ kg})(1.61 \times 10^{-19} \text{ J})} - \sqrt{2(6.647 \times 10^{-27} \text{ kg})(1.59 \times 10^{-19} \text{ J})}$$

$$= 0.029 \times 10^{-23} \text{ kg m s}^{-1}$$

The Δp of the helium atom is larger than the Δp of the electron because of the larger mass of the helium atom. The indeterminacy in the position of the helium atom will be proportionately smaller

$$\Delta x \geq \frac{h/4\pi}{\Delta p} = \frac{6.626 \times 10^{-34} \text{ J s/}}{4\pi(0.029 \times 10^{-23} \text{ kg m s}^{-1})}) = \boxed{1.9 \times 10^{-10} \text{ m} = 0.019 \text{ Å}}$$

4.59 **a)** In the different universe ordinary objects in motion at moderate speeds, such as baseballs, have substantial wavelengths

$$\lambda_{\text{baseball}} = \frac{h}{mv} = \frac{1 \text{ J s}}{0.145 \text{ kg} \times 20 \text{ m s}^{-1}} = \boxed{0.34 \text{ m}}$$

---

[5]See text Table 19.1.

b) Take the uncertainty in the velocity of the baseball Δv to be 2 m s^{-1}. Then

$$\Delta p_{\text{baseball}} = (0.145 \text{ kg})(2 \text{ m s}^{-1}) = \boxed{0.29 \text{ kg m s}^{-1}}$$

$$\Delta p_{\text{baseball}}\Delta x_{\text{baseball}} \geq \frac{h}{4\pi} = \frac{1 \text{ J s}}{4\pi} \quad \text{hence} \quad \Delta x_{\text{baseball}} \geq \frac{1 \text{ J s}}{4\pi(0.29 \text{ kg m s}^{-1})} = \boxed{0.27 \text{ m}}$$

This would make the baseball hard to catch.

c) The Bohr radius in the different universe is larger

$$a_0 = \frac{\epsilon_0 h^2}{\pi e^2 m_e} = \frac{(8.854 \times 10^{-12} \text{ C}^2 \text{ J}^{-1} \text{ m}^{-1})(1 \text{ J s})^2}{\pi(1 \text{ C})^2(0.001 \text{ kg})} = 2.82 \times 10^{-9} \text{ m} = \boxed{28 \text{ Å}}$$

4.61

$$E_{n_x,n_y,n_z} = \frac{h^2}{8m}\left[\frac{n_x^2}{(2L)^2} + \frac{n_y^2}{L^2} + \frac{n_z^2}{L^2}\right]$$

Systematically substitute various combinations of the three quantum numbers n_x, n_y, and n_z to discover the five lowest energies. The problem does not give numerical values for L or m. The following lists the energies in terms of these variables:

$$E_{111} = \frac{h^2}{8\,mL^2}(1/4+1+1) = \frac{9}{4}\frac{h^2}{8mL^2} \qquad E_{211} = \frac{h^2}{8\,mL^2}(4/4+1+1) = \frac{12}{4}\frac{h^2}{8mL^2}$$

$$E_{311} = \frac{h^2}{8\,mL^2}(9/4+1+1) = \frac{17}{4}\frac{h^2}{8mL^2} \qquad E_{112} = E_{121} = \frac{h^2}{8\,mL^2}(1/4+4+1) = \frac{21}{4}\frac{h^2}{8mL^2}$$

$$E_{411} = E_{221} = E_{212} = \frac{h^2}{8\,mL^2}(6) = \frac{24}{4}\frac{h^2}{8mL^2} \qquad E_{321} = E_{312} = \frac{h^2}{8\,mL^2}(9/4+4+1) = \frac{29}{4}\frac{h^2}{8mL^2}$$

The (111) state is the ground state; the others are excited states. Note the double degeneracy of two of the excited states and the triple degeneracy of a third.

Chapter 5

Quantum Mechanics and Atomic Structure

The Hydrogen Atom

5.1　**a)** $\boxed{\text{Not allowed}}$; ℓ must be *less* than n.　　　**b)** $\boxed{\text{Allowed}}$. This specifies a $3p$ electron.

c) Has $m > \ell$, which is $\boxed{\text{not allowed}}$.　　　**d)** Has $\ell < 0$, $\boxed{\text{not allowed}}$.

Tip. Use the physical significance of the rules as a memory aide: $(n-1)$ equals the total number of nodes possessed by the wave-function, and ℓ equals the number of angular nodes. Obviously the number of angular nodes cannot exceed the number of all nodes, so ℓ has $(n-1)$ as its maximum. The *minimum* for ℓ is zero because "-1 angular nodes" has no physical meaning. The quantum number m governs the spatial orientation of the angular nodes. It has $2\ell + 1$ possible values: the positive integers *up* to and including ℓ, the negative integers *down* to and including $-\ell$, and zero. The quantum number m_s must equal either $+\frac{1}{2}$ or $-\frac{1}{2}$, reflecting the fact that the spin angular momentum of the electron has only two quantum states.

Tip. Radial nodes are associated with the radial part of a wave-function. Radial nodes are spherical. Angular nodes are associated with the angular part of a wave-function. They are flat surfaces (planes) or curved surfaces other than spheres. The radial parts and angular parts of various H-atom wave-functions appear in text Table 5.2. A whole wave-function consists of the product of a radial part and an angular part.

5.3　**a)** $4p$　　**b)** $2s$　　**c)** $6f$

5.5　Construct the full $2s$ wave-function from the parts listed in text Table 5.2. It is the product of an angular part Y_s and a radial part R_{2s}. The problem deals with a hydrogen atom, so set $Z = 1$ in these mathematical expressions. Also set $a_0 = 1$. Doing this amounts to choosing to express all distances as multiples of the distance a_0 (rather than multiples of 1 m or any other unit of distance) and all volumes as multiples of a_0^3. It also causes σ in Table 5.2 to equal r. The results are a $2s$ wave-function and its square that look much less forbidding than otherwise

$$\psi_{2s} = \sqrt{\frac{1}{4\pi}}\,\frac{1}{2\sqrt{2}}\,(2-r)\exp(-r/2) \quad \text{and} \quad (\psi_{2s})^2 = \frac{1}{32\pi}\,(4 - 4r + r^2)\exp(-r)$$

The units of squared wave-functions such as $(\psi_{2s})^2$ are always reciprocal volumes. Evaluating a squared wave-function at a specific point and then multiplying by a small volume V gives the approximate probability, a unit-less number between 0 and 1, of finding the electron within such a volume centered at the specific point. See text equation 5.6. In this case the specific point is the origin

$(r = 0)$ so that

$$(\psi_{2s})^2 = \frac{1}{8\pi} \, (a_0)^{-3}$$

The problem specifies a suitable small volume. Express this volume in units of $(a_0)^3$

$$V = 0.75 \times 10^{-36} \text{ m}^3 \times \left(\frac{(1 \, a_0)^3}{(0.529 \times 10^{-10} \text{ m})^3} \right) = 5.066 \times 10^{-6} \, (a_0)^3$$

Multiplying this volume by $(\psi_{2s})^2$ gives the desired estimate of probability at and near $r = 0$:

$$(\psi_{2s})^2 \, V = \left(\frac{1}{8\pi} \, (a_0)^{-3} \right) \, (5.066 \times 10^{-6} \, (a_0)^3) = \boxed{2.0 \times 10^{-7}}$$

Moving the small volume out to a distance of 105.8 pm requires a new calculation of $(\psi_{2s})^2$. As it happens, the new r equals 2.00 a_0. The expression for the square of the wave-function is still

$$(\psi_{2s})^2 = \frac{1}{32\pi} \, (4 - 4r + r^2) \, \exp(-r)$$

Inserting 2.00 in place of r gives

$$(\psi_{2s})^2 = \frac{1}{32\pi} \, \left(4 - 4(2.00) + (2.00)^2 \right) \, \exp(-r) = 0.00$$

The wave-function goes to zero at this distance. This means that the probability of finding the electron in the small volume at this distance is $\boxed{0.00}$. This probability does not depend on the direction in which the movement occurs because the $2s$ wave-function has no dependence on the angular coordinates θ and φ, It is spherically symmetrical. Mention of the $+z$-axis is superfluous.

5.7   The total number of nodes is one less than the quantum number n. The number of angular nodes equals the quantum number ℓ, which is obtained by decoding the s, p, d, f notation, as explained in text Section 5.1.

a) $4p$: 2 radial and 1 angular.   **b)** $2s$: 1 radial and 0 angular.   **c)** $6f$: 2 radial and 3 angular.

5.9   The wave function ψ_{2p_z} is the product of a *radial* part R_{2p} and an *angular* part Y_{p_z}. The two parts of the functions are given in text Table 5.2. Multiply the two parts and then simplify

$$\psi_{2p_z} = (Y_{p_z}) \cdot (R_{2p}) = \left[\left(\frac{3}{4\pi} \right)^{1/2} \cos\theta \right] \cdot \left[\frac{1}{2\sqrt{6}} \left(\frac{Z}{a_0} \right)^{3/2} \left(\frac{Zr}{a_0} \right) \exp(-Zr/2a_0) \right]$$

$$= \left(\frac{1}{32\pi} \right)^{1/2} \left(\frac{Z}{a_0} \right)^{3/2} \cos\theta \left(\frac{Zr}{a_0} \right) \exp(-Zr/2a_0)$$

The radial part of the function contributes the dependence on r; the angular part contributes the dependence on $\cos\theta$. This particular orbital has no dependence on the third coordinate φ. The probability p of finding an electron in the close vicinity of a point[1] specified by the coordinates r, θ and φ is given by the square of this function

$$p(r, \theta, \varphi) = \left(\psi_{2p_z} \right)^2 = \frac{1}{32\pi} \left(\frac{Z}{a_0} \right)^3 \cos^2\theta \left(\frac{Z^2 r^2}{a_0^2} \right) \exp(-Zr/a_0)$$

The question now becomes: what values of r and θ make $p(r, \theta, \varphi) = 0$? Clearly, this happens when $r = 0$ (at the nucleus). It also happens when $\theta = \pi/2 = 90°$ (because $\cos^2 \pi/2 = 0$), and when $\theta = 3\pi/2 = 270°$ (because $\cos^2 3\pi/2 = 0$). Text Figure 5.1 (on text page 179) shows that θ equals

---

[1]More precisely, in the infinitesimal element of volume $dV = r^2 \sin\theta \, dr d\theta d\varphi$ surrounding the point.

$\pm 90°$ at all locations in the xy plane. Therefore, $p(r, \theta, \varphi)$ equals zero at all points in the xy plane. The $\boxed{xy \text{ plane is a nodal plane of the } \psi_{2p_z} \text{ orbital}}$. It is the only node of this wave-function.

Writing out and squaring the whole wave function is not necessary. The angular part of the wave function by itself controls the number and arrangement of the angular nodes. For example, the square of every possible d_{xz} orbital (the $3d_{xz}$, $4d_{xz}$, $5d_{xz}$, ...) depends on the two angular coordinates according to $\sin^2 \theta \cos^2 \theta \cos^2 \varphi$. This function equals zero whenever $\theta = \pi/2$ or $\theta = 3\pi/2$ (in the xy plane) and whenever $\varphi = \pi/2$ or $\varphi = 3\pi/2$ (in the yz plane). This proves that the $\boxed{xy \text{ plane}}$ and $\boxed{yz \text{ plane}}$ are the two angular nodes of all d_{xz} orbitals.

The square of a $d_{x^2-y^2}$ orbital has a $\sin^4 \theta \cos^2 2\varphi$ angular dependence. This trigonometric function goes to zero at these values of φ

$$\varphi = \pi/4 \ (45°) \qquad \varphi = 3\pi/4 \ (135°) \qquad \varphi = 5\pi/4 \ (225°) \qquad \varphi = 7\pi/4 \ (315°)$$

The first and third values of φ define a plane containing the z axis and lying at a $45°$ angle to $+x$ and $+y$ (and also to $-x$ and $-y$). The second and fourth values of φ define a plane also containing the z axis and oriented at right angles to the first plane. These two planes are the angular nodes of all $d_{x^2-y^2}$ orbitals.

The probability of finding a $d_{x^2-y^2}$ electron at a location with $\theta = 0$ or $\theta = \pi$ ($180°$) is also zero. This happens only at points on the z axis, the intersection of the two angular nodes that were just identified.

Shell Model for Many-Electron Atoms

5.11   Text equation 5.9 gives a formula for the average distance $\bar{r}$ between the electron and the nucleus in a hydrogen atom (or in a hydrogen-like ion)

$$\bar{r}_{n\ell} = \frac{n^2 a_0}{Z} \left\{ 1 + \frac{1}{2} \left[1 - \frac{\ell(\ell+1)}{n^2} \right] \right\}$$

For a $2s$ electron in hydrogen, $Z = 1$, $n = 2$, and $\ell = 0$. Substitution gives

$$\bar{r}_{2,0} = \frac{2^2 a_0}{1} \left\{ 1 + \frac{1}{2} \left[1 - \frac{0(0+1)}{2^2} \right] \right\} = 6a_0 = 6(0.529 \text{ Å}) = \boxed{3.17 \text{ Å}}$$

For a $2p$ electron in hydrogen, $Z = 1$, $n = 2$, and $\ell = 1$

$$\bar{r}_{2,1} = \frac{2^2 a_0}{1} \left\{ 1 + \frac{1}{2} \left[1 - \frac{1(1+1)}{2^2} \right] \right\} = 5a_0 = 5(0.529 \text{ Å}) = \boxed{2.64 \text{ Å}}$$

Tip. Compare these answers to the distances marked by the black arrows in text Figure 5.14 on text page 194. It is also worthwhile to confirm that the average distance $\bar{r}$ of a $3p$ electron in hydrogen is $12\frac{1}{2}a_0$ and that the average distance of a $3d$ electron in hydrogen is $10\frac{1}{2}a_0$. These distances are indicated by black arrows in the bottom two graphs in the figure.

5.13   Text equation 5.11 on text page 199 gives an approximate formula for the energy of an electron in Hartree orbital n in an atom

$$\epsilon_n \approx -\frac{[Z_{\text{eff}}(n)]^2}{n^2} \qquad \text{in rydbergs}$$

The problem gives $Z_{\text{eff}}(n) = 1.28$ for a $2s$ ($n = 2$) electron in lithium. Substitute these values

$$\epsilon_{2s} \text{ in Li} \approx -\frac{[Z_{\text{eff}}(n)]^2}{n^2} = -\frac{[1.28]^2}{2^2} = -0.4096 \text{ Ry} = \boxed{-0.410 \text{ Ry}}$$

It is helpful to obtain the answer in other units of energy as well:

$$\approx -0.4096 \text{ Ry} \left(\frac{2.1799 \times 10^{-18} \text{ J}}{1 \text{ Ry}} \right) = \boxed{-8.93 \times 10^{-19} \text{ J}}$$

$$\approx -0.4096 \text{ Ry} \left(\frac{13.607 \text{ eV}}{1 \text{ Ry}} \right) = \boxed{-5.57 \text{ eV}}$$

$$\approx -0.4096 \text{ Ry} \left(\frac{1312 \text{ kJ mol}^{-1}}{1 \text{ Ry}} \right) = \boxed{-537 \text{ kJ mol}^{-1}}$$

Tip. The *observed* energy of the electron in the $2s$ orbital of a lithium atom is -513.3 kJ mol^{-1}. The effective nuclear charge quoted in the problem gives an energy within about five percent of this. See problem **5.55**. Rounding to three significant figure in all four answers is correct.

The approximate average distance of an electron in Hartree orbital n, ℓ is given by

$$\bar{r}_{n,\ell} \approx \frac{n^2 a_0}{Z_{\text{eff}}(n)} \left\{ 1 + \frac{1}{2} \left[1 - \frac{\ell(\ell+1)}{n^2} \right] \right\}$$

For the $2s$ orbital in lithium, $n = 2$ and $\ell = 0$. Hence

$$\bar{r}_{n,\ell} \approx \frac{2^2 a_0}{1.28} \left\{ 1 + \frac{1}{2} \left[1 - \frac{0(0+1)}{2^2} \right] \right\} = \frac{2^2 a_0}{1.28} \left\{ \frac{3}{2} \right\} = 4.69 \, a_0 = \boxed{2.48 \text{ Å}}$$

5.15 The $1s$, $2s$ and $3s$ orbitals in H, Li, and Na respectively contain the outermost electron (when the atoms are in their ground states). The energies of these electrons are

$$\text{For hydrogen} \qquad \epsilon_{1s} = -\frac{Z^2}{n^2} = -\frac{1^2}{1^2} = \boxed{-1 \text{ Ry exactly}}$$

$$\text{For lithium} \qquad \epsilon_{2s} \approx -\frac{[Z_{\text{eff}}(n)]^2}{n^2} = -\frac{[1.28]^2}{2^2} = \boxed{-0.401 \text{ Ry}}$$

$$\text{For sodium} \qquad \epsilon_{3s} \approx -\frac{[Z_{\text{eff}}(n)]^2}{n^2} = -\frac{[1.84]^2}{3^2} = \boxed{-0.376 \text{ Ry}}$$

The energies become algebraically larger as n increases, that is, it becomes progressively easier to remove the outermost electron of the atom moving from hydrogen to sodium.

Aufbau Principle and Electron Configurations

5.17 The ground-state electron configurations are

a) C $\quad 1s^2 2s^2 2p^2$ **b)** Se $\quad 1s^2 2s^2 2p^6 3s^2 3p^6 3d^{10} 4s^2 4p^4$ **c)** Fe $\quad 1s^2 2s^2 2p^6 3s^2 3p^6 3d^6 4s^2$

The use of the bracketed symbol of a noble gas to represent the electron configuration of that element shortens the notation for most configurations

a) C $\quad$ [He]$2s^2 2p^2$ **b)** Se $\quad$ [Ar]$3d^{10} 4s^2 4p^4$ **c)** Fe $\quad$ [Ar]$3d^6 4s^2$

5.19 The ground-state configuration of an ion derives from the ground-state configuration of the atom. In the case of a negative ion, add electrons to available orbitals in order of ascending energy. In the case of positive ions, remove electrons starting with the ones in the highest-energy orbitals

Be$^+$	$1s^2 2s^1$	C$^-$	$1s^2 2s^2 2p^3$	Ne^{2+}	$1s^2 2s^2 2p^4$	Mg$^+$ [Ne]$3s^1$
P^{2+}	[Ne]$3s^2 3p^1$	Cl$^-$	[Ne]$3s^2 3p^6$	As$^+$	[Ar]$3d^{10} 4s^2 4p^2$	
I$^-$	[Kr]$4d^{10} 5s^2 5p^6$					

All of these electron configurations are ground-state (lowest energy) configurations. Be$^+$, C$^-$, Ne^{2+}, Mg$^+$, P^{2+} and As$^+$ all have at least one unpaired electron (they have incomplete subshells) and should be paramagnetic. $\boxed{\text{The Cl}^- \text{ and I}^- \text{ ions are diamagnetic; the others are paramagnetic}}$.

5.21 **a)** The atom has 49 electrons (36 represented by [Kr]; 13 represented by superscripts). It is $\boxed{\text{In}}$.

b) The ion has 18 electrons, and a charge of -2. Its atomic number Z must be 16; it is $\boxed{\text{S}^{2-}}$.

c) The ion has 21 electrons, and a charge of $+4$. Its Z must be 25. it is $\boxed{\text{Mn}^{4+}}$ ion.

5.23 As a halogen this element has a ground-state electron configuration of the form $\ldots ns^2 np^5$. The next p-subshell after the $6p$ (used in the sixth row of the periodic table) is the $7p$. Accordingly, the electron configuration of the element would be $[\text{Rn}] 5f^{14} 6d^{10} 7s^2 7p^5$ where [Rn] stands for the configuration of the first 86 electrons. Since the configuration represents 117 electrons, Z equals $\boxed{117}$.

5.25 If only one electron could occupy each orbital in many-electron atoms, then the configurations $1s^1$ and $1s^1 2s^1 2p^3$ and $1s^1 2s^1 2p^3 3s^1 3p^3$ would be closed-shell electron configurations. Atoms with $Z = \boxed{1, 5, 9}$ respectively would have these ground-state electron configurations.

Shells and the Periodic Table: Photoelectron Spectroscopy

5.27 The photoelectron spectroscopy experiment measures the kinetic energies of electrons that are ejected from atoms by the absorption of high-energy photons. The difference between the energy of the incoming photon and the kinetic energy of an outgoing electron equals the ionization energy (IE) of that electron. Electrons in different orbitals in many-electron atoms have different ionization energies, as illustrated in text Figure 5.20. The electrons detached from the Hg atoms in this problem have 11.7 eV of kinetic energy. Hence

$$IE = h\nu - \frac{1}{2} m_e v^2 = \frac{hc}{\lambda} - \frac{1}{2} m_e v^2$$
$$= \frac{(6.626 \times 10^{-34} \text{ J s})(2.9979 \times 10^8 \text{ m s}^{-1})}{584.4 \times 10^{-10} \text{ m}} - (11.7 \text{ eV})\left(\frac{1.602 \times 10^{-19} \text{ J}}{1 \text{ eV}}\right)$$
$$= \boxed{1.52 \times 10^{-18} \text{ J}}$$

This answer is equivalent to 9.52 eV or 0.699 Ry.

5.29 In one printing of the text the speeds of the electrons in the four peaks are incorrect. The correct speeds are

Peak	Speed of Electron /m s^{-1}	Peak	Speed of Electron / m s^{-1}
a.	7.9924×10^6	c.	2.0712×10^7
b.	2.0421×10^7	d.	2.0956×10^7

a) In the photoelectron spectroscopy (PES) experiment, the ionization energy (IE) of an electron equals the energy of the radiation that detaches it from the atom minus the kinetic energy that it carries away

$$IE = h\nu - \tfrac{1}{2} m_e v^2$$

The energy of the x-rays used to irradiate the Na atoms in this experiment is

$$E_{\text{x-rays}} = h\nu = \frac{hc}{\lambda} = \frac{(6.62607 \times 10^{-34} \text{ J s})(2.99792 \times 10^8 \text{ m s}^{-1})}{9.890 \times 10^{-10} \text{ m}} = 2.0085 \times 10^{-16} \text{ J}$$
$$= 2.0085 \times 10^{-16} \text{ J} \left(\frac{1 \text{ eV}}{1.6022 \times 10^{-19} \text{ J}}\right) = 1253.6 \text{ eV}$$

Subtract the kinetic energies of the electrons in the four peaks from this value. The slowest speed corresponds to the most tightly bound electron, the one with the highest *IE*

$$IE_a = 1253.6 \text{ eV} - \frac{(9.10938 \times 10^{-31} \text{ kg})(7.9924 \times 10^6 \text{ m s}^{-1})^2}{2} \left(\frac{1 \text{ eV}}{1.6022 \times 10^{-19} \text{ J}}\right) = \boxed{1072.0 \text{ eV}}$$

$$IE_b = 1253.6 \text{ eV} - \frac{(9.10938 \times 10^{-31} \text{ kg})(2.0421 \times 10^7 \text{ m s}^{-1})^2}{2} \left(\frac{1 \text{ eV}}{1.6022 \times 10^{-19} \text{ J}}\right) = \boxed{68.1 \text{ eV}}$$

$$IE_c = 1253.6 \text{ eV} - \frac{(9.10938 \times 10^{-31} \text{ kg})(2.0712 \times 10^7 \text{ m s}^{-1})^2}{2} \left(\frac{1 \text{ eV}}{1.6022 \times 10^{-19} \text{ J}}\right) = \boxed{34.0 \text{ eV}}$$

$$IE_d = 1253.6 \text{ eV} - \frac{(9.10938 \times 10^{-31} \text{ kg})(2.0956 \times 10^7 \text{ m s}^{-1})^2}{2} \left(\frac{1 \text{ eV}}{1.6022 \times 10^{-19} \text{ J}}\right) = \boxed{5.2 \text{ eV}}$$

b) The ground-state electron configuration of sodium is $1s^2\,2s^2\,2p^6\,3s^1$. Peak a corresponds to removal of a $1s$ electron, one of the atom's most tightly bound electrons. Peaks b, c, and d correspond to removal of a $2s$, a $2p$, and the $3s$ electron respectively.

Tip. The individual ionization energies in a PES experiment are *not* the same as the successive ionization energies listed in text Table 3.1 (text page 72). In the PES experiment, electrons are knocked out of different orbitals on *neutral* atoms. In text Table 3.1, only the IE_1's involve the removal of electrons from neutral atoms. The IE_2's are for removal from +1 ions, the IE_3's are for removal from +2 ions, and so forth. This point explains why the answer for peak d, the removal of the $3s$ electron, equals IE_1 for sodium (which text Table 3.1 quotes, with greater precision, as 5.14 eV), but the answer for peak c (34.0 eV) differs sharply from IE_2 for sodium (47.29 eV).

5.31   Write text equation 5.11, which is an approximate equation, and solve it for Z_{eff}

$$\epsilon_n \text{ (in rydbergs)} \approx \frac{-[Z_{\text{eff}}(n)]^2}{n^2} \quad \text{which gives} \quad Z_{\text{eff}}(n) \approx \sqrt{n^2(-\epsilon_n \text{ (in rydbergs)})}$$

Now, insert the given energies of $1s$, $2s$, and $2p$ orbitals in fluorine and figure out the three Z_{eff}'s. The data have to be converted from electron-volts to rydbergs. Do this by dividing each by 13.607 eV Ry^{-1}

$$Z_{\text{eff}}(1s, \text{F}) \approx \sqrt{1^2 \left(\frac{-(-689 \text{ eV})}{13.607 \text{ eV Ry}^{-1}}\right)} = \boxed{7.12}$$

$$Z_{\text{eff}}(2s, \text{F}) \approx \sqrt{2^2 \left(\frac{-(-34 \text{ eV})}{13.607 \text{ eV Ry}^{-1}}\right)} = \boxed{3.2}$$

$$Z_{\text{eff}}(2p, \text{F}) \approx \sqrt{2^2 \left(\frac{-(-12 \text{ eV})}{13.607 \text{ eV Ry}^{-1}}\right)} = \boxed{1.9}$$

Tip. These Z_{eff}'s compare poorly to the Z_{eff}'s developed by a much more advanced computational method and listed for fluorine in Table 5.3 on text page 201. On the other hand, Slater's rules (text page 200), give these Z_{eff}'s

$$Z_{\text{eff}}(1s, \text{F}) = 9 - (1 \times 0.35) = 8.65$$
$$Z_{\text{eff}}(2s, \text{F}) = 9 - (6 \times 0.35) - (2 \times 0.85) = 5.20$$
$$Z_{\text{eff}}(2p, \text{F}) = 9 - (6 \times 0.35) - (2 \times 0.85) = 5.20$$

The agreement between Slater's results and the values in text Table 5.3 is impressive. It shows the quality of Slater's work, which was first published in 1930.

Tip. Of course all of the Z_{eff}'s are less than 9, the charge on the fluorine nucleus, but greater than 1, the Z_{eff} that would result from perfect screening of the fluorine nucleus by the first eight electron for the benefit of the ninth.

Periodic Properties and Electronic Structure

5.33   **a)** A ground-state $\boxed{\text{K atom}}$ should have a larger radius than a ground-state Na atom. In K atoms the outermost electron occupies a $4s$ orbital. In ground-state Na atoms the outermost electron occupies a closer-in $3s$ orbital.

    b) The $\boxed{\text{Cs atom}}$ is larger than the Cs^+ ion. As a Cs^+ ion gains an electron to produce a Cs atom, the electron is accommodated in the more distant $n = 6$ shell.

    c) The Rb^+ ion and the Kr atom are isoelectronic. The larger species is the one with smaller nuclear charge: $\boxed{\text{Kr}}$.

    d) A Ca atom has two $4s$ electrons and a K atom has one $4s$ electron. The outermost electrons are in the same shell but Ca has a larger nuclear charge, contracting the electron cloud. Hence the atom of $\boxed{\text{K}}$ is larger.

    e) The Cl^- ion and the Ar atom are isoelectronic. The larger species is the one with smaller nuclear charge: $\boxed{Cl^-}$.

5.35   **a)** The $\boxed{S^{2-}}$ ion should be larger than the O^- ion. Its outermost electrons occupy the $n = 3$ level whereas in O^- ion the outermost electrons are in the closer $n = 2$ level.

    b) The $\boxed{Ti^{2+}}$ ion is larger than the Co^{2+} ion. The two have their outermost electrons in the same subshell, but Ti^{2+} has a smaller nuclear charge.

    c) The $\boxed{Mn^{2+}}$ ion is larger than the Mn^{4+} ion because the outermost electrons (those farthest away) are lost in going from the $+2$ to $+4$ ion.

    d) The $\boxed{Sr^{2+}}$ ion is larger than the Ca^{2+} ion according to the trend to larger size going down the periodic table.

5.37   **a)** The first ionization energy of an element is the minimum energy necessary to remove a single electron from a neutral gaseous atom of the element. The ionization energy of He is particularly high because the removal must overcome the attraction of a large effective nuclear charge and get the electron from the $1s$ orbital, which has the smallest radius of all orbitals.

    b) The element $\boxed{\text{Li}}$ should have the highest *second* ionization energy. The electron is lost from the Li^+ ion, which has an electron configuration like that of helium, and helium has the largest first ionization energy of any atom.

    Tip. Review the ionization energies given in text Table 3.1 (text page 72).

    c) The photons of the radiation must supply at least enough energy to equal IE_1 of He

$$\lambda_{\max} = \frac{hc}{IE_1} = \frac{(6.626 \times 10^{-34} \text{ J s})(2.9979 \times 10^8 \text{ m s}^{-1})}{2370 \text{ kJ mol}^{-1}} \times \left(\frac{1 \text{ kJ}}{1000 \text{ J}} \right) \left(\frac{6.022 \times 10^{23} \text{ molecule}}{1 \text{ mol}} \right)$$

$$= \boxed{5.05 \times 10^{-8} \text{ m}} = 50.5 \text{ nm}$$

5.39   **a)** The Ca^{2+} ion is smaller than the Ar atom because the two are isoelectronic (18 electrons each) and Ca^{2+} has a larger nuclear charge. The Mg^{2+} ion is smaller than the Ca^{2+} ion because Mg is above Ca in the same column of the periodic table. Compare Ar to Br^- by using Cl^- as a bridge. Ar is smaller than Cl^- because the two are isoelectronic and Ar has a larger nuclear charge. But Cl^- is smaller than Br^- ion because Cl lies above Br in a column of the periodic table. Hence

$$\boxed{Mg^{2+} < Ca^{2+} < Ar < Br^-}$$

    b) Ne and Na^+ have the same electron configuration, but Na^+ has a positive charge, making its ionization energy greater; Ne has a larger ionization energy than O based on the periodic trend in the second period; Na has a lower ionization energy than Li, based on periodic trends, and Li has a lower ionization energy than O. Hence

$$\boxed{\text{Na} < \text{O} < \text{Ne} < \text{Na}^+}$$

c) Al is metallic and thus electropositive. The three non-metals should increase in electronegativity moving to the right in the periodic table. Hence

$$\boxed{\text{Al} < \text{H} < \text{O} < \text{F}}$$

5.41   Convert the first ionization energy of cesium from kilojoules per mole to joules per atom by multiplying it by 1000 J kJ^{-1} (to get to joules per mole) and then dividing by 6.022×10^{23} mol^{-1} (Avogadro's number). The result is 6.239×10^{-19} J. Proceed as in problem **5.37**

$$\lambda_{\max} = \frac{hc}{IE_1} = \frac{(6.626 \times 10^{-34} \text{ J s})(2.9979 \times 10^8 \text{ m s}^{-1})}{6.239 \times 10^{-19} \text{ J}} = 3.184 \times 10^{-7} \text{ m} = \boxed{318.4 \text{ nm}}$$

This wavelength is in the $\boxed{\text{near ultraviolet}}$ region of the electromagnetic spectrum.

ADDITIONAL PROBLEMS

5.43   The energy of the photon emitted in the $2p \longrightarrow 1s$ transition in iron is

$$E_{\text{photon}} = \Delta E_{\text{atom}} = \frac{hc}{\lambda} = \frac{(6.626 \times 10^{-34} \text{ J s})(2.9979 \times 10^8 \text{ m s}^{-1})}{0.193 \times 10^{-9} \text{ m}} = \boxed{1.029 \times 10^{-15} \text{ J}}$$

This is about 630 times larger than 16.2×10^{-19} J, which is quoted as the energy spacing between the $2p$ and $1s$ levels in hydrogen.

In iron, the $1s$ orbital experiences the (almost) completely unshielded attraction of the nucleus ($Z = 26$), while the $2p$ orbital experiences a smaller effective nuclear charge because electrons in other orbitals screen it from the nucleus. This separates the $1s$ from the $2p$ orbital in terms of energy much more than in hydrogen ($Z = 1$), where there is no shielding of any orbital.

5.45   The total change in the energy of the atom is the same regardless of whether it relaxes from its excited state in two steps or one

$$\Delta E_{\text{total}} = \Delta E_{\text{Step1}} + \Delta E_{\text{Step2}}$$

The energy change is inversely proportional to the wavelength of the emitted photon that is emitted ($\Delta E = hc/\lambda$) so

$$\frac{hc}{\lambda_{\text{total}}} = \frac{hc}{\lambda_1} + \frac{hc}{\lambda_2} \qquad \text{Division by } hc \text{ gives} \qquad \boxed{\frac{1}{\lambda_{\text{total}}} = \frac{1}{\lambda_1} + \frac{1}{\lambda_2}}$$

The ΔE for any step equals $h\nu$. Therefore $\boxed{\nu_{\text{total}} = \nu_1 + \nu_2}$.

5.47   The $3d_{xy}$ orbital in O^{7+} ion has the $\boxed{\text{same shape}}$ as the $3d_{xy}$ orbital in an H atom. Both have two nodal planes at right angles to each other. The $3d_{xy}$ orbital differs in O^{7+} by being $\boxed{\text{much smaller}}$. The shrinkage results from the larger nuclear charge in O^{7+}.

Tip. Use text equation 5.9 to check the average distance between the nucleus and an electron in the $3d_{xy}$ orbital in O^{7+} ion and in the H atom:

$$\bar{r}_{3,2} \approx \frac{3^2 a_0}{Z} \left[1 + \frac{1}{2} \left(1 - \frac{2(2+1)}{3^2} \right) \right] = \frac{9a_0}{Z} \left[\frac{21}{18} \right] = \frac{10.5 a_0}{Z} = \frac{10.5(0.529 \text{ Å})}{Z}$$

For H ($Z = 1$), $\bar{r}_{3,2} = 5.55$ Å;   for O^{7+} ($Z = 8$), $\bar{r}_{3,2} = 0.694$ Å.

5.49 This atom of sodium is in an $\boxed{\text{excited state}}$. It can lose energy in a variety of ways to end up ultimately in its ground state, which is represented $[Ne]3s^1$.

5.51 In chromium(IV) oxide, the Cr^{4+} ion has the ground-state electron configuration: $\boxed{[Ar]3d^2}$. The neutral Cr atom (with ground-state configuration $[Ar]3d^5\,4s^1$) has lost its $4s$ electron and three of its five $3d$ electrons. The two remaining $3d$ electrons are unpaired so CrO_2 has $\boxed{\text{two}}$ unpaired spins per Cr atom.

Tip. All of the electrons in the ground-state O^{2-} ion are paired.

5.53 The smallest by far is the hydrogen-like ion Co^{25+}. The rest of the order follows from periodic trends.

$$\boxed{Co^{25+} < F^+ < F < Br < K < Rb < Rb^-}$$

5.55 The first ionization energy of lithium is 520.2×10^3 J mol^{-1}. Dividing this by N_A puts it on a per atom basis: 8.638×10^{-19} J per atom. It requires this much energy to extract the $2s$ electron from a ground-state lithium atom. Therefore

$$\epsilon_{2s}\,(\text{Li}) = -8.638 \times 10^{-19}\text{ J}$$

The energy of an electron in a Hartree orbital n in an atom is given by

$$\epsilon_n \approx -\frac{[Z_{\text{eff}}(n)]^2}{n^2}\text{Ry} = \left(-2.17987 \times 10^{-18}\text{ J}\right)\frac{[Z_{\text{eff}}(n)]^2}{n^2}$$

where $Z_{\text{eff}}(n)$ is the effective nuclear charge acting on the electron and n is the orbital's principal quantum number. For the lithium $2s$ electron

$$-8.524 \times 10^{-19}\text{ J} \approx \left(-2.17987 \times 10^{-18}\text{ J}\right)\frac{[Z_{\text{eff}}(2)]^2}{2^2}\quad\text{from which}\quad Z_{\text{eff}}(2) \approx \boxed{1.26}$$

The true Z of lithium is 3. The inner two electrons spend their time mainly between the $2s$ electron and the nucleus and so reduce the influence of the nucleus on the $2s$ electron from 3 down to 1.26. The screening is very substantial, but perfect screening would lower $Z_{\text{eff}}(n)$ all the way to 1.

For Na ($Z = 11$), the $3s$ electron is lost. Modify the preceding computation as follows

$$\left(-2.17987 \times 10^{-18}\text{ J}\right)\frac{[Z_{\text{eff}}(3)]^2}{3^2} \approx \frac{-496 \times 10^3\text{ J mol}^{-1}}{6.022 \times 10^{23}\text{ mol}^{-1}}\quad\text{from which}\quad [Z_{\text{eff}}(3)] \approx \boxed{1.84}$$

For K ($Z = 19$), the $4s$ electron is lost

$$\left(-2.17987 \times 10^{-18}\text{ J}\right)\frac{[Z_{\text{eff}}(4)]^2}{4^2} \approx \frac{-419 \times 10^3\text{ J mol}^{-1}}{6.022 \times 10^{23}\text{ mol}^{-1}}\quad\text{from which}\quad Z_{\text{eff}}(4) \approx \boxed{2.26}$$

5.57 The $2p \longrightarrow 1s$ transition in the Fe atom occurs with emission of x-rays of wavelength 0.193 nm (0.193×10^{-9} m). The energy of these x-rays is

$$E_{\text{x-rays}} = \frac{hc}{\lambda} = \frac{(6.626 \times 10^{-34}\text{ J s})(2.9979 \times 10^8\text{ m s}^{-1})}{0.193 \times 10^{-9}\text{ m}} = 1.03 \times 10^{-15}\text{ J}$$

The difference in the energy of the two states in the Fe atom consequently is

$$E_{1s} - E_{2p} = \boxed{-1.03 \times 10^{-15}\text{ J}}$$

where the negative sign reflects the fact that the $1s$ orbital lies at lower energy than the $2p$. This difference exceeds the difference between the $1s$ and $2p$ states in H by a large factor

$$\frac{(E_{1s} - E_{2p})\text{ in Fe}}{(E_{1s} - E_{2p})\text{ in H}} = \frac{-1.03 \times 10^{-15}\text{ J}}{-16.2 \times 10^{-19}\text{ J}} = 636$$

The much larger spacing between the $2p$ and $1s$ states in Fe results from the larger nuclear charge ($Z = 26$ for Fe versus $Z = 1$ for H). The larger Z lowers the energies of both the $2p$ and $1s$ in Fe compared to H, but lowers the energy of the $1s$ much more. Note that the ratio of the energy differences (636) is roughly equal to $(Z_{\text{Fe}}/Z_{\text{H}})^2 = (26/1)^2 = 676$.

5.59 The similarities in chemical properties among the alkali metals strongly suggest similar ground-state valence electron configurations. This configuration must have the form ns^1 because no $(n-1)d$ orbitals are occupied in ground-state Li and Na atoms. Therefore the ns orbital must have lower energy (greater stability) than the $(n-1)d$ orbital in K, Rb, Cs. A similar argument can be made with respect to the alkaline-earth elements.

Chapter 6

Quantum Mechanics and Molecular Structure

Quantum Picture of the Chemical Bond

6.1 Nodes are surfaces across which wave-functions (orbitals) change their sign. The diagrams in text Figure 6.5 (text page 227) show the shapes of the eight lowest-energy molecular orbitals (MO's) of the H_2^+ ion. In these diagrams, color indicates sign: red for positive, blue for negative. Changes in color show the orientation and location of nodal surfaces.[1] Looking at the diagrams and counting the nodes in each gives these results

Molecular Orbital	$1\sigma_g$	$1\sigma_u^*$	$2\sigma_g$	$2\sigma_u^*$	$3\sigma_g$	$3\sigma_u^*$	$1\pi_u$	$1\pi_u^*$
No. of nodes	0	1	1	2	2	3	1	2

The node-count equals $n-1$ (where n is the quantum number appearing in the labels) for the three σ bonding MO's and n for the three σ^* antibonding MO's. The pattern with the π MO's is similar, but stepped up one: The $1\pi_u$ MO has n nodes and the $1\pi_u^*$ MO has $n+1$ nodes. The problem asks about the number of nodes encountered while traversing the z-axis, which lies in the plane of the paper. Counting each encounter with a nodal surface separately the answer for the six σ orbitals is $\boxed{0,1,2,3,2,3}$. Some nodes are crossed twice.

Tip. The last two entries in the preceding table partially answer problem 6.2.

Tip. Figure 6.55 (text page 280) repeats some of the diagrams in Figure 6.5 in conjunction with contour plots and line scans that help in visualization of the shapes of the orbitals and the locations of the nodes.[2]

6.3 The electron probability density in sigma (σ) molecular orbitals is cylindrically symmetrical about the internuclear axis. The cross-section of a cylinder is a circle. The desired shapes are consequently $\boxed{\text{all circles.}}$ What are the relative sizes of these circles? Text Figure 6.55 (c) provides a way to obtain estimates. The midpoint between the two H nuclei in all of the graphs lies at $z = 0$ Å. The points that are 1/4 of the way between the nuclei lie at $z = \pm 0.265$ Å.[3] Read the values of ψ off the graph at $z = 0$ and $z = \pm 0.265$ Å.[4] Then square them to obtain the relative radii of the circles. Text Figure

---

[1]The $2\sigma_g$ MO in the figure has an inner negative region (blue) that is completely surrounded by an outer positive region (red) that prevents the inner blue from being seen.

[2]The scans along the z axis in Figure 6.55 (c) touch the axes at the far left and right ends. This is an artifact of the printing process. These wave-functions never actually reach zero in these regions, but only approach zero asymptotically. These regions are not nodes.

[3]The internuclear distance in H_2^+ is 1.060 Å, see text Section 6.2.

[4]The numbers on the ψ axis in the plots are fractions of the maximum amplitude of the wave-function. They therefore have no units.

58

6.55 exhibits eight molecular orbitals. Of these, six are σ MO's. They account for the six entries in the following table. Note that the lowest energy MO appears at the bottom of the table, as in text Figure 6.55

Molecular Orbital	at $z = 0$		at $z = \pm 0.265$	
	ψ	ψ^2	ψ	ψ^2
$3\sigma_u^*$	0	0	∓ 0.50	0.25
$3\sigma_g$	-0.60	0.36	-0.40	0.16
$2\sigma_u^*$	0	0	± 0.20	0.04
$2\sigma_g$	-0.95	.90	-0.98	0.96
$1\sigma_u^*$	0	0	± 0.30	0.09
$1\sigma_g$	0.71	0.50	0.75	0.56

For the $1\sigma_u^*$, $2\sigma_u^*$, and $3\sigma_u^*$ MO's the cross-section in the $z = 0$ plane consists of a point (a circle of zero radius). This occurs because these three MO's have nodes at $z = 0$.

Tip. Values for ψ^2 at $z = 0$ and $z = \pm 0.265$ Å can also be obtained by reading the height of the ψ^2 curve above the z axis at these points in the graphs in column (c) of text Figure 6.56.

6.5   The electron density in the $1\sigma_g$ molecular orbital in H_2^+ ion is concentrated between the H nuclei. Therefore, the $\boxed{1\sigma_g}$ MO in text Figure 6.5 describes the bond in the H_2^+ ion. An electron in the $1\sigma_u^*$ MO in H_2^+ ion would spend most of its time in the regions *not* between the H nuclei (as in text Figure 3.22c). It would be an antibonding electron; because it would oppose the bond between the nuclei.

Tip. Interestingly, an electron in the $2\sigma_g$ MO in H_2^+ ion, which is at higher energy than the $1\sigma_u^*$, also maintains bonding in H_2^+ ion. The classical model cannot account for the existence of such a bound state.

MO Theory and the LCAO Approximation for H_2^+: Homonuclear Diatomic Molecules

6.7   The state $\boxed{(\sigma_{g1s})(\sigma_{u1s}^*)^2}$ has the higher energy because it has two electrons in antibonding orbitals. The state $(\sigma_{g1s})^1(\sigma_{g1s}^*)^1$ has one electron in a lower-energy bonding molecular orbital

Tip. The state $(\sigma_{g1s})^1(\sigma_{g1s}^*)^1$ is non-bonding, but the state $(\sigma_{g1s})(\sigma_{g1s}^*)^2$ is actively repulsive.

6.9   The H_2 molecule, which has the ground state electron configuration $(\sigma_{g1s})^2$, has two electrons in bonding molecular orbitals. The He_2^+ ion, which has the ground state electron configuration $(\sigma_{g1s})^2(\sigma_{u1s}^*)^1$, also has 2 electrons in bonding MO's but 1 electron in an antibonding MO. The bond order in He_2^+ is $\frac{1}{2}$; the bond order in H_2 is 1. $\boxed{H_2}$ has the larger bond energy.

6.11   The species $\boxed{He_2^+}$ has a longer bond distance because it has a lower bond order (see problem **6.9**).

6.13   The sketch graph of the potential energy curve for He_2^+ should have a shallower minimum at a longer internuclear distance than the one for H_2, as in the following

Qualitative Potential Energy Curves for H_2 and He_2^+

6.15 H_2 gives H_2^+ when a single electron is removed. The electron comes from a bonding molecular orbital. The product is therefore less strongly bonded than the original molecule. H_2^+ has a [smaller bond energy] and a [longer bond length].

Homonuclear Diatomic Molecules: Second-Period Atoms

6.17 **a)** Fluorine (F_2) is a homonuclear diatomic molecule. It has 18 electrons of which 14 are valence electrons. The F_2^+ ion has lost a valence electron and so has 13 remaining. Consult the correlation diagram for the F_2 molecule in text Figure 6.18b to obtain the energetic order of the MO's. Also see text Example 6.3. Remove an electron from the highest energy MO to get the configuration for the F_2^+ ion from the configuration for the F_2 molecule

$$F_2 \quad (\sigma_{g2s})^2(\sigma_{u2s}^*)^2(\sigma_{g2p})^2(\pi_{u2p})^4(\pi_{g2p}^*)^4$$
$$F_2^+ \quad (\sigma_{g2s})^2(\sigma_{u2s}^*)^2(\sigma_{g2p})^2(\pi_{u2p})^4(\pi_{g2p}^*)^3$$

b) The F_2 molecule has two more bonding than antibonding electrons. Its bond order is $\boxed{1}$; F_2^+ ion has *three* more bonding than antibonding electrons. Its bond order is $\boxed{\frac{3}{2}}$.

c) The F_2 molecule has zero unpaired electrons. Accordingly, F_2 is diamagnetic. The F_2^+ ion has an odd number of electrons. Because at least one electron (a π_{g2p}^* electron in this case) is unpaired, $\boxed{F_2^+}$ is paramagnetic.

d) The $\boxed{F_2^+}$ ion has a larger bond order and therefore requires [more energy] to dissociate than does the F_2 molecule.

6.19 The ground-state valence-electron configuration of the S_2 molecule should duplicate that of O_2 except in using $n = 3$ orbitals. The valence-electron configuration of O_2 appears in text Figure 6.15 (page 235). Assume that the S_2 molecule is in its ground state despite the high temperature. Then the valence electron configuration is

$$\boxed{(\sigma_{g3s})^2(\sigma_{u3s}^*)^2(\sigma_{g3p_z})^2(\pi_{u3p_x})^2(\pi_{u3p_y})^2(\pi_{g3p_x}^*)^1(\pi_{g3p_y}^*)^1}$$

The bond order is $\boxed{2}$; the molecule should be [paramagnetic] (two unpaired electrons).

6.21 In each case, count the valence electrons. This result together with the charge on the species identifies the column of the periodic table in which the element is located. All the configurations involve MO's from the $n = 2$ shell and therefore involve elements in the second row of the periodic table. The bond order is half the number of bonding electrons minus half the number of antibonding electrons:

a) F_2, bond order 1.   **b)** N_2^+, bond order $\frac{5}{2}$.   **c)** O_2^-, bond order $\frac{3}{2}$.

6.23 Check for unpaired valence electrons:

    a) F_2 is diamagnetic.   **b)** N_2^+ is paramagnetic.   **c)** O_2^- is paramagnetic.

6.25 Copy text Figure 6.21, the correlation diagram for heteronuclear diatomic molecules. Nitrogen is more electronegative than carbon. The energies of its atomic orbitals are *lower* than the energies of the corresponding orbitals in the carbon atom. It is "atom B" (on the right side of the correlation diagram); carbon is "atom A" on the left. The CN molecule has 9 valence electrons and is isoelectronic with the BO molecule. The figure displays the ground-state valence electron configuration of BO explicitly. The ground-state valence-electron configuration of CN is identical

$$\text{CN}\quad (\sigma_{2s})^2(\sigma_{2s}^*)^2(\pi_{2p})^4(\sigma_{2p})^1$$

The bond order accordingly is $\boxed{\frac{5}{2}}$; the unpaired electron causes $\boxed{\text{paramagnetism}}$.

Tip. The valence-electron configurations of CN and BO are the same because the *pattern* of relative energies of the MO's is the same in the two molecules. The actual, numerical energies of the MO's differ in the two molecules.

6.27 The molecule CF is a heteronuclear diatomic molecule. Refer to the correlation diagram in text Figure 6.21. Fluorine is more electronegative than carbon; regard F as the atom on the right of the correlation diagram and C as the one on the left. The CF molecule has 11 valence electrons; the CF^+ molecular ion has 10 valence electrons. The ground-state valence electron configurations of the two are

$$\text{CF}\quad (\sigma_{2s})^2(\sigma_{2s}^*)^2(\pi_{2p})^4(\sigma_{2p})^2(\pi_{2p}^*)^1 \quad\text{and}\quad \text{CF}^+\quad (\sigma_{2s})^2(\sigma_{2s}^*)^2(\pi_{2p})^4(\sigma_{2p})^2(\pi_{2p}^*)^0$$

The electron that a CF molecule loses to form a CF^+ ion comes from the π_{2p}^* orbital. The loss of this antibonding electron increases the bond order from $\frac{5}{2}$ to 3. The bond $\boxed{\text{strengthens}}$.

Tip. The CF^+ ion is isoelectronic with the N_2 molecule (both have 10 valence electrons). Finding a triple bond in CF^+ is reasonable; N_2 has a triple bond too.

6.29 The ground-state electron configuration for HeH^- would be $(\sigma_{1s})^2(\sigma_{1s}^*)^2$. The ion has a bond order of $\boxed{\text{zero}}$ and should be $\boxed{\text{unstable}}$.

6.31 Count the valence electrons in each of the four diatomic species: 11 for CF; 10 for CN^-; 5 for CH; 4 for CH^+. Text Figure 6.21 gives the energetic order of the valence MO's for heteronuclear diatomic molecules. Feed the correct number of electrons into the MO's in order of increasing energy. Put a maximum of two electrons into any one MO. In the usual notation, Greek letters with subscripts and superscripts represent MO's of different types, MO's of the same energy are grouped in parentheses, and right superscripts tell the number of electrons in each MO or group of MO's. The ground-state configurations for CF and CN^- are

$$\text{CF}\quad \boxed{(\sigma_{2s})^2(\sigma_{2s}^*)^2(\pi_{2px},\pi_{2py})^4(\sigma_{2pz}^*)^2(\pi_{2px}^*,\pi_{2py}^*)^1}$$

$$\text{CN}^-\quad \boxed{(\sigma_{2s})^2(\sigma_{2s}^*)^2(\pi_{2px},\pi_{2py})^4(\sigma_{2pz}^*)^2}$$

CH^+ and CH have 4 and 5 valence electrons respectively. If the σ MO and π^{nb} MO's differed a lot in energy in CH^+ and CH, as they do in the case of HF in text Figure 6.24, then the ground-state configurations would be CH $(\sigma^{nb})^2\sigma^2(\pi_x^{nb},\pi_y^{nb})^1$ and CH^+ $(\sigma^{nb})^2\sigma^2(\pi_x^{nb},\pi_y^{nb})^0$. But this is not the case. The $\Delta(EN)$ (electronegativity difference) for C—H is only 0.30, much less than for C—F, for which $\Delta(EN)$ is 1.43. See Text Figure 3.18. This greatly reduces the difference in energy between the σ and (π_x^{nb},π_y^{nb}) MO's in CH and CH^+, and the electrons remain unpaired in the ground-state configurations: CH^+ $\boxed{(\sigma^{nb})^2\sigma^1(\pi_x^{nb},\pi_y^{nb})^1}$ and CH $\boxed{(\sigma^{nb})^2\sigma^1(\pi_x^{nb},\pi_y^{nb})^2}$.

Tip. In their ground-states, CH^+ has 2 unpaired electrons and CH has 3. Configurations for CH^+ and CH having 0 and 1 unpaired electrons respectively are possible. They are excited states of the molecule and the molecule-ion, not ground-states.

6.33 CF^+ has 10 valence electrons (4 from C, 7 from F less 1 to form the +1 ion). NO has 11 valence electrons, and CF^- has 12. Text Figure 6.21 applies to all three of these heteronuclear diatomic species. Use it to arrive at ground-state valence electron configurations for the three:

$$CF^+ \quad (\sigma_{2s})^2(\sigma_{2s}^*)^2(\pi_{2px}, \pi_{2py})^4(\sigma_{2pz}^*)^2(\pi_{2px}^*, \pi_{2py})^0$$

$$NO \quad (\sigma_{2s})^2(\sigma_{2s}^*)^2(\pi_{2px}, \pi_{2py})^4(\sigma_{2pz}^*)^2(\pi_{2px}^*, \pi_{2py})^1$$

$$CF^- \quad (\sigma_{2s})^2(\sigma_{2s}^*)^2(\pi_{2px}, \pi_{2py})^4(\sigma_{2pz}^*)^2(\pi_{2px}^*, \pi_{2py})^2$$

Now use the formula

$$\text{bond order} = \frac{\text{No. bonding e}^- - \text{No. antibonding e}^-}{2}$$

to figure out the bond orders of each of the species. CF^+ has 8 valence electrons in bonding MO's and 2 in antibonding MO's. Its bond order is 3. NO has 8 valence electrons in bonding MO's and 3 in antibonding MO's. Its bond order is 2.5. CF^- has 8 electrons in bonding MO's and 2 in an antibonding MO. Its bond order is 2. The observed downward trend in bond dissociation energy $CF^+ > NO > CF^-$ $\boxed{\text{follows the decrease in bond order}}$.

Photoelectron Spectroscopy

6.35   The binding energy of the electron in each case equals the energy of the ionizing radiation minus the kinetic energy that the ejected electron carried

$$\text{Binding energy} = h\nu_{\text{photon}} - \tfrac{1}{2}m_e v^2$$

$$= 21.22 \text{ eV} - 5.63 \text{ eV} = \boxed{15.59 \text{ eV}}$$

$$= 21.22 \text{ eV} - 4.53 \text{ eV} = \boxed{16.69 \text{ eV}}$$

Electrons with the 15.59 eV binding energy are from the $\boxed{\sigma_{g2p_z}}$ molecular orbital, which is the highest occupied molecular orbital in ground-state N_2. See the top portion of text Figure 6.18a. Electrons with the 16.69 eV binding energy come from a $\boxed{\pi_{u2p}}$ molecular orbital (either the π_{u2p_y} or a π_{u2p_x}, the two possess the same energy, as the correlation diagram shows).

6.37   The requested diagram should resemble text Figure 6.18a except that it should be correctly scaled on the vertical, or energy, axis. The $n = 0$ peaks in a photoelectron spectrum are the ones without vibrational excitation; they appear at *lowest* energy in each group of closely spaced peaks arising when electrons are detached from a specific orbital. The experimental photoelectron spectrum for N_2 on text page 251 and the results of problem **6.35**, combine to give the energies of the valence MO's in N_2: -36 eV, -18.8 eV, -16.69 eV, -15.59 eV. Plot them to scale as in the following

This experimental energy-level diagram confirms the electron configuration shown in text Table 6.3.

6.39 Take the order of the MO's in HBr to be the same as the order of the MO's in HF. Text Figure 6.24 shows this order. The smaller ionization energy (11.88 eV) is the energy required to remove a electron from the highest occupied molecular orbital in H—Br. This is a non-bonding molecular orbital, either the $\boxed{4\pi_x^{\text{nb}} \text{ or the } 4\pi_y^{\text{nb}}}$. The "4" replaces the "2" in these symbols because Br is in the fourth period of the periodic table, two rows below F. These two MO's, which are equivalent except for their orientation in space, are localized mainly on the Br atom and derive only a little of their character from the H atom. The larger ionization energy (15.2 eV) is the energy required to remove one of the two electrons in the $\boxed{\sigma}$ orbital, which is a bonding orbital.

Localized Bonds: The Valence Bond Model

6.41 The valence-bond (VB) wave-function for Li_2 is constructed by overlap of the half-filled $2s$ orbital on the first lithium (atom A) and the half-filled $2s$ orbital on the second lithium (atom B)

$$\boxed{\psi_\sigma^{\text{bond}}(1,2) = c_1\left[2s^A(1)\,2s^B(2) + 2s^A(2)\,2s^B(1)\right]}$$

The 1 and 2 in parentheses are shorthand references to the coordinates of the two valence electrons in the Li_2 molecule. The value of the wave-function ψ depends on these coordinates. The wave-function also depends on R_{AB}, the distance between the two nuclei. This could be signified explicitly by including an R in the parentheses on the left, if desired. This wave-function closely resembles the VB wave-function for the bond in H_2 (text equation 6.14). Just as with H_2, combining the two terms on the right using a minus instead of a plus sign (the *ungerade* combination) gives repulsion between the nuclei at all distances. Li_2 is predicted to have a single bond $\boxed{\text{bond order 1}}$. This prediction is $\boxed{\text{the same as the LCAO prediction}}$.

The electron configurations of the C atoms before they bond in C_2 are $2s^2 2p_x^1 2p_y^1$. The filled $2s$ orbitals on the two atoms cannot overlap, but the four half-filled $2p$ orbitals on the atoms can. Construct the VB wave-functions by analogy to text equation 6.19

$$\psi_\pi^{\text{bond}}(1,2) = c_1\left[2p_x^A(1)\,2p_x^B(2) + 2p_x^A(2)\,2p_x^B(1)\right]$$
$$\psi_\pi^{\text{bond}}(3,4) = c_2\left[2p_y^A(3)\,2p_y^B(4) + 2p_y^A(4)\,2p_y^B(3)\right]$$

C_2 is thus predicted to have $\boxed{\text{bond order 2}}$. This prediction is the $\boxed{\text{same as the LCAO prediction}}$.

Tip. Electrons 1 and 2 do not have to be associated with the p_x orbitals; electrons 3 and 4 do not have to be associated with the p_y orbitals. The set of labels could be reversed. The point is to assert a difference, not a specific identification.

Tip. The two π bonds are equivalent and exist concurrently in the same region of space. Consequently $c_2 = c_1$ and a combined VB wave-function can be written for the double bond

$$\psi_{\pi\pi}^{\text{bond}}(1,2,3,4) = c_1\left[2p_x^A(1)\,2p_x^B(2)\right]\left[2p_y^A(3)\,2p_y^B(4)\right] + c_1\left[2p_x^A(2)\,2p_x^B(1)\right]\left[2p_y^A(4)\,2p_y^B(3)\right]$$

6.43 In the simple VB model, the first Be atom has no unpaired electrons to overlap with orbitals on the second (and vice versa). The simple VB model thus predicts no bonding between two Be atoms. The LCAO approach predicts the same thing: two of the valence electrons in Be_2 occupy a bonding molecular orbital, but the other two are obliged to occupy an antibonding molecular orbital, leading to a bond order of zero.

6.45 The simple VB model incorrectly predicts that B (valence electron configuration $2s^2 2p^1$) and H (valence electron configuration $1s^1$) form the $\boxed{\text{linear diatomic molecule B—H}}$ with one electron each from the B and H using the function

$$\psi_\sigma^{\text{bond}}(1,2) = c_1\left[1s^H(1)\,2p_z^B(2)\right] + c_2\left[1s^H(2)\,2p_z^B(1)\right]$$

Modified VB theory (using the concept of hybridization) predicts formation of BH_3, as explained beginning on text page 272.

6.47 The simple VB model predicts that N (valence electron configuration $2s^2\,2p_x^1 2p_y^1 2p_z^1$) forms single bonds with each of the three H atoms. These atoms are designated H1, H2, and H3 in the following valence-bond wave functions, which come from identical overlap with the $2p_x$, $2p_y$, and $2p_z$ orbitals on the N with the respective $1s$ orbitals on the H's

$$N-H1 \quad \psi_\sigma^{bond}(1,2) = c_1\left[1s^{H1}(1)\,2p_x^N(2)\right] + c_2\left[1s^{H1}(2)\,2p_x^N(1)\right]$$

$$N-H2 \quad \psi_\sigma^{bond}(1,2) = c_1\left[1s^{H2}(3)\,2p_y^N(4)\right] + c_2\left[1s^{H2}(4)\,2p_y^N(3)\right]$$

$$N-H3 \quad \psi_\sigma^{bond}(1,2) = c_1\left[1s^{H3}(5)\,2p_z^N(6)\right] + c_2\left[1s^{H3}(6)\,2p_z^N(5)\right]$$

The model incorrectly predicts a $\boxed{\text{trigonal pyramidal}}$ geometry for the NH_3 molecule, one in which all three H—N—H angles equal $90°$.

Orbital Hybridization; Predicting Molecular Structures and Shapes

6.49 Eight valence electrons (5 from the N, 1 each from the H's and 1 from elsewhere) surround the central N atom in NH_2^-. The valence orbitals of the N atom are $\boxed{sp^3}$ hybridized. Two of these hybrid orbitals overlap with $1s$ orbitals on the two H atoms to form two σ bonds. The other two contain lone pairs. The molecular ion should be $\boxed{\text{bent}}$ with an H—N—H angle less than $109.5°$.

Tip. Experimentally, the angle equals $106.7°$.

6.51 In all of these species, the hybridization on the central atom follows from the number of lone pairs plus the number of bonded atoms that surround the central atom (this sum equals the steric number *SN*). The molecular geometry depends on the hybridization, but the shapes of molecules are named only with reference to actual atoms.

a)   b)   c)   d)   e)

tetrahedral   linear   bent   pyramidal   linear

a) The central C in CCl_4 has *SN* 4. This atom is $\boxed{sp^3}$ hybridized, and the molecule is $\boxed{\text{tetrahedral}}$.

b) The central C in CO_2 has *SN* 2 and is $\boxed{sp}$ hybridized. The molecule is $\boxed{\text{linear}}$.

c) The central O in OF_2 has *SN* 4 and is $\boxed{sp^3}$ hybridized. Two of the hybrid orbitals on the O accommodate lone pairs of electrons, and two overlap with orbitals on the fluorine atoms. The molecule is $\boxed{\text{bent}}$.

d) The central C in the CH_3^- ion has *SN* 4 and is $\boxed{sp^3}$ hybridized. One of the four hybrid orbitals contains a lone pair of electrons. The other three overlap with $1s$ orbitals of the three H atoms. The molecular ion is $\boxed{\text{pyramidal}}$.

e) The central Be in BeH_2 has *SN* 2 and is $\boxed{sp}$ hybridized. The molecule is $\boxed{\text{linear}}$.

6.53 The ClO_3^+ and ClO_2^+ ions have 24 and 18 valence electrons respectively. The central Cl atom in ClO_3^+ has *SN* 3 and therefore three $\boxed{sp^2}$ hybrid orbitals overlapping with orbitals from the oxygen atoms. It has a $\boxed{\text{trigonal planar}}$ geometry. The central Cl atom in ClO_2^+ likewise has a set of three $\boxed{sp^2}$ hybrid orbitals, but only two overlap with orbitals on oxygen atoms. The third sp^2 orbital contains a lone pair. The ClO_2^+ molecular ion is $\boxed{\text{bent}}$. The central chlorine atoms in the following Lewis structures are shown with expanded octets. Other resonance structures can be drawn; these particular structures minimize formal charges. Compare to problem **3.91**.

6.55   The central nitrogen atom in the orthonitrate ion can attain an octet by forming four single bonds, one to each of the four oxygen atoms. Expected is $\boxed{sp^3}$ hybridization on the N atom and a $\boxed{\text{tetrahedral}}$ geometry.

Tip. The Lewis structure for single-bonded NO_4^{3-} puts formal charges of -1 on all four oxygen atoms and a formal charge of $+1$ on the central N. Such a build-up of formal charge is undesirable. Expansion of the octet on the central N would moderate the build-up, but elements in the second row of the periodic table resist expansion of their octets for energetic reasons. Elements in the third row of the table expand their octets more readily. This explains why orthophosphate ion PO_4^{3-} is well-known, but orthonitrate ion NO_4^{3-} is difficult to make and keep and is unfamiliar.

Using the LCAO and and Valence Bond Methods Together

6.57   The NF_2 molecule is bent and has 19 valence electrons. Two N—F σ bonds come from overlap of sp^2 hybrid orbitals on the central N atom with $2p$ orbitals on the two F atoms. The two bonds use four electrons. Ten electrons occupy orbitals that are not properly oriented for overlap with other atoms' orbitals: two in the $2s$ orbital on F1; two in the $2s$ orbital on F2; two in a $2p$ orbital on F1, two in a $2p$ orbital on F2; two in the third sp^2 hybrid orbital on the N. This leaves five electrons and three $2p$ orbitals (one each on three atoms). These orbitals overlap to form a π molecular orbital system that accommodates electrons as follows: $(\pi)^2(\pi^{nb})^2(\pi^*)^1$. See text Figure 6.47. The bond order of the whole molecule based on σ bonding is 2 and the π system adds $\frac{1}{2}$ to this making the total bond order of the molecule $2\frac{1}{2}$. The equivalent N—F bonds are both $\frac{5}{4}$ bonds.

6.59   The azide ion N_3^- is linear and has 16 valence electrons just like the CO_2 molecule, which is discussed in some detail on text page 272. Two N—N σ bonds result from overlap of sp hybrid orbitals on the central N atom with $2p_z$ orbitals on the two outer N atoms. These bonds use 4 electrons. Lone pairs in each of the $2s$ orbitals of the outer N atoms account for another 4 electrons. The remaining six p orbitals (two each on three atoms) overlap to form a π molecular orbital system to accommodate the remaining 8 valence electrons. The correlation diagram for CO_2 (text Figure 6.51) gives the relative energies of the MO's in this system. Four of the eight electrons thus go into the low-lying π_x and π_y orbitals. The remaining four go into the two π^{nb} orbitals. The π configuration is $(\pi)^4(\pi^{nb})^4$. This means a total of two π bonds and an overall bond order for the molecule of 4: (2 σ bonds plus 2 π bonds). The two N-to-N linkages are identical; each has bond order 2. All of the electrons are paired so the compound is $\boxed{\text{diamagnetic}}$.

The N_3 molecule has 15 valence electrons. It derives from N_3^- by the loss of an electron. The loss comes from the highest energy molecular orbital which is a nonbonding MO. N_3 is $\boxed{\text{bound}}$ with an overall bond order of 4, just like N_3^-. Unlike N_3^-, N_3 has an unpaired electron and is $\boxed{\text{paramagnetic}}$.

The N_3^+ ion has 14 valence-electrons. It derives from N_3^- by the loss of two nonbonding π electrons. The N_3^+ molecular ion is therefore $\boxed{\text{bound}}$ with bond order 4. There are two unpaired electrons in the set of π^{nb} MO's so N_3^+ is $\boxed{\text{paramagnetic}}$, too.

6.61   Draw the Lewis structure for NO_2^- and use VSEPR theory to determine the steric number and structure. The best two Lewis structures are

VSEPR theory assigns $\boxed{SN\ 3}$ to the central N. The O atoms occupy two of the three sites, and the lone pair the third. The molecular ion is therefore $\boxed{\text{bent}}$. The hybridization at the nitrogen atom is $\boxed{sp^2}$. Two of the three sp^2 hybrid orbitals form the σ bonds to the oxygen atoms, and the third accommodates the lone pair. The unhybridized $2p_z$ atomic orbital on the N atom is oriented perpendicular to the plane of the molecule. It overlaps with the $2p_z$ atomic orbitals of the two oxygens to form a π system. See text Figure 6.52. Two electrons occupy the bonding π orbital in this system and two electrons occupy the nonbonding (π^{nb}) orbital. The antibonding (π^*) orbital remains empty. Adding the σ MO's to the bonding contributed by the π system gives an overall bond order of 3, which amounts to $\boxed{\frac{3}{2}}$ for each bond. In a localized-orbital scheme, two resonance forms are necessary to represent the bonding in NO_2^-.

Tip. The bonding in the nitrite ion is discussed on text page 273.

6.63 The molecules of pyridine C_5H_5N and benzene C_6H_6 have great similarities in their bonding. Both have 30 valence electrons; both have six π molecular orbitals that arise as a combination of the six $2p_z$ orbitals of the six atoms that comprise their rings. However, the MO's that put electron density onto the N atom in pyridine are lower in energy than comparable MO's in benzene because N is more electronegative than C. In other terms, MO's that have the $p_z(N)$ orbital among their "parents" are lower in energy. Draw the structure of benzene and let an N atom replace the C at position 1 in a numbering scheme that goes around the ring

Text Figure 6.54 shows the six lowest-energy π MO's of benzene. The most strongly bonding and strongly antibonding (the highest and lowest in energy among those shown) both have parentage that includes the $2p_z$ orbital on atom 1. These two molecular orbitals are therefore *lowered* in energy in pyridine relative to benzene. One of the two weakly bonding molecular orbitals in benzene has $2p_z$(atom 1) parentage, but the other does not. Its parentage includes p_z orbitals from C atoms at positions 2, 3, 5, and 6 only. The first of these two weakly bonding MO's (on the right in Figure 6.54) is therefore lowered in energy in pyridine relative to benzene, but the energy of the second is (almost completely) unaffected. Similarly, the two weakly antibonding MO's in benzene become split in energy when an N goes in at position 1. The one that has some $2p_z$(N) parentage (on the right in Figure 6.54) is lowered, but the other is (almost completely) unchanged. The result is an energy-level diagram for pyridine with six different π orbital energies, four lower than the corresponding π orbitals in benzene and two (almost completely) unchanged in energy.

ADDITIONAL PROBLEMS

6.65 **a)** Refer to text Figure 6.18 for pictures of the shapes of the five occupied MO's in ground-state N_2. This figure also gives the ground-state electron configuration of N_2, which is

$$N_2 \quad (\sigma_{g2s})^2(\sigma_{u2s}^*)^2(\pi_{u2p_x})^2(\pi_{u2p_y})^2(\sigma_{g2p_z})^2$$

The highest-energy occupied orbital is a σ_g MO, derived from $2p_z$-$2p_z$ overlap. Its shape is shown in the lower part of Figure 6.18a. The next two highest occupied MO's are a pair of π_{u2p} bonding MO's of equal energy. The pair have identical shapes but differ in orientation, as shown in the lower part of the figure. One comes from $2p_x$-to-$2p_x$ overlap; the other comes from $2p_y$-to-$2p_y$ overlap. Occupation of these two π MO's by four electrons furnishes a cylindrical muff of π electron density to surround the σ bond between the two N atoms. At lower energy lies a σ^*_{u2s} orbital. This MO derives from antibonding overlap of the 2s orbitals. Then at the bottom comes a σ_{g2s} orbital from bonding overlap of the same orbitals.

Tip. It is instructive actually to copy the diagrams by hand with a pencil on a piece of scratch paper.

b) Since the highest occupied molecular orbital of N_2 is a bonding orbital, the removal of one electron from N_2 decreases the bond order, and $\boxed{\text{lengthens}}$ the N-to-N bond.

6.67 Using the correlation diagram in text Figure 6.18a for B_2, which has six valence electrons, gives the ground-state configuration

$$(\sigma_{g2s})^2 (\sigma^*_{u2s})^2 (\pi_{u2p_x})^1 (\pi_{u2p_y})^1$$

There are two unpaired electrons in this configuration. The correlation diagram in Figure 6.18b on the other hand gives the ground-state configuration

$$(\sigma_{g2s})^2 (\sigma^*_{u2s})^2 (\sigma_{g2p_z})^2$$

There are no unpaired electrons in this configuration. Use of the diagram in 6.18b is inconsistent with the fact that B_2 is paramagnetic.

6.69 **a)** Look at the MO correlation diagrams for H_2 and O_2 (text Figures 6.12 and 6.18b respectively). The ionization energy equals the minimum energy required to remove the highest energy electron from a gaseous molecule or atom. In the case of H compared to H_2, the 1s electron of the ground-state atom lies higher in energy than a σ_{g1s} electron in the ground-state molecule. It consequently requires less energy to remove the atom's 1s electron than to remove one of the molecule's σ_{g1s} electrons.

The O-to-O_2 comparison is different. In the diagram, a 2p electron of a ground-state O atom lies lower in energy than the π^*_{g2p} electron, which is the highest-energy electron in the ground-state O_2 molecule.[5] Consequently, it requires more energy to ionize O than O_2.

b) The highest occupied molecular orbital of F_2 is the π^*_{g2p} orbital (see text Figure 6.18b). It exceeds the 2p atomic orbital in energy. The prediction on this basis is that the $\boxed{F_2 \text{ has a lower } IE}$ than the F atom.

6.71 The molecular orbital and the square of the molecular orbital for the ground state of the heteronuclear molecule are

$$\psi = C_A \psi_A + C_B \psi_B \quad \text{and} \quad \psi^2 = C_A^2 \psi_A^2 + 2 C_A C_B \psi_A \psi_B + C_B^2 \psi_B^2$$

where the C's are constants. The *square* of the wave-function is given because its value in any small region of space is proportional to the probability of finding the electron within that region. Neglecting the overlap of the two orbitals means neglecting the cross-term in the squared wave-function

$$\psi^2 \approx C_A^2 \psi_A^2 + C_B^2 \psi_B^2$$

If the electron spends 90 percent of its time in orbital ψ_A, then $C_A^2 = 9 C_B^2$. Also, the electron must be either on atom A or atom B so $C_A^2 + C_B^2 = 1$. Solution of the two simultaneous equations gives $C_A = \boxed{0.949}$ and $C_B = \boxed{0.316}$.

6.73 **a)** Nitramide H_2NNO_2 has 24 valence electrons. It must have one double bond somewhere if the octet rule is obeyed. If the structure is non-planar, this double bond is strongly localized to the —NO_2 portion of the molecule. The two electrons occupy a π orbital derived from $2p_z$ orbitals on the N atom and the two O atoms bonded to it. Participation in this π system by orbitals on the

---

[5]Remember that O_2 has 12 valence electrons, not 14, the number shown in the diagram in Figure 6.18b, which is furnished with electrons for the case of F_2.

other N atom would require coplanarity of the H_2N- and $-NO_2$ portions of the molecule. If the two portions are not coplanar, then overlap and effective mixing of p orbitals are not possible, and the N—N bond order is $\boxed{1}$. The following Lewis resonance structures represent this situation

The N—O bonds have bond order $\frac{3}{2}$, according to these structures

b) If the spectroscopic results are faulty and the nitramide molecule really is planar, the four $2p_z$ orbitals available on the two nitrogen atoms and two oxygen atoms after completion of the framework of σ bonding could overlap to form one π, two π^{nb}, and one π^* MO's. Four electrons would occupy this π system: two would occupy the bonding orbital, and two would occupy the non-bonding orbitals. The resulting π system would possess a total of two bonding electrons across three bonds among the four non-H atoms. The bond order of the N—N bond would be 1 (from the σ interaction) plus 1/3 (from the π system) or $\boxed{\frac{4}{3}}$. In terms of Lewis structures, this situation would be represented

Chapter 7

Bonding in Organic Molecules

Petroleum Refining and the Hydrocarbons

7.1 Gasoline having an octane number of 100 burns in a test engine with the same amount of knocking as "isooctane" (2,2,4-trimethylpentane), which was selected years ago as a reference compound. Octane was selected as a reference because it is an excellent anti-knock gasoline. This does not make the best possible anti-knock gasoline. If a gasoline turns out to burn with *less* knocking than isooctane, then its octane number exceeds 100. An octane rating exceeding 100 $\boxed{\text{is possible}}$. Such fuels are known.

7.3 The alkane is gaseous, and one mole of it generates two moles of gaseous carbon dioxide (under the same condition of temperature and pressure) when it burns. This means that each molecule of the alkane contains two atoms of carbon. This alkane cannot be cyclic because it has only two carbon atoms per molecule. Non-cyclic alkanes all have the general formula C_nH_{2n+2}. This alkane is $\boxed{C_2H_6, \text{ ethane}}$. The equation for its combustion is

$$\boxed{2\,C_2H_6(g) + 7\,O_2(g) \longrightarrow 4\,CO_2(g) + 6\,H_2O(g)}$$

Tip. If 1 mol of gaseous alkane generated 3 mol of carbon dioxide, then the problem would have two possible answers: propane (C_3H_8) and cyclopropane (C_3H_6); if 1 mol of gaseous alkane generated 4 mol of carbon dioxide, then the problem would have four answers: normal butane (C_4H_{10}), 2-methylpropane (C_4H_{10}), cyclobutane (C_4H_8), and methylcyclopropane (C_4H_8).

7.5 **a)** The catalytic cracking reaction is

b) An isomer of the above alkene, which is named 1-pentene, is $\boxed{\text{2-pentene}}$. In it the double bond is located between the second and third carbon atoms in the chain as follows

69

Tip. 1-pentene and 2-pentene are the only isomeric pentenes: "3-pentene" is the same as 2-pentene, and "4-pentene" is the same as 1-pentene.

7.7   The structural formulas are

Tip. Note the absence of C's at the corners of the triangle in structure (c). Unlabelled vertices in organic structures are understood to indicate a C atom.

7.9   The *cis* isomer of 3-heptene is at the left and the *trans* is at the right:

7.11   a) 1,2-hexadiene   **b)** 1,3,5-hexatriene   **c)** 2-methyl-1-hexene   **d)** 3-hexyne

7.13   Count the atoms linked directly to the C atom in question, and add the number of lone pairs.[1] The results equals the steric number SN of that C atom. If SN equals 2, the C is sp hybridized; if SN equals 3, the C is sp^2 hybridized; if SN equals 4, the C is sp^3 hybridized.

---
[1]Lone pairs are not common on C atoms in organic compounds.

a) From the left, the hybridization of the C atoms is: sp^2, sp, sp^2, sp^3, sp^3, and sp^3 .

b) All C atoms are sp^2 hybridized.

c) The two C atoms involved in the double bond are sp^2 hybridized. The rest are sp^3 hybridized.

d) From the left, the hybridization is sp^3, sp^3, sp, sp, sp^3, and sp^3 .

Fullerenes

7.15 Figure out the number of two-electron bonds required to hold together the 60 carbon atoms that comprise fullerene *and* simultaneously obey the octet rule on all atoms. The octet rule requires 4 bonds for each C. Each bond is shared by two C's (one at each end), so the answer is 120 (which equals $(4 \times 60)/2$).

The structure of fullerene has 60 vertices (the 60 C atoms), 32 faces (20 hexagonal and 12 pentagonal), and 90 edges. The text gives these numbers on page 303.[2] The 90 edges must be spanned by 60 single bonds and 30 double bonds. Only this combination accounts for 120 shared pairs of electrons. The most symmetrical way of placing the 30 double bonds is to position them on all edges that join two hexagonal faces .

Functional Groups and Organic Reactions

7.17 The balanced equation for the conversion is $C_2H_4 + Cl_2 \longrightarrow C_2H_4Cl_2$. Compute the mass of ethene required to make the 6.26×10^9 kg of 1,2-dichloroethane as follows

$$m_{C_2H_4} = 6.26 \times 10^9 \text{ kg } C_2H_4Cl_2 \times \left(\frac{1 \text{ mol } C_2H_4Cl_2}{0.09896 \text{ kg } C_2H_4Cl_2} \right) \left(\frac{1 \text{ mol } C_2H_4}{1 \text{ mol } C_2H_4Cl_2} \right)$$
$$\times \left(\frac{0.02805 \text{ kg } C_2H_4}{1 \text{ mol } C_2H_4} \right) = 1.77 \times 10^9 \text{ kg } C_2H_4$$

This is 11.2% of the 15.87×10^9 kg total annual production of ethene.
The mass of the chlorine required for this conversion is

$$m_{Cl_2} = 6.26 \times 10^9 \text{ kg } C_2H_4Cl_2 \times \left(\frac{1 \text{ mol } C_2H_4Cl_2}{0.09896 \text{ kg } C_2H_4Cl_2} \right) \left(\frac{1 \text{ mol } Cl_2}{1 \text{ mol } C_2H_4Cl_2} \right)$$
$$\times \left(\frac{0.07091 \text{ kg } Cl_2}{1 \text{ mol } Cl_2} \right) = \boxed{4.49 \times 10^9 \text{ kg } Cl_2}$$

Tip. Check against errors in arithmetic whenever possible. Here, adding the mass of chlorine and ethene put into the process gives the mass of 1,2-dichloroethane that comes out. This strongly suggests that the arithmetic is correct.

7.19 **a)** This is an esterification. It formally resembles an acid-base neutralization with the alcohol playing the part of the base. The organic product is butyl acetate.

$$CH_3COOH + HO{-}CH_2CH_2CH_2CH_3 \longrightarrow CH_3COOCH_2CH_2CH_2CH_3 + H_2O$$

b) The reaction is a dehydration: $H_3C{-}COO^- NH_4^+ \longrightarrow H_3C{-}CO{-}NH_2 + H_2O$.

---

[2]Confirm the number of edges, faces, and vertices by inspecting a soccer ball, which has the correct pattern of hexagons and pentagons marked on its surface. The 12 black pentagons are easy to see and count. Each pentagon is adjoined by five hexagons. Simply multiplying 5×12 counts each hexagon three times, because each hexagon is adjoined by three pentagons. The correct number of hexagons is $(5 \times 12)/3 = 20$. Adding the 12 pentagonal faces makes a total of 32 faces. Then use the formula $V + F = E + 2$, which relates the number of vertices V, edges E, and faces F of any polyhedron, to obtain the number of edges.

c) The H atom on the O atom and one of the H atoms on the neighboring C atom are removed and combined to give H_2: $\boxed{H_3CCH_2CH_2OH \longrightarrow H_3CCH_2CHO + H_2}$. The organic product is propanal (also called propionaldehyde). The reaction is a dehydrogenation.

d) $\boxed{H_3CCH_2CH_2CH_2CH_2CH_2CH_3 + 11\,O_2 \longrightarrow 7\,CO_2 + 8\,H_2O}$.

7.21   **a)** Brominate ethene to give 1,2-dibromoethane. Then dehydrobrominate the 1,2-dibromoethane. This means: treat ethene with bromine so that the bromine molecule adds across the double bond. Then treat the product in such a way that hydrogen bromide is abstracted:

$$CH_2{=}CH_2 + Br_2 \longrightarrow CH_2BrCH_2Br \quad \text{then} \quad CH_2BrCH_2Br \longrightarrow CH_2{=}CHBr + HBr$$

b) Treat 1-butene with water (in the presence of H_2SO_4)

$$CH_3CH_2CH{=}CH_2 + H_2O \longrightarrow CH_3CH_2CH(OH)CH_3$$

c) Treat propene with water to give 2-propanol, and then dehydrogenate over a catalyst (such as metallic copper)

$$CH_3CH{=}CH_2 + H_2O \longrightarrow CH_3CH(OH)CH_3 \quad \text{and} \quad CH_3CH(OH)CH_3 \longrightarrow CH_3COCH_3 + H_2$$

7.23

7.25   The balanced chemical equation for the conversion is $C_{18}H_{32}O_2 + 2\,H_2(g) \longrightarrow C_{18}H_{36}O_2(g)$. Compute the chemical amount of H_2 that is needed

$$n_{H_2} = 500.0 \text{ g } C_{18}H_{32}O_2 \times \left(\frac{1 \text{ mol}}{280.45 \text{ g } C_{18}H_{32}O_2} \right) \times \left(\frac{2 \text{ mol } H_2}{1 \text{ mol } C_{18}H_{32}O_2} \right) = 3.5657 \text{ mol } H_2$$

Then use the ideal-gas equation

$$V_{H_2} = \frac{n_{H_2}RT}{P} = \frac{(3.5657 \text{ mol})(0.082057 \text{ L atm mol}^{-1}\text{K}^{-1})(273 \text{ K})}{1 \text{ atm}} = \boxed{79.9 \text{ L}} \text{ at STP}$$

7.27   The Lewis diagram for acetaldehyde is

This structure shows a total of 18 valence electrons. The SN's (steric numbers) of the carbon atoms are 4 (for the methyl carbon) and 3 (for the carbonyl carbon). The SN's of the other atoms are immaterial because each of them forms only one bond. The SN 4 on the methyl C means it has $\boxed{sp^3}$ hybridization; the SN 3 on the carbonyl C means it has $\boxed{sp^2}$ hybridization. Constructing the single bond framework of the molecule uses 12 valence electrons. At this point a $2p$ orbital on the carbonyl C contains a single electron and the three $2p$ orbitals on the oxygen contain 5 electrons. The $2p$ orbital on the C and a $2p$ orbitals on the O overlap to form a π bonding orbital. Two electrons occupy this orbital. The remaining 4 electrons remain as lone pairs on the O. A π^* antibonding orbital is created simultaneously with the π orbital spanning C and O, but it remains empty. The three groups bonded to the carbonyl C lie in a plane with bond angles near $\boxed{120°}$. The geometry at the methyl C atom is approximately tetrahedral, with all six H—C—H and H—C—C angles near $\boxed{109.5°}$.

7.29   Lewis diagrams for HCOOH and HCOO⁻ are respectively

One resonance structure suffices for formic acid HCOOH; the formate anion HCOO⁻ requires two. In formic acid, one oxygen atom is doubly bonded to the carbon atom, and the other is singly bonded. In the formate ion, both C—O bonds have intermediate character: partially single and partially double. The carbon atom in HCOOH is sp^2 hybridized (SN 3), and the OH oxygen atom is sp^3 hybridized (SN 4). The immediate surroundings of the carbon atom have trigonal planar geometry, and the C—O—H group is bent. In the HCOO⁻ ion, the carbon atom and both oxygen atoms are sp^2 hybridized (SN 3), possessing a three-center four-electron π system. In HCOOH, π overlap occurs between orbitals on the carbon atom and only one oxygen atom. Both C-to-O bond lengths in the formate ion should lie somewhere between the value for the single bond (1.36 Å) and the value for the double bond (1.23 Å).

Pesticides and Pharmaceuticals

7.31   **a)** The juvenile-hormone insecticide methoprene ($C_{16}H_{28}O_3$) is the isopropyl (2-propyl) ester of an unsaturated carboxylic acid. The structures of the acid and the alcohol are

$$CH_3-O-\overset{\overset{\displaystyle CH_3}{|}}{\underset{\underset{\displaystyle {}^{12}CH_3}{|}}{\overset{11}{C}}}-\overset{\overset{\displaystyle H}{|}}{\underset{\underset{\displaystyle H}{|}}{\overset{10}{C}}}-\overset{\overset{\displaystyle H}{|}}{\underset{\underset{\displaystyle H}{|}}{\overset{9}{C}}}-\overset{\overset{\displaystyle H}{|}}{\underset{\underset{\displaystyle H}{|}}{\overset{8}{C}}}-\overset{\overset{\displaystyle CH_3}{|}}{\underset{\underset{\displaystyle H}{|}}{\overset{7}{C}}}-\overset{\overset{\displaystyle H}{|}}{\underset{\underset{\displaystyle H}{|}}{\overset{6}{C}}}-\overset{5}{C}=\overset{4}{C}-\overset{\overset{\displaystyle CH_3}{|}}{\overset{3}{C}}=\overset{\underset{\underset{\displaystyle H}{|}}{}}{\overset{2}{C}}-\overset{1}{COOH} \qquad H-\overset{\overset{\displaystyle CH_3}{|}}{\underset{\underset{\displaystyle CH_3}{|}}{C}}-OH$$

The alcohol is 2-propanol , also called isopropyl alcohol.

Tip.   The numerals near the carbons in the structure of the acid identify positions on the chain according to standard rules for naming. This chain starts with a carboxylic acid group and has one substituent on C-3, one on C-7, and two on C-11. It is 11-methoxy-3,7,11-trimethyl-2,4-dodecadienoic acid.

b) Replacing the 11-methoxy group (at the far left) by H and the carboxylic acid group (at the far right) by CH_3 gives

$$H-\overset{\overset{\displaystyle CH_3}{|}}{\underset{\underset{\displaystyle {}^{12}CH_3}{|}}{\overset{11}{C}}}-\overset{\overset{\displaystyle H}{|}}{\underset{\underset{\displaystyle H}{|}}{\overset{10}{C}}}-\overset{\overset{\displaystyle H}{|}}{\underset{\underset{\displaystyle H}{|}}{\overset{9}{C}}}-\overset{\overset{\displaystyle H}{|}}{\underset{\underset{\displaystyle H}{|}}{\overset{8}{C}}}-\overset{\overset{\displaystyle CH_3}{|}}{\underset{\underset{\displaystyle H}{|}}{\overset{7}{C}}}-\overset{\overset{\displaystyle H}{|}}{\underset{\underset{\displaystyle H}{|}}{\overset{6}{C}}}-\overset{5}{C}=\overset{4}{C}-\overset{\overset{\displaystyle CH_3}{|}}{\overset{3}{C}}=\overset{\underset{\underset{\displaystyle H}{|}}{}}{\overset{2}{C}}-\overset{1}{CH_3}$$

This name of this compound is 3,7,11-trimethyl-2,4-dodecadiene .

7.33   **a)** Aspirin has a molecular formula of $C_9H_8O_4$.

b) $n_{\text{aspirin}} = 0.325 \text{ g} \times (1 \text{ mol}/180.16 \text{ g}) =$ $1.80 \times 10^{-3} \text{ mol}$.

7.35   The following shows the numbering system that is used to designate the 17 carbons that make up the tetracyclic ring system found in all steroids

This structure is the steroid nucleus. Every numbered point specifies a C-atom. Each C in the nucleus is bonded to two or three other C's that also belong to the nucleus. This bonding does not furnish enough bonds (four for each C) for any of the 17 C's except C-5. Additional bonds make up the deficiency. They link to one or two H-atoms per C or to side-groups such as the —OH group, —CH_3 groups, and —C_8H_{17} group at C-3, C-10 and C-13, and C-17 in the diagram. For clarity, H-atoms are not shown. Thus, C-3 is bonded to an H in addition to the —OH and C-2 and C-4. And C-7 is bonded to two H's in addition to C-6 and C-8.

The specific steroid that is shown is cholesterol ($C_{27}H_{46}O$). Cortisone ($C_{21}H_{28}O_5$) derives from cholesterol by: oxidation of the —OH side-group at C-3 to a ketone (loss of one H by the —OH and one H by C-3); oxidation of C-11 to a ketone (loss of two H's and gain of one O); relocation of a double bond from C-5/C-6 to C-4/C-5 (loss of one H by C-4 and gain of one H by C-6); replacement of the —H at C-17 with an —OH; replacement of the side-group at C-17 with the smaller side-group —$COCH_2OH$.

Tip. Confirm by careful counting that if all of the side-groups on the steroid nucleus shown above are replaced by H atoms, the resulting molecule has the formula $C_{17}H_{26}$.

ADDITIONAL PROBLEMS

7.37   Cyclodecene $C_{10}H_{18}$ has 10 carbon atoms bonded in a cycle. One of the 10 connections is a double bond. In the following, only the H's on the double bonded carbon atoms are shown explicitly. In the *cis* isomer these two H's are on the same side of the ring. The figure shows this by representing the bonds to these H's with thicker lines. In the *trans* isomer, these two H's are or opposite sides of the ring. The nearer bond, which slants up from the plane of the paper, is shown by a thicker line, and the more distant bond, which slants back from the plane of the paper is shown by a dashed line.

7.39   Rewrite the formula of the compound as C_3H_5OH. This recognizes that the compound is an alcohol (contains an —OH group). The fragment C_3H_5 does not contain enough H to be a straight-chain alkane; C_3H_7 would be required. Inserting a double bond would reduce the requirement for H atoms by 2 atoms, but double bonds are ruled out. The only other way to reduce the number of H atoms is formation of a ring. The compound is ⎡cyclopropanol⎤.

7.41 **(a)** The statement of the problem makes it clear that the only reactants and products in this acetylation reaction are X, acetyl chloride, Y, and HCl. The equation for the reaction is therefore

$$C_4H_8O_3 + n\ CH_3COCl \longrightarrow C_8H_{12}O_5 + n\ HCl$$

The only value of n that balances this equation is 2. Therefore compound X contains $\boxed{\text{two}}$ hydroxyl groups.

(b) A possible structure if compound X is an aldehyde is

$$
\begin{array}{ccccccc}
 & H & & OH & & H & & H \\
 & | & & | & & | & & | \\
HO- & C & - & C & - & C & - & C=O \\
 & | & & | & & | & & \\
 & H & & H & & H &
\end{array}
$$

In compound Y, the —OH groups on carbons 3 and 4 are replaced by —OCOCH$_3$ groups.

7.43 The unsaturated hydrocarbon must have a straight-chain skeleton of six carbon atoms because it gives straight-chain hexane when reduced with hydrogen gas. Oxidation at the double bond splits it to a four-carbon acid (butanoic acid) and a two-carbon acid (acetic acid). The double bond is therefore at the 2 position: CH$_3$—CH$_2$—CH$_2$—CH=CH—CH$_3$. This compound is $\boxed{\text{2-hexene}}$. It has *cis* and *trans* isomers, but the available data do not allow a decision about which isomer is present. The balanced equations are

$$C_6H_{12}(l) + H_2(g) \longrightarrow C_6H_{14}(l)$$

$$24\,H_3O^+(aq) + 8\,MnO_4^-(aq) + 5\,C_6H_{12}(l) \longrightarrow 8\,Mn^{2+}(aq) + 5\,C_4H_8O_2(aq) + 5\,C_2H_4O_2(aq) + 36\,H_2O(l)$$

7.45 Both pharmaceuticals and pesticides must be biologically active, preferably in as specific a manner as possible. The strategy for delivery of the compound to the target is similar: fooling the organism's own pathways for intake. Hence, mimicking the organism's natural chemicals is important. The use of both classes of compounds requires close attention to possible side-effects. Pharmaceuticals particularly are expected to do one thing in the body and one thing only. A pesticide on the other hand is expected to kill its target by any means. Consequently more modes of action are possible for pesticides. However, ill-effects of a pesticide on non-target organisms can be more adverse, either in the long-term or short-term, than the deprecations of the pest. Numerous pesticides are so toxic to desirable organisms that they may not be used.

7.47 Cyclohexene absorbs at $\boxed{\text{shorter}}$ wavelengths in the UV region of the spectrum than does benzene. The delocalization of electron density among the three double bonds in benzene lowers the gap in energy between the π and π^* molecular orbitals. This shifts the wavelength of the electronic transition between these states to lower energy (longer wavelength) than in cyclhexene.

Chapter 8

Bonding in Transition Metals and Coordination Complexes

Chemistry of the Transition Metals

8.1   **a)** The expectation is that the more water-soluble of the two compounds will be $\boxed{PtF_4}$ because Pt is in a lower oxidation state and the bonds in the compound are more ionic than the bonds in PtF_6. The fact is that PtF_4 and PtF_6 both react with water.[1] Under certain conditions the reactions are

$$PtF_4 + 4\,H_2O \longrightarrow H_2[PtF_2(OH)_4] + 2\,HF$$
$$2\,PtF_6 + 6\,H_2O \longrightarrow 2\,(H_3O)_2[PtF_6] + O_2$$

In the first reaction, water displaces half of the fluorine atoms from the compound in the form of hydrogen fluoride, but the Pt remains in the +4 oxidation state. In the second, water reduces PtF_6, in which Pt is in a +6 oxidation state, to $[PtF_6]^{2-}$, an ion in which Pt is in a +4 oxidation state. The water undergoes oxidation to oxygen as it does this.[2]

Tip. The portions of the Pt-containing products whose formulas are enclosed in brackets in the preceding equations are coordination complexes, which are discussed in text Section 8.2. Naming them according to the rules given on text page 337 is a nice preview of problem **8.21**. The first is the difluorotetrahydroxoplatinate(IV) ion. The second is the hexafluoroplatinate(IV) ion. Both are anions with charges of -2.

b) The more volatile is $\boxed{PtF_6}$, which is the more covalent of the two.

8.3

$$\boxed{V_{10}O_{28}^{6-}(aq) + 16\,H_3O^+(aq) \longrightarrow 10\,VO_2^+(aq) + 24\,H_2O}$$

The vanadium is in the $\boxed{+5}$ oxidation state both before and after this acid-base reaction. The oxide of vanadium in which V has this oxidation state is $\boxed{V_2O_5}$.

8.5   The reduction of titanium(IV) oxide to titanium(III) oxide with hydrogen is represented

$$\boxed{2\,TiO_2 + H_2 \longrightarrow Ti_2O_3 + H_2O}$$

$\boxed{\text{Titanium(III) oxide}}$ should be more basic than titanium(IV) oxide. It has Ti in a lower oxidation state.

---

[1](a) *Reaction of Platinum Tetrafluoride with Water and Methanol*, L. Kolditz and J. Gisbier, Zeitschrift für Anorganische und Allgemeine Chemie, 366(5-6), 265-73, 1969. (b) *Hydrolysis Reactions of Transition Metal Hexafluorides in Liquid Hydrogen Fluoride: Oxonium Salts with Pt, Ir and Ru*, H. Selig, W. A. Sundek, F. A. Disalvo and W. E. Falconer, Journal of Fluorine Chemistry, 11(2) 39-50, 1978.

[2]Oxidation-reduction reactions are reviewed in text Section 11.4, text page 452.

Introduction to Coordination Complexes

8.7   The oxidation states of manganese are $\boxed{+2}$ in MnO, $\boxed{+4}$ in MnO_2, $\boxed{+3}$ in Mn_2O_3, $\boxed{+7}$ in Mn_2O_7, $\boxed{+3 \text{ and } +2}$ in the ratio of 2 : 1 in Mn_3O_4, $\boxed{+7}$ in MnO_4^-.

Tip. The fractional oxidation state $+8/3$ is sometimes assigned to the Mn "as a whole" in Mn_3O_4.

8.9   The F^- ion is much smaller than the Cl^- ion. Smaller ions have higher charge densities (more charge per unit volume) than larger ions of equal charge. Ions with higher charge density tend to stabilize the higher oxidation states of the transition metals that bond to them because high charge density favors covalent character in the bonds. Chemists (fluorine chemists, anyway) say, "Fluorine gives wings to the metals." The point is that fluorine forms low-boiling, volatile compounds with numerous metals. In these compounds the metal has a high oxidation state, and the bonding has considerable covalent character.

8.11 Al^{3+} is a hard acid that prefers to pair with the hard base O^{2-}, whereas Ni^{2+} and Cu^{2+} are borderline to soft acids that prefer to pair with the soft base S^{2-}.

8.13   Measurement of the melting and boiling point of $TiCl_4$ would provide evidence as to whether the compound is ionic or covalent. $TiCl_4$ is predicted to be a $\boxed{\text{covalent compound}}$ because Ti^{4+} is a high oxidation state cation that tends to accept electrons from Cl^- ions to form covalent (dative) bonds.

8.15   Methylamine is a $\boxed{\text{monodentate}}$ ligand that binds to a central metal ion by donating a lone pair of electrons from the $\boxed{\text{N atom}}$. This is the only lone pair of electrons in the molecule.

8.17   **a)** $[V(NH_3)_4Cl_2]$ The V atom has 6 ligands, four with 0 charges and two with -1 charges. The V atom must be in the $\boxed{+2}$ oxidation state for electrical neutrality.

b) $[Mo_2Cl_8]^{4-}$ The two Mo atoms are in the $\boxed{+2}$ oxidation state. They contribute a net $+4$ charge while the 8 chlorides contribute a net -8 charge. The overall complex thus has a -4 charge.

c) $[Co(H_2O)_2(NH_3)Cl_3]^-$ The Co atom is surrounded by six ligands, four with 0 charges and two with -1 charges. It is in the $\boxed{+2}$ oxidation state.

d) $[Ni(CO)_4]$ The ligand CO (carbon monoxide) is electrically neutral as is the complex itself. The Ni atom must be in the $\boxed{0}$ oxidation state.

8.19   **a)** $Na_2[Zn(OH)_4]$ **b)** $[Co(H_2NCH_2CH_2NH_2)_2Cl_2]NO_3$
   c) $[PtBr(H_2O)_3]Cl$ **d)** $[Pt(NH_3)_4(NO_2)_2]Br_2$

8.21   **a)** Ammonium diamminetetraisothiocyanatochromate(III)
   b) Pentacarbonyltechnetium(I) iodide
   c) Potassium pentacyanomanganate(IV)
   d) Tetraammineaquachlorocobalt(III) bromide

Tip. In part **a)**, the "iso" indicates that the donor atom in the thiocyanate ligand is the N, as in Cr—NCS. If the formula were $NH_4[Cr(NH_3)_2(SCN)_4]$, the linkage would be Cr—SCN and the name would be "thiocyanato." Compare to problem **8.69**.

Structures of Coordination Complexes

8.23   The four substances all dissolve in water to make 0.010 M solutions. The more ions per mole of solute then the greater the conductivity of the solution at a given concentration of solute. Hence in order of increasing conductivity:

$$[Cu(NH_3)_2Cl_2] \quad < \quad KNO_3 \quad < \quad Na_2[PtCl_6] \quad < \quad [Co(NH_3)_6]Cl_3$$
$$\text{0 ions} \qquad\qquad \text{2 ions} \qquad\quad \text{3 ions} \qquad\qquad \text{4 ions}$$

8.25   **a)** $[Pt(NH_3)_2BrCl]$ has two isomers. (*cis* and *trans*). Neither is optically active:

b) The $[Co(CN)_3(H_2O)_2Cl]^-$ ion has three isomers[3]

In one isomer (left), the CN^- ions are mutually *cis* on a triangular face of the octahedron defined by the six ligands' positions around the cobalt. It is a "facial" isomer. In the other two, the three CN^- ions string out along two edges of the octahedron in such a way that one is *cis* to the other two but those two are *trans* to each other. These isomers are "meridional." In one of the meridional isomers (center), the Cl^- ion is *trans* to the middle CN^-. In the other (right), an H_2O is *trans* to the middle CN^-. None of the three isomers is optically active.

Tip. Copy the right-most diagram on a piece of scratch paper and draw in lines to highlight the triangular face of the octahedral that is defined by the OH_2 at the top, the Cl^-, and the CN^- farthest to the right. Think of this as the top face. Draw in thinner lines to define a second triangular face containing the other three ligands. Think of this as the bottom face. Confirm that the three isomers of this complex ion are generated by 120° rotations of the top face relative to the bottom face. (For example, rotating to the left shifts the CN^- to the location of the OH_2, the OH_2 to the location of the Cl^- and the Cl^- to the location of the CN^-.)

c) $[V(C_2O_4)_3]^{3-}$ ion is enantiomeric with two possible optical isomers. In the following the oxalato ligand ($^-OOC—COO^-$) is abbreviated as O—·—·—O

mirror

---

[3]In these sketches the solid lines are chemical bonds. The dotted lines establish a reference plane that partial blocks the view of one of the metal-to-ligand bonds.

The vertical line in the middle of the diagram is a mirror plane viewed edge-on. Each atom in the structure on the left is mirrored by an atom in the structure on the right. The mirroring operation works through the mirror plane.

Tip. In the diagram on the left, the lines representing the backbones of the three oxalato ligands twist to the *right* as they connect from the O's that are nearer to the viewer and marked with small asterisks (*) to the O's that are further away from the viewer. In the diagram on the right, the lines twist to the *left*. Both structures have a built-in twist or helicity, but the twist turns clockwise in the first and counterclockwise in the second.

8.27 Three isomeric $[Fe(en)_2Cl_2]^+$ complexes exist: *trans*-dichlorobis(ethylenediamine)iron(III) ion (at the left below) and the two mirror-image *cis*-dichlorobis(ethylenediamine)iron(III) ions.[4] All involve octahedral coordination at the Fe atom. The NH_2—CH_2—CH_2—NH_2 (abbreviated en) molecules attach to the Fe ion through their two —NH_2 groups, thus serving as bidentate ligands. They can span the edges of the Fe coordination octahedron but are not long enough to attach at opposite corners. In the following the en ligand is represented N—·—·—N

Crystal Field Theory: Optical and Magnetic Properties

8.29 Strong-field octahedral complexes have a large splitting between the t_{2g} and e_g sets of orbitals; weak-field complexes have a small splitting between the t_{2g} and e_g orbitals. When the splitting energy Δ_o exceeds the pairing energy of the electrons, electrons pair up in the t_{2g} level and fill it completely before occupying the e_g level. Otherwise, electrons remain unpaired as long as possible.

a) The electron configuration of Mn^{2+} is $[Ar]3d^5$. It has $\boxed{5}$ unpaired electrons in a weak field and $\boxed{1}$ unpaired electron in a strong field:

b) Zn^{2+} ion has the electron configuration $[Ar]3d^{10}$. It has $\boxed{zero}$ unpaired electrons in both weak and strong fields:

---

[4]The "mirror line" in the figure helps show their left-hand/right-hand relationship.

c) Cr^{3+} ion has configuration $[Ar]3d^3$ and $\boxed{3}$ unpaired electrons in both weak field (left) and strong field (right) octahedral coordination:

d) Mn^{3+} ion has configuration $[Ar]3d^4$. It has $\boxed{4}$ unpaired electrons in weak fields and $\boxed{2}$ in strong fields:

e) Fe^{2+} ion has configuration $[Ar]3d^6$. It has $\boxed{4}$ unpaired electrons in weak fields and $\boxed{zero}$ in strong fields:

8.31 The ground-state Fe^{3+} ion has five d electrons. In the strong octahedral field exerted by six CN^- ligands, its ground-state d electron configuration becomes $(t_{2g})^5(e_g)^0$. All of the d electrons are in the three t_{2g} orbitals; only $\boxed{\text{one electron}}$ can be unpaired. In the weak field exerted by six H_2O ligands, the ion's ground-state d electron configuration is $(t_{2g})^3(e_g)^2$. All $\boxed{\text{five}}$ d electrons remain unpaired.

Each of the five d electrons in t_{2g} levels stabilizes $[Fe(CN)_6]^{3-}$ by $2/5\Delta_o$. The total crystal-field stabilization energy equals $5 \times -2/5\Delta_o$ or $\boxed{-2\Delta_o}$. See text Table 8.5.

The two e_g electrons in the weak-field $[Fe(H_2O)_6]^{3+}$ complex contribute $+3/5\Delta_o$ each to the CFSE, and the three t_{2g} electrons contribute $-2/5\Delta_0$ each for a total CFSE of $\boxed{zero}$.

8.33 In an octahedral field, d^3 systems are particularly stable because the t_{2g} set is half-filled whether the ligands are strong-field or weak-field. (Half-filled configurations enhance stability.) In d^8 octahedral systems, the t_{2g} set is completely filled, and the e_g set is half-filled whether the ligand is strong-field or weak-field. This also promotes stability. Octahedral d^5 systems would be expected to be particularly stable in complexes of weak-field ligands. The configuration of the d electron system is then $(t_{2g})^3(e_g)^2$, in which the t_{2g} and e_g sets of orbitals are half filled. Octahedral d^6 systems would be expected to be particularly stable in complexes of strong-field ligands. The configuration of the d electron system is then $(t_{2g})^6(e_g)^0$ in which the t_{2g} orbitals are a filled set.

Optical Properties and the Spectrochemical Series

8.35 The color perceived in a solution is the complement of the color of light absorbed. A colorless ion (such as $[Zn(H_2O)_6]^{2+}$) $\boxed{\text{does not absorb}}$ a significant amount of visible light.

8.37 A solution of $[Fe(CN)_6]^{3-}$ ion transmits red light. Assuming that it absorbs any visible light, the complex ion must absorb light in the green portion of the spectrum. According to text Figure 4.3, green light has a wavelength of about $\boxed{5 \times 10^{-7} \text{ m}}$. Using the relationship $\Delta E = hc/\lambda$ with Planck's constant and the speed of light in the proper units gives an energy difference of 4×10^{-19} J. This is equivalent to 240 kJ mol^{-1}. Assume that absorption of the light excites a single electron from a low-lying t_{2g} orbital to an e_g level. Then Δ_o equals this energy difference, and is also about $\boxed{240 \text{ kJ mol}^{-1}}$.

8.39 The hexacyanoferrate(III) ion has a d^5 configuration on the central ion that is split by a strong octahedral field. As text Table 8.5 shows, the crystal field stabilization energy for the resulting $(t_{2g})^5$ configuration is $-2\Delta_o$. The value of Δ_o is 240 kJ mol^{-1} (see preceding problem) so the CFSE equals $\boxed{-480 \text{ kJ mol}^{-1}}$.

8.41 **a)** The color complement of blue-violet is $\boxed{\text{orange-yellow}}$.

b) The absorbed light is $\boxed{\text{orange-yellow}}$ with a wavelength λ of maximum absorption near $\boxed{600 \text{ nm}}$. See text Figure 4.3. Experimentally, the absorption turns out to have its maximum at 575 nm.

c) Cyanide ion is a strong-field ligand, and water is an weak-field ligand. Replacing coordinated water molecules with cyanide ions increases the crystal-field splitting. Increasing the splitting increases the frequency of the light that is absorbed, and causes a $\boxed{\text{decrease}}$ in the wavelength of maximum absorption.

8.43 **a)** In an aqueous solution of $Fe(NO_3)_3$, the Fe^{3+} ion is coordinated to six water molecules. The weak field of these ligands allows the high-spin electron configuration $(t_{2g})^3(e_g)^2$ on the central Fe^{3+} ion. In the case of $[Fe(CN)_6]^{3-}$, the strong field exerted by the CN^- ligands forces the electron configuration $(t_{2g})^5(e_g)^0$. Replacing the weak-field ligand water with the weak-field ligand fluoride should not change the $(t_{2g})^3(e_g)^2$ configuration. The absorption of light by the fluoride complex ion should therefore resemble that by the aqua complex. The solution of $K_3[FeF_6]$ should be $\boxed{\text{pale}}$.

b) The ground-state electron configuration of the Hg^{2+} ion is $[Xe]4f^{14}5d^{10}$. The completed subshell of 10 d electrons means that electronic transitions in which electrons are redistributed among d orbitals are not possible. Such transitions are mostly responsible for the colors of coordination complexes. A solution of $K_2[HgI_4]$ should therefore be $\boxed{\text{colorless}}$.

8.45 In $[AuBr_4]^-$ ion, the gold is Au^{+3} with valence-electron configuration $[Xe]4f^{14}5d^8$. In $NiBr_4^{2-}$ ion, the nickel is Ni^{+2} with valence-electron configuration $[Ar]3d^8$. Both complex ions are four-coordinate; both contain d^8 metal ions. Square-planar arrangements occur frequently for d^8 ions coordinated by strong-field ligands because occupancy by eight electrons with no unpaired spins leaves the high-energy $d_{x^2-y^2}$ orbital vacant. This can be confirmed using the diagram in text Figure 8.19. Square-planar arrangements are also favored for d^8 metal ions in the fourth and fifth periods (like Au^{3+} ion) in comparison to those in the third period (like Ni^{2+} ion). Tetrahedral arrangements on the other hand are favored when the central metal ion is small (like the Ni^{2+} ion) and the ligands are bulky (like the Br^- ion). Based on these facts, the best prediction is that $[AuBr_4]^-$ ion is $\boxed{\text{square-planar and low-spin}}$ and that $NiBr_4^{2-}$ ion is $\boxed{\text{tetrahedral and high-spin}}$. The last is predicted because all d^8 complexes in a tetrahedral field are high-spin, as use of text Figure 8.20 (a) shows.

ADDITIONAL PROBLEMS

8.47 $\boxed{\text{Zinc}}$ has the lowest melting point and lowest boiling point of the fourth-period transition metals. This element has a d^{10} configuration. Bonding in metals, including elemental zinc, arises from the combination of valence atomic orbitals into delocalized bonding and antibonding molecular orbitals. A complete d subshell means that the antibonding orbitals deriving from d orbitals are completely occupied. This makes the bonding in the metal weaker, which implies a lower melting point and boiling point than in other metals.

8.49 Electronegativity is a measure of the ability of an atom in a compound to draw electrons to itself. $\boxed{\text{Mo(VI)}}$, the highest oxidation state of Mo has (formally at least) the most positive charge and should therefore have the highest electronegativity.

Tip. The electronegativity values mentioned in the problem were obtained by the method of Pauling. They equal 2.16 for Mo(II), 2.19 for Mo(III), 2.24 for Mo(IV), 2.27 for MO(V), and 2.35 for Mo(VI).[5]

8.51 In $[Ru_2(NH_3)_6Br_3](ClO_4)_2$, all six ammonia ligands are neutral, each bromide ion has a -1 oxidation state and each perchlorate ion has a -1 charge. Because the sum of the oxidation states must equal zero (the overall charge of the complex), the oxidation state of the ruthenium is $\boxed{2.5}$.

Tip. One of the ruthenium atoms may be in the $+2$ oxidation state, and the other in the $+3$ oxidation state.

8.53 Ligands are electron-pair donors. On the basis of simple electrostatics, it is much harder for a positively charged species to donate electron pairs than for a neutral or negatively charged species to do so.

8.55 Water coordinated by the central Cr is more tightly bonded and harder for a dehydrating agent to remove than water held in the solid as water of crystallization. Compound 1 loses two moles of H_2O per mole; it therefore has two waters of crystallization. The other water is in the coordination sphere: $\boxed{[Cr(H_2O)_4Cl_2]Cl\cdot 2H_2O}$. This ion has $\boxed{\text{octahedral}}$ coordination about the central Cr atom. Compound 2 loses only one mole of H_2O per mole so it has only one water of hydration: $\boxed{[Cr(H_2O)_5Cl]Cl_2\cdot H_2O}$. Compound 2 also has an $\boxed{\text{octahedral}}$ structure. Compound 3 loses no water so it must have all six water molecules coordinated: $\boxed{[Cr(H_2O)_6]Cl_3}$; compound 3 has an $\boxed{\text{octahedral}}$ structure.

Solutions of silver nitrate give a prompt precipitate of AgCl only with chloride ions that are not in the coordination sphere. Therefore, compound 1 gives one mole of AgCl per mole of complex; compound 2 gives two moles of AgCl per mole of complex; compound 3 gives three moles of AgCl per mole of complex. For compound 1

$$m_{AgCl} = 100.0 \text{ g} \times \left(\frac{1 \text{ mol}}{266.44 \text{ g}}\right)\left(\frac{1 \text{ mol AgCl}}{1 \text{ mol}}\right)\left(\frac{143.32 \text{ g AgCl}}{1 \text{ mol AgCl}}\right) = \boxed{53.79 \text{ g AgCl}}$$

Similar 100.0 g samples of compounds 2 and 3, which have the same molar mass, give respectively twice and three times as much AgCl: $\boxed{107.6 \text{ g}}$ and $\boxed{161.4 \text{ g}}$.

8.57 In $[CoCl_2(en)]$ the two ends of the ethylenediamine (en) ligand are equivalent and the two Cl ligands are also equivalent, as suggested in the following

---

[5]A. L. Allred, *J. Inorg. Nucl. Chem.*, **1961**, *17*, 215.

Tetrahedral structures do not exhibit *cis-trans* isomerism (geometrical isomerism) because the four corner of a tetrahedron are equidistant from each other. Tetrahedral structure *can* exhibit mirror-image isomerism if they have four different atoms attached to the central atom. The complex $[CoBrCl(NH_3)(NH_2CH_3)]$ would in principle exhibit mirror-image isomerism. But $[CoCl_2(en)]$ does not have four different ligands. Consequently $[CoCl_2(en)]$ $\boxed{\text{cannot exhibit geometrical isomerism}}$ and $\boxed{\text{cannot exhibit optical isomerism.}}$

8.59 **a)** In the preferred formulation, the Pt in the $[Pt(en)_2(SCN)_2]^{2+}$ cation is in the +4 oxidation state, while the Pt in the $[PtBr_2(SCN)_2]^{2-}$ anion is in the +2 oxidation state. These complexes form a 1-to-1 ionic compound with the correct empirical formula. The Pt(IV) has 6 *d* electrons, and is surrounded by 6 donors. If all 6 *d* electrons are in the t_{2g} level (strong-field), the ion has no unpaired electrons and is diamagnetic. The Pt(II) has 8 *d* electrons and is surrounded by 4 donors. The anion can therefore also have all of its electrons paired and be diamagnetic. By contrast, if the molecular formula were $[PtBr(en)(SCN)_2]$, then all Pt atoms would be in the +3 oxidation state and have 7 *d* electrons. Such a compound would be paramagnetic.

b) Bis(ethylenediamine)dithiocyanatoplatinum(IV) dibromodithiocyanatoplatinate(II).

8.61 The cyanide ligand has a −1 charge. This means that in $[Mn(CN)_6]^{5-}$ the oxidation state of the Mn must be $\boxed{+1}$. A +1 added to 6(−1) gives the observed −5 charge on the complex ion. Similarly, the oxidation states of Mn in $[Mn(CN)_6]^{4-}$ and $[Mn(CN)_6]^{3-}$ are $\boxed{+2}$ and $\boxed{+3}$, respectively. The ground-state electron configurations of the manganese ions are

$$Mn^{+1} \quad [Ar]3d^6 \qquad\qquad Mn^{+2} \quad [Ar]3d^5 \qquad\qquad Mn^{+3} \quad [Ar]3d^4$$

The low-spin (strong-field) *d* orbital occupancy diagrams for each complex are

8.63 The size of these +2 ions, if observed in free space, would be expected to contract as the nuclear charge *Z* increases. Additional electrons join the same subshell (the 3*d* subshell) and so are about the same distance from the nucleus while nuclear charge (and with it attractive power for electrons) is steadily increasing.

Ionic Radii of Some Fourth-Period +2 Metal Ions In Oxide Structures

The data show that the situation differs when the M^{2+} ions (the metal ions) are in oxide structures. Minima occur with V^{2+} ion, which has three d electrons, and Ni^{2+} ion, which has eight d electrons. All of these metal oxides have the rock-salt structure. Consequently each M^{2+} ion is surrounded by six O^{2-} ions (and, each O^{2-} is surrounded by six M^{2+} ions). The symmetry of the field is octahedral. If the field be weak, then the metal ions have high-spin configurations, and the 4th and 9th electrons to join the d subshell must go into a higher energy e_g orbital. This makes the Mn^{2+} and Cr^{2+} ions bigger because the e_g orbitals are further away from the nucleus. Exactly such an increase in size is observed. It follows that the metal ions are in weak-field environments and are well described as high-spin octahedral complexes.

8.65 The central cobalt(II) in $[CoCl_4]^{2-}$ ion has a total of 7 d electrons. The Co atom (electron configuration $[Ar]3d^74s^2$) loses its two valence s electrons to become Co^{2+}. The tetrahedral field of the four chloride ligands splits the d orbitals of Co(II) into e_g orbitals (at *lower* energy) and t_{2g} orbitals (at *higher* energy). The ground-state electron configuration of the seven electrons is $(e_g)^4(t_{2g})^3$. This configuration results regardless of the strength of the ligand field: only high-spin states are possible for a d^7 species in a tetrahedral field. The tetrahedral complex has crystal-field stabilization energy (CFSE) equal to $-6/5\Delta_t$; this stabilization favors the tetrahedral complex.

8.67 The cesium ions each have a +1 charge, and the fluoride ions each have a -1 charge. Thus, copper is in the $+4$ oxidation state. There are six monodentate ligands attached to the Cu(IV), so the most likely geometry about the central metal atom will be octahedral. The ground-state electron configuration of Cu(IV) is $[Ar]3d^7$. In a weak octahedral field, the d electron configuration would become $(t_{2g})^5(e_g)^2$. This high-spin configuration is far more likely than the low-spin $(t_{2g})^6(e_g)^1$ configuration because the F^- ion is a weak-field ligand.

8.69 The cyclopentadienyl ion $(C_5H_5)^-$ has a -1 charge and is a six-electron donor. The neutral C_6H_6 molecule is also a six-electron donor. The other ligands are all two-electron donors.

The Co atom in $[Co(C_5H_5)_2]^+$ has six $3d$ electrons and shares 12 more from the two ligands. It sees a total of 18 valence electrons.

The Fe atom in $[Fe(C_5H_5)(CO)_2Cl]$ also sees a total of 18 valence electrons: the Fe(II) starts with six; the $(C_5H_5)^-$ donates six; the Cl^-'s and the CO donate two each.

The Mo in $[Mo(C_5H_5)_2Cl_2]$ sees a total of 18 valence electrons: the Mo(IV) starts with two and each $(C_5H_5)^-$ contributes six; the Cl^- ligands donate two electrons each.

The Mn in the complex $[Mn(C_5H_5)(C_6H_6)]$ sees a total of 18 valence electrons: the Mn(I) ion has six, the $(C_5H_5)^-$ contributes six, and the C_6H_6 also contributes six.

8.71 The rule of 18 holds that special stability is conferred on organometallic compounds having a central metal atom surrounded by 18 electrons. To relate this to ligand-field theory, study the MO diagram in text Figure 8.28. The nine molecular orbitals of lowest energy in the diagram are all either bonding or nonbonding. These MO's can hold as many as 18 electrons. Any complex having electrons in these MO's exclusively has zero antibonding electrons. In $Cr(CO)_6$, the ground-state valence electron configuration is

$$\sigma_s^2\sigma_p^6\sigma_d^4(d_{xy}^{nb})^2(d_{xz}^{nb})^2(d_{yz}^{nb})^2$$

with no electrons in antibonding orbitals. Additional electrons would go into antibonding orbitals and decrease the stability of the species.

8.73 Transition metal coordination compounds play a central role in biology. In hemoglobin, for example, iron is essential in binding O_2 for transport to cells and CO_2 for transport from cells. Cobalt is the central metal atom in vitamin B_{12}, a coordination compound that plays a vital role in metabolism.

Some of these biological complexes are colored because of the closely spaced d levels that allow absorption of visible light. Ligand exchange lets these complexes act as catalysts for chemical reactions in living organisms, in analogy to the catalysis developed in synthetic chemistry. Finally, the existence of more than one oxidation state lets some complexes act as oxidizing or reducing agents in biochemical processes.

CUMULATIVE PROBLEMS

8.75 Manganese and chlorine appear in the same group ("Gruppe VII") in Mendeleev's early periodic table mainly because of similarities in their compounds in the +7 oxidation state. Both elements form heptaoxides (Cl_2O_7 and Mn_2O_7) that are liquids are room conditions (indicating covalent bonding) and that react with water to give strong acids (perchloric acid $HClO_4$ and permanganic acid $HMnO_4$). Both of these acids are powerful oxidizing agents, and many perchlorate salts resemble permanganate salts closely.

Among the lower oxides of manganese, MnO is basic, Mn_2O_3 is weakly basic, MnO_2 is feebly acidic, and MnO_3 is more strongly acidic. A similar trend is found among the oxides of chlorine. Manganic acid (H_2MnO_4) resembles chloric acid $HClO_3$ in that it disproportionates to an acid in a higher oxidation state ($HMnO_4$) and an oxide in a lower oxidation state (MnO_2). However manganates (salts of manganic acid) resemble sulfates and chromates much more than chlorates. Manganese and chlorine both have seven valence electrons, but these include d electrons in the case of Mn. The availability of additional low-lying d orbitals in elemental manganese leads to metallic bonding in that element and enormous differences between it and elemental chlorine.

Chapter 9

The Gaseous State

The Chemistry of Gases

9.1 The decomposition of ammonium hydrosulfide produces ammonia NH_3 and hydrogen sulfide H_2S

$$\boxed{NH_4HS(s) \longrightarrow NH_3(g) + H_2S(g)}$$

Tip. To predict the course of chemical reactions, look for the formulas of stable gaseous substances embedded in more complicated formulas. Heating often drives out such compounds. Look for water (H_2O), carbon dioxide (CO_2), carbon monoxide (CO), ammonia (NH_3), hydrogen chloride and the other hydrogen halides (HBr, HI, HF), oxygen (O_2), nitrogen (N_2), and hydrogen sulfide (H_2S).

9.3 Generate ammonia from ammonium bromide by dissolving the ammonium bromide in a small amount of water and adding a strong base, such as aqueous sodium hydroxide. NH_4Br dissolves in water to form ammonium ion $NH_4^+(aq)$ and bromide ion $Br^-(aq)$. The $NH_4^+(aq)$ ion donates a hydrogen ion to the strong base $OH^-(aq)$ to form gaseous ammonia $NH_3(g)$

$$\boxed{NH_4^+(aq) + OH^-(aq) \longrightarrow NH_3(g) + H_2O(l)}$$

Tip. Heating speeds up the liberation of the ammonia. Most reactions accelerate with increasing temperature. Also ammonia, like most gases, is less soluble in hot water than in cold.

Pressure and Temperature of Gases

9.5 Because water is less dense than mercury, a longer column of water is required to balance the pressure of the atmosphere. A pressure of 1.00 atm is balanced in a barometer by a column of mercury 76.0 cm high. The density of mercury is 13.6 g cm^{-3}; the density of water is only 1.00 g cm^{-3}. The height of the column of water must be greater in inverse proportion to the ratio of the densities

$$h_{H_2O} = h_{Hg}\left(\frac{\rho_{Hg}}{\rho_{H_2O}}\right) = 76.0 \text{ cm}\left(\frac{13.6 \text{ g cm}^{-3}}{1.00 \text{ g cm}^{-3}}\right) = 1.03 \times 10^3 \text{ cm} = \boxed{10.3 \text{ m}}$$

This is nearly 34 feet. Water barometers have been built and used but they are awkwardly large.

Tip. The problem can also be solved by substitution into the equation $P = \rho g h$

$$h_{H_2O} = \frac{P}{\rho_{H_2O}\, g} = \frac{101\,325 \text{ Pa}}{(1.00 \times 10^3 \text{ kg m}^{-3})(9.807 \text{ m s}^{-2})} = \boxed{10.3 \text{ m}}$$

Note the conversion of the pressure to pascals, the SI unit, and the conversion of the density of water to kilograms per cubic meter, a combination of SI base units.

9.7   Convert the pressure (414 atm) to pascals (Pa), and then use the formula $P = \rho g h$ to compute the depth (h) of sea-water that exerts the same pressure. In this computation, g is the acceleration of the earth's gravity and ρ is the density of water. Take the density of sea-water as a constant 1.00×10^3 kg m^{-3}.[1] Take g to equal g_0, its standard value, which is 9.80665 m s^{-2}.[2] The pressure is

$$P = 414 \text{ atm} \times \left(\frac{1.01325 \times 10^5 \text{ Pa}}{1 \text{ atm}} \right) = 4.19 \times 10^7 \text{ Pa}$$

The height of a column of water that exerts this pressure at its base is

$$h = \frac{P}{\rho g} = \frac{4.19 \times 10^7 \text{ kg m}^{-1}\text{s}^{-2}}{(1.0 \times 10^3 \text{ kg m}^{-3})\, 9.807 \text{ m s}^{-2}} = 4.3 \times 10^3 \text{ m}$$

Text Table B.2 confirms that a pascal equals a kg m^{-1}s^{-2}. One meter equals 3.28 feet; a depth of 4300 m equals $\boxed{1.40 \times 10^4 \text{ feet}}$.

9.9   The pascal (Pa) is a newton per square meter (N m^{-2}). One standard atmosphere is defined as 1.01325×10^5 Pa. Convert using unit factors as follows

$$P = 172.00 \text{ MPa} \times \left(\frac{10^6 \text{ Pa}}{\text{MPa}} \right) \left(\frac{1 \text{ atm}}{1.01325 \times 10^5 \text{ Pa}} \right) = \boxed{1.6975 \times 10^3 \text{ atm}}$$

Convert the pressure from pascals to bars by multiplying by the proper unit factor

$$P = (1.7200 \times 10^8 \text{ Pa}) \times \left(\frac{1 \text{ bar}}{10^5 \text{ Pa}} \right) = \boxed{1.7200 \times 10^3 \text{ bar}}$$

All of the unit factors in the preceding conversions come from definitions and therefore have an unlimited number of significant digits.

9.11   Assume that the N_2 in the tank behaves ideally. Since neither the temperature nor the amount of the N_2 changes during the expansion, use Boyle's law in the form $P_1 V_1 = P_2 V_2$. The initial pressure P_1 is 3.00 atm, the initial volume V_1 is 2.00 L, P_2 is the final pressure (this is the desired answer), and V_2 is the final volume. If the volumes of the valve and associated plumbing are negligibly small, $V_2 = 2.00 + 5.00 = 7.00$ L. Then

$$P_2 = \frac{P_1 V_1}{V_2} = (3.00 \text{ atm}) \left(\frac{2.00 \text{ L}}{7.00 \text{ L}} \right) = \boxed{0.857 \text{ atm}}$$

9.13   Apply Charles's law in the form $V_1 T_2 = V_2 T_1$. The V_1 is given (4.00 L), and the absolute temperature is doubled, that is, $T_2 = 2T_1$. Accordingly, $V_2 = \boxed{8.00 \text{ L}}$.

9.15   Use Charles's Law. In this problem, $V_1 = 17.4$ gill, and V_2 is required. The temperatures are given on the Fahrenheit scale and must be converted to an absolute scale for use with Charles's law. The following equations show these as unit conversions

$$T_1 = (100°\text{F} - 32°\text{F}) \left(\frac{5°\text{C}}{9°\text{F}} \right) \left(\frac{1 \text{ K}}{1°\text{C}} \right) + 273.15 \text{ K} = 310.9 \text{ K}$$

$$T_2 = (0°\text{F} - 32°\text{F}) \left(\frac{5°\text{C}}{9°\text{F}} \right) \left(\frac{1 \text{ K}}{1°\text{C}} \right) + 273.15 \text{ K} = 255.4 \text{ K}$$

Insert these T's into the equation for Charles's law:

$$V_2 = \left(\frac{T_2}{T_1} \right) V_1 = \left(\frac{255.4 \text{ K}}{310.9 \text{ K}} \right) 17.4 \text{ gill} = \boxed{14.3 \text{ gill}}$$

Tip. Do not worry about converting gills to more familiar units.

---

[1] The density of the deep sea is in fact substantially affected by dissolved salts and the compression of overlying layers.
[2] The acceleration of gravity at and near the surface of the Earth varies slightly with both latitude and altitude.

9.17 The complete reaction of a set mass of CaC_2 produces a set mass of $C_2H_2(g)$ regardless of T and P. Since the pressure is the same (1 atm) in both cases, this is a Charles's law problem with $V_1 = 64.5$ L, $t_1 = 50°C$, $t_2 = 400°C$, and V_2 unknown. Convert the temperatures to kelvins. Then

$$V_2 = \left(\frac{T_2}{T_1}\right) V_1 = \left(\frac{(400 + 273.15) \text{ K}}{(50 + 273.15) \text{ K}}\right) 64.5 \text{ L} = \boxed{134 \text{ L}}$$

The Ideal Gas Law

9.19 The true pressure inside the bicycle tire is 14.7 psi (1 atm) more than the gauge pressure since the gauge reads zero when the pressure is 1 atm.[3] P is $30.0 + 14.7 = 44.7$ psi. Assume that the air in the tire behaves ideally. Then

$$\frac{P_1 V_1}{n_1 T_1} = \frac{P_2 V_2}{n_2 T_2}$$

where the subscripts designate the values of the variables before and after warming from 0°C to 32°C. Because the tire does not expand, V_1 equals V_2. Also, heating does not change the quantity of air inside the tire, so n_1 equals n_2. Converting the temperatures to the Kelvin scale gives $T_1 = 273$ K and $T_2 = 305$ K; $P_1 = 44.7$ psi as just established. Cancel the V's and n's, rearrange and substitute

$$P_2 = P_1\left(\frac{T_2}{T_1}\right) = P_1\left(\frac{305 \text{ K}}{273 \text{ K}}\right) = 44.7 \text{ psi}\left(\frac{305}{273}\right) = 49.9 \text{ psi}$$

Gauge pressure is always 1 atm (14.7 psi) less than the actual pressure. The gauge pressure of the tire at 32°C is thus $49.9 - 14.7 = \boxed{35.2 \text{ psi}}$.

Tip. Expressing the temperatures on an absolute scale is essential. Using temperatures in degrees Celsius here leads to division by zero!

9.21 **a)** Let state 1 be the original state of the air, and let state 2 be the state of the air after the compression. Assume that the ideal-gas law applies. Neither the chemical amount of air nor its temperature changes between state 1 and state 2. Hence Boyle's law applies

$$P_2 = \frac{P_1 V_1}{V_2} = \frac{(1.01 \text{ atm})(20.6 \text{ L})}{1.05 \text{ L}} = \boxed{19.8 \text{ atm}}$$

b) Let state 3 be the state of the bottled air in the European laboratory. The pressure in the bottle is bigger because T_3 (294 K) exceeds T_2 (253 K). Note the conversion of the temperatures to an absolute scale. Solve the ideal-gas equation for n/V and write it for states 2 and 3

$$\frac{P_2}{RT_2} = \frac{n_2}{V_2} \quad \text{and} \quad \frac{P_3}{RT_3} = \frac{n_3}{V_3}$$

Neither the volume of the bottle nor the chemical amount of the air it contains changes during the trip to Europe. This means

$$\frac{n_2}{V_2} = \frac{n_3}{V_3} \quad \text{and consequently} \quad \frac{P_2}{T_2} = \frac{P_3}{T_3}$$

Solve this last equation for P_3, and substitute the known values of the other quantities

$$P_3 = P_2\left(\frac{T_3}{T_2}\right) = (19.8 \text{ atm})\left(\frac{294 \text{ K}}{253 \text{ K}}\right) = \boxed{23.0 \text{ atm}}$$

---

[3]A completely flat tire (at sea level) still has about 1 atm of air pressure inside it.

9.23 The information about the density of the gas is nearly worthless because gas densities depend strongly on P and T, but neither is specified. One might assume room P and T, but other assumptions are plausible. Using the fact that $n = m/\mathcal{M}$, the ideal-gas equation can be rewritten as

$$PV = \frac{m}{\mathcal{M}}RT \quad \text{which gives} \quad \frac{m}{V} = \frac{P\mathcal{M}}{RT} \quad \text{which becomes} \quad \rho = \frac{P\mathcal{M}}{RT}$$

because the density ρ of *anything* equals its mass divided by its volume. Solve the last equation for T and insert the density (6.234 g L^{-1}) and molar mass (129.615 g mol^{-1}) of the H$_2$Te(g)

$$T = \frac{P\mathcal{M}}{R\rho} = \frac{(1.00 \text{ atm})(129.615 \text{ g mol}^{-1})}{(0.08206 \text{ L atm mol}^{-1}\text{K}^{-1})(6.234 \text{ g L}^{-1})} = 253.4 \text{ K} = \boxed{-19.8°\text{C}}$$

9.25 **a)** The reaction also generates sodium chloride $\boxed{2\,\text{Na}(s) + 2\,\text{HCl}(g) \longrightarrow \text{H}_2(g) + 2\,\text{NaCl}(s)}$.

b) Calculate the chemical amount of H$_2(g)$ produced by the complete reaction of 6.24 g of Na(s)

$$n_{\text{H}_2} = 6.24 \text{ g Na} \times \left(\frac{1 \text{ mol Na}}{22.99 \text{ g Na}}\right)\left(\frac{1 \text{ mol H}_2}{2 \text{ mol Na}}\right) = 0.1357 \text{ mol H}_2$$

Express the temperature in kelvins (323 K), rearrange the ideal-gas equation to give V explicitly, and substitute the known values

$$V_{\text{H}_2} = \frac{n_{\text{H}_2}RT}{P} = \frac{(0.1357 \text{ mol})(0.08206 \text{ L atm mol}^{-1}\text{K}^{-1})(323 \text{ K})}{0.850 \text{ atm}} = \boxed{4.23 \text{ L}}$$

9.27 Calculate the chemical amount of NaCl being reacted

$$n_{\text{NaCl}} = 2.5 \times 10^6 \text{ g NaCl} \times \left(\frac{1 \text{ mol NaCl}}{58.44 \text{ g NaCl}}\right) = 4.28 \times 10^4 \text{ mol NaCl}$$

According to the balanced equation, 1 mol of HCl forms per 1 mol of NaCl consumed. Therefore, 4.28×10^4 mol NaCl in theory produces 4.28×10^4 mol HCl. Convert the temperature from the Celsius to the Kelvin scale and then use the ideal-gas equation to compute the volume of this amount of gaseous HCl under the specified conditions

$$V_{\text{HCl}} = \frac{n_{\text{HCl}}RT}{P} = \frac{(4.28 \times 10^4 \text{ mol})(0.08206 \text{ L atm mol}^{-1}\text{K}^{-1})(823 \text{ K})}{0.970 \text{ atm}} = \boxed{3.0 \times 10^6 \text{ L}}$$

9.29 According to the balanced chemical equation, a 3-to-2 molar ratio exists between the O$_2$ formed and KClO$_3$ consumed. This fact furnishes a crucial unit factor in the following series

$$n_{\text{O}_2} = 87.6 \text{ g KClO}_3 \times \left(\frac{1 \text{ mol KClO}_3}{122.54 \text{ g KClO}_3}\right)\left(\frac{3 \text{ mol O}_2}{2 \text{ mol KClO}_3}\right) = 1.072 \text{ mol O}_2$$

Now, use the ideal-gas equation to compute the volume, not forgetting to convert the temperature from Celsius to absolute

$$V_{\text{O}_2} = \frac{n_{\text{O}_2}RT}{P} = \frac{(1.072 \text{ mol})(0.08206 \text{ L atm mol}^{-1}\text{K}^{-1})(13.2 + 273.15 \text{ K})}{1.04 \text{ atm}} = \boxed{24.2 \text{ L}}$$

9.31 **a)** The problem is similar to text Example 9.5 and to problem **9.25**. Calculate the theoretical amount of H$_2$S needed in this reaction to give 2.00 kg of S

$$n_{\text{H}_2\text{S}} = (2.00 \times 10^3 \text{ g S}) \times \left(\frac{1 \text{ mol S}}{32.066 \text{ g S}}\right)\left(\frac{2 \text{ mol H}_2\text{S}}{3 \text{ mol S}}\right) = 41.58 \text{ mol H}_2\text{S}$$

Use the ideal-gas equation to compute the volume that the gaseous H_2S occupies under the stated conditions

$$V_{H_2S} = \frac{n_{H_2S}RT}{P} = \frac{(41.58 \text{ mol})(0.08206 \text{ L atm mol}^{-1}\text{K}^{-1})(273.15 \text{ K})}{1.00 \text{ atm}} = \boxed{932 \text{ L}}$$

Note once again that the temperature has been expressed on an absolute scale. The final answer has three significant figures because the mass of sulfur was given to only three significant figures.

Tip. Deviations from ideality make the use of more than three significant figures misleading in any case.

b) Use the approach that worked in the preceding part

$$n_{SO_2} = (2.00 \times 10^3 \text{ g S}) \times \left(\frac{1 \text{ mol S}}{32.066 \text{ g S}}\right)\left(\frac{1 \text{ mol SO}_2}{3 \text{ mol S}}\right) = 20.79 \text{ mol SO}_2$$

$$m_{SO_2} = 20.79 \text{ mol SO}_2 \times \left(\frac{64.06 \text{ g SO}_2}{1 \text{ mol SO}_2}\right) = \boxed{1.33 \times 10^3 \text{ g SO}_2}$$

$$V_{SO_2} = \frac{n_{SO_2}RT}{P} = \frac{(20.79 \text{ mol})(0.08206 \text{ L atm mol}^{-1}\text{K}^{-1})(273.15 \text{ K})}{1.00 \text{ atm}} = \boxed{466 \text{ L}}$$

Another way to get V_{SO_2} is to use Avogadro's principle,[4] which requires that V_{SO_2} equal half of V_{H_2S}.

Mixtures of Gases

9.33   Apply the definition of mole fraction to the SO_3

$$X_{SO_3} = \frac{n_{SO_3}}{n_{tot}} = \frac{17.0 \text{ mol}}{26.0 \text{ mol} + 83.0 \text{ mol} + 17.0 \text{ mol}} = \frac{17.0 \text{ mol}}{126.0 \text{ mol}} = \boxed{0.135}$$

The partial pressure of the SO_3 equals its mole fraction times the total pressure

$$P_{SO_3} = X_{SO_3}P_{tot} = (0.135)(0.950 \text{ atm}) = \boxed{0.128 \text{ atm}}$$

9.35   An ideal gas at a given temperature and pressure occupies volume in direct proportion to the number of molecules that comprises the gas. Hence the mole percentage (or mole fraction) of N_2 in Martian air equals its volume percentage (or volume fraction). In this case $X_{N_2} = \boxed{0.027}$. The partial pressure of N_2 equals the total pressure multiplied by the mole fraction of N_2

$$P_{N_2} = X_{N_2}P_{tot} = (0.027)(5.92 \times 10^{-3} \text{ atm}) = \boxed{1.6 \times 10^{-4} \text{ atm}}$$

9.37   **a)** The mole fraction of CO in the CO/CO_2 mixture is

$$X_{CO} = \frac{n_{CO}}{n_{tot}} = \frac{10.0 \text{ mol}}{10.0 \text{ mol} + 12.5 \text{ mol}} = \frac{10.0 \text{ mol}}{22.5 \text{ mol}} = \boxed{0.444}$$

b) The balanced chemical equation shows that formation of 3.0 mol of CO_2 consumes 3.0 moles of CO. Therefore, the chemical amount of CO in the mixture at the "certain point" is 7.0 mol. The chemical amount of O_2 at this point equals $12.5 \text{ mol} - \frac{1}{2}(3.0 \text{ mol}) = 11.0 \text{ mol}$. The $\frac{1}{2}$ comes from the 1-to-2 molar ratio between O_2 and CO_2 in the balanced equation. The mixture of gases consists of 7.0 mol of CO, 11.0 mol of O_2, and 3.0 mol of CO_2. The mole fraction of CO is

$$X_{CO} = \frac{n_{CO}}{n_{tot}} = \frac{7.0 \text{ mol}}{7.0 \text{ mol} + 11.0 \text{ mol} + 3.0 \text{ mol}} = \frac{7.0 \text{ mol}}{21.0 \text{ mol}} = \boxed{0.33}$$

---

[4]Text Section 1.3

9.39 **a)** Treat the saturated air as an ideal mixture of ideal gases, that is, assume that Dalton's law and the ideal-gas equation apply. The volume of the mixture is 1.0 cm³, its temperature is $(20 + 273.15)$ K, and the partial pressure of water vapor equals 0.0230 atm. Solve the ideal-gas equation for n and apply it to the water vapor by inserting suitable subscripts

$$n_{H_2O} = \frac{P_{H_2O}V}{RT} = \frac{(0.0230 \text{ atm})(1.0 \text{ cm}^3)}{(82.057 \text{ cm}^3 \text{ atm K}^{-1}\text{mol}^{-1})(20 + 273.15) \text{ K}} = 9.56 \times 10^{-7} \text{ mol}$$

A mole of H_2O contains N_A molecules. It follows that the number of molecules of water in 1.00 cm³ of saturated air is

$$N_{H_2O} = 9.56 \times 10^{-7} \text{ mol H}_2\text{O} \times \left(\frac{6.022 \times 10^{23} \text{ molecules H}_2\text{O}}{1 \text{ mol H}_2\text{O}}\right) = \boxed{5.8 \times 10^{17} \text{ molecules H}_2\text{O}}$$

b) 1.00 cm³ of air holds only about 10^{-6} mol of water. It requires much *more* than 1.00 cm³ of air to hold 0.500 mol of water

$$V_{\text{sat. air}} = 0.500 \text{ mol H}_2\text{O} \times \left(\frac{1.00 \text{ cm}^3 \text{ sat. air}}{9.56 \times 10^{-7} \text{ mol H}_2\text{O}}\right) \times \left(\frac{1 \text{ L}}{1000 \text{ cm}^3}\right) = \boxed{523 \text{ L}}$$

The Kinetic Theory of Gases

9.41 **a)** The root-mean-square speed of the molecules in a gas is given by

$$u_{\text{rms}} = \sqrt{\frac{3RT}{\mathcal{M}}} = \sqrt{\frac{3k_BT}{m}}$$

where T is the absolute temperature, m is the molecular mass, R is the gas constant, k_B is the Boltzmann constant, and $\mathcal{M}$ is the molar mass of the gas. For H_2 at 300 K

$$u_{\text{rms}} = \sqrt{\frac{3RT}{\mathcal{M}}} = \sqrt{\frac{3(8.3145 \text{ J K}^{-1}\text{mol}^{-1})(300 \text{ K})}{0.002016 \text{ kg mol}^{-1}}} = \boxed{1.93 \times 10^3 \text{ m s}^{-1}}$$

b) For sulfur hexafluoride SF_6 at 300 K:

$$u_{\text{rms}} = \sqrt{\frac{3RT}{\mathcal{M}}} = \sqrt{\frac{3(8.3145 \text{ J K}^{-1}\text{mol}^{-1})(300 \text{ K})}{0.14605 \text{ kg mol}^{-1}}} = \boxed{226 \text{ m s}^{-1}}$$

The more massive SF_6 molecules have a rms speed about 8.5 times slower than the less massive H_2 molecule at the same temperature. The ratio of the speeds equals the square root of the reciprocal of the ratio of the molar masses.

Tip. The analysis of the units is worth separate study

$$\sqrt{\frac{(\text{ J K}^{-1}\text{mol}^{-1}) \text{ K}}{\text{kg mol}^{-1}}} = \sqrt{\frac{\text{kg m}^2 \text{ s}^{-2} \text{ K}^{-1} \text{ mol}^{-1} \text{ K}}{\text{kg mol}^{-1}}} = \sqrt{\text{m}^2 \text{ s}^{-2}} = \text{m s}^{-1}$$

If the equivalent equation

$$u_{\text{rms}} = \sqrt{\frac{3k_BT}{m}}$$

is used, the units work out this way

$$\sqrt{\frac{(\text{J K}^{-1}) \text{ K}}{\text{kg}}} = \sqrt{\frac{(\text{kg m}^2 \text{ s}^{-2} \text{ K}^{-1}) \text{ K}}{\text{kg}}} = \sqrt{\text{m}^2 \text{ s}^{-2}} = \text{m s}^{-1}$$

The first equation treats things on a "per mole" basis; the second on a "per molecule" basis.

9.43   The rms speed of He atoms at the surface of the sun (where $T = 6000$ K) is

$$u_{rms} = \sqrt{\frac{3RT}{\mathcal{M}}} = \sqrt{\frac{3(8.3145 \text{ J K}^{-1}\text{mol}^{-1})(6000 \text{ K})}{0.004003 \text{ kg mol}^{-1}}} = \boxed{6100 \text{ m s}^{-1}}$$

In an interstellar cloud at 100 K

$$u_{rms} = \sqrt{\frac{3RT}{\mathcal{M}}} = \sqrt{\frac{3(8.3145 \text{ J K}^{-1}\text{mol}^{-1})(100 \text{ K})}{0.004003 \text{ kg mol}^{-1}}} = \boxed{790 \text{ m s}^{-1}}$$

"Comparison" may mean to take the ratio of the two rms speeds rather than to calculate the actual speeds. Getting the ratio is simpler that the previous calculation because $3R$ and $\mathcal{M}$ cancel out

$$\frac{u_{rms} \text{ (near sun)}}{u_{rms} \text{ (interstellar)}} = \sqrt{\frac{6000 \text{ K}}{100 \text{ K}}} = 7.7$$

The rms speed of the molecules of a gas rises by a factor of only about eight from 100 K and 6000 K.

9.45   For a Maxwell-Boltzmann speed distribution, the quantity $f(u)\Delta u$ equals the fraction of the molecules in a gas having speeds between u and $u + \Delta u$.[5] This fraction equals the desired probability and is

$$\frac{\Delta N}{N} = f(u)\Delta u = 4\pi \left(\frac{\mathcal{M}}{2\pi RT}\right)^{3/2} u^2 \exp\left(-\mathcal{M}u^2/2RT\right)\Delta u$$

The gas in this case is O_2, for which $\mathcal{M} = 0.0320$ kg mol^{-1}. The temperature equals 300 K, and Δu equals 10 m s^{-1} (from 500 to 510 m s^{-1}). Evaluate $f(u)$ at $u = 500$ m s^{-1}

$$f(u) = 4\pi \left(\frac{0.0320 \text{ kg mol}^{-1}}{2\pi(8.3145 \text{ J K}^{-1}\text{mol}^{-1})(300 \text{ K})}\right)^{3/2} (500 \text{ m s}^{-1})^2 \exp\left(\frac{-0.0320 \text{ kg mol}^{-1}(500 \text{ m s}^{-1})^2}{2(8.3145 \text{ J K}^{-1}\text{mol}^{-1})(300 \text{ K})}\right)$$
$$= 1.843 \times 10^{-3} \text{ s m}^{-1}$$

The value of $f(u)$ changes over the range of u. The hint proposes a way to deal with this change, which amounts to a 2.5 percent decrease as u rises from 500 to 510 meters per second. This decrease is shown in the following table

$u\,/\,\mathrm{m\,s^{-1}}$	$f(u)\,/\,\mathrm{s\,m^{-1}}$	$u\,/\,\mathrm{m\,s^{-1}}$	$f(u)\,/\,\mathrm{s\,m^{-1}}$
500	1.843×10^{-3}	506	1.816×10^{-3}
501	1.839×10^{-3}	507	1.812×10^{-3}
502	1.834×10^{-3}	508	1.807×10^{-3}
503	1.830×10^{-3}	509	1.802×10^{-3}
504	1.825×10^{-3}	510	1.797×10^{-3}
505	1.821×10^{-3}	-	-

The desired probability $f(u)\Delta u$ equals the area under the distribution curve between 500 and 510 m s^{-1}. This area has a width of 10 m s^{-1} and a smoothly changing height. Approximate it by 10 narrow columns of width 1 m s^{-1} and heights in s m^{-1} given by the first ten values of $f(u)$ in the table. The procedure is illustrated in text Figure C.3 (in Appendix C). The sum of these ten areas is 1.823×10^{-2} (no units). The desired probability is thus approximately $\boxed{1.82 \text{ percent}}$.

9.47   The molecular speeds in $ClO_2(g)$ follow the Maxwell-Boltzmann distribution because the gas is at thermal equilibrium. If 35.0% of the molecules have speeds exceeding 400 m s^{-1}, then obviously 65.0% have speeds between 0 and 400 m s^{-1}. Since over half of the molecules have speeds less than

---

[5]See the text's discussion of the Maxwell-Boltzmann distribution, which is text equation 9.19.

400 m s^{-1}, that speed lies well to the high side of the hump in the Maxwell-Boltzmann distribution.[6] An increase in temperature shifts the hump toward higher speeds and also flattens it out. Because the temperature increase is slight, the first effect predominates. The percentage of molecules having speeds in excess of 400 m s^{-1} will $\boxed{\text{increase}}$.

Tip. The answer can be confirmed mathematically. The area under this Maxwell-Boltzmann curve from $v = 0$ to $v = 400 \text{ m s}^{-1}$ equals 0.65 of the total area under the curve. Integrate the Maxwell-Boltzmann function and evaluate from $v = 0$ to $v = 400$; then compute the T that makes this integral equal 0.65. The answer is 397 K. If the temperature is raised slightly, the area under the curve between 0 and 400 m s^{-1} changes as the hump in the distribution shifts to the right. Suppose the slightly higher new temperature is 407 K. The integral of the $T = 407$ Boltzmann-Maxwell function from 0 to 400 m s^{-1} is 0.64. The percentage of molecules having speeds exceeding 400 m s^{-1} rises from 35% to about 36% when ClO_2 is heated from 397 to 407 K.

Real Gases: Intermolecular Forces

9.49   Solve the van der Waals equation for P, substitute for n, V, T, a, and b, and complete the arithmetic. The values $a = 1.360 \text{ atm L}^2\text{mol}^{-2}$ and $b = 0.03183 \text{ L mol}^{-1}$ come from text Table 9.3. The chemical amount of O_2 equals 212.5 mol, which is its mass divided by its molar mass.

$$P = \frac{nRT}{V - nb} - a\frac{n^2}{V^2} = \frac{(212.5 \text{ mol})(0.08206 \text{ L atm mol}^{-1}\text{K}^{-1})(273.15 + 20) \text{ K}}{28.0 \text{ L} - (212.5 \text{ mol})(0.03183 \text{ L mol}^{-1})} - a\frac{n^2}{V^2}$$

$$= 240.7 \text{ atm} - (1.360 \text{ atm L}^2\text{mol}^{-2})\frac{(212.5)^2 \text{ mol}^2}{(28.0)^2 \text{ L}^2} = 240.7 \text{ atm} - 78.33 \text{ atm} = \boxed{162 \text{ atm}}$$

This pressure is equivalent to $\boxed{2380 \text{ psi}}$, since 14.696 psi equals 1 atm.

9.51   The problem provides a comparison between the ideal-gas pressure and the van der Waals pressure of a typical gas under ordinary conditions. The data for this sample of CO_2 are

$$n = 50.0 \text{ g}/44.0 \text{ g mol}^{-1} = 1.136 \text{ mol} \qquad T = 298.15 \text{ K} \qquad V = 1.00 \text{ L}$$

a) Solve the ideal-gas equation for P and substitute

$$P = \frac{(1.136 \text{ mol})(0.08206 \text{ L atm mol}^{-1}\text{K}^{-1})(298.15 \text{ K})}{1.00 \text{ L}} = \boxed{27.8 \text{ atm}}$$

b) The van der Waals equation includes terms (a and b) that depend on the identity of the gas. For CO_2, $a = 3.592 \text{ atm L}^2\text{mol}^{-2}$ and $b = 0.04267 \text{ L mol}^{-1}$.[7] Solve the van der Waals equation for P and substitute

$$P = \frac{nRT}{V - nb} - a\frac{n^2}{V^2} = \frac{(1.136 \text{ mol})(0.08206 \text{ L atm mol}^{-1}\text{K}^{-1})(298.15 \text{ K})}{1.00 \text{ L} - (1.136 \text{ mol})(0.04267 \text{ L mol}^{-1})} - a\frac{n^2}{V^2}$$

$$= 29.2 \text{ atm} - (3.592 \text{ atm L}^2\text{mol}^{-2})\left(\frac{(1.136)^2 \text{ mol}^2}{(1.00)^2 \text{ L}^2}\right) = 29.2 \text{ atm} - 4.64 \text{ atm} = \boxed{24.6 \text{ atm}}$$

The van der Waals P is less than the ideal-gas P. The effect of the b term in the van der Waals equation was to increase P from 27.8 to 29.2 atm; the effect of a was to decrease P by 4.64 atm. The a term in this case is more influential than the b term. That is, $\boxed{\text{attractive forces dominate}}$.

Tip. The a correction and b correction oppose each other. Consequently the ideal-gas equation gives fair approximations for P over a larger range of conditions that it would otherwise.

---

[6]See text Figure 9.18.
[7]See text Table 9.3

A Deeper Look...Molecular Collisions and Rate Processes

9.53   Figure the chemical amount of air that leaked into the 500 cm^3 bulb during the one-hour period immediately after it is sealed. Do this by using the pressure observed at the one-hour time and the known volume and temperature in the ideal-gas equation

$$n_{air} = \frac{PV}{RT} = \frac{(1.00 \times 10^{-7} \text{ atm})(0.500 \text{ L})}{(0.082057 \text{ L atm mol}^{-1}\text{K}^{-1})(300 \text{ K})} = 2.03 \times 10^{-9} \text{ mol}$$

The rate of the leak is

$$\text{rate} = \frac{2.03 \times 10^{-9} \text{ mol}}{1 \text{ h}} \times \left(\frac{6.022 \times 10^{23} \text{ molecule}}{\text{mol}} \right) \left(\frac{1 \text{ h}}{3600 \text{ s}} \right) = \frac{3.40 \times 10^{11} \text{ molecule}}{\text{s}}$$

Outside air enters the vessel when its molecules "collide" with the area of the tiny hole. Work with the formula for the rate of wall collisions by a gas.[8] The rate at which molecules exit the bulb through the hole is surely negligible, so the observed rate of the leak equals Z_w in this formula. To use the formula, one must first have the density of the outside air. Compute it using the ideal-gas equation

$$\left(\frac{n}{V} \right)_{air} = \left(\frac{P}{RT} \right)_{air} = \left(\frac{1.00 \text{ atm}}{(0.082057 \text{ L atm mol}^{-1}\text{K}^{-1})(300 \text{ K})} \right) = 4.062 \times 10^{-2} \text{ mol L}^{-1}$$

Multiplying this molar density by N_A converts it to a number density

$$\left(\frac{N}{V} \right)_{air} = \frac{4.062 \times 10^{-2} \text{ mol}}{1 \text{ L}} \times \left(\frac{6.022 \times 10^{23} \text{ molec.}}{\text{mol}} \right) \left(\frac{1000 \text{ L}}{1 \text{ m}^3} \right) = 2.45 \times 10^{25} \text{ molecule m}^{-3}$$

Solve text equation 9.25 for A, the area of the wall

$$A = 4 \frac{1}{(N/V)} \sqrt{\frac{\pi \mathcal{M}}{8RT}} Z_w$$

Insert numbers for the several quantities on the right, taking care with units

$$A = 4 \left(\frac{1}{2.45 \times 10^{25} \text{ m}^{-3}} \right) \sqrt{\frac{\pi (0.0288 \text{ kg mol}^{-1})}{8(8.3145 \text{ J K}^{-1}\text{mol}^{-1})(300 \text{ K})}} (3.40 \times 10^{11} \text{ s}^{-1}) = 1.18 \times 10^{-16} \text{ m}^2$$

Because the hole is circular, its radius is

$$r = \sqrt{\frac{A}{\pi}} = \sqrt{\frac{1.18 \times 10^{-16} \text{ m}^2}{3.1416}} = \boxed{6.1 \times 10^{-9} \text{ m}}$$

9.55   The ratio of the rates of effusion of two gases A and B is given by Graham's law[9]

$$\frac{\text{rate of effusion of A}}{\text{rate of effusion of B}} = \frac{N_A/V}{N_B/V} \sqrt{\frac{\mathcal{M}_B}{\mathcal{M}_A}}$$

Call the unknown gas X. Insert the molar mass of methane (CH_4) and the observed rates of effusion of methane and X into Graham's law

$$\frac{\text{rate}_{CH_4}}{\text{rate}_X} = \frac{1.30 \times 10^{-8} \text{ mol s}^{-1}}{5.41 \times 10^{-9} \text{ mol s}^{-1}} = \frac{N_{CH_4}/V}{N_X/V} \sqrt{\frac{\mathcal{M}_X}{16.04 \text{ g mol}^{-1}}}$$

---

[8]Text equation 9.25
[9]Text equation 9.26.

But $N_{CH_4}/V = N_X/V$ because the gases in the two effusion experiments are held at the same temperature and pressure and in the same volume. These two terms cancel out in the preceding to give

$$\frac{1.30 \times 10^{-8} \text{ mol s}^{-1}}{5.41 \times 10^{-9} \text{ mol s}^{-1}} = \sqrt{\frac{\mathcal{M}_X}{16.04 \text{ g mol}^{-1}}} \quad \text{from which} \quad \mathcal{M}_X = \boxed{92.6 \text{ g mol}^{-1}}$$

9.57 Enrichment in gaseous diffusion always occurs for the gas having the less massive molecules. One pass of a mixture of $^{235}UF_6$ and $^{238}UF_6$ through a diffusion apparatus enriches the product mixture in $^{235}UF_6$ by a factor of $\sqrt{352.038/349.028}$ or 1.0043. This result is computed in text Example 9.10. This problem calls for enrichment from 0.72 percent $^{235}UF_6$ to 95 percent $^{235}UF_6$. It is understood that these are number percentages, not mass percentages. Let (N_{235}/N_{238}) equal the ratio of the number of molecules of the lighter gas $^{235}UF_6$ to the number of molecules of the heavier gas $^{238}UF_6$ after any number of diffusion passes. Then

$$\left(\frac{N_{235}}{N_{238}}\right)_{\text{before}} = \frac{0.73}{99.27} = 0.007354 \quad \text{and} \quad \left(\frac{N_{235}}{N_{238}}\right)_{\text{after}} = \frac{95}{5} = 19$$

Each pass multiplies the light-heavy ratio from the previous pass by 1.0043. Let x equal the number of passes. Then

$$(0.007354)(1.0043)^x = 19$$

Dividing through by 0.007354 and taking the logarithm of both sides of the equation gives

$$x \log(1.0043) = \log\left(\frac{19}{0.007354}\right) \quad \text{from which} \quad x = \boxed{1831}$$

9.59 Recognize that the pressure of the krypton is directly proportional to its number density. This follows from the ideal-gas law

$$P = \left(\frac{n}{V}\right) RT \quad \text{from which} \quad P = \frac{1}{N_A}\left(\frac{N}{V}\right) RT$$

because N, the number of molecules, divided by N_A, Avogadro's number, equals the number of moles of any substance. The mean free path (λ) of the krypton molecules is

$$\lambda = \frac{1}{\sqrt{2}\pi d^2 (N/V)} \quad \text{which implies} \quad \frac{N}{V} = \frac{1}{\sqrt{2}\pi d^2 \lambda}$$

The d in these equations is the molecular diameter of the krypton atom. Substituting this equation into the preceding gives

$$P = \frac{1}{N_A}\left(\frac{1}{\sqrt{2}\pi d^2 \lambda}\right) RT$$

Next, obtain the diameter of the spherical vessel. The volume equals 1.00 L (1.00×10^{-3} m^3). For a sphere, $V = \frac{4}{3}\pi r^3$. Solving for r gives the radius as 0.0620 m. Its diameter is therefore 0.124 m.

Set λ equal to 0.124 m, because the mean free path must be comparable to the diameter of the vessel. From the statement of the problem, T is 300 K and d is 3.16×10^{-10} m. The gas constant equals 8.206×10^{-5} m^3 atm mol^{-1}K^{-1} (note the carefully chosen units). Substitution gives

$$P = \frac{1}{6.022 \times 10^{23} \text{ mol}^{-1}}\left(\frac{1}{\sqrt{2}\pi(3.16 \times 10^{-10} \text{ m})^2(0.124 \text{ m})}\right)\left(\frac{8.206 \times 10^{-5} \text{ m}^3 \text{ atm}}{\text{mol K}}\right)(300 \text{ K})$$

$$= \boxed{7.4 \times 10^{-7} \text{ atm}}$$

The number density (N/V) of the krypton is needed to calculate the diffusion constant

$$\left(\frac{n}{V}\right)_{\mathrm{Kr}} = \frac{P}{RT} = \frac{7.4 \times 10^{-7}\ \mathrm{atm}}{(0.08206\ \mathrm{L\ atm\ mol^{-1}K^{-1}})(300\ \mathrm{K})} = 3.01 \times 10^{-8}\ \mathrm{mol\ L^{-1}}$$

$$\left(\frac{N}{V}\right)_{\mathrm{Kr}} = \frac{3.01 \times 10^{-8}\ \mathrm{mol}}{\mathrm{L}} \times \left(\frac{6.022 \times 10^{23}\ \mathrm{molecule}}{\mathrm{mol}}\right)\left(\frac{1000\ \mathrm{L}}{1\ \mathrm{m^3}}\right) = \frac{1.81 \times 10^{19}\ \mathrm{molecule}}{\mathrm{m^3}}$$

Substitute this intermediate result into the formula for the diffusion constant of a gas

$$D = \frac{3}{8}\sqrt{\frac{RT}{\pi\mathcal{M}}}\left(\frac{1}{d^2 N/V}\right)$$

$$= \frac{3}{8}\sqrt{\frac{8.3145\ \mathrm{J\ K^{-1}mol^{-1}}(300\ \mathrm{K})}{\pi(0.08380\ \mathrm{kg\ mol^{-1}})}}\left(\frac{1}{(3.16 \times 10^{-10}\ \mathrm{m})^2(1.81 \times 10^{19}\ \mathrm{m^{-3}})}\right) = \boxed{20\ \mathrm{m^2\,s^{-1}}}$$

Tip. Check the units separately

$$\sqrt{\frac{\mathrm{J\ K^{-1}mol^{-1}\ K}}{\mathrm{kg\ mol^{-1}}}}\left(\frac{1}{\mathrm{m^2\,m^{-3}}}\right) = \sqrt{\frac{\mathrm{kg\,m^2\,s^{-2}\,K^{-1}\,mol^{-1}\ K}}{\mathrm{kg\,mol^{-1}}}}\left(\frac{1}{\mathrm{m^{-1}}}\right) = \sqrt{\mathrm{m^2\,s^{-2}}}\,(\mathrm{m}) = \mathrm{m^2\,s^{-1}}$$

ADDITIONAL PROBLEMS

9.61   The solid Earth is surrounded by an ocean of air that exerts an average pressure of 730 mm Hg all over its surface. Imagine the air to be replaced by an ocean of liquid mercury. To exert an equal pressure the mercury would need to be only 730 mm deep. The volume of the mercury would equal, to a close approximation, the surface area of the Earth times the depth of the ocean d. Insert $r = 6.370 \times 10^6$ m (converted from km) and $d = 730 \times 10^{-3}$ m into the formula for this volume

$$V_{\mathrm{Hg}} = 4\pi r^2 d = 4(3.1416)(6.370 \times 10^6\ \mathrm{m})^2(730 \times 10^{-3}\ \mathrm{m}) = 3.72 \times 10^{14}\ \mathrm{m^3}$$

Compute the mass of the ocean of mercury by multiplying its volume by its density. Look up and use the density of mercury at room temperature

$$m_{\mathrm{Hg}} = (3.72 \times 10^{14}\ \mathrm{m^3}) \times \left(\frac{10^6\ \mathrm{cm^3}}{1\ \mathrm{m^3}}\right)\left(\frac{13.546\ \mathrm{g}}{\mathrm{cm^3}}\right)\left(\frac{1\ \mathrm{kg}}{1000\ \mathrm{g}}\right) = \boxed{5.04 \times 10^{18}\ \mathrm{kg}}$$

This mass also equals the mass of the atmosphere because the atmosphere exerts the same pressure at the surface of the earth as the hypothetical ocean of mercury.

Tip. The mass of the whole Earth (including its atmosphere) is 5.97×10^{24} kg. This is about 1.2 million times greater than the mass of the atmosphere, which is close to 5×10^{18} kg. Any answer to this problem remotely as large as 10^{24} kg would be quite suspect.

9.63   From the equation $P = \rho g h$ the height h of the column of liquid in a barometer is inversely proportional at a given pressure P to the density of the liquid that is in it: the denser the liquid, the less height it needs. Hence, the height that the column of Hg would have at 0.0°C is

$$h_{0.0°\mathrm{C}} = h_{35°\mathrm{C}} \times \left(\frac{13.5094\ \mathrm{g\ cm^{-3}}}{13.5955\ \mathrm{g\ cm^{-3}}}\right) = 760.0\ \mathrm{mm} \times 0.993667 = 755.19\ \mathrm{mm}$$

This height of mercury in a mercury barometer[10] means P equals 755.19 torr, or $\boxed{0.9937\ \mathrm{atm}}$.

Tip. Substitution into the formula $P = \rho g h$ gives the same answer.

---

[10]At 0.0°C because mercury expands slightly with increasing temperature.

9.65 By analogy to the textbook statements of Boyle's law (text equation 9.2) and Charles's law (text equation 9.6), Amontons's law must state that at constant volume for a fixed amount of gas, the pressure of the gas is directly proportional to its absolute temperature

$$\boxed{P \propto T \quad \text{(fixed volume and fixed amount of gas)}}$$

Another version of Amontons's law can be obtained by setting $V_1 = V_2$ and $n_1 = n_2$ in text equation 9.8, canceling out equal quantities, and re-arranging to put the P's on one side of the equation and the T's on the other

$$\frac{P_1}{P_2} = \frac{T_2}{T_2} \quad \text{at constant V and } n$$

9.67 **a)** 1005 mol of helium displaces 1005 mol of air since the pressure and temperature of the gases inside and outside of the balloon are the same.[11] The masses of these gases are

$$m_{\text{He}} = 1005 \text{ mol He} \left(\frac{4.003 \text{ g He}}{1 \text{ mol He}} \right) = 4\,023 \text{ g} \qquad m_{\text{air}} = 1005 \text{ mol air} \left(\frac{29.0 \text{ g air}}{1 \text{ mol air}} \right) = 29\,145 \text{ g}$$

The answer is the difference between these two masses. It equals $\boxed{25\,100 \text{ g}}$.

b) The balloon still contains 1005 mol of He after the ascent, and the T and P inside the balloon still equal the T and P outside of it. Therefore the balloon, which has increased greatly in volume, still displaces 1005 mol of air. Therefore, the answer is again $\boxed{25\,100 \text{ g}}$.

9.69 One way to solve this problem is to think in terms of proportions. The ideal-gas law states that a given volume contains moles of gas in *inverse* proportion to their absolute temperature, as long as the pressure is constant. This means the higher the temperature, the lower the amount of gas. Also, the amount of products of a chemical reaction is in *direct* proportion to the amount of reactants. Raising the temperature at which HCl is collected from 323.15 K to 773.15 K (from 50°C to 500°C) multiplies T by 2.392. The number of moles of HCl produced in the high-temperature experiment therefore equals the number produced in the low-temperature experiment divided by this factor. Since the number of moles of HCl(g) equals the number of moles of NaCl reacted, the amount of NaCl used in the high-temperature experiment is less in the same proportion. The answer is simply 10.0 kg divided by 2.392. It equals $\boxed{4.18 \text{ kg NaCl}}$.

A second way is to compute the "certain volume" and then find the number of moles of gas that it contains at both 323.15 K and 773.15 K. Find the theoretical yield of HCl from 10.0 kg of NaCl

$$n_{\text{HCl}} = (10.0 \times 10^3 \text{ g NaCl}) \times \left(\frac{1 \text{ mol NaCl}}{58.44 \text{ g NaCl}} \right) \left(\frac{1 \text{ mol HCl}}{1 \text{ mol NaCl}} \right) = 171.1 \text{ mol HCl}$$

The "certain volume" occupied by this HCl at 50°C (323.15 K) is

$$V_{\text{HCl}} = \frac{nRT}{P} = \frac{171.1 \text{ mol} \, (0.08206 \text{ L atm mol}^{-1}\text{K}^{-1})(323.15 \text{ K})}{1.00 \text{ atm}} = 4537 \text{ L}$$

At 500°C (773.15 K) this volume contains fewer moles

$$n_{\text{HCl, 500°C}} = \frac{PV}{RT} = \frac{(1.00 \text{ atm})(4537 \text{ L})}{(0.08206 \text{ L atm mol}^{-1}\text{K}^{-1})(773.15 \text{ K})} = 71.51 \text{ mol HCl}$$

Finally, compute the mass of NaCl required to produce this amount of HCl

$$m_{\text{NaCl}} = 71.51 \text{ mol HCl} \times \left(\frac{1 \text{ mol NaCl}}{1 \text{ mol HCl}} \right) \left(\frac{58.44 \text{ g NaCl}}{1 \text{ mol NaCl}} \right) \left(\frac{1 \text{ kg NaCl}}{1000 \text{ g NaCl}} \right) = \boxed{4.18 \text{ kg NaCl}}$$

---

[11]This follows from Avogadro's principle, text Section 1.3.6

Tip. The second calculation is slower, but attractive in its concreteness. But suppose that the problem had stated that the pressure was the same in the two experiments but withheld the actual value. None of the intermediate numbers in the second method could be computed, but the first method would still work.

9.71 **a)** Balance the equations for the two reactions

$$CaCO_3(s) \longrightarrow CaO(s) + CO_2(g) \qquad \text{and} \qquad CaO(s) + H_2O(l) \longrightarrow Ca(OH)_2(s)$$

Determine the chemical amount of CO_2 needed to produce 8.47 kg of $Ca(OH)_2$

$$n_{CO_2} = 8.47 \text{ kg Ca(OH)}_2 \times \left(\frac{1 \text{ mol Ca(OH)}_2}{0.07409 \text{ kg Ca(OH)}_2} \right) \left(\frac{1 \text{ mol CO}_2}{1 \text{ mol Ca(OH)}_2} \right) = 114.3 \text{ mol CO}_2$$

Use the ideal-gas law to find the volume of the gaseous CO_2

$$V_{CO_2} = \frac{nRT}{P} = \frac{(114.3 \text{ mol})(0.08206 \text{ L atm mol}^{-1}\text{K}^{-1})(1223 \text{ K})}{0.976 \text{ atm}} = \boxed{1.18 \times 10^4 \text{ L}}$$

The final answer is rounded to three significant figures because the mass of $Ca(OH)_2$ was given to only three significant figures.

9.73 The total volume of the diagrammed system is 12.00 L, the sum of the volumes of the three containers (neglecting the volume of the connecting tube). If the three gases behave ideally in their containers, their chemical amounts are

$$n_{O_2} = \frac{2.51 \times 5.00 \text{ L atm}}{RT} \qquad n_{N_2} = \frac{0.792 \times 4.00 \text{ L atm}}{RT} \qquad n_{Ar} = \frac{1.23 \times 3.00 \text{ L atm}}{RT}$$

The total pressure of the gas mixture after the stopcocks are opened is also given by the ideal-gas equation, assuming Dalton's law holds. In the expression for P_{tot}, the total chemical amount of the mixed gas equals the sum of the chemical amounts of the three components

$$P_{tot} = n_{tot} \frac{RT}{V} = (n_{O_2} + n_{N_2} + n_{Ar}) \frac{RT}{V}$$

$$= \left(\frac{12.55 \text{ L atm}}{RT} + \frac{3.168 \text{ L atm}}{RT} + \frac{3.69 \text{ L atm}}{RT} \right) \left(\frac{RT}{12.00 \text{ L}} \right) = \boxed{1.62 \text{ atm}}$$

9.75 Assume the gases behave ideally before the reaction starts and after the reaction reaches equilibrium. The reaction itself is profoundly non-ideal behavior. Imagine that the temperature and volume of the system are such that the total chemical amount of gases equals 1.00 mol at the start. The reaction

$$C_2H_2(g) + 2 H_2(g) \longrightarrow C_2H_6(g)$$

then must decrease the total chemical amount of gas to 0.42 mol. This is true because chemical amount is directly proportional to pressure if T and V do not change. Let x represent the original chemical amount of $C_2H_2(g)$ and y the original chemical amount of $H_2(g)$. Before the reaction starts, there is no $C_2H_6(g)$, so

$$x + y = 1.00 \text{ mol}$$

The reaction produces x mol of $C_2H_6(g)$ as it consumes $2x$ mol of $H_2(g)$ and x mol of $C_2H_2(g)$. Since $C_2H_2(g)$ is the limiting reagent, the reaction stops when the x mol of C_2H_2 gas is gone. At the end of the reaction, the vessel contains x mol of $C_2H_6(g)$, the product, and $(y - 2x)$ mol of left-over $H_2(g)$. Hence

$$x + (y - 2x) = 0.42 \text{ mol}$$

Solving the two simultaneous equations gives $x = 0.029$ mol. The original amount of $C_2H_2(g)$ is 0.29 mol; the original mole fraction of $C_2H_2(g)$ is 0.29 mol/1.00 mol = $\boxed{0.29}$.

Tip. If the system had been assumed to hold, say, 100 mol of gases, then all of the numbers in this computation, except the answer, would have been 100 times bigger.

9.77   **a)** The average kinetic energy of the atoms of deuterium, which is given in the problem, depends only on the absolute temperature (assuming ideal-gas behavior) $\bar{E} = \frac{3}{2}k_B T$. Solve for T, the temperature, and substitute

$$T = \frac{2\bar{E}}{3 k_B} = \frac{2(8 \times 10^{-16} \text{ J})}{3(1.38 \times 10^{-23} \text{ J K}^{-1})} = \boxed{4 \times 10^7 \text{ K}}$$

The relative atomic mass of ^{2}H is not needed in this part of the problem.

b) The average kinetic energy of the particles in a gas equals $\frac{1}{2}m\bar{u^2}$. Write this relationship for ^{1}H and divide it by a similar relationship for ^{2}D

$$\frac{\bar{E}_H}{\bar{E}_D} = \frac{\frac{1}{2}m_H \bar{u_H^2}}{\frac{1}{2}m_D \bar{u_D^2}} = \frac{32 \times 10^{-16} \text{ J}}{8 \times 10^{-16} \text{ J}}$$

Solve for the ratio of the rms speeds

$$\frac{u_{\text{rms,H}}}{u_{\text{rms,D}}} = \sqrt{\frac{\bar{u_H^2}}{\bar{u_D^2}}} = \sqrt{\frac{32}{8}}\sqrt{\frac{2.015}{1.0078}} = \boxed{2.8}$$

9.79   **a)** From the geometry shown in the figure, $r\cos\theta = \Delta\ell/2$ from which $\boxed{\Delta\ell = 2r\cos\theta}$.

b) A molecule is following the path marked by the lines with arrows on them. Its momentum is mv. The component of its momentum perpendicular to the wall at a point of collision is $mv\cos\theta$. After collision this component of momentum is $-mv\cos\theta$. The change in momentum during a collision is $\Delta p = -mv\cos\theta - (mv\cos\theta) = -2mv\cos\theta$. The change in momentum of the wall is equal and opposite: $\boxed{\Delta p_{\text{wall}} = +2mv\cos\theta}$.

c) Let the speed of the molecule equal v. The time Δt between collisions with the wall is $\Delta\ell/v$. The force on the wall by the molecule F_{wall} equals the momentum change per collision multiplied by the rate of collisions. The rate of collisions is $1/\Delta t$, so

$$F_{\text{wall}} = \Delta p_{\text{wall}}\left(\frac{1}{\Delta t}\right) = \Delta p_{\text{wall}}\left(\frac{v}{\Delta\ell}\right) = (2m\,v\cos\theta)\left(\frac{v}{2r\cos\theta}\right) = \boxed{\frac{m\,v^2}{r}}$$

Note that this force is the negative of the centrally directed force needed to maintain an object of mass m in uniform circular motion of radius r at speed v. A molecule just skating around the circumference of the container (such that 2θ approaches $180°$) executes such motion.

d) F_{wall} in part **c)** is the force exerted on the wall by a single molecule. The force from N molecules is the sum of N such contributions

$$F = \sum_N \frac{m\,v^2}{r} = N\frac{m\bar{v^2}}{r}$$

The overbar signifies the average of the squares of the individual speeds of the N molecules. The volume V of the sphere is $\frac{4}{3}\pi r^3$; the surface area A of the sphere is $4\pi r^2$. The pressure exerted by N molecules is

$$P = \frac{F}{A} = \frac{Nm\bar{v^2}}{4\pi r^2(r)} \qquad \text{Multiplying } P \text{ by } V \text{ gives} \quad PV = \left(\frac{Nm\bar{v^2}}{4\pi r^2(r)}\right)\left(\frac{4}{3}\pi r^3\right) = \frac{1}{3}Nm\bar{v^2}$$

9.81   The Lennard-Jones potential V_{LJ} and its derivative follow:

$$V_{\text{LJ}} = 4\epsilon\left[\left(\frac{\sigma}{R}\right)^{12} - \left(\frac{\sigma}{R}\right)^{6}\right]$$

$$F(R) = \frac{-dV_{\text{LJ}}}{dR} = \frac{4\epsilon}{\sigma}\left[12\left(\frac{\sigma}{R}\right)^{13} - 6\left(\frac{\sigma}{R}\right)^{7}\right] = \frac{24\epsilon}{\sigma}\left[2\left(\frac{\sigma}{R}\right)^{13} - \left(\frac{\sigma}{R}\right)^{7}\right]$$

Substitute $\sigma = 3.40$ Å and $\epsilon = 1.654 \times 10^{-21}$ J, the LJ parameters for Ar into the second of these expressions

$$F(R) = (1.168 \times 10^{-10} \text{ J m}^{-1}) \left[2 \left(\frac{3.40 \text{ Å}}{R} \right)^{13} - \left(\frac{3.40 \text{ Å}}{R} \right)^{7} \right]$$

at $R = 3.0$ Å, $F(R) = 9.08 \times 10^{-10}$ J m^{-1} at $R = 3.4$ Å, $F(R) = 1.17 \times 10^{-10}$ J m^{-1}
at $R = 3.8$ Å, $F(R) = 1.40 \times 10^{-12}$ J m^{-1} at $R = 4.2$ Å, $F(R) = -1.16 \times 10^{-11}$ J m^{-1}

The force is repulsive at the first three distances, attractive at the fourth.

9.83   The rate of collision of methane molecules with the wall is

$$Z_{\text{w}} = \frac{1}{4} \frac{N}{V} \sqrt{\frac{8RT}{\pi \mathcal{M}}} A \qquad \text{or} \qquad Z_{\text{w}} = \frac{1}{4} \frac{n}{V} \sqrt{\frac{8RT}{\pi \mathcal{M}}} A$$

The first equation (text equation 9.25) gives the rate in molecules per second. The second gives the rate in moles per second. Combine the second version with the ideal-gas law

$$Z_{\text{w}} = \frac{1}{4} \left(\frac{P}{RT} \right) \sqrt{\frac{8RT}{\pi \mathcal{M}}} A = \frac{1}{4} P \sqrt{\frac{8}{\pi \mathcal{M} RT}} A$$

All of the quantities on the right-hand side are known, including

$$P = 2000 \text{ psi} \left(\frac{1 \text{ atm}}{14.696 \text{ psi}} \right) \left(\frac{101325 \text{ Pa}}{1 \text{ atm}} \right) = 1.38 \times 10^7 \text{ Pa}$$

$$\mathcal{M} = 16.04 \times 10^{-3} \text{ kg mol}^{-1} \qquad A = 1.0 \text{ mm}^2 = 1.0 \times 10^{-6} \text{ m}^2$$

Substitution gives the collision rate

$$Z_{\text{w}} = \frac{1}{4} (1.38 \times 10^7 \text{ Pa}) \sqrt{\frac{8}{\pi (0.01604 \text{ kg mol}^{-1})(8.315 \text{ J K}^{-1}\text{mol}^{-1})(293 \text{ K})}} (1.0 \times 10^{-6} \text{ m}^2)$$

$$= 0.88 \text{ mol s}^{-1}$$

Assume that this rate does not drop significantly because of the loss of gas through the hole. Then, the mass of natural gas (methane) that leaks out in one day is

$$m_{\text{CH}_4 \text{ lost}} = (1 \text{ d}) \left(\frac{0.88 \text{ mol}}{\text{s}} \right) \left(\frac{86\,400 \text{ s}}{\text{d}} \right) \left(\frac{16.04 \text{ g CH}_4}{\text{mol}} \right) = \boxed{1.2 \times 10^6 \text{ g}}$$

The pressure and temperature of the methane in the tank are 1.38×10^7 Pa (as just computed) and 293 K. Now obtain the volume and chemical amount of the methane

$$V_{\text{CH}_4} = \pi r^2 h = \pi (20 \text{ ft})^2 (50 \text{ ft}) \left(\frac{0.3048^3 \text{ m}^3}{\text{ft}^3} \right) = 1.78 \times 10^3 \text{ m}^3$$

$$n_{\text{CH}_4} = \frac{P_{\text{CH}_4} V_{\text{CH}_4}}{RT_{\text{CH}_4}} = \frac{(1.38 \times 10^7 \text{ Pa})(1.78 \times 10^3 \text{ m}^3)}{(8.315 \text{ J K}^{-1}\text{mol}^{-1})(293 \text{ K})} = 1.01 \times 10^7 \text{ mol}$$

The fraction of the methane that escapes in one day is

$$f_{\text{CH}_4 \text{ lost}} = \frac{n_{\text{CH}_4 \text{ lost}}}{n_{\text{CH}_4}} = \frac{(1 \text{ d})(0.88 \text{ mol s}^{-1})(86\,400 \text{ s d}^{-1})}{1.01 \times 10^7 \text{ mol}} = 0.0075$$

9.85 The problem concerns the diffusion of smelly and poisonous ammonia through air after a spill at a temperature of 20°C.

a) Treat the air and the ammonia (NH_3) as if they were ideal gases. Calculate the number density of an ideal gas under the conditions described in the problem

$$\frac{N}{V} = N_A \frac{P}{RT} = \left(\frac{6.022 \times 10^{23}}{\text{mol}}\right) \frac{1.00 \text{ atm}}{(0.08206 \text{ L atm mol}^{-1}\text{K}^{-1})293 \text{ K}} = 2.50 \times 10^{22} \text{ L}^{-1}$$

$$= 2.50 \times 10^{22} \text{ L}^{-1} \left(\frac{1 \text{ L}}{\text{dm}^3}\right) \left(\frac{10^3 \text{ dm}^3}{\text{m}^3}\right) = 2.50 \times 10^{25} \text{ m}^{-3}$$

The problem suggests treating the various molecular species in air (nitrogen, oxygen and so forth) as if they had the same molar mass ($0.017 \text{ kg mol}^{-1}$) as ammonia and the same diameter (3.0×10^{-10} m) as molecules of ammonia. The justification is that the goal is to get an estimate, not an exact answer. This simplification allows the use of text equation 9.33, which really applies only to the case of self-diffusion.[12] Substitute the reciprocal of N/V and the other numbers into this equation

$$D = \frac{3}{8}\sqrt{\frac{RT}{\pi \mathcal{M}}}\left(\frac{1}{d^2 N/V}\right) = \frac{3}{8}\sqrt{\frac{RT}{\pi \mathcal{M}}}\left(\frac{V}{d^2 N}\right)$$

$$= \frac{3}{8}\sqrt{\frac{(8.314 \text{ J mol}^{-1}\text{K}^{-1})(293 \text{ K})}{\pi (0.017 \text{ kg mol}^{-1})}}\left(\frac{1 \text{ m}^3}{(3.0 \times 10^{-10} \text{ m})^2 (2.50 \times 10^{25})}\right)$$

$$= \frac{3}{8}(213.57 \text{ m s}^{-1})\left(4.44 \times 10^{-7} \text{ m}\right) = \boxed{3.56 \times 10^{-5} \text{ m}^2\text{ s}^{-1}}$$

b) Solve text equation 9.32 for t and insert the estimated diffusion constant of ammonia and the 100 m distance

$$t = \frac{\overline{\Delta r^2}}{6D} = \frac{(100 \text{ m})^2}{6(3.56 \times 10^{-5} \text{ m}^2\text{ s}^{-1})} = \boxed{4.68 \times 10^7 \text{ s}} \approx 1.5 \text{ yr}$$

Clearly, true diffusive mixing is S-L-O ... and wait for the W. It is also non-existent in environmental settings.

9.87 **a)** Assume that interstellar gas consists exclusively of atomic hydrogen. Convert its number density to a density in moles per liter

$$\frac{n}{V} = \frac{10 \text{ atoms}}{1 \text{ cm}^3} \times \frac{1 \text{ mol}}{6.022 \times 10^{23} \text{ atoms}} \times \frac{1000 \text{ cm}^3}{1 \text{ L}} = 1.66 \times 10^{-20} \text{ mol L}^{-1}$$

Substitute this value into the ideal-gas equation $P = (n/V)RT$. $P = \boxed{1.4 \times 10^{-19} \text{ atm}}$.

b) Calculate the average speed of the H atoms

$$v_{\text{avr}} = \sqrt{\frac{8RT}{\pi \mathcal{M}}} = \sqrt{\frac{8(8.3145 \text{ J K}^{-1} \text{ mol}^{-1})(100 \text{ K})}{\pi (0.0010 \text{ kg mol}^{-1})}} = 1.45 \times 10^3 \text{ m s}^{-1}$$

Multiplying this speed by the average time between collisions gives the average distance between collisions. It is $\boxed{1.46 \times 10^{12} \text{ m}}$. This is roughly 10.5 times the distance between Earth and Sun.

9.89 **a)** Text equation 9.29 gives the rate at which a single molecule in a gas collides with others. Substitute the various constants, the given T, P and the molar mass of methane. Then insert 3.82×10^{-10} m,

---

[12]The diffusion of labelled molecules through an atmosphere of otherwise identical unlabelled molecules.

the Lennard-Jones σ for methane, in place of the collision diameter d

$$Z_1 = 4\frac{N}{V}d^2\sqrt{\frac{\pi RT}{\mathcal{M}}} = 4\left(\frac{N_A P}{RT}\right)d^2\sqrt{\frac{\pi RT}{\mathcal{M}}}$$

$$= 4\left(\frac{6.022 \times 10^{23}\ \text{mol}^{-1}(101\,325\ \text{Pa})}{(8.3145\ \text{J K}^{-1}\ \text{mol}^{-1})(298\ \text{K})}\right)d^2\sqrt{\frac{\pi(8.3145\ \text{J K}^{-1}\ \text{mol}^{-1})(298\ \text{K})}{0.01604\ \text{kg mol}^{-1}}}$$

$$= 4\,(2.463 \times 10^{25}\ \text{m}^{-3})\,(3.82 \times 10^{-10}\ \text{m})^2\,(696.6\ \text{m s}^{-1}) = \boxed{1.00 \times 10^{10}\ \text{s}^{-1}}$$

b) At $P = 1.0 \times 10^{-7}$ atm the collision rate is smaller by a factor of 10^7. It is $\boxed{1.0 \times 10^3\ \text{s}^{-1}}$.

CUMULATIVE PROBLEMS

9.91 See the solution to problem 1.43, which is nearly identical. The molecular formula is $\boxed{B_9Cl_9}$.

9.93 **a)** The reaction is $Rb_2SO_3 + 2\,HBr \longrightarrow 2\,RbBr + H_2O + SO_2$. Initially

$$n_{Rb_2SO_3} = \frac{6.24\ \text{g}}{251.0\ \text{g mol}^{-1}} = 0.02486\ \text{mol}$$

$$n_{HBr} = \frac{PV}{RT} = \frac{(0.953\ \text{atm})(1.38\ \text{L})}{(0.08206\ \text{L atm mol}^{-1}\text{K}^{-1})(348.15\ \text{K})} = 0.04603\ \text{mol}$$

1 mol of Rb_2SO_3 consumes 2 mol of HBr; the HBr is used up first. $\boxed{\text{HBr is the limiting reactant}}$.

b) Each mole of HBr that reacts should give one mole of RbBr. The theoretical yield of RbBr is

$$m_{RbBr} = (0.04603\ \text{mol RbBr})(165.37\ \text{g mol}^{-1}) = \boxed{7.61\ \text{g}}$$

c) The percentage yield of RbBr is $(7.32\ \text{g}/7.61\ \text{g}) \times 100\% = \boxed{96.2\%}$.

Chapter 10

Solids, Liquids, and Phase Transitions

Bulk Properties of Gases, Liquids, and Solids: Molecular Interpretation

10.1   The substance is most likely to be a $\boxed{\text{gas}}$. Its large compressibility and large coefficient of thermal expansion are typical of gaseous materials.

10.3   **a)** Compute the density of the material

$$d = \left(\frac{2.71 \text{ kg}}{258 \text{ cm}^3} \right) \times \left(\frac{1000 \text{ g}}{1 \text{ kg}} \right) = 10.5 \text{ g cm}^{-3}$$

The material is certainly $\boxed{\text{condensed}}$, on the basis of its high density, which greatly exceeds 1 g cm^{-3}, the density of water.

b)

$$\text{molar volume} = V_{\text{m}} = \left(\frac{1 \text{ cm}^3}{10.5 \text{ g}} \right) \times \left(\frac{108 \text{ g}}{1 \text{ mol}} \right) = \boxed{10.3 \text{ cm}^3 \text{ mol}^{-1}}$$

10.5   The volume increase amounts to only 0.3%. Warming a gas from 293 K to 313 K would, according to Charles's law, cause an expansion of

$$\frac{(313 - 293) \text{ K}}{293 \text{ K}} \times 100\% = 6.8\%$$

A non-ideal gas would also expand by several percent during this change in temperature. The substance is accordingly $\boxed{\text{condensed}}$.

10.7   The huge increase in volume is due to $\boxed{\text{vaporization}}$, the transition of water from a condensed phase (liquid water) to a gaseous phase (steam) at 100°C.

10.9   Non-directional ion-ion interactions maintain the structure of solid sodium chloride. Weaker dispersion forces maintain carbon tetrachloride as a solid. Indentation requires breaking bonds in the solid. Solid NaCl is $\boxed{\text{harder}}$ than solid CCl_4 because the ion-ion interactions in NaCl are stronger than the dispersion interactions in CCl_4.

10.11   In all three phases, the diffusion constant should $\boxed{\text{decrease}}$ as the density of the phase is increased. At higher densities molecules are closer to each other. In gases they will collide more often and travel shorter distances between collisions. In liquids and solids, there will be less space for molecules to move around each other.

Intermolecular Forces: Origins in Molecular Structure

10.13   Ion-dipole forces arise from molecular or ionic properties that are always present. An example is the interaction of the permanent charge of a potassium ion with the permanent dipole of a water molecule. Induced dipole forces arise when a permanent dipole possessed by a molecule or the charge on an ion induces a *temporary* separation of electric charge (a temporary dipole) in an otherwise non-polar species Such an induced dipole goes away when the ion or dipole that provoked it is removed. An example is the attraction between an Fe^{3+} ion and a molecule of oxygen.

10.15 **a)** Potassium and fluorine differ considerably in electronegativity. The bonding in the compound potassium fluoride is consequently expected to be ionic, as explained in text Section 3.6. The $\boxed{\text{ion-ion}}$ attractions predominate in this compound; dispersion forces are also present.

b) $\boxed{\text{Dipole-dipole}}$ attractions predominate in the interactions between molecules in hydrogen iodide. The positive (H) end of one molecule is attracted by the negative (I) end of another, but repelled by its positive (H) end. Dispersion forces are also present.

c) $\boxed{\text{Dispersion forces}}$ are the only forces operating among the atoms in a sample of radon. Single Rn atoms have completely symmetrical (spherically symmetrical) distributions of charge. Two neighboring Rn atoms induce temporary dipoles in each other that cause them to attract each other.

d) $\boxed{\text{Dispersion forces}}$ are the only intermolecular forces possible between molecules of N_2.

10.17 A sodium ion should be most strongly attracted to a $\boxed{\text{bromide ion}}$. The attraction between ions of unlike charge such as these is stronger than the ion-dipole attraction between Na^+ and HBr and the ion-induced dipole attraction between Na^+ and Kr.

10.19 **a)** Read along the horizontal axis in text Figure 10.9 to find the locations of the minima in the potential energy curves. These locations give the bond distances. In Cl_2 the bond is about $\boxed{2.0 \times 10^{-10} \text{ m}}$ long; in molecular KCl, the bond is about $\boxed{2.5 \times 10^{-10} \text{ m}}$ long.

b) The bond in Cl_2 is shorter than the bond in KCl, but is distinctly weaker, as shown by the depth of the minimum in the potential energy curve for Cl—Cl (about -225 kJ mol^{-1}) compared to the depth for K—Cl (about -490 kJ mol^{-1}).

Tip. Potential energy curves such as the ones in text Figure 10.9 are much more informative than over-generalizations such as the one quoted in the problem.

Intermolecular Forces in Liquids

10.21 The boiling point and melting point of a substance depend on the strength of the attractive forces operating among the particles that comprise the liquid or solid substance. These forces tend to increase with increasing molar mass in a group of related substances. The halogens are certainly closely related chemically. Therefore their boiling and melting points tend to rise with increasing molar mass.

Tip. The observed normal melting points and boiling points of the halogens are

Substance	$\mathcal{M}/$ g mol^{-1}	m.p. / °C	b.p. / °C
F$_2$ fluorine	38	−219.6	−187.9
Cl$_2$ chlorine	71	−101	−34.05
Br$_2$ bromine	160	−7.2	58.2
I$_2$ iodine	254	113.6	184.5

10.23 Substances with the strongest intermolecular forces require the highest temperature to make them boil. Liquid RbCl has strong Coulomb (electrostatic) forces holding its ions together. It has the highest boiling point. Liquid NH_3 has dipole-dipole attractions, as does liquid NO. In liquid NH_3, these are particularly strong. They are hydrogen bonds. In liquid NO, the dipole-dipole attractions are weaker. Liquid NH_3 boils at a higher temperature than liquid NO. Induced dipole-induced dipole forces are the only intermolecular attractions in liquid neon. Consequently, it has the lowest boiling point of all: $\boxed{\text{Ne} < \text{NO} < \text{NH}_3 < \text{RbCl}}$.

10.25 The facts suggest that the $(CH_3OH)_4$ molecule has a ⟨cyclic⟩ structure. Why would a straight chain of molecules stop at exactly four links? The ring is probably maintained by hydrogen bonding between the —OH hydrogen atom and the O of a neighboring molecule. The ring would be puckered and consist of a total of eight atoms—four H's and 4 O's in alternation

10.27 Although the two substances have comparable molar masses, the boiling point of hydrazine N_2H_4 should ⟨exceed⟩ the boiling point of ethylene C_2H_4 because N—H $\cdots$ N hydrogen bonds are present in hydrazine but no hydrogen bonds occur in ethylene.

10.29 Compute the number of molecules of water present in the sample

$$N_{H_2O} = 1.0 \text{ kg } H_2O \times \left(\frac{1 \text{ mol } H_2O}{0.018 \text{ kg } H_2O} \right) \left(\frac{6.022 \times 10^{23} \text{ molecules}}{1 \text{ mol } H_2O} \right) = 3.35 \times 10^{25} \text{ molecules}$$

Each water molecule participates in a maximum of 4 hydrogen bonds, and each hydrogen bond connects 2 water molecules. Therefore the maximum number of H-bonds in the sample is 4/2 or twice the number of molecules: 6.7×10^{25}.

Phase Equilibria

10.31 Rearrange the ideal-gas equation and use it as follows

$$V_{H_2} = \frac{n_{H_2}RT}{P} = \frac{(1.00 \text{ mol})(0.08206 \text{ L atm mol}^{-1}\text{K}^{-1})(16.0 \text{ K})}{0.213 \text{ atm}} = 6.16 \text{ L}$$

This volume of a mole of H_2 vapor at this low temperature is 27% of the volume of a mole of gaseous hydrogen at STP.

10.33 Assume that the Hg vapors behave ideally. Compute the number of moles of Hg per unit volume in the space above the surface of the mercury

$$\left(\frac{n}{V} \right)_{Hg} = \frac{P}{RT} = \frac{2.87 \times 10^{-6} \text{ atm}}{(0.08206 \text{ L atm mol}^{-1}\text{K}^{-1})(300.15 \text{ K})} = 1.165 \times 10^{-7} \text{ mol L}^{-1}$$

Multiply by Avogadro's number to find the number of Hg atoms per unit volume

$$\left(\frac{N}{V} \right)_{Hg} = 1.165 \times 10^{-7} \text{ mol L}^{-1} \times \left(\frac{6.022 \times 10^{23} \text{ atom}}{1 \text{ mol}} \right) \left(\frac{1 \text{ L}}{1000 \text{ cm}^3} \right) = 7.02 \times 10^{13} \text{ atom cm}^{-3}$$

10.35 The pressure on the interior walls of the vessel containing the collected acetylene comes from collisions by molecules of H_2O as well as molecules of C_2H_2. Therefore, subtract the partial pressure of the water vapor from the total pressure inside the container. This gives the pressure that the acetylene would exert if it were present by itself. The partial pressure of water vapor depends solely on the temperature. The required value is given, so $P_{C_2H_2} = P_{\text{total}} - P_{\text{water}} = 0.9950 - 0.0728 = 0.9222$ atm.

Use the ideal-gas law to compute the chemical amount of acetylene per unit volume of collected gases that exerts a pressure of 0.9222 atm at a T of 40°C (313.15 K)

$$\frac{n_{C_2H_2}}{V} = \frac{P_{C_2H_2}}{RT} = \frac{0.9222 \text{ atm}}{(0.08206 \text{ L atm mol}^{-1}\text{K}^{-1})(313.15 \text{ K})} = 0.03589 \text{ mol L}^{-1}$$

The molar mass of acetylene equals 26.038 g mol^{-1}. Multiplying the chemical amount by this molar mass gives $\boxed{0.9345 \text{ g L}^{-1}}$ as the mass of acetylene present per unit volume of gas.

10.37 Determine the partial pressure of the CO_2, and then use the ideal-gas law to obtain the chemical amount of CO_2. From the balanced equation, the number of moles of $CaCO_3$ that reacts equals the number of moles of CO_2 that forms. Convert this amount of $CaCO_3$ to a mass

$$P_{CO_2} = P_{total} - P_{water} = 0.9963 \text{ atm} - 0.0231 \text{ atm} = 0.9732 \text{ atm}$$

$$n_{CO_2} = \frac{P_{CO_2}V_{mixture}}{RT} = \frac{(0.9732 \text{ atm})(0.722 \text{ L})}{(0.08206 \text{ L atm mol}^{-1}\text{K}^{-1})(293.15 \text{ K})} = 0.0292 \text{ mol CO}_2$$

$$m_{CaCO_3} = 0.0292 \text{ mol CO}_2 \times \left(\frac{1 \text{ mol CaCO}_3}{1 \text{ mol CO}_2}\right)\left(\frac{100.09 \text{ g CaCO}_3}{1 \text{ mol CaCO}_3}\right) = \boxed{2.92 \text{ g CaCO}_3}$$

Phase Transitions

10.39 Reading from text Figure 10.18, the vapor pressure of water at 90°C is approximately $\boxed{0.70 \text{ atm}}$. Because the whole of the earth's atmosphere exerts a pressure of 1 atm, this means that 70% of the earth's atmosphere is above the level of the explorer's camp; $\boxed{30\%}$ of the atmosphere is below that level.

10.41 The boiling and melting points indicate that the interatomic attractions are much stronger in iridium than in sodium. The surface tension of molten $\boxed{\text{iridium}}$ should therefore exceed that of molten sodium.

Phase Diagrams

10.43 Compression favors the denser phase, but the denser phase (the liquid) is already present. The sample of Pu $\boxed{\text{stays liquid}}$.

10.45

Phase Diagram for NH₃ (not to scale)

10.47 Check whether the specified point is within the boundaries of the solid, liquid, or gas region on the phase diagram of argon that appears in text Figure 10.24.

 a) liquid **b)** gas **c)** solid **d)** gas.

10.49 **a)** The temperature of the triple point of acetylene lies above $-84.0°C$. For a pure substance, T's at which liquid and gas are in equilibrium equal or exceed T's at which liquid and solid are in equilibrium.

 b) Note that 0.8 atm is *less* than 760 torr, which is the vapor pressure of solid acetylene at $-84°C$. If solid acetylene is heated at $P = 0.8$ atm, it therefore passes directly into the vaporous (gaseous) state without ever existing as a liquid. It sublimes at some temperature below $-84.0°C$.

10.51 The nitrogen confined in the glass tube becomes supercritical as it is heated past the critical temperature. This is assured because the size of the container and the amount of nitrogen that it contains assure that the critical density of nitrogen is exceeded. Originally, a meniscus separates the liquid from gaseous nitrogen within the tube. The meniscus disappears at 126.19 K as the two phases merge into a single fluid phase of uniform density: the distinction between liquid and gas ceases to exist.

ADDITIONAL PROBLEMS

10.53 Candle wax at room temperature is a solid , although a low-melting one. (It melts at about 60°C, depending on its source.) Natural rubber, a polymer of an organic compound called isoprene (C_8H_8), is a solid at room conditions. Both are incompressible compared to a gas. Neither flows readily, compared to ordinary liquids and gases.

10.55 The HCl molecule is polar. When liquid hydrogen chloride dissolves an ionic compound, its molecules tend to line up with their positive ends (the H ends) closest to negatively charged ions and with their negative ends (the Cl atoms) closest to positively charged ions.

 Tip. The diagram shows four H—Cl molecules interacting with a +1 ion and four more interacting with a −1 ion, but the number can vary depending on the size and charge of the ions. Positive ions are often smaller than negative ions, as shown.

10.57 The changes under discussion occur in the average *internal* kinetic energy and average *internal* potential energy. The internal kinetic energy derives from the motions of the molecules and atoms within the sample; the internal potential energy derives from the relative positions and attractions (or repulsions) among the atoms and molecules within the sample. Neither has to do with the movements or position of the sample as a whole. The changes occur in four steps:

1. The sample starts as a solid (ice) at $T = 10$ K and $P = 1$ atm. Heating this very cold ice causes its internal kinetic energy to rise. The motions of the molecules consist mostly of vibrations around fixed positions in the solid. The temperature, which is a measure of the average kinetic energy of a sample's molecules, rises as the intensity of these vibrations increases. The average internal potential energy, which comes as a result of the attractions among the molecule neighbors, increases somewhat as the cold ice expands against the constant outside pressure.

2. At $T = 273.15$ K and $P = 1$ atm, the ice starts to melt. During this phase-change, the temperature of the sample stays constant, which means that the average kinetic energy of its molecules remains constant. The added energy goes toward increasing the average internal potential energy.

3. Once the sample is entirely liquefied, fresh additions of heat go mostly to raising its internal kinetic energy. The temperature of the sample starts to rise again. As in step 1, some energy goes to increase the internal potential energy. Liquid water expands by roughly four percent between 273.15 and 373.15 K

4. At 373.15 K and $P = 1$ atm the sample starts to boil. During this phase-change, its average internal kinetic energy remains constant. The average internal potential energy increases greatly as the molecules, which attract each other, are repositioned from being fairly close neighbors to locations extremely remote from each other.

10.59   "Saturation" means that the vapor pressure of the water in the room has reached its maximum—the humidity is 100% and drops of water are about to start condensing on the walls. The partial pressure of water vapor in the room is very close to the vapor pressure of pure water at the given temperature.[1] This vapor pressure equals 0.03126 atm. The rest of the pressure in the room comes from oxygen, nitrogen, and the other components of the air. Compute the volume of the room

$$V_{\text{room}} = 110 \text{ m}^3 \times \left(\frac{10^3 \text{ L}}{\text{m}^3}\right) = 1.10 \times 10^5 \text{ L}$$

Combine this volume with the known partial pressure of water to obtain the chemical amount and the mass of the water vapor in the air in the room. Assume that the ideal-gas equation applies

$$n_{\text{H}_2\text{O}} = \frac{PV}{RT} = \frac{0.03126 \text{ atm}(1.10 \times 10^5 \text{ L})}{(0.08206 \text{ L atm mol}^{-1}\text{K}^{-1})(298.15 \text{ K})} = 140.5 \text{ mol}$$

$$m_{\text{H}_2\text{O}} = 140.5 \text{ mol H}_2\text{O} \times \left(\frac{18.02 \text{ g H}_2\text{O}}{1 \text{ mol H}_2\text{O}}\right) = \boxed{2530 \text{ g H}_2\text{O}}$$

This is roughly 2.5 L of liquid water.

Tip. Converting R to m^3 atm mol^{-1} K^{-1} (to fit the units of volume as given) leads to the same answer. A standard reference[2] reports that the measured mass of water in one cubic meter of saturated air at 1 atm and 298.15 K to equal 23.14 g. This corresponds to 2545 g of water in 110 m^3 of air, which is close to the answer just obtained.

10.61   The only substance within the sealed can is water. If liquid water and water vapor coexist at 60°C, then the pressure of the water vapor must be approximately $\boxed{0.20 \text{ atm}}$, reading from text Figure 10.18.

10.63   Why don't spacecraft just boil away in the vacuum of space? The term "boiling" implies an active or even violent event. Spacecraft do tend to lose individual atoms from their hulls into the vacuum of space. This process is however so very slow that no change is apparent.

10.65   When chunks of solid CO_2 are added to room-temperature ethanol in an open beaker, portions of the solid sublime off as gaseous CO_2. Much bubbling and roiling accompany the escape of this gas. The process will in theory chill the ethanol to $\boxed{-78.5°\text{C}}$, the sublimation temperature of the solid CO_2, but no lower. Since ethanol at $P = 1$ atm requires a temperature below $-114.5°$C to freeze, it stays liquid in this experiment. Once the ethanol is good and cold, small amounts of further sublimation take place to counteract heat flowing in from the surroundings; the mixture will always fizz a little if its surroundings are at room temperature.

10.67   Substitution of the appropriate "VDW" (van der Waals) a's and b's into the three equations quoted in the problem gives the results that appear in the columns to the right of the vertical line in following

---

[1] The difference derives from the interaction of the water with the air.
[2] Kaye and Laby, *Tables of Physical and Chemical Constants*, 16th edition, page 38.

table. Because the different VDW constants are in atm L^2 mol^{-2} and L mol^{-1} respectively, a value of 0.082057 L atm $mol^{-1}K^{-1}$ was used for R. The answer columns include the observed values (in parentheses) for comparison.

Gas	a (atm L^2 mol^{-2})	b (L mol^{-1})	T_c (K)	P_c (atm)	$(V/n)_c$ (L mol^{-1})
O_2	1.360	0.03183	154.3 (154.6)	49.72 (49.8)	0.09549 (0.0734)
CO_2	3.592	0.04267	303.9 (304.2)	73.07 (72.9)	0.1280 (0.0940)
H_2O	5.464	0.03049	647.1 (647.1)	217.7 (217.6)	0.09147 (0.0567)

Tip. The experimental T_c's and P_c's in parentheses are more precise than can be obtained by reading the scale in text Figure 10.24. Of course, molar volumes cannot be read from *P-T* phase diagrams at all.

10.69 a)

b) Such a diagram cannot be drawn because two different phases (liquid and solid) can have the same temperature and molar volume.

10.71 The problem gives the boiling points of fluorides of elements in the second row of the periodic table. The high boiling points of LiF and BeF_2 result from the strong ion-ion attractions in the liquids. The large decrease in boiling point going from BeF_2 to BF_3 suggests a changeover from ion-ion forces to much weaker dipole-dipole or induced-dipole forces. The continued decrease in boiling point of the chlorides across the rest of the second period corresponds to decreasing ionic character in the bonds and a parallel decrease in the strength of the intermolecular attractions.

CUMULATIVE PROBLEMS

10.73 The bonds in SbF_5 have less ionic character than in SbF_3. Ionic character tends to decrease with increasing oxidation number of the metal. The bonds in AsF_5 should have less ionic character than those in SbF_5 because Sb lies below As in the periodic table. The bond in F_2 has the least ionic character of all, since it is a covalent bond. Therefore the trend in ionic character is

$$\text{least} \quad F_2 < AsF_5 < SbF_5 < SbF_3 \quad \text{most}$$

If the boiling points increase with increasing ionic character, then

$$\boxed{F_2 < AsF_5 < SbF_5 < SbF_3}$$

Tip. The experimental boiling points of the first three substances at 1 atm pressure respectively equal: -188.14, -53, and $149.5°C$. SbF_3 sublimes at this pressure at $319°C$.

Chapter 11

Solutions

Composition of Solutions

11.1 **a)** The molar mass of cholesterol is 386.64 g mol^{-1}. One liter (L) equals 10 deciliters (dL); one gram (g) equals 1000 milligrams (mg). Use unit-factors derived from these facts as follows

$$c_{\text{cholesterol}} = \frac{214 \text{ mg}}{1 \text{ dL}} \times \left(\frac{1 \text{ g}}{1000 \text{ mg}} \right) \left(\frac{10 \text{ dL}}{1 \text{ L}} \right) \left(\frac{1 \text{ mol}}{386.64 \text{ g}} \right) = \boxed{0.00553 \text{ mol L}^{-1}}$$

b) Assume that blood, like water, has a density of 1.0 g mL^{-1}. One liter (1000 mL) of blood then has a mass of 1.0 kg, of which 2.14 g (0.00214 kg) is cholesterol. Assume that the remaining 0.998 kg is solvent water. Insert these numbers into the definition of molality

$$m_{\text{cholesterol}} = \frac{\text{moles of cholesterol}}{\text{kilograms of solvent}} = 2.14 \text{ g} \times \left(\frac{1 \text{ mol}}{386.64 \text{ g}} \right) \times \left(\frac{1}{0.998 \text{ kg}} \right) = \boxed{0.0055 \text{ mol kg}^{-1}}$$

c) If there is 2.14 g of cholesterol per liter of blood, then one liter of blood contains 2.14 g of cholesterol. This simple turn-about gives the unit-factor in the following

$$V_{\text{blood}} = 8.10 \text{ g cholesterol} \times \left(\frac{1 \text{ L blood}}{2.14 \text{ g cholesterol}} \right) = \boxed{3.79 \text{ L blood}}$$

Tip. The second answer has only two significant figures because the assumption about the density of blood is weak. Whole blood in fact has a density of 1.06 g mL^{-1}. This means that 1000 mL of blood weighs 1.06 kg. Taking this into account alters the calculation of the molality of the cholesterol

$$m = \frac{0.00553 \text{ mol cholesterol}}{1.0579 \text{ kg solvent}} = 0.0052 \text{ mol kg}^{-1}$$

Clearly, the subtraction of the 2.14 g is superfluous in both calculations.

11.3 To compute the various quantities, obtain the masses and chemical amounts of HCl and H$_2$O in some set quantity of solution. Then use the definitions. Exactly 100.0 g of solution contains 38.00 g of HCl and 62.00 g of H$_2$O. Its volume is

$$V_{\text{solution}} = 100 \text{ g solution} \times \left(\frac{1 \text{ mL solution}}{1.1886 \text{ g solution}} \right) = 84.133 \text{ mL}$$

Use the molar masses of HCl and H$_2$O to compute their chemical amounts from their masses

$$n_{\text{HCl}} = \frac{38.00 \text{ g HCl}}{36.4606 \text{ g mol}^{-1}} = 1.0422 \text{ mol HCl} \qquad n_{\text{H}_2\text{O}} = \frac{62.00 \text{ g H}_2\text{O}}{18.0153 \text{ g mol}^{-1}} = 3.4415 \text{ mol H}_2\text{O}$$

The molarity of the HCl equals the number of moles of HCl divided by the number of liters of solution.

$$c_{HCl} = \frac{1.0422 \text{ mol}}{84.133 \text{ mL}} \times \left(\frac{1000 \text{ mL}}{1 \text{ L}} \right) = \boxed{12.39 \text{ mol L}^{-1}}$$

The molality of the HCl equals the number of moles of HCl divided by the number of kilograms of solvent

$$m_{HCl} = \frac{1.0422 \text{ mol}}{62.00 \text{ g}} \times \left(\frac{1000 \text{ g}}{1 \text{ kg}} \right) = \boxed{16.81 \text{ mol kg}^{-1}}$$

The mole fraction of water equals the number of moles of water divided by the total number of moles of all components of the solution

$$X_{H_2O} = \frac{3.4415 \text{ mol}}{1.0422 + 3.4415 \text{ mol}} = 0.7676$$

Only two components are present; X_{HCl} is simply $1 - 0.7676 = \boxed{0.2324}$.

Tip. In the preceding m_{HCl} denotes the molality of the HCl in the solution. In other contexts the same symbol denotes the *mass* of HCl. The official recommendation for avoiding possible confusion is to indicate molality with a b and reserve m for mass. However chemists rarely do this.

11.5   Compute the mass of acetic acid in 1 kg of the 6.0835 M solution. To do this, construct suitable unit-factors from the molarity of the acetic acid and the density of the solution

$$m_{C_2H_4O_2} = 1 \text{ kg solution} \times \left(\frac{1 \text{ L solution}}{1.0438 \text{ kg solution}} \right) \left(\frac{6.0835 \text{ mol C}_2\text{H}_4\text{O}_2}{1 \text{ L sol'n}} \right) \left(\frac{60.052 \text{ g C}_2\text{H}_4\text{O}_2}{\text{mol C}_2\text{H}_4\text{O}_2} \right)$$

$$= 350.00 \text{ g C}_2\text{H}_4\text{O}_2$$

By subtraction, 1 kg of solution contains 650.00 g (0.65000 kg) of water. The 350.00 g of acetic acid equals 5.8282 mol of acetic acid. The molality of acetic acid equals the number of moles of acetic acid divided by the mass of the solvent in kilograms

$$\text{molality}_{C_2H_4O_2} = \frac{5.8283 \text{ mol C}_2\text{H}_4\text{O}_2}{0.65000 \text{ kg}} = \boxed{8.9665 \text{ mol kg}^{-1}} = 8.9665 \text{ molal}$$

11.7   Water is the solute, and liquid nitrogen is the solvent. Nevertheless, the definition of mole fraction works the same. Use it to write the equation

$$X_{H_2O} = 1.00 \times 10^{-5} = \frac{n_{H_2O}}{n_{N_2} + n_{H_2O}} = \frac{n_{H_2O}}{35.6972 + n_{H_2O}}$$

where 35.6972 is the number of moles of N_2 in 1.00 kg of N_2 (non-significant figures are carried along deliberately). Solving the equation for n_{H_2O} is simplified by noting that n_{H_2O} can be neglected in the denominator. Thus, the 1.00 kg of $N_2(l)$ contains 3.5697×10^{-4} mol of dissolved H_2O. This amounts to $\boxed{0.00643 \text{ g}}$ of H_2O.

11.9   **a)** According to the problem, 100.00 g of commercial $H_3PO_4(aq)$ contains 90.00 g of pure H_3PO_4 and 10.00 g of H_2O. This is the first factor in the following. The two following unit-factors come from the molar mass of H_3PO_4 and the concentration of the solution

$$\rho = \frac{100.00 \text{ g solution}}{90.00 \text{ g H}_3\text{PO}_4} \times \left(\frac{97.995 \text{ g H}_3\text{PO}_4}{1.0000 \text{ mol H}_3\text{PO}_4} \right) \left(\frac{12.2 \text{ mol H}_3\text{PO}_4}{1 \text{ L solution}} \right) = \frac{1.33 \times 10^3 \text{ g solution}}{1 \text{ L solution}}$$

This is a correct answer, but densities are more frequently given in grams per milliliter. The answer in these units is $\boxed{1.33 \text{ g mL}^{-1}}$. This unit is the same as a g cm^{-3}.

b) The 2.00 L of 1.00 M $H_3PO_4(aq)$ must contain 2.00 mol of H_3PO_4. The volume of the 12.2 M $H_3PO_4(aq)$ solution that provides 2.00 mol of H_3PO_4 is

$$V_{\text{solution}} = 2.00 \text{ mol } H_3PO_4 \times \left(\frac{1 \text{ L solution}}{12.2 \text{ mol } H_3PO_4} \right) = \boxed{0.164 \text{ L solution}}$$

To make 2.00 L of a 1.00 M $H_3PO_4(aq)$ solution, put 0.164 L (164 mL) of 12.2 M H_3PO_4 in a 2-liter volumetric flask and then add water to bring the total volume up to the 2.00 L mark.

11.11   First calculate how many moles of NaOH were in the solution before any solid NaOH was added

$$n_{\text{NaOH}} = 1.50 \text{ L solution} \times \left(\frac{2.40 \text{ mol NaOH}}{1 \text{ L solution}} \right) = 3.60 \text{ mol NaOH}$$

The molar mass of NaOH is 40.00 g mol^{-1}, so the additional 25.0 g of NaOH equals 0.625 mol. After the addition, the total amount of NaOH in the container equals $3.60 + 0.625 = 4.23$ mol. Meanwhile, the final volume of the solution is 4.00 L. The final concentration of NaOH is therefore

$$c_{\text{NaOH}} = \frac{4.23 \text{ mol}}{4.00 \text{ L}} = \boxed{1.06 \text{ mol L}^{-1}}$$

Nature of Dissolved Species

11.13   Break up each soluble reactant and product into its component ions and rewrite the chemical equation in terms of these species. Identify ions that appear unchanged on the two sides of the equation. Cancel out these spectator ions.

a) $Ag^+(aq) + Cl^-(aq) \longrightarrow AgCl(s)$

b) $K_2CO_3(s) + 2 H^+(aq) \longrightarrow 2 K^+(aq) + CO_2(g) + H_2O(l)$[1]

c) $2 Cs(s) + 2 H_2O(l) \longrightarrow 2 Cs^+(aq) + 2 OH^-(aq) + H_2(g)$

d) $2 MnO_4^-(aq) + 16 H^+(aq) + 10 Cl^-(aq) \longrightarrow 5 Cl_2(g) + 2 Mn^{2+}(aq) + 8 H_2O(l)$

Reaction Stoichiometry in Solution: Acid-Base Titrations

11.15   The balanced chemical equation that appears in the problem indicates a 4-to-2 molar relationship between HNO_3 and PbO_2 in the reaction. Use this fact to construct a unit-factor. The given molarity furnishes another unit-factor, and the molar mass of PbO_2 (239.2 g mol^{-1}) furnishes a third. The computation using these three factors goes as follows

$$V_{\text{solution}} = 15.9 \text{ g } PbO_2 \times \left(\frac{1 \text{ mol } PbO_2}{239.2 \text{ g } PbO_2} \right) \left(\frac{4 \text{ mol } HNO_3}{2 \text{ mol } PbO_2} \right) \left(\frac{1 \text{ L solution}}{7.91 \text{ mol } HNO_3} \right) = \boxed{0.0168 \text{ L}}$$

11.17   The carbon dioxide in this problem is a gas (with volume measured in liters), and the potassium carbonate is in aqueous solution (with volume *also* measured in liters). According to the balanced equation, the $CO_2(g)$ and $K_2CO_3(aq)$ react in a 1-to-1 molar ratio. The following uses this fact together with the concentration of the $K_2CO_3(aq)$ to determine the chemical amount of $CO_2(g)$ that reacts

$$n_{CO_2} = 187 \text{ L solution} \times \left(\frac{1.36 \text{ mol } K_2CO_3}{1 \text{ L solution}} \right) \left(\frac{1 \text{ mol } CO_2}{1 \text{ mol } K_2CO_3} \right) = 254.3 \text{ mol } CO_2$$

The volume occupied by this amount of $CO_2(g)$ depends on T and P. Assume that the CO_2 behaves ideally and insert n_{CO_2} and the given T and P in the ideal-gas equation

$$V_{CO_2} = \frac{n_{CO_2}RT}{P} = \frac{(254.3 \text{ mol})(0.08206 \text{ L atm mol}^{-1}\text{K}^{-1})(323.15 \text{ K})}{1.00 \text{ atm}} = \boxed{6.74 \times 10^3 \text{ L}}$$

---

[1]Some printings of the text have a typographical error in the given equation. Correct is $K_2CO_3(s) + 2 HCl(aq) \longrightarrow 2 KCl(aq) + CO_2(g) + H_2O(l)$.

11.19 When a salt forms in an acid-base reaction, the cation derives from the base, and the anion from the acid.

a) $Ca(OH)_2(aq) + 2\,HF(aq) \longrightarrow CaF_2(s) + 2\,H_2O(l)$
calcium hydroxide + hydrofluoric acid $\longrightarrow$ calcium fluoride + water

b) $2\,RbOH(aq) + H_2SO_4(aq) \longrightarrow Rb_2SO_4(aq) + 2\,H_2O(l)$
rubidium hydroxide + sulfuric acid $\longrightarrow$ rubidium sulfate + water

c) $Zn(OH)_2(aq) + 2\,HNO_3(aq) \longrightarrow Zn(NO_3)_2(aq) + 2\,H_2O(l)$
zinc hydroxide + nitric acid $\longrightarrow$ zinc nitrate + water

d) $KOH(aq) + CH_3COOH(aq) \longrightarrow KCH_3COO(aq) + H_2O(l)$
potassium hydroxide + acetic acid $\longrightarrow$ potassium acetate + water

11.21 The reaction is $H_2S + 2\,NaOH \longrightarrow Na_2S + 2\,H_2O$. $\boxed{\text{Sodium sulfide}}$ is the salt produced by this neutralization reaction.

Tip. Without the hint, NaHS (sodium hydrogen sulfide) is a possible answer.

11.23 **a)** Phosphorus trifluoride is PF_3; phosphorous acid is H_3PO_3, and hydrofluoric acid is HF. The equation is easily balanced by inspection. Assign 1 as the coefficient for PF_3. All of the fluorine ends up in HF. This means the coefficient for HF is 3. All of the oxygen ends up in H_3PO_3, making 3 the coefficient for the H_2O: $\boxed{PF_3 + 3\,H_2O \longrightarrow H_3PO_3 + 3\,HF}$.

b) First determine the chemical amount of $PF_3(g)$ in 1.94 L of gaseous PF_3 at 25°C (298 K) and 0.970 atm. Assume ideal-gas behavior by the PF_3. Then

$$n_{PF_3} = \frac{PV}{RT} = \frac{(0.970\text{ atm})(1.94\text{ L})}{(0.08206\text{ L atm mol}^{-1}\text{K}^{-1})(298\text{ K})} = 0.07695\text{ mol}$$

According to the balanced equation, 1 mol of PF_3 reacts to give 1 mol of H_3PO_3 and 3 mol of HF. This means 0.07695 mol of H_3PO_3 and 0.2309 mol of HF are produced from 0.07695 mol of PF_3. Both acids dissolve as they are formed. Enough water is present to give a final volume of 872 mL (0.872 L). The acids are mixed with each other, but their respective concentrations are computed by *separately* dividing the chemical amounts by the final volume

$$c_{H_3PO_3} = \frac{0.07695\text{ mol}}{0.872\text{ L}} = \boxed{0.0882\text{ M}} \qquad c_{HF} = \frac{0.2309\text{ mol}}{0.872\text{ L}} = \boxed{0.265\text{ M}}$$

11.25 The problem is very similar to text Example 11.16. Each mole of KOH dissolves in water to give one mole of $K^+(aq)$ ion and one mole of $OH^-(aq)$ ion. Therefore, the chemical amount of $OH^-(aq)$ in 37.85 mL of 0.1279 M aqueous KOH is

$$n_{OH^-} = 37.85\text{ mL} \times \left(\frac{0.1279\text{ mmol}}{1\text{ mL}}\right) = 4.841\text{ mmol}$$

Nitric acid furnishes one mole of $H^+(aq)$ ion per mole dissolved. Also, the stoichiometric ratio in the acid-base reaction between HNO_3 and KOH is 1-to-1. This means that the chemical amount of HNO_3 in the 100.0 mL sample before the reaction was also 4.841 mmol. The concentration of HNO_3 in the original solution was

$$c_{HNO_3} = [HNO_3] = \frac{4.841\text{ mmol}}{100.0\text{ mL}} = \frac{0.04841\text{ mmol}}{1\text{ mL}} = \boxed{0.04841\text{ mol L}^{-1}}$$

Tip. Note the use of the convenient unit, the millimole (mmol). The concentration of a dissolved substance in mmol mL^{-1} is numerically equal to its concentration in mol L^{-1}.

Reaction Stoichiometry in Solutions: Oxidation-Reduction Titrations

11.27 **a)** $2 \overset{+3}{P} F_2 I(l) + 2 \overset{0}{Hg}(l) \longrightarrow \overset{+2}{P_2} F_4(g) + \overset{+1}{Hg_2} I_2(s)$

b) $2 K \overset{+5 \; -2}{Cl O_3}(s) \longrightarrow 2 K \overset{-1}{Cl}(s) + 3 \overset{0}{O_2}(g)$

c) $4 \overset{-3}{N H_3}(g) + 5 \overset{0}{O_2}(g) \longrightarrow 4 \overset{+2 \; -2}{N O}(g) + 6 H_2 \overset{-2}{O}(g)$

d) $2 \overset{0}{As}(s) + 6 Na \overset{+1}{O H}(l) \longrightarrow 2 Na_3 \overset{+3}{As} O_3(s) + 3 \overset{0}{H_2}(g)$

11.29 Refer to the conventions for assigning oxidation numbers in text Section 3.12 (page 112). Neither hydrogen (oxidation state $+1$) nor oxygen (oxidation state -2) changes oxidation state in this reaction. The gold loses $3\,e^-$ per atom, passing from the zero to the $+3$ oxidation state; $\boxed{Au}$ is oxidized. The Se atom in H_2SeO_4 passes from the $+6$ to the $+4$ oxidation state; it gains $2\,e^-$, and so $\boxed{H_2SeO_4}$ is reduced. Note that only half of the H_2SeO_4 reacting is actually reduced.

11.31 The problem requires *completion* as well as balancing. This means the insertion of H_2O, OH^- ion and H_3O^+ ion on one or both sides of the equation. Follow the procedure outlined in text Section 11.4.2

a) The following gives the results of each step. Indications of state such as (aq) or (s) are omitted until the end.

1.	$VO_2^+ \longrightarrow VO^{2+}$	$SO_2 \longrightarrow SO_4^{2-}$
2.	$VO_2^+ \longrightarrow VO^{2+}$	$SO_2 \longrightarrow SO_4^{2-}$
3.	$VO_2^+ \longrightarrow VO^{2+} + H_2O$	$2 H_2O + SO_2 \longrightarrow SO_4^{2-}$
4.	$2 H_3O^+ + VO_2^+ \longrightarrow VO^{2+} + 3 H_2O$	$6 H_2O + SO_2 \longrightarrow SO_4^{2-} + 4 H_3O^+$
5.	$e^- + 2 H_3O^+ + VO_2^+ \longrightarrow VO^{2+} + 3 H_2O$	$6 H_2O + SO_2 \longrightarrow SO_4^{2-} + 4 H_3O^+ + 2 e^-$
6.	$2 e^- + 4 H_3O^+ + 2 VO_2^+ \longrightarrow 2 VO^{2+} + 6 H_2O$	$6 H_2O + SO_2 \longrightarrow SO_4^{2-} + 4 H_3O^+ + 2 e^-$

$$\boxed{2 VO_2^+(aq) + SO_2(g) \longrightarrow 2 VO^{2+}(aq) + SO_4^{2-}(aq)}$$

Step 5 shows that VO_2^+ is reduced (electrons had to be put in on the left side of the half-equation to balance charge) and that SO_2 is oxidized (electrons had to be put in on the right). In the final line, addition of the oxidation half-equation to the reduction half-equation has led, as planned, to the cancellation of e^-'s. The algebraic combining of terms removes six H_2O's and four H_3O^+'s.

Use the same method in the other parts of the problem.

b) $Br_2(l) + SO_2(g) + 6 H_2O(l) \longrightarrow 2 Br^-(aq) + SO_4^{2-}(aq) + 4 H_3O^+(aq)$

c) $Cr_2O_7^{2-}(aq) + 3 Np^{4+}(aq) + 2 H_3O^+(aq) \longrightarrow 2 Cr^{3+}(aq) + 3 NpO_2^{2+}(aq) + 3 H_2O(l)$

d) $5 HCOOH(aq) + 2 MnO_4^-(aq) + 6 H_3O^+(aq) \longrightarrow 5 CO_2(g) + 2 Mn^{2+}(aq) + 14 H_2O(l)$

e) $3 Hg_2HPO_4(s) + 2 Au(s) + 8 Cl^-(aq) + 3 H_3O^+(aq) \longrightarrow 6 Hg(l) + 3 H_2PO_4^-(aq) + 2 AuCl_4^-(aq) + 3 H_2O(l)$

Tip. The determination of oxidation numbers[2] is *not* necessary in balancing redox equations. However, oxidation numbers can come in handy in checking results. For example, in part **a)**, vanadium is reduced from the $+5$ to the $+4$ state and sulfur is oxidized from the $+4$ to the $+6$ state. This requires a gain of one electron per vanadium atom and a loss of two electrons per sulfur atom, confirming the results of step 5.

11.33 Follow the six-step method given in the text. The steps are the same as those used in problem **11.31** except that OH^- and H_2O are used to balance hydrogen in step 4 (instead of H_3O^+ and H_2O).

---

[2]Oxidation numbers are discussed text Section 3.12.

a) The following lists the results of each step. Indications of state such as (aq) or (s) are omitted until the end.

1. $Cr(OH)_3 \longrightarrow CrO_4^{2-}$ $Br_2 \longrightarrow Br^-$
2. $Cr(OH)_3 \longrightarrow CrO_4^{2-}$ $Br_2 \longrightarrow 2\,Br^-$
3. $H_2O + Cr(OH)_3 \longrightarrow CrO_4^{2-}$ $Br_2 \longrightarrow 2\,Br^-$
4. $5\,OH^- + H_2O + Cr(OH)_3 \longrightarrow CrO_4^{2-} + 5\,H_2O$ $Br_2 \longrightarrow 2\,Br^-$
5. $5\,OH^- + H_2O + Cr(OH)_3 \longrightarrow CrO_4^{2-} + 5\,H_2O + 3\,e^-$ $2\,e^- + Br_2 \longrightarrow 2\,Br^-$
6. $10\,OH^- + 2\,H_2O + 2\,Cr(OH)_3 \longrightarrow 2\,CrO_4^{2-} + 10\,H_2O + 6\,e^-$ $6\,e^- + 3\,Br_2 \longrightarrow 6\,Br^-$

$$\boxed{10\,OH^-(aq) + 2\,Cr(OH)_3(s) + 3\,Br_2(aq) \longrightarrow 2\,CrO_4^{2-}(aq) + 6\,Br^-(aq) + 8\,H_2O(l)}$$

In the answer, addition of the oxidation half-equation to the reduction half-equation has led, as planned, to the cancellation of the e^-'s. Cancellation of like terms then removes two H_2O's.

Use the same method in the other parts of the problem.

b) $ZrO(OH)_2(s) + 2\,SO_3^{2-}(aq) \longrightarrow Zr(s) + 2\,SO_4^{2-}(aq) + H_2O(l)$

c) $7\,HPbO_2^-(aq) + 2\,Re(s) \longrightarrow 7\,Pb(s) + 2\,ReO_4^-(aq) + H_2O(l) + 5\,OH^-(aq)$

d) $4\,HXeO_4^-(aq) + 8\,OH^-(aq) \longrightarrow 3\,XeO_6^{4-}(aq) + Xe(g) + 6\,H_2O(l)$

e) $N_2H_4(aq) + 2\,CO_3^{2-}(aq) \longrightarrow N_2(g) + 2\,CO(g) + 4\,OH^-(aq)$

11.35   The problem requires a reversal of the steps for the completion and balancing of redox equations.

a) $Fe^{2+}(aq) \longrightarrow Fe^{3+}(aq) + e^-$ (oxidation)

$H_2O_2(aq) + 2\,H_3O^+(aq) + 2e^- \longrightarrow 4\,H_2O(l)$ (reduction)

b) $5\,H_2O(l) + SO_2(aq) \longrightarrow HSO_4^-(aq) + 3\,H_3O^+(aq) + 2\,e^-$ (oxidation)

$MnO_4^-(aq) + 8\,H_3O^+(aq) + 5\,e^- \longrightarrow Mn^{2+}(aq) + 12\,H_2O(l)$ (reduction)

c) $ClO_2^-(aq) \longrightarrow ClO_2(g) + e^-$ (oxidation)

$ClO_2^-(aq) + 4\,H_3O^+(aq) + 4\,e^- \longrightarrow Cl^-(aq) + 6\,H_2O(l)$ (reduction)

Tip. The reaction in part **c)** is a disproportionation.

11.37   The HNO_2 is both oxidized and reduced in this reaction, which is a disproportionation. The two half-equations are

$$HNO_2(aq) + 4\,H_2O(l) \longrightarrow NO_3^-(aq) + 3\,H_3O^+(aq) + 2\,e^- \quad \text{(oxidation)}$$
$$e^- + HNO_2(aq) + H_3O^+(aq) \longrightarrow NO(g) + 2\,H_2O(l) \quad \text{(reduction)}$$

Eliminate the e^-'s between the two half-equations by doubling the second and adding it to the first. The result is

$$\boxed{3\,HNO_2(aq) \longrightarrow NO_3^-(aq) + 2\,NO(g) + H_3O^+(aq)}$$

11.39   The potassium dichromate solution contains 5.134 g of solute per 1000 mL of solution. 34.26 mL of it brings the titration to the endpoint. The chemical amount of $K_2Cr_2O_7$ that reacts is

$$n_{K_2Cr_2O_7} = 34.26 \text{ mL solution} \times \left(\frac{5.134 \text{ g } K_2Cr_2O_7}{1000 \text{ mL solution}}\right)\left(\frac{1 \text{ mol } K_2Cr_2O_7}{294.18 \text{ g } K_2Cr_2O_7}\right) = 5.979 \times 10^{-4} \text{ mol}$$

In aqueous solution, 1 mol of $Cr_2O_7^{2-}(aq)$ forms for every 1 mol of $K_2Cr_2O_7$ that dissolves. Also, according to the balanced equation (which is a net ionic equation), 1 mol of $Cr_2O_7^{2-}(aq)$ reacts with 6 mol of $Fe^{2+}(aq)$. Cast these facts as unit-factors to compute the chemical amount of Fe^{2+}

$$n_{Fe^{2+}} = 5.979 \times 10^{-4} \text{ mol } K_2Cr_2O_7 \times \left(\frac{1 \text{ mol } Cr_2O_7^{2-}}{1 \text{ mol } K_2Cr_2O_7}\right)\left(\frac{6 \text{ mol } Fe^{2+}}{1 \text{ mol } Cr_2O_7^{2-}}\right) = 0.003587 \text{ mol}$$

This is the amount of Fe^{2+} in 500.0 mL of solution. The amount per liter (1000.0 mL) is twice as much. The concentration of Fe^{2+} in the sample is $\boxed{0.007175 \text{ mol L}^{-1}}$.

Phase Equilibrium in Solutions: Nonvolatile Solutes

11.41 Figure out the molar masses of acetone and benzophenone, and use them to compute the chemical amounts of the two in the solution

$$n_{C_3H_6O} = 50.0 \text{ g C}_3\text{H}_6\text{O} \times \left(\frac{1 \text{ mol C}_3\text{H}_6\text{O}}{58.08 \text{ g C}_3\text{H}_6\text{O}} \right) = 0.86088 \text{ mol C}_3\text{H}_6\text{O}$$

$$n_{C_{13}H_{10}O} = 15.0 \text{ g C}_{13}\text{H}_{10}\text{O} \times \left(\frac{1 \text{ mol C}_{13}\text{H}_{10}\text{O}}{182.22 \text{ g C}_{13}\text{H}_{10}\text{O}} \right) = 0.0823 \text{ mol C}_{13}\text{H}_{10}\text{O}$$

The mole fraction of benzophenone in the solution equals

$$X_{C_{13}H_{10}O} = \frac{0.0823}{0.0823 + 0.86088} = 0.08728$$

The change in vapor pressure of the acetone due to the presence of the benzophenone is

$$\Delta P_{C_3H_6O} = -X_{C_{13}H_{10}O} P^\circ_{C_3H_6O} = -0.08728(0.3270 \text{ atm}) = -0.02854 \text{ atm}$$

The final vapor pressure of the acetone equals its original vapor pressure P° plus the change. This is 0.3270 atm minus 0.02854 atm or $\boxed{0.2985 \text{ atm}}$.

11.43 The boiling-point elevation of a solvent caused by a single nonvolatile solute is proportional to the molality of the solute

$$\Delta T_b = K_b m \qquad \text{hence} \qquad K_b = \frac{\Delta T_b}{m}$$

To obtain the molality m, use its definition.[3] The chemical amount of anthracene equals its mass divided by its molar mass, which is 178.2 g mol^{-1}

$$n_{\text{anthracene}} = \frac{7.80 \text{ g}}{178.2 \text{ g mol}^{-1}} = 0.04376 \text{ mol}$$

The molality of the anthracene in solvent toluene is this number of moles divided by the mass of the toluene in kilograms

$$m_{\text{anthracene}} = \frac{0.04376 \text{ mol}}{0.1000 \text{ kg}} = 0.4376 \text{ mol kg}^{-1}$$

The change in the boiling temperature is clearly $112.06 - 110.60 = 1.46°C$. Then

$$K_b = \frac{\Delta T_b}{m} = \frac{1.46°C}{0.4376 \text{ mol kg}^{-1}} = 3.34°C \text{ kg mol}^{-1}$$

This can also be expressed as $\boxed{3.34 \text{ K kg mol}^{-1}}$.

Tip. Convert both temperatures from °C to K and subtract T_1 from T_2 to check this last statement.

11.45 Compute the molality of the aqueous solution of the nonvolatile, non-dissociating solute sugar using the formula for boiling-point elevation and taking K_b for water from text Table 11.2

$$m_{\text{sugar}} = \frac{\Delta T_b}{K_b} = \frac{0.30 \text{ K}}{0.512 \text{ K kg mol}^{-1}} = 0.5859 \text{ mol kg}^{-1}$$

This means that 200.0 g of water contains 0.1172 mol of sugar. The mass of the sugar is given as 39.8 g. Its molar mass equals 39.8 mol/0.1172 mol = $\boxed{340 \text{ g mol}^{-1}}$.

Tip. The answer is close to 342.3 g mol^{-1}, the molar mass of sucrose (table sugar).

---

[3]The symbol m also represents mass. The International Union of Pure and Applied Chemistry recommends using the symbol b for molality to avoid possible confusion. This is seldom done.

11.47   Assume that the unknown is nonvolatile. The molality of the unknown then is related to the change in the freezing point of the camphor as follows

$$m_{\text{unknown}} = -\frac{\Delta T_f}{K_f} = -\frac{(170.8 - 178.4)°\text{C}}{37.7°\text{C kg mol}^{-1}} = 0.20 \text{ mol kg}^{-1}$$

Note the switch in the temperature units in the freezing-point depression constant from K to °C.[4] There is 0.20 mol of unknown per kilogram of camphor, but the problem deals with 25.0 g (0.0250 kg) of camphor. Compute the amount of unknown in the 25.0 g of camphor

$$n_{\text{unknown}} = 0.20 \text{ mol kg}^{-1} \times 0.0250 \text{ kg} = 0.0050 \text{ mol}$$

The molar mass of a substance equals its mass divided by its chemical amount

$$\mathcal{M}_{\text{unknown}} = 0.840 \text{ g}/0.0050 \text{ mol} = \boxed{1.7 \times 10^2 \text{ g mol}^{-1}}$$

11.49   The ice-cream mixture contains 34 g of sucrose for every 66 g of water. Use this fact in a unit-factor to figure out how much sucrose is present in 1000 g (1 kilogram) of water

$$m_{\text{sucrose}} = 1000 \text{ g water} \times \left(\frac{340 \text{ g sucrose}}{660 \text{ g water}}\right) = 515 \text{ g sucrose}$$

Divide this mass by 342.3 g mol^{-1}, the molar mass of sucrose, to convert to moles. The result is 1.50 mol. Since there is 1.50 mol of sucrose per 1000 g of water, the molality of the aqueous sucrose is 1.50 mol kg^{-1}. The change in the freezing point is

$$\Delta T = -K_f m = (-1.86 \text{ K kg mol}^{-1})(1.50 \text{ mol kg}^{-1}) = -2.8 \text{ K}$$

The freezing point of the mixture equals this change added to the freezing point of the pure solvent (0°C). It is $\boxed{-2.8°\text{C}}$.

As pure ice freezes out, the remaining solution becomes more and more concentrated in sucrose, and the $\boxed{\text{freezing point is depressed further}}$.

11.51   Calculate the effective molality from the change in the freezing point

$$m = -\frac{\Delta T}{K_f} = -\frac{(-4.218 \text{ K})}{1.86 \text{ K kg mol}^{-1}} = 2.268 \text{ mol kg}^{-1}$$

The ratio of the effective molality to the actual molality is 2.268/0.8402 = 2.70. Thus each Na_2SO_4 unit dissociates effectively into $\boxed{2.70 \text{ particles}}$.

Tip. This is less than the theoretical value of 3 (corresponding to two Na^+ and one SO_4^{2-} per formula unit) because the positive and negative ions in this rather concentrated solution tend to associate, reducing the effective number of particles in solution.

11.53   The osmotic pressure π of this solution is related to the concentration of the unknown solute by the equation

$$\pi = cRT$$

Solve this equation for c and insert the known values of the other quantities

$$c_{\text{unknown}} = \frac{\pi}{RT} = \frac{0.0105 \text{ atm}}{(0.08206 \text{ L atm mol}^{-1}\text{K}^{-1})(300 \text{ K})} = 4.265 \times 10^{-4} \text{ mol L}^{-1}$$

---

[4]The switch is legitimate because the problem concerns only a *change* in temperature. See problem **11.43**.

This solution was obtained by dissolving 200 mg (0.200 g) of the unknown in 25.0 mL of solution, a procedure that gives the same concentration as dissolving 8000 mg of solute in 1.00 L of water. Thus 8.00 g (8000 mg) of the unknown equals 4.265×10^{-4} mol. Accordingly, the molar mass is

$$\mathcal{M}_{\text{unknown}} = \frac{8.00 \text{ g}}{4.265 \times 10^{-4} \text{ mol}} = \boxed{1.88 \times 10^4 \text{ g mol}^{-1}}$$

11.55   Text Figure 11.22 shows the experimental set-up. The difference h between the level of the solution in the tube and the level outside the tube is proportional to the osmotic pressure of the solution. The problem gives h as 15.2 cm (0.152 m) of solution. To get the osmotic pressure, substitute the density ρ of the solution and the acceleration g of gravity in the formula $\pi = \rho g h$.[5] The density of the solution is 1.00 g cm^{-3}, which is equivalent to 1.00×10^3 kg m^{-3}. Then

$$\pi = \rho g h = (1.00 \times 10^3 \text{ kg m}^{-3})(9.807 \text{ m s}^{-2})(0.152 \text{ m}) = 1.49 \times 10^3 \text{ kg m}^{-1}\text{s}^{-2}$$

This equals 1.49×10^3 Pa.[6] Converting to atm

$$\pi = 1.49 \times 10^3 \text{ Pa} \times \left(\frac{1 \text{atm}}{101\,325 \text{ Pa}} \right) = 0.0147 \text{ atm}$$

Now, calculate the concentration of the polymer in the solution

$$c = \frac{\pi}{RT} = \frac{0.0147 \text{ atm}}{(0.08206 \text{ L atm mol}^{-1}\text{K}^{-1})(288.15 \text{ K})} = 6.22 \times 10^{-4} \text{ mol L}^{-1}$$

The solution holds 6.22×10^{-4} mol of polymer per liter. It also hold 4.64 g of polymer per liter. Therefore

$$\mathcal{M} = \frac{4.64 \text{ g}}{6.22 \times 10^{-4} \text{ mol}} = \boxed{7.46 \times 10^3 \text{ g mol}^{-1}}$$

Tip. It is possible to compute the concentration of the polymer without changing the pressure to atmospheres. Use SI units as follows

$$c = \frac{\pi}{RT} = \frac{1.49 \times 10^3 \text{ Pa}}{(8.3145 \text{ J K}^{-1}\text{mol}^{-1})(288.15 \text{ K})} = 0.622 \text{ mol m}^{-3}$$

This answer is the same because there are 1000 L in a cubic meter. Confirm the conversion of units using the equivalencies in text Appendix B.

Phase Equilibrium in Solutions: Volatile Solutes

11.57   **a)** The partial pressure of gaseous CO_2 above the aqueous solution of CO_2 is 5.0 atm. Henry's law relates the mole fraction of dissolved CO_2 to this partial pressure

$$P_{CO_2} = k_{CO_2} X_{CO_2} \qquad \text{from which} \qquad X_{CO_2} = \frac{P_{CO_2}}{k_{CO_2}} = \frac{5.00 \text{ atm}}{(1.65 \times 10^3 \text{ atm})} = 0.0030$$

This fraction means that there is 0.0030 mol of CO_2 in solution for every 0.9970 mol of water. But there is 55.5 mol of water per liter of solution if the solution (which is dilute) has the same density as water. Hence

$$c_{CO_2} = \frac{0.0030 \text{ mol } CO_2}{0.9970 \text{ mol } H_2O} \times \left(\frac{55.5 \text{ mol } H_2O}{1.00 \text{ L solution}} \right) = 0.17 \text{ mol L}^{-1}$$

The amount of CO_2 per liter is $\boxed{0.17 \text{ mol}}$.

---

[5]Note that this equation is identical to text equation 9.1, the formula for the "regular" barometric pressure exerted by a gas.
[6]Text Table B.2 in text Appendix B.

b) Before the cap is removed, gaseous CO_2 in the small space above the liquid is in equilibrium with the dissolved CO_2. The equilibrium is dynamic. This means that CO_2 molecules are constantly moving from the gas phase to the aqueous phase and back, and the rate of transfer of CO_2 out of solution equals the rate of transfer into solution. When the cap is removed, gaseous CO_2 escapes from the bottle because the partial pressure of CO_2 in the atmosphere is much less than 1 atm. Equilibrium is re-established with a far smaller concentration of CO_2 in the solution.

11.59   Determine the chemical amount of the methane that was dissolved in the 1.00 kg of solution before the boiling. Assume that all the methane was expelled and that the expelled gas behaves ideally

$$n_{CH_4} = \frac{PV}{RT} = \frac{(1.00 \text{ atm})(3.01 \text{ L})}{(0.08206 \text{ L atm mol}^{-1}\text{K}^{-1})(273.15 \text{ K})} = 0.1343 \text{ mol}$$

The molar mass of CH_4 is 16.04 g mol^{-1}, so the expelled methane has a mass of 2.154 g. The 1.00 kg of solution contained only water and methane. The mass of the water left after removal of methane is therefore 0.9978 kg. This mass of H_2O equals 55.39 mol of water.[7] Now, all the information needed to calculate the mole fraction of CH_4 is available

$$X_{CH_4} = \frac{n_{CH_4}}{n_{CH_4} + n_{H_2O}} = \frac{0.1343 \text{ mol}}{0.1343 + 55.39 \text{ mol}} = 0.002419$$

This fraction of CH_4 was present with 1.00 atm of CH_4 above the solution. Henry's law applies, and the Henry's law k equals

$$k_{CH_4} = \frac{P_{CH_4}}{X_{CH_4}} = \frac{1.00 \text{ atm}}{0.002419} = \boxed{413 \text{ atm}}$$

11.61   Write Raoult's law for both the benzene and toluene

$$P_{\text{benz}} = X_{\text{benz}}P^{\circ}_{\text{benz}} \qquad \text{and} \qquad P_{\text{tol}} = X_{\text{tol}}P^{\circ}_{\text{tol}}$$

Since equal numbers of moles of benzene and toluene were mixed, $X_{\text{benz}} = X_{\text{tol}} = 0.500$. Use these mole fractions and the vapor pressures of the pure substances to compute the partial pressure of each substance above the mixture

$$P_{\text{benz}} = 0.500(0.0987 \text{ atm}) = 0.04935 \text{ atm}$$
$$P_{\text{tol}} = 0.500(0.0289 \text{ atm}) = 0.01445 \text{ atm}$$

The number of moles of a particular gas in an ideal gaseous mixture is directly proportional to its partial pressure. Assume ideality. Then

$$n_{\text{benz}} = P_{\text{benz}}\frac{V}{RT} \qquad \text{and} \qquad n_{\text{tol}} = P_{\text{tol}}\frac{V}{RT}$$

The mole fraction of benzene in the vapor is

$$X_{\text{benz,vap}} = \frac{n_{\text{benz}}}{n_{\text{benz}} + n_{\text{tol}}} = \frac{P_{\text{benz}}(V/RT)}{P_{\text{benz}}(V/RT) + P_{\text{tol}}(V/RT)} = \frac{P_{\text{benz}}}{P_{\text{benz}} + P_{\text{tol}}}$$
$$= \frac{0.04935 \text{ atm}}{(0.04935 + 0.01445) \text{ atm}} = \boxed{0.774}$$

Tip. Half of the molecules in the liquid are benzene, but over three quarters of the molecules in the vapors above the liquid are benzene. Such enrichment in the more volatile component is the basis for separation by distillation.

---

[7]Using $\mathcal{M} = 18.0153$ g mol^{-1} for water.

11.63   **a)** Convert the masses of CCl_4 and $C_2H_4Cl_2$ to chemical amounts by dividing by their respective molar masses

$$n_{CCl_4} = \frac{30.0 \text{ g}}{153.82 \text{ g mol}^{-1}} = 0.1950 \text{ mol} \qquad n_{C_2H_4Cl_2} = \frac{20.0 \text{ g}}{98.96 \text{ g mol}^{-1}} = 0.2021 \text{ mol}$$

Compute the mole fraction of CCl_4 in the solution from these values

$$X_{CCl_4} = \frac{0.1950 \text{ mol}}{(0.1950 + 0.2021) \text{ mol}} = \boxed{0.491}$$

b) The total vapor pressure above the solution equals the sum of the partial pressures of the two components in the vapors above the solution. Raoult's law gives these partial pressures. Therefore

$$P_{tot} = P_{CCl_4} + P_{C_2H_4Cl_2} = X_{CCl_4}P^{\circ}_{CCl_4} + X_{C_2H_4Cl_2}P^{\circ}_{C_2H_4Cl_2}$$
$$= (0.491)(0.293 \text{ atm}) + (1 - 0.491)(0.209 \text{ atm}) = \boxed{0.250 \text{ atm}}$$

where the mole fraction of CCl_4 comes from the preceding part, and the vapor pressures of the pure components are given in the problem. The two mole fractions add up to 1 because there are only two components in the solution.

c) According to Dalton's law of partial pressures, the mole fraction of a component in a gaseous mixture equals its partial pressure divided by the total pressure

$$X_{CCl_4,vap} = \frac{P_{CCl_4}}{P_{tot}}$$

According to Raoult's law, the partial pressure of $CCl_4(g)$ in the vapor above the solution equals its mole fraction in the solution times its vapor pressure when pure

$$X_{CCl_4,vap} = \frac{P_{CCl_4}}{P_{tot}} = \frac{X_{CCl_4}P^{\circ}_{CCl_4}}{P_{tot}}$$

But $P^{\circ}_{CCl_4}$ is given in the problem and X_{CCl_4} was found in part **a)**. Substitution gives

$$X_{CCl_4,vap} = \frac{0.491(0.293 \text{ atm})}{0.250 \text{ atm}} = \boxed{0.575}$$

Tip. The mole fraction of CCl_4 is 0.491 in the solution but 0.575 in the vapors above the solution. As in problem **11.61**, the vapors are enriched in the more volatile component.

ADDITIONAL PROBLEMS

11.65   **a)** The element iodine exists in the Donovan's solution in different chemical forms. However, the total mass of iodine in the solution, which is prepared by mixing pure compounds that dissolve completely, depends only on the masses of the compounds and the fraction of each that iodine contributes. The total mass of iodine in 100 mL of Donovan's solution is thus the fraction by mass of elemental iodine in AsI_3 ($\mathcal{M} = 455.6$ g mol^{-1}) multiplied by the 1.00 g of AsI_3 plus the fraction by mass of iodine in HgI_2 ($\mathcal{M} = 454.4$ g mol^{-1}) multiplied by the 1.00 g of HgI_2

$$m_I = \left(\frac{(3)(126.9 \text{ g I})}{455.6 \text{ g AsI}_3}\right) 1.00 \text{ g AsI}_3 + \left(\frac{(2)(126.9 \text{ g I})}{454.4 \text{ g HgI}_2}\right) 1.00 \text{ g HgI}_2 = 1.39 \text{ g}$$

where 126.9 is the relative atomic mass of iodine. The mass of iodine per liter (which is 10 times 100 mL) of solution is 10 times this answer or $\boxed{13.9 \text{ g L}^{-1}}$.

b) The 0.100 M AsI_3 solution contains 45.56 g of AsI_3 per liter and therefore furnishes 4.556 g of AsI_3 per 100 mL. To make 3.50 L of Donovan's solution, 35.0 g of AsI_3 is needed. Measure out $(35.0/4.556) \times 100$ mL $= 768$ mL of the AsI_3 solution. Add to it 35.0 g of $HgI_2(s)$ and 31.5 g of $NaHCO_3(s)$. Then add enough water to bring the total volume to 3.50 L.

11.67   **a)** Bubbling a mixture of gases that contains the base ammonia through aqueous acetic acid causes the following reaction

$$\boxed{NH_3(g) + HOAc(aq) \longrightarrow NH_4^+(aq) + OAc^-(aq)}$$

The original solution contains the weak electrolyte acetic acid and no other electrolytes and is a poor conductor of electricity. The final solution contains a good concentration of ions and is a good conductor of electricity. The electrical conductivity $\boxed{\text{increases significantly}}$ as the gas is absorbed.

The concentrations of the ions in the final solution equal their chemical amounts divided by the volume of the solution. The acetic acid is "just neutralized" by the ammonia, so the chemical amounts of the two product ions equal each other

$$n_{NH_4^+} = (1.50 \text{ L})(0.200 \text{ mol L}^{-1}) = 0.300 \text{ mol} \qquad n_{OAc^-} = 0.300 \text{ mol}$$

The concentrations of the two in the final solution are also equal

$$c_{NH_4^+} = \frac{0.300 \text{ mol}}{1.50 \text{ L}} = \boxed{0.200 \text{ mol L}^{-1}} \qquad c_{OAc^-} = \boxed{0.200 \text{ mol L}^{-1}}$$

b) Compute the initial chemical amount of gases in the mixture in the flask

$$n_{\text{gas,initial}} = \frac{PV}{RT} = \frac{3.00 \text{ atm} \times 5.0 \text{ L}}{(0.08206 \text{ L atm mol}^{-1}\text{K}^{-1})300.2 \text{ K}} = 0.609 \text{ mol}$$

The pressure in the flask falls from 3.00 atm to 1.00 atm when some gas escapes. The amount that remains is $(1.00/3.00)(0.609)$ mol because n in the ideal gas law is directly proportional to P and neither T nor V changes. Therefore $(2.00/3.00)(0.609)$ mol of mixed gas escapes. The chemical amount of ammonia in the gas that escapes is readily computed because the ammonia neutralizes 0.300 mol of acetic acid

$$n_{NH_3} = (1.50 \text{ L})(0.200 \text{ mol HOAc L}^{-1}) \times \left(\frac{1 \text{ mol NH}_3}{1 \text{ mol HOAc}}\right) = 0.300 \text{ mol NH}_3$$

The escaped gas contained 0.300 mol of NH_3. The rest was nitrogen

$$n_{N_2} = n_{\text{gas}} - n_{NH_3} = \left(\frac{2.00}{3.00}\right)(0.609 \text{ mol}) - 0.300 \text{ mol} = 0.106 \text{ mol}$$

The mass fraction of NH_3 in the escaped gas is

$$\frac{0.300 \text{ mol}(17.0 \text{ g mol}^{-1})}{0.300 \text{ mol}(17.0 \text{ g mol}^{-1}) + 0.106 \text{ mol}(28.0 \text{ g mol}^{-1})} = 0.63$$

The mass percentage of NH_3 in the gas is therefore $\boxed{63\%}$.

11.69   The vanadium(III) sulfate (vanadic sulfate), a reducing agent, is oxidized according to the balanced half-equation

$$V_2(SO_4)_3(aq) + 8 \text{ OH}^-(aq) \longrightarrow 2 \text{ V(OH)}_4^+(aq) + 3 \text{ SO}_4^{2-}(aq) + 4 \text{ } e^-$$

Compute the chemical amount of X from the titration data together with a unit-factor from this half-equation

$$n_X = 15.0 \text{ mL} \times \left(\frac{0.200 \text{ mmol V}_2(SO_4)_3}{1 \text{ mL}}\right)\left(\frac{4 \text{ mmol } e^-}{1 \text{ mmol V}_2(SO_4)_3}\right)\left(\frac{1 \text{ mmol X}}{1 \text{ mmol } e^-}\right) = 12.0 \text{ mmol X}$$

The balanced half-equation provides the second factor in parentheses in this equation. The statement of the problem provides the first and the third. The molar mass of X is

$$\mathcal{M}_X = \frac{540 \text{ mg X}}{12.0 \text{ mmol}} = 45.0 \text{ mg mmol}^{-1} = \boxed{45.0 \text{ g mol}^{-1}}$$

If each molecule of X accepted three electrons, then the third factor in parentheses in the preceding would have a 3 in the denominator instead of a 1. This would reduce the chemical amount of X by a factor of 3 (from 12.0 mmol to 4.00 mmol) and would increase the molar mass of X by a factor of 3, to $\boxed{135 \text{ g mol}^{-1}}$.

11.71 The careful wording of the problem assures the reader that no Cl is lost at any point during the transformation $NaCl \longrightarrow Cl_2 \longrightarrow HCl$. For every mole of Cl present in the original 150 mL of 10.00% aqueous NaCl, one mole of Cl is formed in the 250 mL of HCl(aq). Compute this number of moles

$$n_{NaCl} = 150 \text{ mL solution} \times \left(\frac{1.0726 \text{ g sol'n}}{1 \text{ mL sol'n}} \right) \left(\frac{10.0 \text{ g NaCl}}{100 \text{ g sol'n}} \right) \left(\frac{1 \text{ mol NaCl}}{58.44 \text{ g NaCl}} \right) = 0.2753 \text{ mol}$$

The second term in parentheses exhibits the use of the mass percentage of NaCl in the solution as a unit-factor. The 0.2753 mol of NaCl implies 0.2753 mol of Cl, because the two elements exist in a 1 : 1 ratio in NaCl. All of the Cl ends up in the form of HCl, so 0.2753 mol of HCl is present in the 250 mL of solution that is formed. The concentration of the HCl is

$$c_{HCl} = \frac{0.2753 \text{ mol}}{0.250 \text{ L}} = \boxed{1.10 \text{ mol L}^{-1}}$$

11.73 The change in the vapor pressure ΔP is $0.3868 - 0.3914 = -0.0046$ atm. The mole fraction of the sulfur present in the sulfur-CS_2 system is therefore

$$X_{sulfur} = -\frac{\Delta P}{P^\circ_{CS_2}} = -\frac{-0.0046 \text{ atm}}{0.3914 \text{ atm}} = 0.0117$$

The solution contains 1.00 kg of CS_2 which is 13.13 mol of CS_2.[8] Hence

$$X_{sulfur} = 0.0117 = \frac{n_{sulfur}}{n_{sulfur} + n_{CS_2}} = \frac{n_{sulfur}}{n_{sulfur} + 13.13 \text{ mol}}$$

Solving for the chemical amount of sulfur gives 0.155 mol. Because this amount of sulfur is simultaneously 40.0 g of sulfur, the molar mass of sulfur as it exists in this solution is 40.0 g/0.155 mol = $\boxed{257 \text{ g mol}^{-1}}$. This is almost exactly eight times larger than 32 g mol^{-1}, the molar mass of S. The molecular formula of the sulfur in the solution must be $\boxed{S_8}$.

11.75 The soft drink is a solution of CO_2 (and other substances) in water. When the cap is on, gaseous CO_2 in the space above the fluid is held at a pressure exceeding 1 atm. Henry's law requires a higher concentration of dissolved CO_2 than if the pressure of CO_2 were only 1 atm. The dissolved CO_2 depresses the freezing point of the solution. When the cap is popped off, the partial pressure of CO_2 over the solution suddenly drops to far less than 1 atm. Gaseous CO_2 bubbles out of solution (tiny bubbles). The freezing point of the soft drink rises. If the new, higher freezing point exceeds the temperature of the solution, the solution will freeze.

11.77 Raoult's law allows calculation of the effective mole fraction of $CaCl_2$ in the solution at 25°C using the vapor-pressure lowering

$$X_2 = -\frac{P_1 - P_1^0}{P_1^0} = -\frac{(0.02970 - 0.03126) \text{ atm}}{0.03126 \text{ atm}} = 0.0499$$

---

[8] $\mathcal{M}_{CS_2} = 76.14 \text{ g mol}^{-1}$

$CaCl_2$ dissociates in water to form one mole of $Ca^{2+}(aq)$ cations and two moles of $Cl^-(aq)$ anions per mole dissolved. The X_2 just calculated is therefore not the true mole fraction of the solute, but is an *effective* mole fraction that is larger that the true mole fraction because of the dissociation of $CaCl_2$ into ions. Now, calculate the effective molality of the $CaCl_2$

$$m_{CaCl_2,\,eff} = \frac{0.0499 \text{ mol solute}}{(1.000 - 0.0499) \text{ mol H}_2\text{O}} \times \left(\frac{1 \text{ mol H}_2\text{O}}{0.018015 \text{ kg H}_2\text{O}}\right) = 2.92 \text{ mol kg}^{-1}$$

Put this effective molality into the usual formula for freezing-point depression

$$\Delta T = -K_f\, m_{CaCl_2,\,eff} = -(1.86 \text{ K kg mol}^{-1})(2.92 \text{ mol kg}^{-1}) = -5.43 \text{ K} = -5.43°\text{C}$$

Recall that the kelvin and the degree Celsius are equal in size. The freezing point of the solution is the original freezing point plus the change:

$$0.00°\text{C} + (-5.43°\text{C}) = \boxed{-5.43°\text{C}}$$

Tip. It is assumed that the effective number of particles of solute is unchanged by cooling from $25°\text{C}$, where the vapor pressure was recorded, to $-5.43°\text{C}$.

11.79 The salt $GaCl_2$ would be expected to dissociate in water according to

$$GaCl_2(aq) \longrightarrow Ga^{2+}(aq) + 2\,Cl^-(aq)$$

If the "$GaCl_2$" were actually $Ga[GaCl_4]$,[9] then the dissociation would be

$$Ga[GaCl_4](aq) \longrightarrow Ga^+(aq) + [GaCl_4]^-(aq)$$

In the first case, dissociation gives three ions; in the case, it gives only two ions. Measurement of a colligative property should distinguish between the two cases. For example, imagine that enough compound is dissolved in water to make a solution that is $0.0100 \text{ mol kg}^{-1}$ in $GaCl_2$. This solution would have a freezing point of $-0.056°\text{C}$ if the formula $GaCl_2$ were correct. This freezing point is predicted using an effective molality of $0.0300 \text{ mol kg}^{-1}$ in the formula for freezing-point depression. The effective molality is triple m_{GaCl_2} because three moles of ions are formed by dissociation of one mole of $GaCl_2$. Now try the formula $Ga[GaCl_4]$. The identical aqueous solution has a $m_{Ga[GaCl_4]}$ of $0.00500 \text{ mol kg}^{-1}$, because the new formula for the solute corresponds to a molar mass that is twice as large, and an effective molality of $0.0100 \text{ mol kg}^{-1}$, because two ions are formed upon dissociation. The predicted freezing point of $-0.0186°\text{C}$ differs measurably.

11.81 Determine the effective molality of the NaCl solution that freezes at $-0.406°\text{C}$

$$m_{NaCl} = -\frac{\Delta T_f}{K_f} = -\frac{-0.406 \text{ K}}{1.86 \text{ K kg mol}^{-1}} = 0.218 \text{ mol kg}^{-1}$$

This is also the effective molality of the contents of the red blood cell because the cells neither swell nor shrink. Assume that the molarity and molality are equal and so use 0.218 M to calculate the osmotic pressure

$$\pi = c\,RT = (0.218 \text{ mol L}^{-1})(0.08206 \text{ L atm mol}^{-1}\text{K}^{-1})(298.15 \text{ K}) = \boxed{5.33 \text{ atm}}$$

11.83 Write Henry's law for a solution of benzene in water

$$P_{benz} = k_{benz}X_{benz} = (301 \text{ atm})X_{benz}$$

---

[9]Brackets are often used to set off complex ions. See text Section 8.2

The chemical amount of benzene in the solution described in the problem is 0.0256 mol, obtained by dividing 2.0 g by 78.11 g mol^{-1}, the molar mass of benzene. The mole fraction of benzene is

$$X_{\text{benz}} = \frac{0.0256 \text{ mol}}{0.0256 \text{ mol} + (55.5 \times 10^3) \text{ mol}} = 4.6 \times 10^{-7}$$

The large amount of water drowns out the contribution of the benzene to the denominator of this fraction. Insert the mole fraction of benzene and the given Henry's law constant into the equation for Henry's law

$$P_{\text{benz}} = k_{\text{benz}} X_{\text{benz}} = (301 \text{ atm})(4.6 \times 10^{-7}) = \boxed{1.4 \times 10^{-4} \text{ atm}}$$

Then substitute this pressure and the temperature in kelvins into the rearranged ideal-gas equation

$$c_{\text{benz}} = \frac{n_{\text{benz}}}{V} = \frac{P_{\text{benz}}}{RT} = \frac{1.4 \times 10^{-4} \text{ atm}}{(0.082057 \text{ L atm K}^{-1}\text{mol}^{-1})(298.15 \text{ K})} = 5.7 \times 10^{-6} \text{ mol L}^{-1}$$

This result is the concentration of the benzene in the vapor above the solution. Convert to molecules per cubic centimeter (which is a *number density*) as follows

$$\left(\frac{N}{V}\right)_{\text{benz}} = \frac{5.7 \times 10^{-6} \text{ mol}}{1 \text{ L}} \times \left(\frac{1 \text{ L}}{1000 \text{ cm}^3}\right)\left(\frac{6.022 \times 10^{23} \text{ molecule}}{1 \text{ mol}}\right) = \boxed{3.4 \times 10^{15} \frac{\text{molecule}}{\text{cm}^3}}$$

11.85 The difference between a solution and a colloidal suspension lies with the size of the dispersed particles. In a solution, the solute is dispersed at the molecular (or ionic) level. Each particle is surrounded by a cage of several solvent molecules. Examples are solutions of NaCl or alcohol in water. In a colloidal suspension, the dispersed particles are aggregates of hundreds to thousands of solute molecules. The aggregates are frequently surrounded by interacting solvent molecules that prevent them from sticking together to form a visible precipitate. The particles do not settle on the bottom of the container because the agitation caused by collisions of neighboring molecules is strong enough to keep them up. An example of a colloid is homogenized milk. The white opacity of milk is caused by tiny particles of fat that are too small to be filtered. In some cases, it is difficult to classify a mixture definitively as solution or suspension. If the particles are aggregates of only small numbers of molecules, the properties of the mixture will be similar to those of a solution, but deviate somewhat toward those of a colloidal suspension.

11.87 First obtain the empirical formula of the compound. Compute the chemical amounts of C and H in the sample

$$n_{\text{C}} = 5.46 \text{ g CO}_2 \times \left(\frac{1 \text{ mol CO}_2}{44.0 \text{ g CO}_2}\right)\left(\frac{1 \text{ mol C}}{1 \text{ mol CO}_2}\right) = 0.1241 \text{ mol C}$$

$$n_{\text{H}} = 2.23 \text{ g H}_2\text{O} \times \left(\frac{1 \text{ mol H}_2\text{O}}{18.015 \text{ g H}_2\text{O}}\right)\left(\frac{2 \text{ mol H}}{1 \text{ mol H}_2\text{O}}\right) = 0.2476 \text{ mol H}$$

These amounts correspond to 0.2495 g of H and 1.490 g of C, so the mass of oxygen in the combustion sample equals

$$m_{\text{O}} = m_{\text{tot}} - m_{\text{H}} - m_{\text{C}} = (2.40 - 0.2495 - 1.490) \text{ g} = 0.6605 \text{ g}$$

The chemical amount of O is

$$n_{\text{O}} = 0.6605 \text{ g O} \times \left(\frac{1 \text{ mol O}}{15.9994 \text{ g O}}\right) = 0.0413 \text{ mol}$$

The three elements are present in the molar ratio C$_{0.124}$H$_{0.248}$O$_{0.0413}$, which gives the empirical formula C$_3$H$_6$O.

The observed depression of the freezing point gives the molality of the solution

$$m = -\frac{\Delta T_f}{K_f} = -\frac{-0.97 \text{ K}}{1.86 \text{ K kg mol}^{-1}} = 0.522 \text{ mol kg}^{-1}$$

The 0.281 kg of solvent therefore contains 0.146 mol of the compound. The mass of this 0.146 mol of compound equals 8.69 g. Hence the molar mass of the compound is approximately 59 g mol^{-1}. The molecular formula is clearly $\boxed{C_3H_6O}$, which has a molar mass of 58.08 g mol^{-1}.

Chapter 12

Thermodynamic Processes and Thermochemistry

The First Law of Thermodynamics: Internal Energy, Work, and Heat

12.1 The work done *on* a gas in a change of volume at constant pressure is given by $w = -P_{ext}\Delta V$.[1] The problem gives a value for the external pressure and values for the final and initial volumes. Substitute them into the equation to obtain

$$w = -P_{ext}\Delta V = -(50.0 \text{ atm})(974 \text{ L} - 542 \text{ L}) = -2.16 \times 10^4 \text{ L atm}$$

As ever, the change in a quantity (in this case the volume) is the final value minus the initial. To convert to joules, multiply by the proper unit factor

$$w = -2.16 \times 10^4 \text{ L atm} \times \left(\frac{101.325 \text{ J}}{1 \text{ L atm}}\right) = \boxed{-2.19 \times 10^6 \text{ J}}$$

Tip. -2.19×10^6 J of work is performed on the nitrogen by the surroundings; $+2.19 \times 10^6$ J is performed by the surroundings on the nitrogen. The difference in sign indicates a difference in point of view. Does one sit with the nitrogen looking at the surroundings or in the surroundings looking at the nitrogen?

12.3 A ball of mass m falls a distance Δh under the influence of gravity. It experiences a change in potential energy equal to $mg\Delta h$, where Δh is the change in height and g is the acceleration of gravity. The ball stops dead when it hits the ground (it may be made of clay). According to the problem, the total energy of the ball does not change at impact. All of the potential energy instead is converted into internal energy. This energy goes to heat up the ball. This can be expressed mathematically as

$$mc_s\Delta T + Mg\Delta h = 0$$

where c_s is the specific heat capacity of the ball. Cancel out the m's and solve for Δh

$$\Delta h = -\frac{c_s\Delta T}{g}$$

In this problem, ΔT equals 1.00°C (which equals 1.00 K) and c_s equals 0.850×10^3 J K^{-1} kg^{-1}. Note that c_s is put on a per-kilogram basis to aid the cancellation of units. Also, g is 9.81 m s^{-2}. Substituting gives

$$\Delta h = -\frac{(0.850 \times 10^3 \text{ J K}^{-1} \text{ kg}^{-1})(1.00 \text{ K})}{9.81 \text{ m s}^{-2}} = -86.6 \text{ J kg}^{-1}\text{m}^{-1}\text{s}^2$$

---

[1]This is text equation 12.1.

126

By its definition a joule equals a kg m^{2}s^{-2}. Therefore, in the above cluster of units all but the meter cancel out: Δh is -86.6 m. The negative sign means that the final height of the ball is less than the initial height. The ball falls down (not up) a distance of $\boxed{86.6 \text{ m}}$.

12.5   The molar heat capacity of a substance equals its specific heat capacity multiplied by its molar mass. Here is the calculation of this quantity for lithium

$$c_\text{P} = c_\text{s}\mathcal{M} = 3.57 \text{ J K}^{-1}\text{g}^{-1}(6.94 \text{ g mol}^{-1}) = 24.8 \text{ J K}^{-1}\text{mol}^{-1}$$

The full set of values in the group

Element	Li(s)	Na(s)	K(s)	Rb(s)	Cs(s)
c_P / J K^{-1}mol^{-1}	24.8	28.3	29.6	31.0	32.2

Beyond sodium there is a steady increase of about 1.3 J K^{-1}mol^{-1} for every element. Extrapolation of the trend assigns francium a molar heat capacity of about $\boxed{33.5 \text{ J K}^{-1}\text{mol}^{-1}}$.

Tip. The periodic trend is distinct enough, but small. Indeed, the molar heat capacities of the metallic elements are all pretty close to 25 J K^{-1}g^{-1}. This is the law of Dulong and Petit (see problem **12.7**).

12.7   Again, the molar heat capacity of a substance equals its specific heat capacity multiplied by its molar mass. The calculations proceed as in problem **12.5** with these results

Element	Ni(s)	Zn(s)	Rh(s)	W(s)	Au(s)	U(s)
c_P / J K^{-1}mol^{-1}	26.1	25.4	25.0	24.3	25.4	27.6

12.9   **a)** During the heating step, heat flows into the system. Therefore, $\boxed{q_\text{sys} > 0}$. The container is rigid. It can neither expand nor contract. Hence $\Delta V_\text{sys} = 0$. Consequently, no pressure-volume work is performed on the system. No other type of work is possible, so $\boxed{w_\text{sys} = 0}$. Finally, by the first law, $\boxed{\Delta U_\text{sys} > 0}$.

b) During the cooling step, heat flows out of the system $\boxed{q_\text{sys} < 0}$. Again, no pressure-volume work can be performed on the system: $\boxed{w_\text{sys} = 0}$. By the first law, the internal energy of the system is lowered $\boxed{\Delta U_\text{sys} < 0}$.

c) No work was absorbed in either step 1 or step 2. Hence

$$\boxed{(w_{\text{sys}, 1} + w_{\text{sys}, 2}) = 0}$$

Positive heat is absorbed in step 1 and negative heat is absorbed in step 2. Nothing in the problem indicates that the system ends up in its original thermodynamic state after being cooled back to its original temperature. The final state could have more internal energy than the original state or less or it could even have the same internal energy.

$$\boxed{\text{The sign of } (\Delta U_{\text{sys}, 1} + \Delta U_{\text{sys}, 2}) \text{ cannot be determined}}$$

Combine a statement of the first law for the system with the now-established fact that the system absorbs zero work

$$(\Delta U_{\text{sys}, 1} + \Delta U_{\text{sys}, 2}) = (q_{\text{sys}, 1} + q_{\text{sys}, 2}) + (w_{\text{sys}, 1} + w_{\text{sys}, 2})$$
$$(\Delta U_{\text{sys}, 1} + \Delta U_{\text{sys}, 2}) = (q_{\text{sys}, 1} + q_{\text{sys}, 2}) + 0$$

$$\boxed{\text{The sign of } (q_{\text{sys}, 1} + q_{\text{sys}, 2}) \text{ cannot be determined}}.$$

Tip. A trap in this problem is to assume, without justification, that the material in the container is an ideal gas (for which the internal energy depends only on the temperature). A related trap is to assume that any changes brought on by the rise in temperature are exactly reversed by the drop in temperature. This is not true when an egg is boiled and re-cooled. Why should it be true here?

Heat Capacity, Enthalpy, and Calorimetry

12.11 Imagine the system under consideration to consist of two sub-systems: the metal and the water. If the mixing of hot metal and cool water takes place inside a well-insulated container (which prevents leaks of heat), then the heat absorbed by the system equals zero. The system is the sum of the two sub-systems. Therefore

$$q_{sys} = 0 = q_{metal} + q_{water}$$

For both sub-systems, the amount of heat gained equals the specific heat capacity times the mass times the temperature change

$$q_{metal} + q_{water} = m_{water}c_{s,water}\Delta T_{water} + m_{metal}c_{s,metal}\Delta T_{metal} = 0$$

Solving for the specific heat capacity of the metal:

$$c_{s,metal} = \frac{-m_{water}c_{s,water}\Delta T_{water}}{m_{metal}\Delta T_{metal}} = \frac{-(100.0 \text{ g}) \, 4.18 \text{ J K}^{-1}\text{g}^{-1}(6.39°\text{C})}{(61.0 \text{ g})(-93.61°\text{C})} = \boxed{0.468 \text{ J K}^{-1}\text{g}^{-1}}$$

Tip. Don't bother to convert °C to K. The Kelvin and Celsius scales differ only in the location of their zero points. A temperature *change* of 1°C is identical to a *change* of 1 K.

12.13 Body 1 and body 2 are originally at different temperatures. They are brought into thermal contact with each other and held in thermal isolation from other objects. Then

$$q_1 + q_2 = m_1 c_{s1}\Delta T_1 + m_2 c_{s2}\Delta T_2 = 0$$

If the masses of the two bodies are equal, then $m_1 = m_2$, and

$$c_{s1}\Delta T_1 = -c_{s2}\Delta T_2 \qquad \text{from which} \qquad \boxed{\frac{c_{s1}}{c_{s2}} = -\frac{\Delta T_2}{\Delta T_1}}$$

The last equation shows that the specific heat capacities of the two bodies are inversely proportional to the temperature changes they undergo in this experiment.

Tip. The minus sign in the answer reflects the fact that the ΔT's of body 1 and body 2 are always of opposite signs; one warms up while the other cools down.

12.15 The difference in temperature ΔT between water at its boiling point and melting point is 100°C. The heat needed to bring 1.00 g of water at 0°C to 100°C equals

$$q = mc_s\Delta T = (1.00 \text{ g})\left(4.18 \text{ J} \, (°\text{C})^{-1}\text{ g}^{-1}\right)\left((100.00 - 0.00)°\text{C}\right) = 418 \text{ J}$$

The amount of heat needed to melt 1.00 g of ice is, according to the statement of Lavoisier and Laplace, 3/4 of this amount or $\boxed{314 \text{ J}}$. More recent experiments set the amount of heat to melt 1.00 g of ice at 333 J.

Illustrations of the First Law of Thermodynamics in Ideal Gas Processes

12.17 The 0.500 mol of neon expands against a constant pressure of 0.100 atm. Neon is a monatomic gas. Assume that it is also an ideal gas. Before the expansion, the volume of the neon (which is the system) is 11.20 L (calculated using the ideal-gas equation with n equal 0.500 mol at 1.00 atm and 273 K). The expanded volume is 43.08 L (calculated from the ideal-gas equation with $P = 0.200$ atm,

$n = 0.500$ mol, and $T = 210$ K). The gas expands against a constant pressure (of 0.100 atm). The work done on the neon is

$$w = -P_{\text{ext}} \Delta V = -0.100 \text{ atm}(43.08 - 11.20) \text{ L} = \boxed{-3.19 \text{ L atm}}$$

The neon cools from 273 to 210 K. Since it is an ideal monatomic gas, the change in its internal energy is directly proportional to the change in its temperature; the constant of proportionality is $n(\frac{3}{2})R$, the heat capacity at constant volume

$$\Delta U = nc_v \Delta T = n \left(\frac{3}{2} R \right) \Delta T$$

Substituting gives

$$\Delta U = 0.500 \text{ mol} \left(\frac{3}{2} 0.08206 \text{ L atm mol}^{-1} \text{K}^{-1} \right) (-63 \text{ K}) = \boxed{-3.88 \text{ L atm}}$$

By the first law:

$$q = \Delta U - w = -3.88 \text{ L atm} - (-3.19 \text{ L atm}) = \boxed{-0.69 \text{ L atm}}$$

The three answers can also be given in joules (1 L atm = 101.325 J)

$$w = -323 \text{ J} \qquad \Delta U = -393 \text{ J} \qquad q = -70 \text{ J}$$

12.19 **a)** The statement of the problem gives the initial amount (2.00 mol), pressure (3.00 atm), and temperature (350 K) of the ideal monatomic gas. The initial volume of the gas is $V = nRT/P = 19.15$ L. The final volume is *twice* this original volume or $\boxed{38.3 \text{ L}}$. The change in volume ΔV is $38.30 - 19.15 = 19.15$ L.

b) The adiabatic expansion occurs against a *constant* pressure of 1.00 atm. Under that circumstance, the work done on the gas is

$$w = -P\Delta V = -1.00(19.15) \text{ L atm} \times \left(\frac{101.325 \text{ J}}{1 \text{ L atm}} \right) = \boxed{-1.94 \times 10^3 \text{ J}}$$

The expansion is adiabatic so $\boxed{q = 0}$ by definition, and

$$\Delta U = q + w = 0 - 1.94 \times 10^3 \text{ J} = \boxed{-1.94 \times 10^3 \text{ J}}$$

c) Any change in the internal energy of an ideal gas causes a change in temperature in direct proportion

$$\Delta U = nc_v \Delta T$$

Solve for ΔT and substitute the various values

$$\Delta T = \frac{\Delta U}{nc_v} = \frac{-1.94 \times 10^3 \text{ J}}{2.00 \text{ mol } (3/2) 8.3145 \text{ J K}^{-1}\text{mol}^{-1}} = -77.8 \text{ K}$$

Thus, T_2, the final temperature, is $T_1 + \Delta T = 350 + (-77.8) = \boxed{272 \text{ K}}$.

12.21 The system consists of the 6.00 mol of argon. The "change in energy" means the change in internal energy. For this monatomic gas (assuming ideality) it is

$$\Delta U = nc_v \Delta T = (6.00 \text{ mol}) \left(\frac{3}{2} 8.3145 \text{ J K}^{-1}\text{mol}^{-1} \right) (150 \text{ K}) = \boxed{+11.2 \times 10^3 \text{ J}}$$

The change is adiabatic which means that $q = 0$. From the first law

$$w = \Delta U - q = 11.2 \times 10^3 \text{ J} - 0 = \boxed{+11.2 \times 10^3 \text{ J}}$$

The work done on the argon is $\boxed{11.2 \times 10^3 \text{ J}}$, *all* of which goes to increase its internal energy.

Molecular Contributions to Internal Energy and Heat Capacity

12.23   Diatomic molecules (such as those of $HI(g)$ and $I_2(g)$) possess three degrees of freedom in their translational motion, two degrees of freedom in their rotational motion, and one degree of freedom in their vibrational motion. Each translational and rotational degree of freedom contributes $R/2$ to the heat capacity of the gas; each vibrational degree of freedom contributes $(2R/2)$. The predicted heat capacity of both $HI(g)$ and $I_2(g)$ is

$$c_V = \underbrace{3 \times (R/2)}_{\text{translation}} + \underbrace{2 \times (R/2)}_{\text{rotation}} + \underbrace{1 \times (2R/2)}_{\text{vibration}} = 7(R/2)$$

This result is shown in text Table 12.3 for several diatomic molecules. It applies at constant *volume*. At constant *pressure* (and assuming idea-gas behavior), an additional $2(R/2)$ of heat capacity becomes available because the container can expand. The predicted heat capacity at constant pressure for the two gases is therefore $9(R/2)$. That is,

$$c_P = \frac{9}{2}(8.3145 \text{ J K}^{-1}\text{mol}^{-1}) = \boxed{37.41 \text{ J K}^{-1}\text{mol}^{-1}}$$

The experimental c_P's of $I_2(g)$ and $HI(g)$ at 298 K differ from this prediction. According to text Appendix D, they are 29.16 and 36.90 J K^{-1}mol^{-1} respectively. The following table shows the disagreement in terms of R

Gas	Predicted c_P	Measured c_P at 298 K	Difference
HI	$4.500\,R$	$3.507\,R$	$0.993\,R$
I$_2$	$4.500\,R$	$4.438\,R$	$0.062\,R$

The disagreement arises because some modes of motion become inaccessible for the storage of thermal energy when the temperature is too low. Vibrational motions require higher temperatures for access than do rotational and translational motions. In the case of $HI(g)$, the vibrational mode is almost completely inaccessible at 298 K. It contributes only $0.007\,R$ of the experimentally measured $3.507\,R$. This is just $\boxed{0.2\%}$ of the measured value. The vibrational mode of $HI(g)$ is almost completely "frozen out" at 298 K. The vibrational mode in $I_2(g)$ is more accessible at this temperature. It contributes $0.938\,R$, which is $\boxed{21.1\%}$ of the measured c_P.

Tip. Lower temperatures freeze out $I_2(g)$'s vibrational mode as well. Very low temperatures freeze out the vibrational modes *and* the rotational modes of both gases, reducing their c_P's to $5R/2$ if they do not liquefy them first.

12.25   **a)** Argon is a monatomic gas. Its atoms have just three degree of freedom, all translational. Theory predicts $c_V = 3(R/2)$ and $c_P = 5(R/2)$ if the argon behaves ideally. Then

$$\Delta H = nc_P \Delta T = (2.00 \text{ mol}) \left(\frac{5}{2}(8.3145 \text{ J K}^{-1}\text{mol}^{-1}) \right)(573 \text{ K} - 298 \text{ K}) = \boxed{11.4 \times 10^3 \text{ J}}$$

b) Molecules of ethylene (C_2H_4) have six atoms and a total of $3 \times 6 = 18$ modes of motion. See text Table 12.3. The predicted heat capacity of gaseous ethylene at constant volume, its c_V, is

$$c_V = \underbrace{3 \times (R/2)}_{\text{translation}} + \underbrace{3 \times (R/2)}_{\text{rotation}} + \underbrace{12 \times (2R/2)}_{\text{vibration}} = 30(R/2)$$

Assuming ideal-gas behavior, c_p exceeds c_V by $2(R/2)$. Therefore

$$\Delta H = nc_p\Delta T = (2.00 \text{ mol})\left(\frac{32}{2}\, 8.3145 \text{ J K}^{-1}\text{mol}^{-1}\right)(573 \text{ K} - 298 \text{ K})$$

$$= (2.00 \text{ mol})(133.0 \text{ J K}^{-1}\text{mol}^{-1})(275 \text{ K}) = \boxed{73.2 \times 10^3 \text{ J}}$$

Tip. Appendix D lists the experimental c_P of gaseous argon at 298 K as 20.79 J K^{-1}mol^{-1}. This equals the value obtained and used in part **a)**. The experimental c_P of gaseous ethylene at 298 K is 43.56 J K^{-1}mol^{-1}. This is only about 40% of 133.0 J K^{-1}mol^{-1}, the c_P obtained and used in part **b)**. As with hydrogen iodide in problem **12.23**, the vibrational modes of ethylene are substantially frozen out, inaccessible for the storage of energy, at 298 K.

Thermochemistry

12.27 The three balanced equations tell the enthalpy change taking place during the production or consumption of specific numbers of moles of product or reactant. Put these enthalpy changes on a basis of mass.

a)
$$\Delta H = \frac{-828 \text{ kJ}}{2 \text{ mol Na}_2\text{O}} \times \left(\frac{1.00 \text{ mol Na}_2\text{O}}{62.0 \text{ g Na}_2\text{O}}\right) = \boxed{-6.68 \text{ kJ g}^{-1}}$$

b)
$$\Delta H = \frac{302 \text{ kJ}}{1 \text{ mol MgO}} \times \left(\frac{1.00 \text{ mol MgO}}{40.31 \text{ g MgO}}\right) = \boxed{+7.49 \text{ kJ g}^{-1}}$$

c)
$$\Delta H = \frac{33.3 \text{ kJ}}{2 \text{ mol CO}} \times \left(\frac{1.00 \text{ mol CO}}{28.01 \text{ g CO}}\right) = \boxed{+0.594 \text{ kJ g}^{-1}}$$

12.29 Only 119.0 J of the measured 121.3 J of heat comes from the reaction of the 0.00288 mol of $Br_2(l)$. The rest of the heat (2.34 J) is added mechanically[2] during the course of breaking the capsule and stirring the liquid. Dissolution of 1.00 mol of bromine would evolve considerably more heat

$$q_{\text{evolved}} = 1.00 \text{ mol} \times \left(\frac{119.0 \text{ J}}{2.88 \times 10^{-3} \text{ mol}}\right) = \boxed{41.3 \times 10^3 \text{ J}}$$

Tip. Recall that ΔH of a process equals the amount of heat that is *absorbs* as it proceeds at constant pressure. The ΔH for the dissolution reaction is therefore -41.3×10^3 J mol^{-1}.

12.31 Represent the vaporization as $CO(l) \longrightarrow CO(g)$. Table 12.4 lists ΔH_{vap} as 6.04 kJ mol^{-1}. The following series of unit-factors then provides the answer

$$\Delta H = 2.38 \text{ g CO} \times \left(\frac{1 \text{ mol CO}}{28.01 \text{ g CO}}\right) \times \left(\frac{6.04 \text{ kJ}}{1 \text{ mol CO}}\right) = \boxed{0.513 \text{ kJ}}$$

12.33 A 36.0 g ice cube is also a 2.00 mol ice cube because 18.0 g of H_2O equals 1.00 mol of H_2O. The ice cube is put in contact with 360 g (20.0 mol) of 20°C water. At −10°C, the ice is well below its melting point. It must be warmed up before it can start to melt. Warming the ice from −10°C to 0°C requires an input of heat. call it q_1

$$q_1 = nc_p\Delta T = (2.00 \text{ mol})(38 \text{ J K}^{-1}\text{mol}^{-1})(273.15 \text{ K} - 263.15 \text{ K}) = 760 \text{ J}$$

---

[2]See text Figure 12.8.

Melting the ice at 0°C to obtain water at 0°C also requires heat

$$q_2 = n\Delta H_{\text{fus}} = (2.00 \text{ mol})(6007 \text{ J mol}^{-1}) = 12\,014 \text{ J}$$

On the other hand, taking 360 g (20.0 mol) of water from 20°C to 0°C requires cooling, which is the same as an input of negative heat:

$$q_3 = nc_{\text{p}}(T_{\text{f}} - T_{\text{i}}) = (20.0 \text{ mol})(75 \text{ J K}^{-1}\text{mol}^{-1})(-20 \text{ K}) = -30\,000 \text{ J}$$

Notice that q_3 exceeds the *sum* of q_1 and q_2 in magnitude. This means that the 360 g of warm water never cools down to freezing because all of the ice melts first. The same thing happens with any small ice-cube in a big glassful of warm water. When all changes are finished this system must consist entirely of liquid at a final temperature T_{f} somewhere between 0° and 20°C.

Assume that no heat is lost to or gained from the surroundings. Then

$$q_1 + q_2 + q_4 + q_5 = 0$$

The new quantity q_4 equals the heat absorbed in warming the 2.00 mol of melt-water from 0°C to to T_{f} and the new quantity q_5 equals the heat absorbed in cooling the 20.0 mol of water to T_{f}. The two new q's depend on the final temperature as follows

$$q_4 = (2.00 \text{ mol})(75 \text{ J K}^{-1}\text{mol}^{-1})(T_{\text{f}} - 0)$$
$$q_5 = (20.0 \text{ mol})(75 \text{ J K}^{-1}\text{mol}^{-1})(T_{\text{f}} - 20.0)$$

q_1 and q_2 are already known. Doing the addition gives

$$760 \text{ J} + 12\,014 \text{ J} + (2.00 \text{ mol})(75 \text{ J K}^{-1}\text{mol}^{-1})(T_{\text{f}} - 0)$$
$$+ (20.0 \text{ mol})(75 \text{ J K}^{-1}\text{mol}^{-1})(T_{\text{f}} - 20.0) = 0$$

Solving gives $T_{\text{f}} = \boxed{10.4°\text{C}}$.

Tip. Avoid the wrong concept that ice is always "ice-cold" (has a temperature of 0°C). Like all other materials, ice comes to the temperature of its surroundings.

12.35 Reverse the chemical equation for the combustion of ketene. The ΔH of "un-combustion" of ketene is −1 times the ΔH of combustion.

$$2 CO_2(g) + H_2O(g) \longrightarrow CH_2CO(g) + 2 O_2(g) \qquad \Delta H_1 = 981.1 \text{ kJ}$$

Double the coefficients in the equation for the combustion of methane. The ΔH of the resulting bigger combustion is 2 times the ΔH of the original

$$2 CH_4(g) + 4 O_2(g) \longrightarrow 2 CO_2(g) + 4 H_2O(g) \qquad \Delta H_2 = -1604.6 \text{ kJ}$$

Adding these two equations gives the desired equation:

$$2 CH_4(g) + 2 O_2(g) \longrightarrow CH_2CO(g) + 3 H_2O(g) \qquad \Delta H_3 = \boxed{-623.5 \text{ kJ}}$$

$\Delta H_3 = \Delta H_1 + \Delta H_2$ by Hess's law.

Tip. How do you know which equations to reverse or double in problems like this? Manipulate to put the correct number of moles of each substance on the correct side of the final equation. Thus, the ketene equation had to be reversed because ketene appears among the products in the target equation.

12.37 The conversion $C(gr) \longrightarrow C(dia)$ is endothermic (positive ΔH). Therefore, one pound of diamonds contains more enthalpy than one pound of graphite. Both diamond and graphite give the same product (gaseous carbon dioxide) when burned. When burned, the pound of $\boxed{\text{diamonds}}$ will give off more heat.

12.39 A standard enthalpy of reaction is calculated by summing the standard enthalpies of formation of the products and subtracting the standard enthalpies of formation of the reactants

$$N_2H_4(l) + 3\,O_2(g) \longrightarrow 2\,NO_2(g) + 2\,H_2O(l)$$

$$\Delta H^\circ = 2 \underbrace{(33.18)}_{NO_2(g)} + 2 \underbrace{(-285.83)}_{H_2O(l)} - 1 \underbrace{(50.63)}_{N_2H_4(l)} - 3 \underbrace{(0)}_{O_2(g)} = \boxed{-555.93 \text{ kJ}}$$

In the preceding equation, all of the ΔH_f°'s were found in Appendix D and are in kJ mol^{-1}. Each is multiplied by the number of moles of the substance represented in the balanced equation.

Tip. This procedure is not restricted to ΔH°'s (standard enthalpy changes). It also works for the ΔH's of reactions involving substances in non-standard states. However enthalpies of formation for substances in non-standard states are rarely tabulated.

12.41 **a)** As in problem **12.39**

$$2\,ZnS(s) + 3\,O_2(g) \longrightarrow 2\,ZnO(s) + 2\,SO_2(g)$$

$$\Delta H^\circ = 2 \underbrace{(-348.28)}_{ZnO(s)} + 2 \underbrace{(-296.83)}_{SO_2(g)} - 2 \underbrace{(-205.98)}_{ZnS(s)} - 3 \underbrace{(0)}_{O_2(g)} = \boxed{-878.26 \text{ kJ}}$$

b) Compute the chemical amount of ZnS (in moles) and multiply it by the molar ΔH° to get the amount of heat absorbed in the roasting of the 3.00 metric tons of ZnS. It is known that 2 mol of ZnS(s) has a ΔH° of -878.26 kJ. Hence

$$q_p = \Delta H^\circ = 3.00 \text{ ton ZnS} \times \left(\frac{10^6 \text{ g}}{\text{ton}}\right)\left(\frac{1 \text{ mol}}{97.456 \text{ g}}\right)\left(\frac{-878.26 \text{ kJ}}{2 \text{ mol ZnS}}\right) = \boxed{-1.35 \times 10^7 \text{ kJ}}$$

12.43 **a)** The balanced equation is $CaCl_2(s) \longrightarrow Ca^{2+}(aq) + 2\,Cl^-(aq)$. Combine the standard enthalpies of formation as follows:

$$\Delta H^\circ = 2 \underbrace{(-167.16)}_{Cl^-(aq)} - 1 \underbrace{(542.83)}_{Ca^{2+}(aq)} - 1 \underbrace{(-795.8)}_{CaCl_2(s)} = \boxed{-81.4 \text{ kJ}}$$

b) Compute ΔH° for the dissolution of 20.0 g of $CaCl_2(s)$

$$\Delta H^\circ = 20.0 \text{ g CaCl}_2 \times \left(\frac{1 \text{ mol CaCl}_2}{110.98 \text{ g CaCl}_2}\right)\left(\frac{-81.35 \text{ kJ}}{1 \text{ mol CaCl}_2}\right) = -14.66 \text{ kJ}$$

The process of dissolution absorbs -14.66 kJ. The immediate surroundings of the dissolution (the water) therefore must absorb $+14.66$ kJ. The temperature change of the water equals the heat it absorbs divided by its heat capacity

$$\Delta T = \frac{q}{c_p M} = \frac{14.66 \times 10^3 \text{ J}}{418 \text{ J K}^{-1}} = 35.1 \text{ K} = 35.1°C$$

The final temperature is $T_f = 20.0°C + 35.1°C = \boxed{55.1°C}$.

12.45 The balanced equation is $C_6H_{12}(l) + 9\,O_2(g) \longrightarrow 6\,CO_2(g) + 6\,H_2O(l)$. Set up a calculation of a standard enthalpy of this combustion reaction in terms of standard enthalpies of formation of the products and reactants. The standard enthalpy of combustion is known, but one of the ΔH_f°'s is not known

$$\Delta H^\circ = -3923.7 \text{ kJ} = 6\underbrace{(-393.51)}_{CO_2(g)} + 6\underbrace{(-285.83)}_{H_2O(g)} - 1\underbrace{(\Delta H_f^\circ)}_{(C_6H_{12}(l))} - 9\underbrace{(0)}_{O_2(g)}$$

The standard enthalpies of formation are all in kJ mol^{-1}. All are therefore multiplied by the number of moles of each substance appearing in the balanced equation. Solving gives the ΔH_f° of liquid cyclohexane as $\boxed{-152.3 \text{ kJ mol}^{-1}}$.

12.47 a) The equation is $\boxed{C_{10}H_8(s) + 12\,O_2(g) \longrightarrow 10\,CO_2(g) + 4\,H_2O(l)}$.

b) The amount of heat evolved $(-q)$ in the combustion of 0.6410 g of naphthalene was observed to equal 25.79 kJ. Since the combustion was performed at constant volume, no work was done on the system ($w = 0$). Therefore, $\Delta U = q + w = -25.79 \text{ kJ} + 0 = -25.79 \text{ kJ}$. Put this ΔU on a molar basis to correspond to the 1 mol of naphthalene appearing in the balanced equation

$$\Delta U = \left(\frac{-25.79 \text{ kJ}}{0.6410 \text{ g } C_{10}H_8}\right) \times \left(\frac{128.17 \text{ g } C_{10}H_8}{1 \text{ mol } C_{10}H_8}\right) = -5157 \text{ kJ mol}^{-1}$$

The temperature is 25°C both before and after the reaction. Therefore for the equation written above (which shows 1 mol of naphthalene) $\Delta U^\circ = \boxed{-5157 \text{ kJ}}$.

c) To calculate ΔH° use the definition

$$\Delta H^\circ = \Delta U^\circ + \Delta(PV)$$

Assume that the gases are ideal and that the volumes of the solids are negligible. Then $\Delta(PV) = (\Delta n_g)RT$, and

$$\Delta H^\circ = \Delta U^\circ + (\Delta n_g)RT$$

The Δn_g is the change in the number of moles of gases during the reaction. The combustion of 1 mol of naphthalene produces 10 mol of gas, while consuming 12 mol of gas. Accordingly

$$(\Delta n_g)RT = (-2 \text{ mol})(8.3145 \text{ J K}^{-1}\text{mol}^{-1})(298.15 \text{ K}) = -4.96 \text{ kJ}$$

$$\Delta H^\circ = \Delta U^\circ + (\Delta n_g)RT = -5157 \text{ kJ} - 4.96 \text{ kJ} = \boxed{-5162 \text{ kJ}}$$

d) Specialize text equation 12.12 to apply to the combustion of naphthalene

$$\Delta H^\circ = -5162 \text{ kJ} = 10\underbrace{(-393.51)}_{CO_2(g)} + 4\underbrace{(-285.83)}_{H_2O(l)} - 12\underbrace{(0)}_{O_2(g)} - 1\underbrace{\Delta H_f^\circ}_{C_{10}H_8(s)}$$

Each term on the right in this equation consists of a ΔH_f° in kJ mol^{-1} multiplied by the number of moles in the balanced equation. Solving gives the ΔH_f° (at 25°C) of solid naphthalene as $\boxed{+84 \text{ kJ mol}^{-1}}$.

12.49 Write an equation for the formation of 1 mole of $CCl_3F(g)$ from gaseous atoms that are well-separated in space from all other atoms. Such "naked atoms." result from the process of atomization of a substance. The equation is

$$C(g) + 3\,Cl(g) + F(g) \longrightarrow CCl_3F(g)$$

The bond enthalpies in text Table 12.5 allow estimation of the ΔH° for this reaction as follows:

$$\Delta H^\circ \approx 1 \text{ mol}\underbrace{(-441 \text{ kJ mol}^{-1})}_{C-F} + 3 \text{ mol}\underbrace{(-328 \text{ kJ mol}^{-1})}_{C-Cl} = -1425 \text{ kJ}$$

Next, write equations that show the preparation of the naked atoms from the elements in their standard states. Each of these atomization processes has an enthalpy that can be computed from the data in text Table 12.5. Each is multiplied by the number of moles of the atom involved.

$$C(s) \longrightarrow C(g) \qquad \Delta H° = 1 \text{ mol} \, (716.7 \text{ kJ mol}^{-1}) = 716.7 \text{ kJ}$$
$$3/2 \, Cl_2(g) \longrightarrow 3 \, Cl(g) \qquad \Delta H° = 3 \text{ mol} \, (121.7 \text{ kJ mol}^{-1}) = 365.1 \text{ kJ}$$
$$1/2 \, F_2(g) \longrightarrow F(g) \qquad \Delta H° = 1 \text{ mol} \, (79.0 \text{ kJ mol}^{-1}) \; = 79.0 \text{ kJ}$$

Combine the four chemical equations in the preceding to obtain the equation for the formation of 1 mole of $CCl_3F(g)$ from its elements in their standard states. Do the same for the $\Delta H°$'s

$$C(g) + 3 \, Cl(g) + F(g) \longrightarrow CCl_3F(g) \qquad \Delta H° \approx -1425 \text{ kJ}$$
$$C(s) \longrightarrow C(g) \qquad \Delta H° = 716.7 \text{ kJ}$$
$$3/2 \, Cl_2(g) \longrightarrow 3 \, Cl(g) \qquad \Delta H° = 365.1 \text{ kJ}$$
$$\underline{1/2 \, F_2(g) \longrightarrow F(g) \qquad \Delta H° = 79.0 \text{ kJ}}$$
$$C(s) + 3/2 \, Cl_2(g) + 1/2 \, F_2(g) \longrightarrow CCl_3F(g) \qquad \Delta H° \approx -264 \text{ kJ}$$

The desired estimate of the molar enthalpy of formation of $CCl_3F(g)$ is $\boxed{-264 \text{ kJ mol}^{-1}}$.

12.51 The reaction is the combustion of propane in oxygen

$$C_3H_8(g) + 5 \, O_2(g) \longrightarrow 3 \, CO_2(g) + 4 \, H_2O(g)$$

As this reaction proceeds, bonds are both broken and formed. Broken are 2 mol of C—C bonds, 8 mol of C—H bonds, and 5 mol of O=O double bonds. It *always* requires enthalpy to break bonds, so these events have positive ΔH's. Formed are 6 mol of C=O double bonds and 8 mol of O—H bonds. Bond-formation events *always* have negative ΔH's. The net enthalpy of the reaction (approximately) equals the sum of the enthalpy changes in all these events

$$\Delta H \approx 6 \underbrace{(-728)}_{C=O} + 8 \underbrace{(-463)}_{O-H} + 5 \underbrace{(498)}_{O=O} + 8 \underbrace{(413)}_{C-H} + 2 \underbrace{(348)}_{C-C} = \boxed{-1.58 \times 10^3 \text{ kJ}}$$

Tip. The answer -1.582×10^3 kJ is correct according to the rules for significant digits.[3] But bond enthalpies are only approximately constant from one compound to the next. For this reason the answer is rounded off to -1.58×10^3 kJ.

12.53 The Lewis structures of the two reactants are in the top row of the following, and the Lewis structure of the two products are in the bottom row

Each row contains three B-Br bonds and three B-Cl bonds. The sum of the *average* bond enthalpies in the product molecules equals the sum of the *average* bond enthalpies in the reactant molecules.

---

[3]Text Appendix A.

The ΔH° of the reaction will differ from zero only if the B—Br and B—Cl bond enthalpies in the four molecules differ from their average values. Slight differences are to be expected because the bonds have different neighbors in the products than in the reactants.

Tip. How close to zero is ΔH°? The ΔH_f°'s of $BBr_2Cl(g)$ and $BBrCl_2(g)$ have been measured and tabulated.[4] Combine them with the ΔH_f°'s of $BBr_3(g)$ and $BCl_3(g)$ from text Appendix D. The resulting ΔH° of reaction is 0.63 kJ mol^{-1}. This is indeed quite small.

Reversible Processes in Ideal Gases

12.55 The system consists of 2.00 mol of ideal gas. In an isothermal change $\Delta T = 0$. The internal energy of an ideal gas depends only on its temperature which means that $\boxed{\Delta U = 0}$. As for the enthalpy:

$$\Delta H = \Delta U + \Delta(PV) = 0 + \Delta(nRT) = 0 + nR\Delta T = 0 + 0 \qquad \text{that is,} \qquad \boxed{\Delta H = 0}$$

The expansion is reversible. Hence

$$w = -nRT\ln\left(\frac{V_2}{V_1}\right) = -(2.00 \text{ mol})\left(\frac{8.3145 \text{ J}}{\text{mol K}}\right)(298 \text{ K})\ln\left(\frac{36.00}{9.00}\right) = \boxed{-6.87 \text{ kJ}}$$

The first law requires that $\Delta U = q + w$. Hence q equals $\boxed{+6.87 \text{ kJ}}$.

12.57 During any adiabatic process $q = 0$. During this *reversible* adiabatic expansion of an ideal gas

$$T_1 V_1^{\gamma-1} = T_2 V_2^{\gamma-1}$$

where γ is c_p/c_v and the subscripts refer to the initial and final states of the gas. In this problem, V_1 is 20.0 L, V_2 is 60.0 L, γ is 5/3, and T_1 is 300 K. Solving for T_2 and substituting gives

$$T_2 = T_1\left(\frac{V_1}{V_2}\right)^{\gamma-1} = (300 \text{ K})\left(\frac{20.0 \text{ L}}{60.0 \text{ L}}\right)^{5/3-3/3} = 300\left(\frac{20.0 \text{ L}}{60.0 \text{ L}}\right)^{2/3} \text{ K} = 144.22 \text{ K} = \boxed{144 \text{ K}}$$

Meanwhile, the ΔU of an ideal gas in any process depends solely on the change in its temperature

$$\Delta U = nc_v\Delta T = (2.00 \text{ mol})\left(\frac{3}{2}\, 8.3145 \text{ J K}^{-1}\text{mol}^{-1}\right)(-155.78 \text{ K}) = \boxed{-3.89 \text{ kJ}}$$

This number also equals w, the work done on the gas, because $\Delta U = q + w$ and q is zero in this adiabatic process. Finally, ΔH of an ideal gas also depends entirely on ΔT

$$\Delta H = nc_p\Delta T = (2.00 \text{ mol})\left(\frac{5}{2}\, 8.3145 \text{ J K}^{-1}\text{mol}^{-1}\right)(-155.78 \text{ K}) = \boxed{-6.48 \text{ kJ}}$$

Note that $\Delta H = \gamma\Delta U$.

Tip. Avoid the common mistake of using the seductively attractive equation $T_1 V_1^{\gamma-1} = T_2 V_2^{\gamma-1}$ carelessly. It works only for *reversible* processes involving *ideal gases*.

Distribution of Energy among Molecules

12.59 The Boltzmann energy distribution $P_n = Ce^{-\epsilon_n/k_B T}$ gives the probability that molecules in a gas occupy a given quantized energy level n. Notice that the *higher* the energy ϵ_n of the level, the *lower* the probability. Call the higher of the energy levels under consideration in this gas n_2 and let it have energy ϵ_2 while the lower level n_1 has energy ϵ_1. The relative population of the two levels is given by ratio of Boltzmann equations written for the two levels

$$\frac{P_2}{P_1} = \frac{Ce^{-\epsilon_2/k_B T}}{Ce^{-\epsilon_1/k_B T}} = \frac{e^{-\epsilon_2/k_B T}}{e^{-\epsilon_1/k_B T}} = e^{-(\epsilon_2-\epsilon_1)/k_B T}$$

---

[4]At http://webbook.nist.gov/chemistry/ for example.

The energy difference and T are given in the problem, and k_B is well-known. Insert the numbers

$$\frac{P_2}{P_1} = \exp\left(\frac{-(0.4 \times 10^{-21} \text{ J})}{1.38 \times 10^{-23} \text{ J K}^{-1}(298 \text{ K})}\right) = \boxed{0.91}$$

Tip. The normalization constant C cancels out because the focus is on *relative* populations.

12.61   The relative populations of two vibrational quantum levels i and j of different energies is given by the following ratio, which is derived by dividing text equation 12.19 as written for the i-th level by the same equation as written for the j-th level

$$\frac{P_i}{P_j} = \exp\left(-(\epsilon_i - \epsilon_j)/k_B T\right)$$

The problem asks for this ratio for the vibrational ground state and first excited state in N_2 at 450 K. Let the j-th state be the $v = 0$ state, the ground state, and let the i-th state be the $v = 1$ state, the first excited vibrational state. The difference between the energies of these states is just h times the fundamental oscillation frequency (vibrational frequency) of the system

$$\epsilon_i - \epsilon_j = h\nu = (6.626 \times 10^{-34} \text{ J s})(7.07 \times 10^{13} \text{ s}^{-1}) = 4.685 \times 10^{-20} \text{ J}$$

Substitute this and the temperature into the first equation

$$\frac{P_i}{P_j} = \exp\left(\frac{-(4.685 \times 10^{-20} \text{ J})}{(1.3808 \times 10^{-23} \text{ J K}^{-1})(450 \text{ K})}\right) = \boxed{0.00053}$$

Tip. At 450 K, which is pretty hot, the lowest excited vibrational level still remains quite sparsely populated relative to the ground state.

12.63   This problem resembles problem **12.61**, but with the complication that the fundamental vibrational frequency of the HF molecule ν_{HF} has to be computed. Getting it requires knowledge of the force constant of the H—F bond, which is given in the problem, and the reduced mass μ of HF. Obtain the reduced mass by using the definition of reduced mass

$$\mu_{HF} = \frac{m_F m_H}{m_F + m_H} = \frac{(18.998 \text{ u})(1.0078 \text{ u})}{(18.998 \text{ u}) + (1.0078 \text{ u})} = 0.9570 \text{ u} = 1.589 \times 10^{-27} \text{ kg}$$

Now put the reduced mass and the force constant into text equation 12.20 to obtain ν_{HF}

$$h\nu_{HF} = \frac{h}{2\pi}\sqrt{\frac{k_{HF}}{\mu_{HF}}}$$

$$\nu_{HF} = \frac{1}{2\pi}\sqrt{\frac{966 \text{ N m}^{-1}}{1.589 \times 10^{-27} \text{ kg}}} = 1.241 \times 10^{14} \text{ s}^{-1}$$

The energy difference between the equally spaced vibrational levels of HF is

$$\epsilon_i - \epsilon_j = h\nu = (6.626 \times 10^{-34} \text{ J s})(1.241 \times 10^{14} \text{ s}^{-1}) = 8.222 \times 10^{-20} \text{ J}$$

Substitute this energy difference into the equation used in problem **12.61** for the relative populations of two energy states

$$\frac{P_i}{P_j} = \exp\left(\frac{-(8.222 \times 10^{-20} \text{ J})}{(1.3807 \times 10^{-23} \text{ J K}^{-1})(300 \text{ K})}\right) = e^{-19.85} = \boxed{2.39 \times 10^{-9}}$$

At 300 K (near room temperature) nearly all of the HF molecules are in the $\nu = 0$ vibrational state, which is the vibrational state of lowest energy.

ADDITIONAL PROBLEMS

12.65   The law of Dulong and Petit states that all metals have a molar heat capacity of approximately 25 $J\ K^{-1}mol^{-1}$. The molar heat capacity equals the specific heat capacity of a substance multiplied by its molar mass. Hence:

$$c = c_s \mathcal{M} \approx 25\ J\ K^{-1}mol^{-1}$$

The experimental specific heat capacity of indium is $0.233\ J\ K^{-1}g^{-1}$. A molar mass of $76\ g\ mol^{-1}$ combines with this number to give a molar heat capacity for indium of only $17.7\ J\ K^{-1}mol^{-1}$. This violates the law of Dulong and Petit badly. The currently accepted value of the molar mass of indium ($114.8\ g\ mol^{-1}$) gives a molar heat capacity consistent with the law of Dulong and Petit.

Tip. For solids and liquids the distinction between c_p and c_v is usually unimportant, especially in an approximate relationship. For this reason the p or v subscript does not appear on c in this problem.

12.67   **a)** The system is the 2.00 mol of argon gas. The work done *on* the system is $-P\Delta V$. Since the gas is ideal and P is constant, $P\Delta V = nR\Delta T$ for the system. In this case ΔT is given as -100 K. Then

$$w = -nR\Delta T = -(2.00\ \text{mol})(8.3145\ J\ K^{-1}mol^{-1})(-100\ K) = \boxed{+1.66 \times 10^3\ J}$$

b) The process goes on at constant pressure so the heat absorbed is q_p

$$q_p = nc_p\Delta T = (2.00\ \text{mol})\left(\frac{5}{2}\ 8.3145\ J\ K^{-1}mol^{-1}\right)(-100\ K) = \boxed{-4.16 \times 10^3\ J}$$

c) Use the first law of thermodynamics: $\Delta U = q + w = -4157 + 1663 = -2494$ J. This rounds off to $\boxed{-2.49\ kJ}$. Note the use of un-rounded answers from parts a) and b) in the addition.

d) The ΔH of a system during a change always equals q_p. Hence, ΔH is $\boxed{-4.16\ kJ}$.

12.69   Because frictional losses and leaks do not occur, the amount of work done by the gas on the paddle mechanism equals the negative of the work absorbed by the gas

$$w = -(-P\Delta V) = (1.00\ \text{atm})(13.00 - 5.00)\ L = 8.00\ L\,\text{atm}$$

$$= 8.00\ L\,\text{atm} \times \left(\frac{101.325\ J}{L\,\text{atm}}\right) = 811\ J$$

All of this work goes to heat up the 1.00 L of water. At the given density, the 1.00 L of water weighs 1.00×10^3 g. Therefore

$$\Delta T_{H_2O} = \frac{q_{H_2O}}{c_{s,\,H_2O}\,m_{H_2O}} = \frac{811\ J}{(4.18\ J\ K^{-1}g^{-1})(1.00 \times 10^3\ g)} = \boxed{0.194\ K}$$

Tip. Notice that details about the gas in the cylinder (ideal or non-ideal, monatomic or polyatomic, and so forth) are immaterial.

12.71   Use the molar mass of glucose ($C_6H_{12}O_6$) as a unit-factor to obtain the chemical amount of glucose in the candy bar. Then use the molar enthalpy of combustion of glucose as a unit-factor to obtain the heat absorbed

$$q = 14.3\ g\ C_6H_{12}O_6 \times \left(\frac{1\ \text{mol}\ C_6H_{12}O_6}{180.16\ g\ C_6H_{12}O_6}\right)\left(\frac{-2820\ kJ}{1\ \text{mol}\ C_6H_{12}O_6}\right) = -223.8\ kJ$$

The heat absorbed by the surroundings of the reaction (which are the person's body) therefore equals $+223.8$ kJ. The change in temperature as 50 kg of water absorbs this amount of heat is

$$\Delta T = \frac{q}{c_s M} = \frac{223.8 \times 10^3\ J}{(4.18\ J\ K^{-1}g^{-1})(50 \times 10^3\ g)} = \boxed{1.1\ K}$$

12.73   Determine whether He(l) or $N_2(l)$ is a better coolant near the boiling point by comparing their specific heat capacities. He(l) absorbs 4.25 J of heat per gram as it heats up by 1 K. $N_2(l)$ absorbs only 1.95 J of heat per gram as it warms by the same amount. Therefore, $\boxed{\text{He}(l)}$ is a better coolant near the boiling point.

At their boiling point, the two liquids cool by vaporization; $\boxed{N_2(l)}$ is better because it absorbs much more heat per gram in vaporization than He(l).

12.75   **a)** The equation for the combustion of isooctane is

$$\boxed{C_8H_{18}(l) + \tfrac{25}{2}\,O_2(g) \longrightarrow 8\,CO_2(g) + 9\,H_2O(l)}$$

b) The combustion of 0.542 g of isooctane is exothermic (isooctane is a fuel). A bomb calorimeter is a sealed thick-walled vessel. Using it means that the reaction takes place at constant volume (barring an explosion). Imagine that this closed system consists of three sub-systems: the calorimeter body, the water inside the calorimeter, and the combustion reaction. As a whole, the system neither gains nor loses heat because the bomb is well-insulated from its surroundings. The amounts of heat gained by the three sub-systems add up to zero

$$q_{\text{calorimeter}} + q_{H_2O} + q_{\text{combustion}} = q_{\text{sys}} = 0 \quad \text{constant } V$$

The heat absorbed by a system (or sub-system) in a change at constant volume can be computed using either

$$q_v = c_v \Delta T \qquad \text{or} \qquad q_v = m c_s \Delta T$$

depending on whether a heat capacity or specific heat capacity is available. The problem gives the heat capacity c_v of the calorimeter (48 J K^{-1}) and the specific heat capacity of water (4.184 J K^{-1}g^{-1}). The change in temperature of the calorimeter equals $28.670 - 20.450 = 8.220°C$, which means that $\Delta T = 8.220$ K. The ΔT of the 750 g of water also equals 8.220 K, because the water and calorimeter are in thermal contact. Insert these numbers for the q's of the water and calorimeter

$$\underbrace{(48 \text{ J K}^{-1})(8.22 \text{ K})}_{\text{calorimeter}} + \underbrace{(750 \text{ g})(4.184 \text{ J K}^{-1}\text{g}^{-1})(8.22 \text{ K})}_{\text{water}} + q_{\text{combustion}} = 0$$

Solving for the last q gives -2.62×10^4 J. This equals the heat absorbed by this combustion reaction at constant volume. At constant volume, zero work is done by or upon the combustion reaction. Hence[5]

$$\Delta U_{\text{combustion}} = q + w = q_v + 0 = \boxed{-2.62 \times 10^4 \text{ J}}$$

c) The molar mass of C_8H_{18} is 114.23 g mol^{-1}. The combustion of an entire mole of isooctane absorbs much more heat than the combustion of 0.542 g

$$\Delta U = \frac{-2.62 \times 10^4 \text{ J}}{0.542 \text{ g}} \times \frac{114.23 \text{ g}}{1 \text{ mol isooctane}} = -5.52 \times 10^6 \text{ J mol}^{-1} = \boxed{\frac{-5520 \text{ kJ}}{1 \text{ mol isooctane}}}$$

d) By definition, $\Delta H = \Delta U + \Delta(PV)$. If the gases in the combustion reaction are ideal and the liquids have negligible volume in comparison to the volume of the gases, then $\Delta(PV) = (\Delta n_g)RT$, where Δn_g is the change in the number of moles of gas. Assume that these two conditions are met. The balanced equation for the combustion of 1 mol of isooctane shows that

$$\Delta n_g = \frac{8 \text{ mol gases}}{1 \text{ mol isooctane}} - \frac{12.5 \text{ mol gases}}{1 \text{ mol isooctane}} = \frac{-4.5 \text{ mol gases}}{1 \text{ mol isooctane}}$$

---

[5]Some printings of the text ask for a calculation of ΔE. ΔU is intended.

Therefore

$$(\Delta n_{\mathrm{g}})RT = \left(\frac{-4.5 \text{ mol gases}}{1 \text{ mol isooctane}}\right)\left(\frac{8.3145 \times 10^{-3} \text{ kJ}}{\text{mol gases K}}\right)(298 \text{ K}) = \frac{-11.15 \text{ kJ}}{1 \text{ mol isooctane}}$$

Although the temperature rises from 20.450° to 28.670°C, taking it as a constant 25°C (298 K) causes little error. Complete the calculation as follows

$$\Delta H = \Delta U + \Delta(PV) = -5520 \text{ kJ mol}^{-1} + (-11.15 \text{ kJ mol}^{-1}) = \boxed{-5530 \text{ kJ mol}^{-1}}$$

Tip. The generally accepted[6] standard enthalpy of combustion of isooctane is -5461.3 kJ mol^{-1}.

e) The standard enthalpy of the combustion of isooctane according to the equation in part **a)** depends on the standard enthalpies of formation of the several reactants and products as follows

$$\Delta H° = 8 \, \Delta H_{\mathrm{f}}°(CO_2(g)) + 9 \, \Delta H_{\mathrm{f}}°(H_2O(l)) - \Delta H_{\mathrm{f}}°(\text{isooctane})$$

The ΔH obtained from the bomb calorimetry experiment does not equal the $\Delta H°$ of this reaction at 298.15 K, but should approximate it closely. Insert the ΔH on the left in the preceding, substitute $\Delta H_{\mathrm{f}}°$'s from text Appendix D on the right, and solve for $\Delta H_{\mathrm{f}}°(\text{isooctane})$

$$-5530 \text{ kJ} = 8 \underbrace{(-393.51)}_{CO_2(g)} + 9 \underbrace{(-285.83)}_{H_2O(l)} - \Delta H_{\mathrm{f}}°(\text{isooctane}) - \frac{25}{2}\underbrace{(0.00)}_{O_2(g)}$$

$$\Delta H_{\mathrm{f}}°(\text{isooctane}) = \boxed{-191 \text{ kJ mol}^{-1}}$$

Tip. The accepted $\Delta H_{\mathrm{f}}°$ of isooctane at 298 K is -259.3 kJ mol^{-1}. As in part **d)**, the difference from the accepted value is 69 kJ mol^{-1}.

12.77   **a)** To get the $\Delta H°$ for the combustion of 1 mol of acetylene, combine $\Delta H_{\mathrm{f}}°$'s as follows

$$\Delta H° = 2 \underbrace{(-393.51)}_{CO_2(g)} + 1 \underbrace{(-241.82)}_{H_2O(g)} - 1 \underbrace{(226.73)}_{C_2H_2(g)} - \frac{5}{2}\underbrace{(0.00)}_{O_2(g)} = \boxed{-1255.57 \text{ kJ}}$$

b) The total heat capacity of the mixture of the two gases equals the molar heat capacity of the first multiplied by the number of moles of the first plus the molar heat capacity of the second multiplied by the number of moles of the second

$$nc_{\mathrm{p}} = (2.00 \text{ mol})\underbrace{(37 \text{ J K}^{-1}\text{mol}^{-1})}_{CO_2} + (1.00 \text{ mol})\underbrace{(36 \text{ J K}^{-1}\text{mol}^{-1})}_{H_2O} = \boxed{110 \text{ J K}^{-1}}$$

c) Assume for convenience that 1.00 mol of $C_2H_2(g)$ is burned. The product gases, which are 2.00 mol of $CO_2(g)$ and 1.00 mol of $H_2O(g)$, absorb 1255.57 kJ of heat. For these gases, which comprise the flame

$$\Delta T = \frac{q}{nc_{\mathrm{p}}} = \frac{1.25557 \times 10^6 \text{ J}}{110 \text{ J K}^{-1}} = 1.14 \times 10^4 \text{ K} = 11\,400°C$$

If the temperature before combustion is 25°C, the maximum flame temperature is 11 425°C, which rounds off to $\boxed{11\,400°C}$.

---

[6]See http://webbook.nist.gov/

12.79 Define the system as the contents of the engine's cylinder. Before the explosive combustion of the octane, the temperature of this system is 600 K, it has a volume is 0.150 L, and its pressure is 12.0 atm. Apply the ideal-gas equation to the mixed contents of the cylinder before the combustion

$$n_{\text{octane}} + n_{\text{air}} = \frac{PV}{RT} = \frac{(12.0 \text{ atm})(0.150 \text{ L})}{(0.08206 \text{ L atm mol}^{-1}\text{K}^{-1})(600 \text{ K})} = 0.03656 \text{ mol}$$

The octane and air in the system are in a 1-to-80 molar ratio

$$80\, n_{\text{octane}} = n_{\text{air}}$$

Solving the preceding pair of simultaneous equations gives

$$n_{\text{octane}} = 0.4514 \text{ mmol} \qquad \text{and} \qquad n_{\text{air}} = 36.11 \text{ mmol}$$

According to the problem, the system does not change its volume during the actual combustion of the fuel, so w is zero. Furthermore, q is zero (the combustion happens so fast that there is no time for heat to be lost or gained). Since w and q both equal zero, ΔU of the system equals zero, by the first law of thermodynamics. Imagine the combustion to occur in two stages: a: the reaction goes at a constant temperature of 600 K; b: the product gases heat up at constant volume. The sum of these two changes is the overall change within the cylinder during combustion. Therefore

$$\Delta U_{\text{sys}} = 0 = \Delta U_a + \Delta U_b \quad \text{which means} \quad \Delta U_a = -\Delta U_b$$

The problem offers data pertaining to enthalpy changes, not energy changes, in the two steps. Deal with this by substituting for the ΔU_a and ΔU_b in terms of ΔH's:

$$\Delta H_a - \Delta(PV)_a = -\left(\Delta H_b - \Delta(PV)_b\right)$$

Step a involves ideal gases, takes place at a constant temperature, and involves change in the chemical amount of gas. Therefore $\Delta(PV)_a$ equals $\Delta n_g RT$, where Δn_g is the change in the chemical amount of gases during the combustion reaction (step a). Step b is the quick after-the-reaction heating of the ideal gases inside the cylinder that goes on, according to the story of the problem, before any expansion or loss of heat by conduction to the outside. The term $\Delta(PV)_b$ therefore equals $n_{\text{after}}R\Delta T$ where "n_{after}" signifies the chemical amount of gases present after the reaction. Finally, for the change in temperature that comprises step b, ΔH_b is equal to $n_{\text{after}}c_p\Delta T$, just as in text equation 12.11, as long as the molar heat capacity c_P is independent of temperature, which it is under the assumption of ideality. Inserting these three relations into the preceding equation gives

$$\Delta H_a - \Delta n_g RT = -(n_{\text{after}}c_p\Delta T - n_{\text{after}}R\Delta T)$$

In this equation T is 600 K, and ΔT is the temperature change during the heating. The plan is to compute all the other quantities and then substitute their values into this equation to get ΔT and, from it, the final temperature.

The cylinder contains 36.11 mmol of air and 0.4514 mmol of octane before the reaction. Air is 78.08% N_2 and 20.95% O_2 on a molar basis. The remaining one percent or so consists of argon, carbon dioxide, and other trace gases. Lump them in with the nitrogen. The resulting approximation is 79.05 mole percent N_2 and 20.95 mole percent O_2. Write down the amounts of the various gases in the cylinder *before* the combustion

$$n_{N_2} = 0.7905(36.11) \text{ mmol} \qquad n_{O_2} = 0.2095(36.11) \text{ mmol} \qquad n_{\text{octane}} = 0.4514 \text{ mmol}$$

The octane burns according to

$$C_8H_{18}(g) + 12\tfrac{1}{2}\, O_2(g) \longrightarrow 8\, CO_2(g) + 9\, H_2O(g)$$

so that *after* the reaction the amounts of the various gases are

$$n_{N_2} = 0.7905(36.11) = 28.54 \text{ mmol} \qquad n_{O_2} = 0.2095(36.11) - 12.5(0.4514) = 1.923 \text{ mmol}$$

$$n_{CO_2} = 8(0.4514) = 3.611 \text{ mmol} \qquad n_{H_2O} = 9(0.4514) = 4.063 \text{ mmol}$$

Notice that N_2 does not react, and that the octane, the limiting reactant, is all used up. Adding up the amounts of gas before and after the reaction gives

$$n_{before} = 36.56 \text{ mmol} \qquad n_{after} = 38.14 \text{ mmol} \quad \text{from which} \quad \Delta n_g = +1.58 \text{ mmol}$$

The enthalpy of formation of gaseous octane at 600 K is given, but the enthalpies of formation of oxygen, water vapor, and carbon dioxide at 600 K are not. Use the enthalpies of formation of these three substances at 298 K (text Appendix D) as approximations for their values at 600 K. The enthalpy of combustion of octane at 600 K is then

$$\underbrace{\Delta H^\circ_{600}}_{\text{(combustion)}} = 9 \underbrace{(-241.8)}_{H_2O(g)} + 8 \underbrace{(-393.5)}_{CO_2(g)} - 1 \underbrace{(-57.4)}_{\text{octane}(g)} - 12.5 \underbrace{(0.00)}_{O_2(g)} = -5266.8 \text{ kJ mol}^{-1}$$

This per-mole value is much larger than ΔH_a, the enthalpy of the combustion reaction in the cylinder. Only 0.4514 mmol of octane burns in the cylinder, not an entire mole. The enthalpy of combustion of the contents of the cylinder is

$$\Delta H_a = -5266.8 \times 10^3 \text{ J mol}^{-1} \times (0.4514 \times 10^{-3} \text{ mol}) = -2377 \text{ J}$$

The composite heat capacity of the contents of the cylinder after the reaction is the sum of the nc_p values for the four product gases, as in problem **12.67b**

$$n_{after}c_p = (0.00158 \text{ mol}) \underbrace{(35.2 \text{ J K}^{-1}\text{mol}^{-1})}_{O_2} + (0.0289 \text{ mol}) \underbrace{(29.8 \text{ J K}^{-1}\text{mol}^{-1})}_{N_2}$$

$$+ (0.00406 \text{ mol}) \underbrace{(38.9 \text{ J K}^{-1}\text{mol}^{-1})}_{H_2O} + (0.00361 \text{ mol}) \underbrace{(45.5 \text{ J K}^{-1}\text{mol}^{-1})}_{CO_2} = 1.24 \text{ J K}^{-1}$$

Now, solve this problem's "master equation"

$$\Delta H_a - \Delta n_g RT = -(n_{after}c_p\Delta T - n_{after}R\Delta T)$$

for ΔT and make the required numerical substitutions

$$\Delta T = \frac{\Delta H_a - \Delta n_g RT}{n_{after}R - n_{after}c_p} = \frac{-2377 \text{ J} - (0.00158 \text{ mol})(8.3145 \text{ J K}^{-1}\text{mol}^{-1})(600 \text{ K})}{(0.03814 \text{ mol})(8.3145 \text{ J K}^{-1}\text{mol}^{-1}) - 1.24 \text{ J K}^{-1}} = 2580 \text{ K}$$

The maximum temperature is 600 K + 2580 K = 3180 K or about $\boxed{2900°C}$.

12.81 **a)** The gases trapped inside the cylinder of the "one-lung" engine have volume V_1 when the piston is fully withdrawn but a smaller volume V_2 when the piston is thrust home. The compression ratio is 8 : 1 so $V_1 = 8V_2$. The area of the base of the engine's cylinder is πr^2, where r is the radius of the base. The volume of a cylinder is the area of its base times its height h

$$V_1 = Ah \quad \text{and} \quad V_2 = A(h - 12.00 \text{ cm})$$

which employs the (given) fact that full compression shortens h by 12.00 cm. Because r is 5.00 cm, the area A is 78.54 cm^2. Substituting for V_1 and V_2 in terms of A and h gives

$$Ah = 8A(h - 12.00 \text{ cm})$$

The A's cancel, allowing solution for h. The result is 13.714 cm. Once h is known, it is easy to compute V_1 and V_2, which equal 1.077 L and 0.1347 L respectively.

The temperature and pressure of the air-fuel mixture are 353 K (80°C) and 1.00 atm when the mixture enters the cylinder with fully withdrawn piston (V_1). Assuming the air-fuel mixture is an ideal gas

$$n_{\text{mixture}} = \frac{(1.00 \text{ atm})(1.077 \text{ L})}{(0.08206 \text{ L atm mol}^{-1}\text{K}^{-1})(353 \text{ K})} = 0.0372 \text{ mol}$$

The molar ratio of air to fuel (C_8H_{18}) is 62.5 to 1, so $n_{\text{air}} = 62.5 \, n_{\text{fuel}}$. and it follows that at the start the cylinder contains 0.0366 mol of air and 5.86×10^{-4} mol of octane fuel.

During the compression stroke, the system undergoes an irreversible adiabatic compression to one-eighth of its initial volume. None of the relationships that govern *reversible* adiabatic processes applies. Assume however, as advised in the problem, that the compression is near to reversible. If it is, and if the mixture of gases in the cylinder is approximately ideal, then

$$T_1V_1^{\gamma-1} \approx T_2V_2^{\gamma-1} \quad \text{where } \gamma = \frac{c_P}{c_V} = \frac{c_P}{c_P - R} = \frac{35 \text{ J K}^{-1}\text{mol}^{-1}}{(35 - 8.315) \text{ J K}^{-1}\text{mol}^{-1}} = 1.31$$

The temperature after the compression stroke is

$$T_2 \approx T_1 \left(\frac{V_1}{V_2}\right)^{\gamma-1} = (353 \text{ K}) \left(\frac{1.077 \text{ L}}{0.1347 \text{ L}}\right)^{0.31} = (353 \text{ K})(8)^{0.31} = \boxed{673 \text{ K}}$$

b) The compressed gases occupy a volume of $\boxed{0.135 \text{ L}}$ just before they are ignited, as calculated above.

c) The pressure of the compressed air-fuel mixture just before ignition is P_2. Compute it by applying the ideal-gas equation to the system with $T_2 = 673$ K, $V_2 = 0.1347$ L, and $n = 0.0372$ mol.

$$P_2 \approx \frac{nRT_2}{V_2} = \frac{0.0372 \text{ mol}(0.082057 \text{ L atm mol}^{-1}\text{K}^{-1})(673 \text{ K})}{0.1347 \text{ L}} = \boxed{15.3 \text{ atm}}$$

Alternatively, estimate P_2 using the formula for a reversible adiabatic change

$$P_2 \approx P_1 \left(\frac{V_1}{V_2}\right)^{\gamma} = 1.00 \text{ atm} \left(\frac{1.077}{0.1347}\right)^{1.31} = \boxed{15.3 \text{ atm}}$$

d) The ΔH for the combustion of gaseous octane at 600 K is -5266.8 kJ mol^{-1}, as estimated in problem **12.79**. This value is preferable to -5530 kJ mol^{-1}, the figure was obtained for the combustion of liquid isooctane in problem **12.75**. The latter is for combustion of a different compound (isooctane) in a different form (liquid, not gaseous) at a different temperature (298 K not 600 K) to give a different product (liquid water, not water vapor). The air-octane mixture inside the cylinder contains 5.86×10^{-4} mol of octane. Consequently, the ΔH of combustion in this system equals

$$\Delta H = \left(\frac{-5266.8 \text{ kJ}}{1 \text{ mol}}\right) \times (5.86 \times 10^{-4} \text{ mol}) = -3.09 \text{ kJ}$$

After the combustion, the cylinder contains $CO_2(g)$, $H_2O(g)$, and unreacted $O_2(g)$ and $N_2(g)$. The balanced chemical equation shows that the combustion consumes 5.86×10^{-4} mol of $C_8H_{18}(g)$ and $12.5 \times (5.86 \times 10^{-4})$ mol of $O_2(g)$ to produce $8 \times (5.86 \times 10^{-4})$ mol of $CO_2(g)$ and $9 \times (5.86 \times 10^{-4})$ mol of $H_2O(g)$. The effect of the reaction is to increase the chemical amount of gases in the cylinder by $3.5 \times (5.86 \times 10^{-4} \text{ mol}) = 0.002051$ mol. This is Δn_g for the reaction. The original amount of gases is 0.0372 mol. After the combustion there is 0.0393 mol of gases. The *energy* (not enthalpy) released from the reaction all goes to heat up the gaseous contents of the cylinder as long as no heat escapes to the cylinder walls and no work is done until the power stroke starts. Therefore

$$\Delta T = \frac{\Delta H_{\text{react}} - \Delta n_{\text{g}} RT}{nR - nc_{\text{p}}}$$

In this equation, which is derived in problem **12.79** and featured there as the "master equation" of the problem, every quantity but ΔT is known:

$$\Delta T = \frac{-3090 \text{ J} - (0.002051 \text{ mol})(8.3145 \text{ J K}^{-1}\text{mol}^{-1})(673 \text{ K})}{(0.0393 \text{ mol})(8.3145 \text{ J K}^{-1}\text{mol}^{-1}) - (0.0393 \text{ mol})(35 \text{ J K}^{-1}\text{mol}^{-1})} = 2960 \text{ K}$$

The temperature inside the cylinder rises by 2960 K to a maximum of $\boxed{3630 \text{ K}}$.

e) Assume that the expansion stroke is not only adiabatic but reversible. Then the formula

$$T_2 = T_1 \left(\frac{V_1}{V_2}\right)^{\gamma - 1}$$

applies. In this case, T_1 is 3630 K. The ratio V_1/V_2 is 1 to 8 because now the initial state is the *small* volume state just before the expansion stroke of the piston. The exponent $\gamma - 1$ is still 0.31, as previously established. Substituting gives

$$T_2 = (3\,630 \text{ K}) \left(\frac{1}{8}\right)^{0.31} = \boxed{1\,900 \text{ K}}$$

This is the temperature of the exhaust gases.

12.83 Write a chemical equation for oxidation of the $CO(g)$

$$CO(g) + 1/2\, O_2(g) \longrightarrow CO_2(g)$$

At 25°C (298 K), the standard enthalpy of this reaction is

$$\Delta H^{\circ}_{298} = \Delta H^{\circ}_{\text{f}}(CO_2(g)) - 1/2\, \Delta H^{\circ}_{\text{f}}(O_2(g)) - \Delta H^{\circ}_{\text{f}}(CO(g))$$
$$= -393.5 \text{ kJ mol}^{-1} - 1/2\, (0 \text{ kJ mol}^{-1}) - (-110.5 \text{ kJ mol}^{-1}) = -283.0 \text{ kJ mol}^{-1}$$

Let the system consist of two subsystems, the reaction and the dry air. Assume that the system neither loses heat to nor gains heat from its surroundings (such as the catalyst and the body of the tube) as the $CO(g)$ reacts. Assume further that the dry air has a mass of 1 g and that ΔH°_{298} equals the actual ΔH for the reaction. Then

$$q_{\text{sys}} = q_{\text{reaction}} + q_{\text{air}}$$
$$0 = n_{\text{CO}} \left(\frac{-283.0 \times 10^3 \text{ J}}{1 \text{ mol CO}}\right) + m_{\text{air}} c_{s,\text{air}} \Delta T$$
$$0 = \frac{m_{\text{CO}}}{28.0 \text{ g mol}^{-1}} \left(\frac{-283.0 \times 10^3 \text{ J}}{1 \text{ mol CO}}\right) + 1 \text{ g} (1.01 \text{ J g}^{-1} \text{K}^{-1}) 3.2 \text{ K}$$

Solving the last equation for the only unknown, which is m_{CO}, the mass of the carbon monoxide, gives 3.2×10^{-4} g. The mass percentage of CO in the air is

$$\frac{3.2 \times 10^{-4} \text{ g CO}}{1 \text{ g air}} \times 100\% = \boxed{0.0032\%}$$

12.85 Solid $Ca(OH)_2$ dissolves in aqueous HCl according to

$$Ca(OH)_2(s) + 2\, H_3O^+(aq) \longrightarrow Ca^{2+}(aq) + 4\, H_2O(l)$$

This event can be viewed as the dissolution of the solid followed by the neutralization of the OH^- ion. Use ΔH_f° values from text Appendix D to compute its ΔH°

$$\Delta H^\circ = 1\, \Delta H_f^\circ(Ca^{2+}(aq)) + 4\, \Delta H_f^\circ(H_2O(l)) - 1\, \Delta H_f^\circ(Ca(OH)_2(s)) - 2\, \Delta H_f^\circ(H_3O^+(aq))$$

$$= -542.83 + 4(-285.83) - (-986.09) - 2(-285.83) = -128.40 \text{ kJ}$$

in which the standard units of mol and kJ mol^{-1} are omitted for compactness. The greatest change in temperature occurs when the system is thermally insulated from its surroundings. Under that condition, the system as a whole absorbs zero heat as the subsystems "reaction" and "solution" exchange heat

$$0 = q_{\text{sys}} = q_{\text{reaction}} + q_{\text{solution}} = \Delta H + mc_s\Delta T$$

The q of the reaction should be quite close to ΔH° because the reactants and products are in standard states. The rise in temperature does affect ΔH°, but the effect is surely slight. The mass of the solution can be computed from its volume and its density. Approximate the density as 1.0 g mL^{-1}, which is the density of pure water at 25°C. Then

$$0 = 0.05 \text{ mol } Ca(OH)_2 \left(\frac{-128.40 \times 10^3 \text{ J}}{1 \text{ mol } Ca(OH)_2}\right) + 1 \text{ L} \left(\frac{1.0 \times 10^3 \text{ g}}{L}\right)(4.184 \text{ J g}^{-1}\text{K}^{-1})\Delta T$$

$$0 = -6420 \text{ J} + (4184 \text{ J K}^{-1})\Delta T$$

Solving gives $\Delta T = 1.5$ K. The maximum temperature is $25° + 1.5° = \boxed{26.5°}$.

CUMULATIVE PROBLEMS

12.87 The chemical amount of the silane can be computed using the T and P stated in the problem, if the silane behaves ideally

$$n_{\text{SiH}_4} = \frac{PV}{RT} = \frac{(0.658 \text{ atm})(0.250 \text{ L})}{(0.08206 \text{ L atm mol}^{-1}\text{K}^{-1})(298 \text{ K})} = 6.727 \times 10^{-3} \text{ mol}$$

The combustion of this much silane at constant volume (in a bomb calorimeter) at 298 K absorbs -9.757 kJ of heat (which is the same as evolving $+9.757$ kJ). That is, $\Delta U = q_v = -9.757$ kJ. This is a standard energy of reaction, a ΔU°, if the reactants begin in standard states and the products end up in standard states. Assume that they do. Moreover, the silane burns at 298 K, so this is a ΔU_{298}°.[7] The standard molar energy of combustion of silane at 298 K is then

$$\Delta U_{298}^\circ = \frac{-9.757 \text{ kJ}}{6.727 \times 10^{-3} \text{ mol}} = -1450 \text{ kJ mol}^{-1}$$

Next, compute the ΔH_{298}° of the combustion of silane. The balanced equation given in the problem shows that 3 mol of gaseous reactants gives 0 mol of gaseous products

$$\Delta H_{298}^\circ = \Delta U_{298}^\circ + RT\Delta n_g$$

$$= -1450 \text{ kJ mol}^{-1} + (0.008315 \text{ kJ K}^{-1}\text{mol}^{-1})(298.15 \text{ K})\left(\frac{-3 \text{ mol gas}}{\text{mol of reaction}}\right)$$

$$= -1458 \text{ kJ mol}^{-1}$$

---

[7]Saying that the silane burns at 298 K does not mean that the temperature *stays* at 298 K throughout the reaction. The temperature rises temporarily but is adjusted to 298 K at the end of the experiment by removing a sufficient quantity of heat.

ΔH°_{298} for the combustion of silane equals the sum of the standard enthalpies of formation of the products minus the sum of the standard enthalpies of formation of the reactants. Use the chemical equation given in the problem in conjunction with values of standard enthalpies of formation from text Appendix D

$$\Delta H^\circ_{298} = -1458 \text{ kJ mol}^{-1} = 1 \underbrace{(-910.94)}_{SiO_2 \text{ quartz}} + 2 \underbrace{(-285.83)}_{H_2O(l)} - 1 \underbrace{(\Delta H^\circ_f)_{298}}_{SiH_4(g)}$$

$$\underbrace{(\Delta H^\circ_f)_{298}}_{SiH_4(g)} = \boxed{-25 \text{ kJ mol}^{-1}}$$

Compute the ΔU°_f of silane at 298 K from the ΔH°_f at 298 K and Δn_g that occurs during the formation of 1 mol of silane from its elements

$$(\Delta U^\circ_f)_{298} = (\Delta H^\circ_f)_{298} - RT\Delta n_g$$

$$= -25 \text{ kJ mol}^{-1} - (0.008315 \text{ kJ K}^{-1}\text{mol}^{-1})(298.15 \text{ K})\left(\frac{-1 \text{ mol gas}}{\text{mol of reaction}}\right)$$

$$= \boxed{-23 \text{ kJ mol}^{-1}}$$

12.89 Substances with the strongest intermolecular forces have the highest enthalpies of vaporization. Liquid KBr has strong ion-ion forces holding its ions together. It has the highest ΔH_{vap}. NH_3 has dipole-dipole attractions, which are stronger than the weak dispersion (van der Waals) forces that maintain the liquid in Ar and He. The dispersion forces should be stronger in Ar than in He because Ar has a larger molar mass. Therefore $\boxed{\text{He} < \text{Ar} < NH_3 < KCl}$.

12.91 **a)** Lewis structures for carbonic acid show two O—H single bonds, two C—O single bonds and one C=O double bond

b) Imagine the reaction $H_2CO_3 \longrightarrow H_2O + CO_2$ to proceed by the breaking of all of the bonds in H_2CO_3 followed by the making of the four bonds in H—O—H plus O=C=O. The enthalpy of bond breaking is positive; the enthalpy of bond making is negative. Take bond enthalpies from text Table 12.5 and combine them accordingly

$$\Delta H = +2 \underbrace{(463)}_{O-H} +2 \underbrace{(351)}_{O-C} +1 \underbrace{(728)}_{C=O} -2 \underbrace{(463)}_{O-H} -2 \underbrace{(728)}_{C=O} = \boxed{-26 \text{ kJ}}$$

Chapter 13

Spontaneous Processes and Thermodynamic Equilibrium

The Nature of Spontaneous Processes

13.1 Deciding what constitutes the contents of a thermodynamical system in a problem or real-life situation is entirely up to the analyst (you). Once the system is defined, the surroundings are automatically "the rest of the universe." A wise choice of system can greatly simplify the analysis of a thermodynamic problem. It also often pays explicitly to recognize the nature of a chosen system's *immediate* surroundings. The following are typical useful choices of system and surroundings.

a) The system is the reaction $NH_4NO_3(s) \longrightarrow NH_4^+(aq) + NO_3^-(aq)$. This means that the system includes solid ammonium nitrate, the water in which it dissolves and the aquated ions that dissolution generates. The inclusion of water in the system is indicated only rather subtly (by the (aq)'s on the formulas of the product ions). The immediate surroundings include the flask or beaker in which the system is held, the air above the system, and other neighboring materials. The dissolution of ammonium nitrate is $\boxed{\text{spontaneous}}$. Before the process can proceed, any physical barrier (such as a glass wall or an air-gap) between the water and the ammonium nitrate must be removed. The parts (sub-systems) of a system need not be physically contiguous.

b) The system is the reaction $H_2(g) + O_2(g) \longrightarrow$ products. Its immediate surroundings are the walls of the bomb and other portions of its environment that might deliver heat or work or absorb heat or work. The reaction of hydrogen with oxygen is $\boxed{\text{spontaneous}}$. Once hydrogen and oxygen are mixed in a closed bomb, $\boxed{\text{no constraint}}$ exists to prevent their reaction. That is, the system just defined is thermodynamically unstable with respect to the explosion. Experimentally this system gives products quite slowly at room temperature (no immediate explosion). It explodes instantly at higher temperatures.

c) The system is the rubber band. The immediate surroundings consist of the weight (visualized as attached to the lower end of the rubber band), a hangar at the top of the rubber band, and the air in contact with the rubber band. The change is $\boxed{\text{spontaneous}}$ once a constraint such as a supporting finger underneath the weight is removed.

d) The system is the gas contained in the chamber. The immediate surroundings are the walls of the chamber and the moveable piston head. The process is spontaneous if the force exerted by the weight on the piston exceeds the force exerted by the collisions of the molecules of the gas on the bottom of the piston.[1] Because slow compression of the gas is observed, the change is $\boxed{\text{spontaneous}}$.

---

[1]The forces due to the mass of the piston itself and friction between the piston and the walls within which it slides are neglected.

e) The system is the drinking glass in the process: glass $\longrightarrow$ fragments. The immediate surroundings are the floor, the air, and the other materials in the room. The change is $\boxed{\text{spontaneous}}$. It occurs when the constraint, which is whatever portion of the surroundings holds the glass above the floor, is removed.

Entropy and Irreversibility: A Statistical Interpretation

13.3 **a)** The number of available microstates equals the number of possible ways for a number to come up on one die times the number of possible ways for a number to come up on the other. Each die has six faces and therefore 6 available microstates. The total number of available microstates is $\boxed{36}$.

b) The probability that the first die will show a six is $1/6$. The same is true for the second die. It follows that the probability that two sixes show up at the same time is $(1/6) \times (1/6) = \boxed{1/36}$.

13.5 The major driving force for

$$H_2O(l) + D_2O(l) \longrightarrow 2\,HOD(l)$$

is the $\boxed{\text{tendency for the entropy to increase}}$. Two moles of HOD have a larger entropy than a mixture of one mol of H_2O and one mol of D_2O because there is a much larger number of ways for the available H's, D's and O's to be assembled into a collection of HOD molecules than into a collection of H_2O's and D_2O's. The change occurs spontaneously, once the reactants are mixed, even if the system is completely separated from its surroundings.

Tip. A mixture of $HOD(l)$, $H_2O(l)$, and $D_2O(l)$ has an larger entropy than either the pure product or the pure reactants. For this reason the reaction does not "go to completion" to generate pure $HOD(l)$. Instead it settles in a final condition in which the three species coexist.

13.7 Before the stopcock is opened, the number of microstates available to a single H_2 (or He) is proportional to the volume of the glass bulb: $\Omega = cV$ where c is a constant. There are N_A molecules of H_2 and N_A atoms of He. The number of possible microstates for each gas is

$$\Omega_{H_2} = (cV)^{N_A} \quad \text{and} \quad \Omega_{He} = (cV)^{N_A}$$

The number of microstates of the entire system, still before the valve is opened, is the product of the Ω's

$$\Omega_{sys} = \Omega_{H_2}\Omega_{He} = (cV)^{2N_A}$$

This is the number of microstates that have all of the H_2 in the first bulb and all of the He in the second. By symmetry it is also the number of microstates that have all of the H_2 in the *second* bulb and all the He in the first. *After* the stopcock is opened, $2N_A$ molecules occupy a volume of $2V$ and

$$\Omega_{sys} = (c\,2V)^{2N_A}$$

The probability p of the "cross-diffused" result, the state in which the H_2 and He trade places, is the number of ways in which it can be constituted divided by the number of ways in which the mixed system can be constituted

$$p = \frac{(cV)^{2N_A}}{(c2V)^{2N_A}} = 2^{-2N_A}$$

Taking the logarithm of both sides of this equation helps in arriving at p

$$\log p = -2N_A \log 2 = -2N_A(0.301) = -3.63 \times 10^{23} \qquad p = \boxed{10^{-3.63 \times 10^{23}}}$$

Tip. The probability is incredibly small. This change will never happen spontaneously, ever though "never" is admittedly a long long time.

13.9 If the number of accessible microstates in a system increases when a process occurs, then the change in entropy ΔS of the system for that process is positive.

a) When NaCl melts it goes from an ordered solid (fewer microstates) to a relatively disordered liquid state (more microstates): $\boxed{\Delta S > 0}$.

b) When a building is demolished its constituent particles go from a situation corresponding to relatively fewer microstates (the arrangements of the particles' positions and momenta that are recognizable as the building) to a situation corresponding to far many more microstates (the arrangements that are recognizable as a heap of rubble: $\boxed{\Delta S > 0}$.

c) The mixture of nitrogen, oxygen, and argon has far more microstates than the three separate volumes, each containing a different gas: $\boxed{\Delta S < 0}$.

Entropy Changes in Reversible Processes

13.11 Define the system to consist of 1 mol of solid tungsten at its melting point of 3410°C (3683 K). Imagine supplying 35.4 kJ of heat infinitely slowly and in such a way that the temperature stays constant but the tungsten melts. 35.4 kJ then equals q_{rev} for the melting. Substitute this value and T into the equation that defines the change of entropy of a system. Because the change occurs at a constant temperature, T may be taken outside the integral sign

$$\Delta S = \int \frac{dq_{rev}}{T} = \frac{1}{T}\int dq_{rev} = \frac{1}{T}q_{rev} = \frac{35.4 \times 10^3 \text{ J mol}^{-1}}{3683 \text{ K}} = \boxed{9.61 \text{ J K}^{-1}\text{mol}^{-1}}$$

Tip. The temperature *must* be an absolute temperature (in kelvins, for example).

13.13 Trouton's rule states that ΔS_{vap}, the entropy of vaporization, equals 88 ± 5 J K^{-1}mol^{-1} for most liquids. The approximate molar enthalpy of vaporization of acetone then is

$$\Delta H_{vap} = T\Delta S_{vap} \approx (329.35 \text{ K})(88 \text{ J K}^{-1}\text{mol}^{-1}) = \boxed{29 \times 10^3 \text{ J mol}^{-1}}$$

Tip. The experimental ΔH_{vap} of acetone equals 30.2×10^3 J mol^{-1} at its boiling point.

13.15 Assume that the 4.00 mol of hydrogen, which is a gas at 400 K, behaves ideally. The internal energy of an ideal gas depends solely on its absolute temperature T. In an isothermal process, T does not change. Hence, $\boxed{\Delta U = 0}$.

To evaluate ΔH, use its definition

$$\Delta H = \Delta U + \Delta(PV) \qquad \text{which implies} \qquad \Delta H = \Delta U + nR\Delta T$$

since $PV = nRT$ for an ideal gas. But ΔT and ΔU equal zero. Hence, $\boxed{\Delta H = 0}$.

The work done *on* the gas during the reversible isothermal expansion from 12.0 L to 30.0 L is

$$w = -nRT \ln\left(\frac{V_2}{V_1}\right) = -4.00 \text{ mol} \left(8.3145 \text{ J K}^{-1}\text{mol}^{-1}\right)(400 \text{ K}) \ln\left(\frac{30.0}{12.0}\right)$$

$$= -12.2 \times 10^3 \text{ J} = \boxed{-12.2 \text{ kJ}}$$

The first law requires that if $\Delta U = 0$, then $q = -w$. This means the gas absorbs 12.2 kJ of heat during its expansion, just enough to account for the 12.2 kJ of work that it performs: $q = \boxed{+12.2 \text{ kJ}}$.

Finally, $\Delta S = q_{rev}/T$ for an isothermal process, and q_{rev} is the q just computed

$$\Delta S = \frac{q_{rev}}{T} = \frac{+12.2 \times 10^3 \text{ J}}{400 \text{ K}} = \boxed{+30.5 \text{ J K}^{-1}}$$

Entropy Changes and Spontaneity

13.17   Break down the overall process into the three steps described in the problem and calculate ΔS_{sys} for each. Then add up the three contributions. The steps are: I, warming of ice; II, melting of ice; III, warming of melted ice. According to text equation 13.8, ΔS for a temperature change at constant pressure can be computed using the equation

$$\Delta S = nc_{\mathrm{p}} \ln\left(\frac{T_2}{T_1}\right) \qquad (\text{constant } P)$$

as long as c_{P} stays constant over the range of the temperature change. Use this formula to obtain ΔS of the system for the first and third steps

$$\Delta S_{\mathrm{I}} = (1.00\ \mathrm{mol})(38\ \mathrm{J\ K^{-1}mol^{-1}}) \ln\left(\frac{273.15}{253.15}\right) = 2.9\ \mathrm{J\ K^{-1}}$$

$$\Delta S_{\mathrm{III}} = (1.00\ \mathrm{mol})(75\ \mathrm{J\ K^{-1}mol^{-1}}) \ln\left(\frac{293.15}{273.15}\right) = 5.3\ \mathrm{J\ K^{-1}}$$

In the second step, T stays at 273.15 K, and ΔS equals the quantity of heat absorbed reversibly by the system (q_{rev}) divided by this temperature

$$\Delta S_{\mathrm{II}} = \frac{6007\ \mathrm{J}}{273.15\ \mathrm{K}} = 21.99\ \mathrm{J\ K^{-1}}$$

The *total* ΔS of the system equals

$$\Delta S_{\mathrm{sys}} = \Delta S_{\mathrm{I}} + \Delta S_{\mathrm{II}} + \Delta S_{\mathrm{III}} = (2.9 + 21.99 + 5.3)\ \mathrm{J\ K^{-1}} = \boxed{+30.2\ \mathrm{J\ K^{-1}}}$$

The entire process is reversible. Consequently, the entropy of the universe remains constant: $\Delta S_{\mathrm{univ}} = \Delta S_{\mathrm{sys}} + \Delta S_{\mathrm{surr}} = \boxed{0}$. This means $\Delta S_{\mathrm{surr}} = \boxed{-30.2\ \mathrm{J\ K^{-1}}}$.

13.19   Let the system consist of two subsystems, the 72.4 g piece of iron and the 100.0 g of water. The hot iron is plunged into cool water, and the final temperature of the system is a uniform 16.5°C (289.65 K). The process is far from reversible. Nevertheless, the ΔS's of both the iron and the water may be computed using the equation

$$\Delta S = nc_{\mathrm{p}} \ln\left(\frac{T_2}{T_1}\right)$$

as long as their c_{P}'s stay constant over the ranges of their temperature changes. This approach succeeds because entropy is a state function. Its change depends only on the original and final states of the system (or subsystem) and not on the path by which the change occurs. The ΔS of the system then equals the sum of the ΔS's of its subsystems.

The iron cools from 100.0°C to 16.5°C. This means that $T_1 = 373.15$ K and $T_2 = 289.65$ K. Substitute these absolute temperatures into the formula for ΔS given in text equation 13.8

$$\Delta S_{\mathrm{Fe}} = nc_{\mathrm{p}} \ln\left(\frac{T_2}{T_1}\right) = \left(\frac{72.4\ \mathrm{g}}{55.85\ \mathrm{g\ mol^{-1}}}\right)(25.1\ \mathrm{J\ K^{-1}mol^{-1}}) \ln\left(\frac{289.65\ \mathrm{K}}{373.15\ \mathrm{K}}\right) = \boxed{-8.24\ \mathrm{J\ K^{-1}}}$$

The water warms from 283.15 K to 289.65 K. Use the same formula

$$\Delta S_{\mathrm{H_2O}} = nc_{\mathrm{p}} \ln\left(\frac{T_2}{T_1}\right) = \left(\frac{100.0\ \mathrm{g}}{18.015\ \mathrm{g\ mol^{-1}}}\right)(75.3\ \mathrm{J\ K^{-1}mol^{-1}}) \ln\left(\frac{289.65\ \mathrm{K}}{283.15\ \mathrm{K}}\right) = \boxed{+9.49\ \mathrm{J\ K^{-1}}}$$

The ΔS_{tot} in the problem refers to the whole system. It is $-8.24 + 9.49 = \boxed{+1.25\ \mathrm{J\ K^{-1}}}$.

The Third Law of Thermodynamics

13.21 **a)** The ΔS°_{298} of the reaction as written equals the standard molar entropies of the products at 298 K, each multiplied by its coefficient in the balanced equation, minus the standard molar entropies of the reactants at 298 K, each multiplied by its coefficient in the balanced equation. Copy the equation

$$N_2H_4(l) + 3\,O_2(g) \longrightarrow 2\,NO_2(g) + 2\,H_2O)(l)$$

and, taking data from text Appendix D, figure the standard entropy *change* of the reaction at 25°C as represented by the equation as follows

$$\Delta S^\circ_{298} = 2\underbrace{(239.95)}_{NO_2(g)} + 2\underbrace{(69.91)}_{H_2O(l)} - 1\underbrace{(121.21)}_{N_2H_4(l)} - 3\underbrace{(205.03)}_{O_2(g)} = \boxed{-116.58 \text{ J K}^{-1}}$$

Tip. The symbol for a standard molar entropy at 298 K is S°_{298}. These quantities are *always* positive numbers. Avoid the common error of taking the S°_{298} of an element to equal zero.

b) The gaseous form of a substance always has a larger molar entropy that its liquid or solid form. This means that for the process $N_2H_4(l) \longrightarrow N_2H_4(g)$, $\Delta S > 0$. which means that the S°_{298} of $N_2H_4(g)$ is more positive than the S°_{298} of $N_2H_4(l)$. This causes ΔS°_{298} for the reaction of $N_2H_4(g)$ with $O_2(g)$ to be $\boxed{\text{algebraically smaller,}}$ or more negative, than ΔS°_{298} for the reaction of $N_2H_4(l)$ with $O_2(g)$.

13.23 The computations use the method of problem **13.21**.

$$\text{For LiCl:}\quad \Delta S^\circ_{298} = 2\underbrace{(59.33)}_{LiCl(s)} - 2\underbrace{(29.12)}_{Li(s)} - 1\underbrace{(222.96)}_{Cl_2(g)} = \boxed{-162.54 \text{ J K}^{-1}}$$

$$\text{For NaCl:}\quad \Delta S^\circ_{298} = 2\underbrace{(72.13)}_{NaCl(s)} - 2\underbrace{(51.21)}_{Na(s)} - 1\underbrace{(222.96)}_{Cl_2(g)} = \boxed{-181.12 \text{ J K}^{-1}}$$

$$\text{For KCl:}\quad \Delta S^\circ_{298} = 2\underbrace{(82.59)}_{KCl(s)} - 2\underbrace{(64.18)}_{K(s)} - 1\underbrace{(222.96)}_{Cl_2(g)} = \boxed{-186.14 \text{ J K}^{-1}}$$

$$\text{For RbCl:}\quad \Delta S^\circ_{298} = 2\underbrace{(95.90)}_{RbCl(s)} - 2\underbrace{(76.78)}_{Rb(s)} - 1\underbrace{(222.96)}_{Cl_2(g)} = \boxed{-184.72 \text{ J K}^{-1}}$$

$$\text{For CsCl:}\quad \Delta S^\circ_{298} = 2\underbrace{(101.17)}_{CsCl(s)} - 2\underbrace{(85.23)}_{CsCl(s)} - 1\underbrace{(222.96)}_{Cl_2(g)} = \boxed{-191.08 \text{ J K}^{-1}}$$

The answers grow more negative moving down the group, but RbCl is an exception.

13.25 By the second law of thermodynamics, $\Delta S_{univ} = \Delta S_{sys} + \Delta S_{surr} > 0$. In this example, $\Delta S_{sys} = -44.7$ J K^{-1}. Therefore, $\boxed{\Delta S_{surr} > +44.7 \text{ J K}^{-1}}$.

13.27 The change is the breakdown of $SiO_2(s)$ to $Si(s)$ and $O_2(g)$. The products consist of a mole of solid and a mole of gas, but the reactant is simply a mole of solid. The products have many more microstates both because there are more particles and because the particles take up more volume.

The Gibbs Free Energy

13.29 **a)** Solid ammonia is held at a constant temperature of 170 K. It is implied that the pressure is a constant 1 atm. The molar Gibbs energy of fusion is

$$\Delta G_{fus} = \Delta H_{fus} - T\Delta S_{fus} = 5.65 \text{ kJ mol}^{-1} - (170 \text{ K})(0.0289 \text{ kJ K}^{-1}\text{mol}^{-1}) = 0.74 \text{ kJ mol}^{-1}$$

The change in the Gibbs energy of 1.00 mol of ammonia when it melts is therefore $\boxed{+0.74 \text{ kJ}}$.

b) This case differs from part **a)** only in the amount of ammonia. Multiply the molar Gibbs energy of fusion by 3.60 mol, the amount of NH_3 that melts. The result is $\boxed{2.65 \text{ kJ}}$.

c) At 170 K, $\Delta G > 0$. Hence the melting of ammonia is $\boxed{\text{not spontaneous}}$ at 170 K (and 1 atm pressure).

d) If solid and liquid NH_3 are in equilibrium, then ΔG equals zero for the fusion process, which is solid $\rightleftharpoons$ liquid. Calculate the T that makes this true. Use the molar enthalpy of fusion and the molar entropy of fusion that are quoted in the problem

$$\text{If} \quad \Delta G = \Delta H_{\text{fus}} - T\Delta S_{\text{fus}} = 0 \quad \text{then} \quad T_{\text{fus}} = \frac{\Delta H_{\text{fus}}}{\Delta S_{\text{fus}}} = \frac{5.65 \times 10^3 \text{ J mol}^{-1}}{28.9 \text{ J K}^{-1}\text{mol}^{-1}} = \boxed{196 \text{ K}}$$

13.31   When 1.00 mol of ethanol is vaporized at its normal boiling point, ΔH equals 38.7 kJ. The vaporization process goes on at constant pressure, so $\boxed{q = \Delta H = 38.7 \text{ kJ}}$. The vaporization is also isothermal and reversible, so q is also q_{rev}. Then

$$\Delta S = \frac{q_{\text{rev}}}{T} = \frac{38.7 \text{ kJ}}{351.1 \text{ K}} = 0.110 \text{ kJ K}^{-1} = \boxed{110 \text{ J K}^{-1}}$$

Now for the calculation of ΔU. From the definition of enthalpy

$$\Delta U = \Delta H - \Delta(PV)$$

At constant pressure $\Delta(PV) = P\Delta V = P(V_2 - V_1)$. In this case, V_2 is the volume of one mole of vaporous ethanol at 351.1 K and V_1 is the volume of one mole of liquid ethanol, also at 351.1 K. The vapor behaves ideally

$$V_2 = \frac{nRT}{P} = \frac{1.00 \text{ mol}(0.08206 \text{ L atm mol}^{-1}\text{K}^{-1})(351.15 \text{ K})}{1.00 \text{ atm}} = 28.8 \text{ L}$$

The volume of one mole of liquid ethanol (V_1) is less than 0.1 L, which makes it negligibly small compared to 28.8 L. Therefore

$$P\Delta V = P(V_2 - V_1) = (1.00 \text{ atm})(28.8 \text{ L}) = 28.8 \text{ L atm} \times \left(\frac{0.101325 \text{ kJ}}{1 \text{ L atm}} \right) = 2.92 \text{ kJ}$$

Substitute these values into the expression for ΔU

$$\Delta U = \Delta H - \Delta(PV) = 38.7 \text{ kJ} - 2.92 \text{ kJ} = \boxed{35.8 \text{ kJ}}$$

By expanding against a constant pressure, the system performs +2.92 kJ of pressure-volume work on its surroundings. This is the only kind of work possible. The work done by the surroundings on the system is the negative of the work done by the system on the surroundings: $\boxed{w = -2.92 \text{ kJ}}$.

For any reversible processes at constant T and P, $\Delta G = 0$. This can be verified in this case

$$\Delta G = \Delta H - T\Delta S = 38.7 \text{ kJ} - (351.15 \text{ K})(0.110 \text{ kJ K}^{-1}) = \boxed{0.0 \text{ kJ}}$$

13.33   Add the two reactions given in the problem and their ΔG's at 1200°C

$$2 \text{ Fe}_2\text{O}_3(s) \longrightarrow 4 \text{ Fe}(s) + 3 \text{ O}_2(g) \qquad \Delta G = +840 \text{ kJ}$$
$$\underline{3 \text{ C}(s) + 3 \text{ O}_2(g) \longrightarrow 3 \text{ CO}_2(g) \qquad \Delta G = -1200 \text{ kJ}}$$
$$2 \text{ Fe}_2\text{O}_3(s) + 3 \text{ C}(s) \longrightarrow 4 \text{ Fe}(s) + 3 \text{ O}_2(g) \qquad \Delta G = \boxed{-360 \text{ kJ}}$$

The last reaction is spontaneous because it has a negative ΔG. The removal of O_2 by reaction with C drives the decomposition of the Fe_2O_3.

13.35   **a)** Calculate ΔH°_{298} and ΔS°_{298} of the reaction $4\,\mathrm{Fe}(s) + 3\,\mathrm{O}_2(g) \longrightarrow 2\,\mathrm{Fe}_2\mathrm{O}_3(s)$ from the data in text Appendix D

$$\Delta H^{\circ}_{298} = 2\underbrace{(-824.2)}_{\mathrm{Fe}_2\mathrm{O}_3(s)} - 4\underbrace{(0.00)}_{\mathrm{Fe}(s)} - 3\underbrace{(0.00)}_{\mathrm{O}_2(g)} = -1648.4\ \mathrm{kJ}$$

$$\Delta S^{\circ}_{298} = 2\underbrace{(87.40)}_{\mathrm{Fe}_2\mathrm{O}_3(s)} - 4\underbrace{(27.28)}_{\mathrm{Fe}(s)} - 3\underbrace{(205.03)}_{\mathrm{O}_2(g)} = -549.41\ \mathrm{J\ K^{-1}}$$

The problem asks for the temperature range in which the reaction is spontaneous. The changeover from spontaneity to non-spontaneity occurs at $\Delta G^{\circ} = 0$. Use the relationship $\Delta G^{\circ} = \Delta H^{\circ} - T\Delta S^{\circ}$ to obtain the temperature T that makes ΔG° equal zero. Take ΔH°_{298} and ΔS°_{298} as good approximations of the actual values of ΔH° and ΔS° at whatever T turns out to be. Remember to convert ΔS°_{298} to kJ K^{-1} (or ΔH°_{298} to J) so that the units cancel out properly

$$T \approx \frac{\Delta H^{\circ}_{298}}{\Delta S^{\circ}_{298}} = \frac{-1648.1\ \mathrm{kJ}}{-0.54941\ \mathrm{kJ\ K^{-1}}} = 3000\ \mathrm{K}$$

Because ΔH° and ΔS° are both negative, the reaction is $\boxed{\text{spontaneous below 3000 K}}$. Above 3000 K the ever-growing $-T\Delta S^{\circ}$ term finally makes ΔG° positive.

b) Perform similar calculations for the reaction $\mathrm{SO}_2(g) + \frac{1}{2}\mathrm{O}_2(g) \longrightarrow \mathrm{SO}_3(g)$

$$\Delta H^{\circ}_{298} = \underbrace{(-395.72)}_{\mathrm{SO}_3(g)} - \underbrace{(-296.83)}_{\mathrm{SO}_2(g)} - 0.5\underbrace{(0.00)}_{\mathrm{O}_2(g)} = -98.89\ \mathrm{kJ}$$

$$\Delta S^{\circ}_{298} = \underbrace{(256.65)}_{\mathrm{SO}_3(g)} - \underbrace{(248.11)}_{\mathrm{SO}_2(g)} - 0.5\underbrace{(205.03)}_{\mathrm{O}_2(g)} = -93.98\ \mathrm{J\ K^{-1}}$$

$$T \approx \frac{\Delta H^{\circ}_{298}}{\Delta S^{\circ}_{298}} = \frac{-98.89\ \mathrm{kJ}}{-0.09398\ \mathrm{kJ\ K^{-1}}} = 1052\ \mathrm{K}$$

Since ΔH° and ΔS° are both negative, the reaction is $\boxed{\text{spontaneous below 1050 K}}$.

c) The reaction is $\mathrm{NH}_4\mathrm{NO}_3(s) \longrightarrow \mathrm{N}_2\mathrm{O}(g) + 2\,\mathrm{H}_2\mathrm{O}(g)$

$$\Delta H^{\circ}_{298} = \underbrace{(82.05)}_{\mathrm{N}_2\mathrm{O}(g)} + 2\underbrace{(-241.82)}_{\mathrm{H}_2\mathrm{O}(g)} - \underbrace{(-365.56)}_{\mathrm{NH}_4\mathrm{NO}_3(s)} = -36.03\ \mathrm{kJ}$$

$$\Delta S^{\circ}_{298} = \underbrace{(219.74)}_{\mathrm{N}_2\mathrm{O}(g)} + 2\underbrace{(188.72)}_{\mathrm{H}_2\mathrm{O}(g)} - \underbrace{(151.08)}_{\mathrm{NH}_4\mathrm{NO}_3(s)} = 446.10\ \mathrm{J\ K^{-1}}$$

$$T \approx \frac{\Delta H^{\circ}_{298}}{\Delta S^{\circ}_{298}} = \frac{-36.03\ \mathrm{kJ}}{0.44610\ \mathrm{kJ\ K^{-1}}} = -80.7\ \mathrm{K}$$

A negative absolute temperature is physically meaningless. In this calculation it signals that the reaction is either always spontaneous (ΔS° positive and ΔH° negative) or never spontaneous (ΔS° negative and ΔH° positive). Since ΔS° is positive, this reaction is $\boxed{\text{spontaneous at all temperatures.}}$

13.37   The reduction reaction is $\boxed{\mathrm{WO}_3(s) + 3\,\mathrm{H}_2(g) \longrightarrow \mathrm{W}(s) + 3\,\mathrm{H}_2\mathrm{O}(g)}$.

Calculate ΔH°_{298} and ΔS°_{298} for this reaction using the data in Appendix D. The method and assumptions are the same as in problem **13.35**

$$\Delta H^{\circ}_{298} = 1\underbrace{(0.00)}_{\mathrm{W}(s)} + 3\underbrace{(-241.82)}_{\mathrm{H}_2\mathrm{O}(g)} - 1\underbrace{(-842.87)}_{\mathrm{WO}_3(s)} - 3\underbrace{(0.00)}_{\mathrm{H}_2(g)} = \boxed{+117.41\ \mathrm{kJ}}$$

$$\Delta S^{\circ}_{298} = 1\underbrace{(32.64)}_{\mathrm{W}(s)} + 3\underbrace{(188.72)}_{\mathrm{H}_2\mathrm{O}(g)} - 1\underbrace{(75.90)}_{\mathrm{WO}_3(s)} - 3\underbrace{(130.57)}_{\mathrm{H}_2(g)} = \boxed{+131.19\ \mathrm{J\ K^{-1}}}$$

$\boxed{\text{Because } \Delta H° \text{ and } \Delta S° \text{ are both positive}}$, the reaction becomes spontaneous at high enough temperature. The changeover temperature is

$$T = \frac{\Delta H°}{\Delta S°} = \frac{117.41 \times 10^3 \text{ J}}{131.19 \text{ J K}^{-1}} = \boxed{895 \text{ K}}$$

Tip. The reaction *does* proceed to some extent at temperatures below 895 K, but reactants predominate.

A Deeper Look...Carnot Cycles, Efficiency, and Entropy

13.39 **a)** The maximum theoretical efficiency ϵ of an engine operating between two temperatures is attained when the engine operates reversibly. This maximum efficiency is given by text equation 13.19

$$\epsilon = \frac{T_h - T_\ell}{T_h} = 1 - \frac{T_\ell}{T_h}$$

In this problem, T_ℓ is 300 K and T_h is 450 K so ϵ is $\boxed{0.333}$.

b) The efficiency of the engine is the ratio of the net work it *performs* to the heat that it *absorbs*

$$\epsilon = \frac{-w_{net}}{q}$$

The minus sign is necessary to adhere to the convention that $+w$ is work absorbed. If 1500 J of heat is absorbed per cycle from the 450 K reservoir and ϵ is 0.333, then w_{net} is -500 J in each turn of the cycle. It follows from the first law that the engine discards 1000 J of heat $\boxed{q = -1000 \text{ J}}$ into the low-temperature reservoir during each cycle.

c) The engine absorbs 1500 J of heat during one portion of its cycle of operation. It must lose this amount of energy by the time it completes the cycle (for which ΔU is zero). Of the 1500 J, 1000 J goes to the 300 K reservoir as heat, Concurrently, $\boxed{500 \text{ J}}$ appears as work performed by the engine.

Tip. Do a check. The net work absorbed by the engine during a full cycle is -500 J, and the net heat absorbed is 1500 J. These numbers give the correct answer for the efficiency of the engine

$$\epsilon = \frac{-w_{net}}{q} = \frac{-(-500 \text{ J})}{1500 \text{ J}} = 0.333$$

ADDITIONAL PROBLEMS

13.41 The liquid and gaseous forms of a substance are in equilibrium at its normal boiling point. "Normal" means that the pressure equals exactly 1 atm. It follows that ΔG for boiling equals zero as long as both liquid and vapor are present at 1 atm.

$$\text{If} \quad \Delta G = \Delta H_{vap} - T\Delta S_{vap} = 0 \quad \text{then} \quad \Delta S_{vap} = \frac{\Delta H_{vap}}{T}$$

Substitution of the values from the problem gives

$$\Delta S_{vap} = \frac{38.74 \times 10^3 \text{ J mol}^{-1}}{351.6 \text{ K}} = \boxed{110.2 \text{ J K}^{-1}\text{mol}^{-1}}$$

The computation is identical to the one in problem **13.31** except that an additional significant figure is available.

Trouton's rule states that ΔS_{vap} is close to 88 J K^{-1}mol^{-1} for most liquids. The ΔS_{vap} for ethanol is 25% higher than predicted by Trouton's rule.

Tip. The ΔS_{vap} and ΔH_{vap} in the preceding computation are *not* the same as ΔH°_{298} and ΔS°_{298} for the vaporization of ethanol. Computing ΔH°_{298} and ΔS°_{298} from the 298.15 K data on $C_2H_5OH(l)$ and $C_2H_5OH(g)$ that appear in text Appendix D confirms this

$$C_2H_5OH(l) \longrightarrow C_2H_5OH(g) \quad \Delta H^\circ_{298} = 42.59 \times 19^3 \text{ J mol}^{-1} \quad \Delta S^\circ_{298} = 121.9 \text{ J K}^{-1}\text{mol}^{-1}$$

The reason for the differences is that boiling takes place at the normal boiling point and not at 25°C (298.15 K). This issue is the point of problem **13.62**. Using the 298.15 K data gives 76.2°C for the normal boiling point of ethanol, which is a somewhat more than 2°C too low.

13.43   **a)** The compression of the oxygen from state 1 to state 2 is reversible and adiabatic. This means $q_{rev} = 0$. Therefore, ΔS_{sys} equals $\boxed{\text{zero}}$.

b) The problem proposes an alternative path for taking the 2.60 mol of oxygen from its state 1, in which $P_1 = 1.00$ atm and $T_1 = 300$ K, to the same state 2 that is attained in the first part of the problem. But state 2 is not explicitly described. Only P_2, which is 8.00 atm, is given. T_2 and V_2 must be computed. Assume that the oxygen behaves ideally. When an ideal gas is compressed reversibly and adiabatically from an initial (P_1, V_1) to a final state (P_2, V_2)

$$\frac{P_1}{P_2} = \left(\frac{V_2}{V_1}\right)^\gamma$$

where γ is the ratio of c_p to c_v of the gas. For an ideal gas, $c_v = c_p - R$. This allow the computation of γ for O_2

$$\gamma = \frac{c_p}{c_v} = \frac{29.4 \text{ J K}^{-1}\text{mol}^{-1}}{(29.4 \text{ J K}^{-1}\text{mol}^{-1} - 8.314 \text{ J K}^{-1}\text{mol}^{-1})} = 1.394$$

The original volume (V_1) of the 2.60 mol of oxygen in this problem is 64.0 L, as computed using the ideal-gas equation with $T_1 = 300$ K and $P_1 = 1.00$ atm. Substitute this value of V_1, the value of P_2 (8.00 atm), the value of P_1 (1.00 atm), and the value of γ in the preceding to obtain V_2

$$\frac{1.00 \text{ atm}}{8.00 \text{ atm}} = \left(\frac{V_2}{64.0 \text{ L}}\right)^{1.394} \quad \text{from which} \quad V_2 = 14.4 \text{ L}$$

The numerical result is obtained by taking the logarithm of both sides of the equation, rearranging, and then taking the anti-logarithm of both sides. Inserting P_2 and V_2 in the ideal-gas equation then gives T_2

$$T_2 = \frac{P_2 V_2}{nR} = \frac{(8.00 \text{ atm})(14.4 \text{ L})}{(2.60 \text{ mol})(0.08206 \text{ L atm mol}^{-1}\text{K}^{-1})} = 540 \text{ K}$$

In summary, states 1 and 2 of the 2.60 mol of O_2 are

$$P_1 = 1.00 \text{ atm} \qquad V_1 = 64.0 \text{ L} \qquad T_1 = 300 \text{ K}$$
$$P_2 = 8.00 \text{ atm} \qquad V_2 = 14.4 \text{ L} \qquad T_2 = 540 \text{ K}$$

The oxygen is first heated to T_2 at constant pressure and then compressed reversibly and isothermally to P_2. Compute the state variables in the *intermediate* state (subscripted i) that is attained after the constant-pressure (isochoric) heating but before the isothermal compression

$$P_i = 1.00 \text{ atm} \qquad T_i = 540 \text{ K} \qquad V_i = 115.2 \text{ L}$$

The value of V_i comes by using the ideal-gas equation with $n = 2.60$ mol. The entropy change during the isochoric heating is

$$\Delta S_{1 \to i} = nc_p \ln\left(\frac{T_i}{T_1}\right) = (2.60 \text{ mol})(29.4 \text{ J K}^{-1}\text{mol}^{-1})\ln\left(\frac{540}{300}\right) = 44.9 \text{ J K}^{-1}$$

The entropy change during the constant-temperature compression is

$$\Delta S_{i \rightarrow 2} = nR \ln\left(\frac{V_2}{V_i}\right) = (2.60 \text{ mol})(8.3145 \text{ J K}^{-1}\text{mol}^{-1}) \ln\left(\frac{14.40}{115.2}\right) = -45.0 \text{ J K}^{-1}$$

The ΔS for the overall process is the sum of these two values. It equals $\boxed{\text{zero}}$, allowing for round-off errors.

13.45   **a)** If the motion of air masses through the atmosphere is adiabatic and reversible, then q_{rev} equals zero, and ΔS equals $\boxed{\text{zero}}$.

b) Upward displacement of an air mass within the troposphere (the lowest portion of the atmosphere) causes its temperature and pressure to drop concurrently. Break down this overall process into two parts: a temperature change at constant pressure (step I) and a pressure change at constant temperature (step II). The initial values of temperature and pressure are T_0 and P_0 and the final values are T and P. For the two steps

$$\Delta S_{\text{I}} = n c_{\text{p}} \ln\left(\frac{T}{T_0}\right) \qquad\qquad \Delta S_{\text{II}} = nR \ln\left(\frac{P_0}{P}\right)$$

In the first step, ΔS is *less* than zero because cooling a system reduces its entropy. In the second step, ΔS *exceeds* zero. The sum of these two ΔS's must be zero because the overall process, the sum of the two steps, is isentropic. Hence

$$\boxed{c_{\text{p}} \ln\left(\frac{T}{T_0}\right) + R \ln\left(\frac{P_0}{P}\right) = 0}$$

c) If $\ln(P/P_0)$ is approximately equal to $-\mathcal{M}gh/RT$, then

$$-\ln\left(\frac{P}{P_0}\right) = \ln\left(\frac{P_0}{P}\right) \approx \frac{+\mathcal{M}gh}{RT}$$

Substitute this result into the boxed expression in part **b)**. Rearrangement then quickly gives

$$T \ln\left(\frac{T}{T_0}\right) \approx \frac{-\mathcal{M}gh}{c_{\text{p}}}$$

All of the quantities on the right side are given in the problem. Insert them to obtain

$$T \ln\left(\frac{T}{T_0}\right) \approx \frac{-(0.029 \text{ kg mol}^{-1})(9.8 \text{ m s}^{-2})(5.9 \times 10^3 \text{ m})}{29 \text{ J K}^{-1}\text{mol}^{-1}} = -57.8 \text{ K}$$

This means that T, the temperature of the parcel of air after its ascension from sea level to the top of Mount Kilimanjaro, fulfills the approximate equation

$$T \ln\left(\frac{T}{(273.15 + 38) \text{ K}}\right) \approx -57.8 \text{ K}$$

where the sea-level temperature ($38°\text{C}$) has replaced T_0. Obtain T by guessing a few trial values and using a calculator. It equals 246 K or $\boxed{-27°\text{C}}$.

13.47   In problem **13.20**, a 1.000 mol piece of iron at $100°\text{C}$ is plunged into a large reservoir of water at $0°\text{C}$. It loses 2510 J to the water as its temperature falls from 373 K to 273 K. Its entropy decreases

$$\Delta S_{\text{Fe}} = n c_{\text{p}} \ln\frac{T_2}{T_1} = (1.00 \text{ mol})(25.1 \text{ J K}^{-1}\text{mol}^{-1}) \ln\left(\frac{273.15}{373.15}\right) = -7.83 \text{ J K}^{-1}$$

a) The piece of iron is first cooled from 100 to 50°C and then from 50 to 0°C using two water reservoirs. It loses 1255 J of heat to the first reservoir and 1255 J of heat to the second. The entropy change of the first reservoir, which *absorbs* 1255 J of heat and which is so big it stays at 323.15 K, is

$$\Delta S_{\text{I}} = \frac{q}{T} = \frac{1255 \text{ J}}{323.15 \text{ K}} = 3.88 \text{ J K}^{-1}$$

The entropy change of the second big reservoir, which also absorbs 1255 J of heat but does it at the cooler temperature of 273.15 K, exceeds that of the first. It is

$$\Delta S_{\text{II}} = \frac{q}{T} = \frac{1255 \text{ J}}{273.15 \text{ K}} = 4.59 \text{ J K}^{-1}$$

These two reservoirs comprise the surroundings of the iron

$$\Delta S_{\text{surr}} = \Delta S_{\text{I}} + \Delta S_{\text{II}} = 3.88 \text{ J K}^{-1} + 4.59 \text{ J K}^{-1} = \boxed{8.47 \text{ J K}^{-1}}$$

The ΔS_{Fe} is still $\boxed{-7.83 \text{ J K}^{-1}}$ because only the path by which it cooled has changed. It starts in the same initial state and ends up in the same final state. Therefore

$$\Delta S_{\text{univ}} = \Delta S_{\text{surr}} + \Delta S_{\text{Fe}} = 8.47 \text{ J K}^{-1} - 7.83 \text{ J K}^{-1} = \boxed{0.64 \text{ J K}^{-1}}$$

b) Each of the four reservoirs absorbs 627.5 J of heat, one-fourth of the total given up by the iron. The entropy changes of the four reservoirs are

$$\Delta S_{\text{I}} = \frac{627.5 \text{ J}}{348.15 \text{ K}} = 1.80 \text{ J K}^{-1} \qquad \Delta S_{\text{II}} = \frac{627.5 \text{ J}}{323.15 \text{ K}} = 1.94 \text{ J K}^{-1}$$

$$\Delta S_{\text{III}} = \frac{627.5 \text{ J}}{298.15 \text{ K}} = 2.10 \text{ J K}^{-1} \qquad \Delta S_{\text{IV}} = \frac{627.5 \text{ J}}{273.15 \text{ K}} = 2.30 \text{ J K}^{-1}$$

ΔS_{surr} is the sum of the ΔS's of the four reservoirs. It is $\boxed{8.14 \text{ J K}^{-1}}$. The ΔS_{Fe} is however *still* the same: $\boxed{-7.83 \text{ J K}^{-1}}$.

$$\Delta S_{\text{univ}} = \Delta S_{\text{surr}} + \Delta S_{\text{Fe}} = 8.14 \text{ J K}^{-1} - 7.83 \text{ J K}^{-1} = \boxed{0.31 \text{ J K}^{-1}}$$

c) Using four reservoirs made the process more nearly reversible as shown by the smaller ΔS_{univ} in part **b)**. Adding additional reservoirs would make the process $\boxed{\text{yet more nearly reversible.}}$

To make the process fully reversible would require an infinite series of reservoirs each one absorbing an infinitesimal quantity of heat from the iron at a temperature infinitesimally lower than the temperature of the previous reservoir. The sum of the ΔS's of all of the reservoirs would equal $+7.83$ J K^{-1}, and ΔS_{univ} would be zero.

Tip. Obviously, full thermodynamic reversibility is unattainable in real processes.

13.49 **a)** Several different ideal gases occupy their own original volumes and all have the same temperature and pressure. Constraints are removed (for example, valves between the containers are opened) and the gases mix. Clearly ΔS is positive for the mixing. To get an expression for ΔS compute ΔS for each of the gases *separately*, and then add up the several contributions. Work with the i-th gas. This gas starts at V_1 and expands to V_2. In both state 1 and state 2, the entropy of the i-th gas depends on its number of microstates Ω

$$S_1 = k_{\text{B}} \ln \Omega_1 \qquad S_2 = k_{\text{B}} \ln \Omega_2$$

The *change* in entropy of the i-th gas is

$$\Delta S_i = S_2 - S_1 = k_{\text{B}} \ln \Omega_2 - k_{\text{B}} \ln \Omega_1 = k_{\text{B}} \ln \left(\frac{\Omega_2}{\Omega_1} \right)$$

In both state 1 and state 2, the number of microstates available to one molecule of the gas is proportional to the volume ($\Omega = cV$). The number of microstates available to *all* of the molecules of the i-th gas is proportional to the volume raised to the power $n_i N_A$. This power equals the total number of molecules of the i-th gas (N_A is Avogadro's number and n_i is the number of moles of gas i). The change in entropy for the i-th gas therefore equals

$$\Delta S_i = k_B \ln\left(\frac{\Omega_2}{\Omega_1}\right) = k_B \ln\left(\frac{(cV_2)^{n_i N_A}}{(cV_1)^{n_i N_A}}\right) = n_i N_A k_B \ln\left(\frac{V_2}{V_1}\right)$$

Now, focus on the ratio (V_2/V_1). By Boyle's law, it equals (P_1/P_2), the ratio of the original pressure of the i-th gas to the final partial pressure of the i-th gas in the mixture. This latter pressure is, by Dalton's law of partial pressures

$$P_2 = X_i P_{tot}$$

where X_i is the mole fraction of the i-th gas. But P_1, the original pressure of the i-th gas, *equals* P_{tot}, because all of the gases started at the same pressure.[2] Therefore

$$\frac{V_2}{V_1} = \frac{P_1}{P_2} = \frac{P_{tot}}{X_i P_{tot}} = \frac{1}{X_i}$$

Substituting this result into the expression for ΔS_i gives

$$\Delta S_i = n_i N_A k_B \ln\left(\frac{1}{X_i}\right) = -n_i N_A k_B \ln X_i$$

Next, substitute the gas constant R for $N_A k_B$ and replace n_i with $X_i n$, where n is the total number of moles of gas

$$\Delta S_i = -(X_i n)(N_A k_B) \ln X_i = -nR X_i \ln X_i$$

Finally, add up the contributions of all of the gases to get the overall ΔS

$$\boxed{\Delta S = \sum_i \Delta S_i = -nR \sum_i X_i \ln X_i}$$

b) Divide 50 g by the respective molar masses of O_2, N_2 and Ar to obtain the chemical amount of each. Then calculate the mole fractions of the three gases in the mixture and substitute into the preceding formula. The results of these calculations are

Gas	Chemical Amount / mol	X	$X \ln X$
O_2	1.563	0.3399	−0.3668
N_2	1.784	0.3879	−0.3673
Ar	1.252	0.2722	−0.3542
TOTAL	4.599	1.0000	−1.0883

$$\Delta S = -nR \sum_i X_i \ln X_i = -4.599 \text{ mol}(8.3145 \text{ J K}^{-1}\text{mol}^{-1})(-1.0883) = \boxed{42 \text{ J K}^{-1}}$$

This is the entropy change of mixing at any temperature and pressure as long as the assumption of ideal-gas behavior holds.

c) Separating the components of air is the reverse of mixing them. The entropy change of separation is therefore the negative of the entropy change of mixing, assuming ideal-gas behavior. To solve the problem, compute ΔS_{sys} for the process of mixing and then change its sign. Text Table 9.1 gives the volume percentages of the various gases in the air. The mole fractions (X's) of the gases equal these numbers divided by 100. The following table gives these mole fractions, and the quantity $X \ln X$ for each gas

---

[2]Opening a valve between a container of ideal gas 1 at 2.0 atm and a container of ideal gas 2 also at 2.0 atm gives a mixture at 2.0 atm.

Gas	X	$X \ln X$
N_2	0.78110	−0.19297
O_2	0.20953	−0.32747
Ar	0.00934	−0.04365
Ne	0.00002	−0.00020
TOTAL	1.00000	−0.56429

Continuing the table to include more trace atmospheric gases does not provide $X \ln X$ values significantly different from zero. The mixture of gases has a volume of 100 L, a temperature of 298.15 K, and a pressure of 1 atm. Assuming ideality, the amount of the mixture of gases is

$$n = \frac{PV}{RT} = \frac{(1 \text{ atm})(100 \text{ L})}{(0.082 \text{ L atm mol}^{-1}\text{K}^{-1})(298.15 \text{ K})} = 4.09 \text{ mol}$$

Now use the formula derived in part **a)**

$$\Delta S = -nR \sum_i X_i \ln X_i = -(4.09 \text{ mol})(8.3145 \text{ J K}^{-1}\text{mol}^{-1})(-0.56429) = 19.2 \text{ J mol}^{-1}$$

The entropy change of separation of the components *of the system* is $\boxed{-19.2 \text{ J K}^{-1}}$.

Tip. The problem asks simply for the "entropy change." The entropy change of the universe cannot be calculated because there is no information about what goes on in the surroundings of the system to effect the separation. It is of course certain that $\Delta S_{\text{surr}} > 19.2 \text{ J K}^{-1}$. In a practical process it would be much greater.

13.51 The absolute entropy is proportional to the area under such a curve. Gold has the $\boxed{\text{higher}}$ absolute entropy at 200 K.

13.53 **a)** Higher temperature makes the conversion of rhombic to monoclinic sulfur a spontaneous process. Based on this fact both $\Delta H°$ and $\Delta S°$ are $\boxed{\text{positive}}$. Based on text equation 13.12, if ΔH_{sys} and ΔS_{sys} (which approximately equal $\Delta H°$ and $\Delta S°$) had unlike signs, then the changeover temperature would be less than 0 K (meaning no changeover). If ΔH_{sys} and ΔS_{sys} were negative, then the conversion would be favored by lower temperature.

b) Equilibrium between rhombic and monoclinic sulfur at constant temperature and pressure means $\Delta G° = \Delta H° - T\Delta S° = 0$. Rearranging and substituting the values from the problem gives:

$$\Delta S° = \frac{\Delta H°}{T} = \frac{400 \text{ J}}{368.5 \text{ K}} = \boxed{1.09 \text{ J K}^{-1}}$$

13.55 One mole of H_2O is undercooled (supercooled) to −10°C and thermally insulated from its surroundings. When freezing occurs in this system, the temperature of the system increases from −10 to 0°C. The final state is a mixture of solid and liquid H_2O at 0°C.

a) Assume that the freezing takes place at constant pressure. Then, the change in enthalpy of the system, ΔH_{sys}, equals q_{sys}, which is known to equal zero (because of the insulation). Imagine the change to take part in two steps. In step 1, the H_2O stays liquid as it warms up from −10 to 0°C. Then, in step 2, some of the H_2O freezes. The enthalpy is a state function, so

$$\Delta H_{\text{sys}} = \Delta H_1 + \Delta H_2$$
$$0 = n_{\text{liq}}c_{\text{P (liq)}}\Delta T + n_{\text{ice}}(-\Delta H_{\text{fus}})$$
$$= 1 \text{ mol } (75.3 \text{ J K}^{-1}\text{ mol}^{-1})(273 - 263) \text{ K} + n_{\text{ice}}(-6020 \text{ J mol}^{-1})$$

where it has been assumed that the heat capacity of the undercooled liquid is constant between −10 and 0°C. Solving the final equation gives $n_{\text{ice}} = 0.125$ mol. This is $\boxed{1/8}$ of the 1 mol system.

b) The determination of ΔS requires imagining a reversible path for the change. The obvious path consists of two steps that are similar to the steps just used: heating the 1 mol of undercooled water reversibly from -10 to $0°$; then letting 0.125 mol of the water reversibly freeze. For this path

$$\Delta S_{sys} = \Delta S_1 + \Delta S_2$$

$$= n_{liq} c_{P\,(liq)} \ln \frac{T_2}{T_1} + n_{ice} \frac{-\Delta H}{T}$$

$$= 1\ mol(75.3\ J\ K^{-1}\ mol^{-1}) \ln \frac{273\ K}{263\ K} + 0.125\ mol \frac{-6020\ J\ mol^{-1}}{273\ K}$$

$$= 2.81\ J\ K^{-1} - 2.76\ J\ K^{-1} = \boxed{0.05\ J\ K^{-1}}$$

Tip. The answer to part **a)** is good only to about 10% because the temperature difference is known only to 1°C in 10.

13.57 Compute the $\Delta H_{298}°$ and $\Delta G_{298}°$ of the reaction $3\ Fe_2O_3(s) \longrightarrow 2\ Fe_3O_4(s) + 1/2\ O_2(g)$

$$\Delta H_{298}° = 2\ mol\ \Delta H_f°(Fe_3O_4(s)) + 1/2\ mol\ \Delta H_f°(O_2(g)) - 3\ mol\ \Delta H_f°(Fe_2O_3(s))$$

$$\Delta G_{298}° = 2\ mol\ \Delta G_f°(Fe_3O_4(s)) + 1/2\ mol\ \Delta G_f°(O_2(g)) - 3\ mol\ \Delta G_f°(Fe_2O_3(s))$$

Substitution of the standard-state values at 298 K from text Appendix D gives

$$\Delta H_{298}° = 2(-1118.4) + 1/2(0) - 3\ (-824.2) = \boxed{235.8\ kJ}$$

$$\Delta G_{298}° = 2(-1015.5) + 1/2(0) - 3\ (-742.2) = \boxed{195.6\ kJ}$$

The $\Delta G_{298}°$ of the reaction as it is written is positive. This means that the reverse reaction is favored when the products and reactants are in standard states at 298.15 K. Therefore $\boxed{Fe_2O_3(s)}$ is favored under the stated conditions.

13.59 The reaction is

$$3\ CO_2(g) + Si_3N_4(s) \longrightarrow 3\ SiO_2(s) + 2\ N_2(g) + 3\ C(s)$$

The problem and text Appendix D supply the necessary $\Delta G_f°$ data

$$\Delta G_{298}° = 3 \underbrace{(-856.67)}_{SiO_2(s)} + 2 \underbrace{(0)}_{N_2(g)} + 3 \underbrace{(0)}_{C(s)} - 3 \underbrace{(-394.36)}_{CO_2(g)} - 1 \underbrace{(-642.6)}_{Si_3N_4(s)} = \boxed{-744.3\ kJ}$$

13.61 **a)** The reaction of interest is

$$2\ CuCl_2(s) \longrightarrow 2\ CuCl(s) + Cl_2(g)$$

Text Appendix D supplies $\Delta H_f°$ and $S°$ values at 298 K for the computation of $\Delta H_{298}°$ and $\Delta S_{298}°$

$$\Delta H_{298}° = 2 \underbrace{(-137.2)}_{CuCl(s)} + 1 \underbrace{(0)}_{Cl_2(g)} - 2 \underbrace{(-220.1)}_{CuCl_2(s)} = 165.8\ kJ$$

$$\Delta S_{298}° = 2 \underbrace{(86.2)}_{CuCl(s)} + 1 \underbrace{(222.96)}_{Cl_2(g)} - 2 \underbrace{(108.07)}_{CuCl_2(s)} = 179.2\ J\ K^{-1}$$

b) $\Delta G_{590}° \approx \Delta H_{298}° - T\Delta S_{298}° = 165.8\ kJ - (590\ K)(0.1792\ kJ\ K^{-1}) = \boxed{60.1\ kJ}$

c) Use the experimental values at 590 K instead of the values at 298.15 K

$$\Delta G_{590} = \Delta H_{590}° - T\Delta S_{590}° = 158.36\ kJ - (590\ K)(0.17774\ kJ\ K^{-1}) = \boxed{53.5\ kJ}$$

The answer using $\Delta H_{298}°$ and $\Delta S_{298}°$ is about $\boxed{12\%}$ larger than the actual ΔG_{590}.

Tip. The temperature dependence of $\Delta H°$ and $\Delta S°$ should not always be neglected. Taking it into consideration becomes important when the temperature differs a lot from 298.15 K.

13.63 **a)** Potassium ions tend to cross the wall in the direction that equalizes the K^+ concentrations on the two sides of the cell wall. In the case K^+ ions will $\boxed{\text{leave the muscle cells}}$ and pass into the surrounding fluids.

b) The problem is to compute ΔG for transporting 1.00 mol of K^+ ions from a concentration c_1 of 0.0050 M to a concentration c_2 of 0.15 M. The answer must be positive because the separation of a uniform concentration of K^+ ions into two regions of differing concentration is clearly non-spontaneous. Assume that the two solutions are ideal solutions and that T is the normal human body temperature of 37°C (310 K). Then use text equation 14.5a, which gives the change in the Gibbs free energy of the system of an ideal solution when its concentration changes from c_1 to c_2

$$\Delta G = nRT \ln \left(\frac{c_2}{c_1} \right)$$

Substitute the numbers

$$\Delta G = (1.00 \text{ mol})(8.3145 \text{ J K}^{-1}\text{mol}^{-1})(310 \text{ K}) \ln \left(\frac{0.15}{0.0050} \right) = \boxed{8770 \text{ J}}$$

13.65 The problem deals with this vaporization

$$2.00 \text{ mol NH}_3(l, 1 \text{ atm}, 240 \text{ K}) \longrightarrow 2.00 \text{ mol NH}_3(g, 1 \text{ atm}, 298 \text{ K})$$

a) The ΔH for the change equals the sum of the ΔH's of two steps: vaporization of NH_3 at 240 K and heating of the vapor from 240 K to 298 K. It is

$$\Delta H = \Delta H_1 + \Delta H_2 = n \, \Delta H_{\text{vap}} + n \, c_p \Delta T$$

$$= 2.00 \text{ mol} \, (23.4 \times 10^3 \text{ J mol}^{-1}) + 2.00 \text{ mol} \left(\frac{38 \text{ J}}{\text{K mol}} \right) (298 - 240) \text{ K}$$

$$= 51.2 \times 10^3 \text{ J}$$

The q for the two-stage change, like all others, depends on the path. Fortunately, this q is a q_P because both changes occur at constant pressure. The q therefore equals the preceding ΔH, that is, $q = \boxed{51.2 \times 10^3 \text{ J}}$. The change absorbs some pressure-volume work. Use text equation 12.1 to compute this work

$$w = -P\Delta V = -P(V_2 - V_1) = -P(V_{\text{gas}} - V_{\text{liq}})$$

The V_{liq} in this equation is the volume of 2.00 mol of liquid ammonia at the starting temperature. According to the problem, this volume can be neglected in comparison to V_{gas}, the volume of the gas at the final temperature. Therefore

$$w = -PV_{\text{gas}} = -P\frac{nRT}{P} = -(2.00 \text{ mol}) (8.3145 \text{ J K}^{-1}\text{mol}^{-1}) (298 \text{ K}) = \boxed{-4.96 \times 10^3 \text{ J}}$$

By the first law of thermodynamics

$$\Delta U = q + w = 51.2 \times 10^3 \text{ J} + (-4.96 \times 10^3 \text{ J}) = \boxed{46.2 \text{ J}}$$

b) The molar entropy of vaporization is

$$\Delta S = \frac{\Delta H_{\text{vap}}}{T} = \frac{23.4 \times 10^3 \text{ J mol}^{-1}}{240 \text{ K}} = \boxed{97.5 \text{ J K}^{-1}\text{mol}^{-1}}$$

This is also the standard molar entropy of the transition at 240 K.

Chapter 14

Chemical Equilibrium

The Empirical Law of Mass Action

14.1 A mass-action expression is written only with reference to a specific balanced equation. For these equations

a) $\dfrac{(P_{H_2O})^2_{eq}}{(P_{H_2})^2_{eq}(P_{O_2})_{eq}}$ b) $\dfrac{(P_{XeF_6})_{eq}}{(P_{Xe})_{eq}(P_{F_2})^3_{eq}}$ c) $\dfrac{(P_{CO_2})^{12}_{eq}(P_{H_2O})^6_{eq}}{(P_{C_6H_6})^2_{eq}(P_{O_2})^{15}_{eq}}$

Tip. If equilibrium partial pressures are inserted directly into these expressions, the result in each case is a K_P. If all P_{eq}'s are divided first by a reference pressure (1 atm, for example), the result is a thermodynamic K (assuming that the gases behave ideally).

14.3 One balanced equation and its associated mass-action expression are

$$P_4(g) + 2\,O_2(g) + 6\,Cl_2(g) \rightleftharpoons 4\,POCl_3(g) \qquad \dfrac{P^4_{POCl_3}}{P_{P_4}P^2_{O_2}P^6_{Cl_2}}$$

The answer omits the subscript eq's. This common practice, which is common, is followed in further answers in this chapter.

Tip. Other answers are possible. For example, if the coefficients in the balanced equation are doubled, then the exponents in the mass-action expression are doubled, and the overall expression is squared

$$2\,P_4(g) + 4\,O_2(g) + 12\,Cl_2(g) \rightleftharpoons 8\,POCl_3(g) \qquad \dfrac{P^8_{POCl_3}}{P^2_{P_4}P^4_{O_2}P^{12}_{Cl_2}}$$

14.5 **a)** The law of mass action applied to the reaction

$$CO(g) + H_2O(g) \rightleftharpoons CO_2(g) + H_2(g)$$

states that the following equation, in which K is a constant, holds at equilibrium

$$\boxed{\dfrac{(P_{CO_2}/P_{ref})(P_{H_2}/P_{ref})}{(P_{CO}/P_{ref})(P_{H_2O}/P_{ref})} = K}$$

b) The problem gives K and three of the four equilibrium partial pressures. Substitute in the mass-action expression

$$\dfrac{(0.70)(P_{H_2}/P_{ref})}{(0.10)(0.10)} = 3.9$$

In this equation, each numerical partial pressure has already been divided by the reference pressure of 1 atm. Solving for P_{H_2} gives $\boxed{0.056 \text{ atm}}$.

162

14.7   **a)** The graph consists of the values in the second column of the following table plotted (on the y axis) versus the values in the first column (on the x axis). The values in the third column are those of the second divided by those of the first

$[N_2O_4]$ / 10^{-3} mol L^{-1}	$[NO_2]^2$ / 10^{-5} mol L^{-1}	$[NO_2]^2/[N_2O_4]$ / 10^{-2} mol L^{-1}
0.190	0.784	4.12
0.686	2.70	3.94
1.54	5.27	3.42
2.55	10.8	4.23
3.75	13.7	3.65
7.86	9.9	3.80
11.9	44.1	3.71

From the mass-action expression for this reaction it follows that

$$[NO_2]^2 = K[N_2O_4]$$

This equation has the form of the equation for a straight line $y = mx + b$, with $y = [NO_2]^2$, $m = K$, $x = [N_2O_4]$ and $b = 0$. Thus, K equals the slope of the line just plotted.

b) The mean of the values in the last column of the table in the preceding part is the mean experimental K. It is $\boxed{3.84 \times 10^{-2}}$.

14.9   The activities of pure solids and liquids equal 1 and may be omitted from the mass-action expressions

$$\textbf{a)}\ \frac{(a_{H_2S})^8}{(a_{H_2})^8} = K \qquad \textbf{b)}\ \frac{a_{COCl_2} a_{H_2}}{a_{Cl_2}} = K \qquad \textbf{c)}\ a_{CO_2} = K \qquad \textbf{d)}\ \frac{1}{(a_{C_2H_2})^3} = K$$

In all four parts, the partial pressures of the gases divided by the standard-state pressure (1 atm) approximate the activity of the gas. This allows rewriting the answers as follows

$$\textbf{a)}\ \frac{(P_{H_2S})^8}{(P_{H_2})^8} = K \qquad \textbf{b)}\ \frac{P_{COCl_2} P_{H_2}}{P_{Cl_2}} = K \qquad \textbf{c)}\ P_{CO_2} = K \qquad \textbf{d)}\ \frac{1}{(P_{C_2H_2})^3} = K$$

14.11   Pure solids and liquids are omitted from the expressions because their activities equal 1.

$$\textbf{a)}\ \frac{a_{Zn^{2+}}}{a_{Ag^+}^2} = K \qquad \textbf{b)}\ \frac{a_{VO_3(OH)^{2-}}\, a_{OH^-}}{a_{VO_4^{3-}}} = K \qquad \textbf{c)}\ \frac{(a_{HCO_3^-})^6}{(a_{As(OH)_6^{3-}})^2 (a_{CO_2})^6} = K$$

In all three parts, the concentrations of the solutes divided by the standard state concentration (1 M) approximates the activity of the solute. Also, the activities of gases can be treated just as in problem **14.9**. This allows rewriting the answers as

a) $\dfrac{[Zn^{2+}]}{[Ag^+]^2} = K$ **b)** $\dfrac{[VO_3(OH)^{2-}][OH^-]}{[VO_4^{3-}]} = K$ **c)** $\dfrac{[HCO_3^-]^6}{[As(OH)_6^{3-}]^2 P_{CO_2}^6} = K$

Thermodynamic Description of the Equilibrium State

14.13   Insert data from text Appendix D into the usual form

$$\Delta G_{298}^\circ = 2\underbrace{(51.29)}_{NO_2(g)} + 3\underbrace{(-228.59)}_{H_2O(g)} - 2\underbrace{(-16.48)}_{NH_3(g)} - 7/2\underbrace{(0)}_{O_2(g)} = \boxed{-550.23\ \text{kJ}}$$

Substitute this answer in the equation

$$\ln K_{298} = \frac{-\Delta G_{298}^\circ}{RT} = \frac{-(-550.23 \times 10^3\ \text{J mol}^{-1})}{8.3145\ \text{J K}^{-1}\text{mol}^{-1}(298.15\ \text{K})} = 221.95$$

Hence, $K = e^{221.95} = \boxed{2.5 \times 10^{96}}$.

Tip. ΔG_{298}° is properly reported in kJ, but the value used in the calculation of K is in J mol^{-1}. The extra per mole refers to "per mole of the reaction as it is written." If the chemical equation were rewritten with all of its coefficients doubled, then ΔG_{298}° would double, and the value of equilibrium constant K would be squared.

14.15   **a)** Calculate the change in the standard Gibbs energy at 25°C (ΔG_{298}°) for the reaction of 1 mol of SO_2 with $\frac{1}{2}$ mol of O_2 to give 1 mol of SO_3. This requires a table of standard molar Gibbs energies of formation (Appendix D). Then compute the equilibrium constant K using the relationship $\Delta G^\circ = -RT \ln K$

$$\Delta G_{298}^\circ = 1\underbrace{(-371.08)}_{SO_3(g)} - 1\underbrace{(-300.19)}_{SO_2(g)} - 1\underbrace{(0.00)}_{O_2(g)} = -70.89\ \text{kJ}$$

$$\ln K_{298} = \frac{-\Delta G_{298}^\circ}{RT} = \frac{-(-70.89 \times 10^3\ \text{J mol}^{-1})}{8.3145\ \text{J K}^{-1}\text{mol}^{-1}(298.15\ \text{K})} = 28.6$$

$$K_{298} = e^{28.6} = \boxed{2.6 \times 10^{12} = \frac{P_{SO_3}}{P_{SO_2}(P_{O_2})^{1/2}}}$$

It is understood that the three partial pressures in the equilibrium expression are to be divided by a reference pressure of 1 atm because the standard-state pressure for the ΔG° is 1 atm.

b)

$$\Delta G_{298}^\circ = 2\underbrace{(-1015.5)}_{Fe_3O_4(s)} + \frac{1}{2}\underbrace{(0.00)}_{O_2(g)} - 3\underbrace{(-742.2)}_{Fe_2O_3(s)} = +195.6\ \text{kJ}$$

$$\ln K_{298} = \frac{195.6 \times 10^3\ \text{J mol}^{-1}}{-8.3145\ \text{J K}^{-1}\text{mol}^{-1}(298.15\ \text{K})} = -78.90$$

$$K_{298} = e^{-78.90} = \boxed{5.4 \times 10^{-35} = (P_{O_2})^{1/2}}$$

c)

$$\Delta G_{298}^\circ = 1\underbrace{(65.49)}_{Cu^{2+}(aq)} + 2\underbrace{(-131.23)}_{Cl^-(aq)} - 1\underbrace{(-175.7)}_{CuCl_2(s)} = -21.27\ \text{kJ}$$

$$\ln K_{298} = \frac{-21.27 \times 10^3\ \text{J mol}^{-1}}{-8.3145\ \text{J K}^{-1}\text{mol}^{-1}(298.15\ \text{K})} = 8.58$$

$$K_{298} = e^{8.58} = \boxed{5.3 \times 10^3 = [Cu^{2+}][Cl^-]^2}$$

Tip. The form of the mass-action expression derives from the set of coefficients that is used in the computation of the $\Delta G°$. In the above, the set was "$1 + \frac{1}{2} \to 1$" and not "$2 + 1 \to 2$" or any of the other sets that balance the equation.

Law of Mass Action for Related and Simultaneous Equilibria

14.17 The reaction is the combustion of carbon disulfide to give carbon dioxide and sulfur dioxide. Equation 2 has all of the coefficients divided by three but represents this as meaningfully as equation 1. The mass-action expression for equation 2 (K_2) therefore equals that for equation 1 (K_1) except with all of the exponents divided by 3. Dividing exponents by 3 corresponds to taking the cube root: $\boxed{K_2 = \sqrt[3]{K_1}}$.

14.19 If two chemical equations add up to give a third, the equilibrium constant associated with the third is the *product* of the equilibrium constants associated with the first two; if an equation is written in reverse, the new equilibrium constant of the new equation is the reciprocal of the original. Writing the first equation in this problem in reverse and adding the second equation gives the equation of interest. Thus, K for the reaction of interest equals $K_2 \times (1/K_1)$ or $\boxed{K_2/K_1}$.

Equilibrium Calculations for Gas-Phase and Heterogeneous Reactions

14.21 Write the mass-action expression corresponding to the chemical equation given in the problem. Then substitute the equilibrium partial pressures and calculate K_{454}

$$3\,Al_2Cl_6(g) \rightleftharpoons 2\,Al_3Cl_9(g) \qquad \frac{(P_{Al_3Cl_9})^2}{(P_{Al_2Cl_6})^3} = K_{454} = \frac{(1.02 \times 10^{-2})^2}{(1.00)^3} = \boxed{1.04 \times 10^{-4}}$$

14.23 The 1,3-di-*t*-butylcyclohexane is a gas at 580 K. In a collection of, say, 10,000 molecules, 642 would be in the chair form and 9358 would be in the boat form at equilibrium. The partial pressures of gases are proportional to the number of molecules present, assuming ideality. That is, $P_{gas} = kN$ where k is a constant of proportionality that depends on the temperature and volume only. Therefore

$$K_{580} = \frac{P_{boat}}{P_{chair}} = \frac{kN_{boat}}{kN_{chair}} = \frac{9358}{642} = \boxed{14.6}$$

14.25 **a)** Calculate the chemical amount of SO_2Cl_2 from the mass of of SO_2Cl_2 that was put into the flask

$$n_{SO_2Cl_2} = 3.174 \text{ g } SO_2Cl_2 \times \left(\frac{1 \text{ mol } SO_2Cl_2}{134.97 \text{ g } SO_2Cl_2}\right) = 0.02352 \text{ mol } SO_2Cl_2$$

Imagine that the SO_2Cl_2 vaporizes in one step and then reacts in a second distinct step. Use the ideal-gas law to compute the partial pressure of the $SO_2Cl_2(g)$ after it fills the flask at 100°C but before it has a chance to react

$$P_{SO_2Cl_2} = n_{SO_2Cl_2}\left(\frac{RT}{V}\right) = 0.02351\left(\frac{(0.08206 \text{ L atm mol}^{-1}\text{K}^{-1})(373.15 \text{ K})}{1.000 \text{ L}}\right) = 0.7199 \text{ atm}$$

The partial pressures of both products equal zero at this point. As the reaction advances toward equilibrium, the three partial pressures change. The SO_2Cl_2 decomposes to generate SO_2 and Cl_2 in equal chemical amounts

	$SO_2Cl_2(g)$	$\rightleftharpoons$	$SO_2(g)$ +	$Cl_2(g)$
Init. pressure (atm)	0.7199		0	0
Change in pressure (atm)	$-x$		$+x$	$+x$
Equil. pressure (atm)	$0.7199 - x$		x	x

The total pressure in the flask at equilibrium is the sum of the three equilibrium partial pressures

$$P_{tot} = 1.30 \text{ atm} = P_{SO_2Cl_2} + P_{Cl_2} + P_{SO_2} = (0.7199 - x) + x + x$$

Solving gives x equal to 0.5801 atm. The equilibrium partial pressures of the two products are accordingly both $\boxed{0.58 \text{ atm}}$, and the equilibrium partial pressure of the reactant is $\boxed{0.14 \text{ atm}}$.

b) K_{373} is computed by substituting equilibrium partial pressures into the appropriate mass-action expression

$$K_{373} = \frac{(P_{SO_2}/1 \text{ atm})(P_{Cl_2}/1 \text{ atm})}{(P_{SO_2Cl_2}/1 \text{ atm})} = \frac{(0.58)(0.58)}{(0.14)} = \boxed{2.4}$$

14.27 **a)** Calculate the chemical amount of the gaseous $C_6H_5CH_2OH$. Then use the ideal-gas law to calculate its initial partial pressure. Abbreviate the formula of benzyl alcohol as BzOH[1]

$$n_{BzOH} = 1.20 \text{ g BzOH} \times \left(\frac{1 \text{ mol BzOH}}{108 \text{ g BzOH}} \right) = 0.0111 \text{ mol BzOH}$$

$$P_{BzOH} = \frac{n_{BzOH}RT}{V} = \frac{(0.0111 \text{ mol})(0.08206 \text{ L atm mol}^{-1}\text{K}^{-1})(523 \text{ K})}{2.00 \text{ L}} = 0.238 \text{ atm}$$

The following three-line table shows how the partial pressures of benzyl alcohol and its products change as equilibrium is approached

	$C_6H_5CH_2OH(g)$	$\rightleftharpoons$	$C_6H_5CHO(g)$ +	$H_2(g)$
Init. pressure (atm)	0.238		0	0
Change in pressure (atm)	$-x$		$+x$	$+x$
Equil. pressure (atm)	$0.238 - x$		x	x

Substitute the final pressures in the mass-action expression

$$\frac{P_{C_6H_5CHO}P_{H_2}}{P_{BzOH}} = 0.558 = \frac{(x)(x)}{(0.238 - x)} \quad \text{from which} \quad x^2 + 0.558x - 0.133 = 0$$

Use the quadratic formula to solve for x

$$x = \frac{-(0.558) \pm \sqrt{(0.558)^2 - 4(1)(-0.133)}}{2(1)} = \frac{-0.558 \pm 0.918}{2}$$

$$x = 0.180 \quad \text{and} \quad x = -0.738$$

Disregard the solution $x = -0.738$ because it leads to impossible partial pressures for all three gases. The answer is $P_{C_6H_5CHO} = \boxed{0.180 \text{ atm}}$.

b) The fraction of the benzyl alcohol dissociated at equilibrium equals the amount dissociated divided by the initial amount. These amounts are respectively proportional to the decrease in partial pressure of the benzyl alcohol and the initial partial pressure of the benzyl alcohol. Hence

$$f = \frac{\text{amount of BzOH dissociated}}{\text{original amount of BzOH}} = \frac{0.180 \text{ atm}}{0.238 \text{ atm}} = \boxed{0.756}$$

14.29 At equilibrium, the bulb contains a three-way mixture of $PCl_3(g)$ and $Cl_2(g)$, the two products, and whatever $PCl_5(g)$, the sole reactant, remains. The total pressure of this mixture is given as 0.895 atm. If this gaseous mixture follows Dalton's law, then its final total pressure is equal to the sum of the partial pressures of the components

$$P_{tot} = P_{PCl_3} + P_{Cl_2} + P_{PCl_5} = 0.895 \text{ atm}$$

The reaction

$$PCl_5(g) \rightleftharpoons Cl_2(g) + PCl_5(g)$$

---

[1] "Bz" is often used for $C_6H_5CH_2-$, the benzyl group.

is the sole reaction taking place because the PCl_5, Cl_2, and PCl_3 are the only chemical species in the vessel at equilibrium. The stoichiometry of the reaction requires that the partial pressures of PCl_3 and Cl_2 remain equal always. Let these two pressures equal x atm. Then the partial pressure of PCl_5 always equals $(0.895 - 2x)$ atm. At equilibrium at 250 K, the partial pressures satisfy the equation

$$K_{250} = 2.15 = \frac{P_{Cl_2} P_{PCl_3}}{P_{PCl_5}} = \frac{(x)(x)}{0.895 - 2x}$$

Solving (using the quadratic formula) gives $x = 0.40866$ and a physically meaningless root ($x = -4.7087$). The equilibrium partial pressures of the Cl_2 and the PCl_3 are both $\boxed{0.409 \text{ atm}}$, and the equilibrium partial pressure of the PCl_5 is $\boxed{0.078 \text{ atm}}$.

14.31 Write the equation for the reaction, and insert the given data in the usual table

	$Br_2(g)$	$+ I_2(g)$	$\rightleftharpoons$	$2 IBr(g)$
Init. pressure (atm)	0.0500	0.0400		0
Change in pressure (atm)	$-x$	$-x$		$+2x$
Equil. pressure (atm)	$0.0500 - x$	$0.0400 - x$		$2x$

$$\frac{P_{IBr}^2}{P_{Br_2} P_{I_2}} = 322 = \frac{(2x)^2}{(0.0500 - x)(0.0400 - x)}$$

Rearrangement leads to the quadratic equation $x^2 - 0.09113x + 2.025 \times 10^{-3} = 0$. Solving (using the quadratic formula) gives $x = 0.0384$ and $x = 0.0527$. The second root is "unphysical" because there was only 0.0400 atm of I_2 at the start. Decreasing by 0.0527 atm is impossible. The correct partial pressures come from the first root

$$P_{IBr} = \boxed{0.0768 \text{ atm}} \qquad P_{I_2} = \boxed{0.0016 \text{ atm}} \qquad P_{Br_2} = \boxed{0.0116 \text{ atm}}$$

Tip. Note the implicit assumption that the reaction given in the problem is the only reaction that takes place. In the laboratory unanticipated reactions can and often do occur.

14.33 Write the equation, the initial partial pressures, the change in the pressures required to reach equilibrium, and the final partial pressures

	$N_2(g)$	$+ O_2(g)$	$\rightleftharpoons$	$2 NO(g)$
Init. pressure (atm)	0.41	0.59		0.22
Change in pressure (atm)	$+x$	$+x$		$-2x$
Equil. pressure (atm)	$0.41 + x$	$0.59 + x$		$0.22 - 2x$

Note the signs in the second line of the table. The x's under the two reactants have positive signs while the x under the product has a negative sign. If x comes out to be a positive number, then the shift in coming to the equilibrium corresponds to the loss of product and the formation of reactants. The mass-action expression relates the equilibrium partial pressures to K, which is known at 25° (298 K)

$$K_{298} = 4.2 \times 10^{-31} = \frac{P_{NO}^2}{P_{N_2} P_{O_2}} = \frac{(0.22 - 2x)^2}{(0.41 + x)(0.59 + x)}$$

Before attempting to solve for x, consider the relative sizes of the numbers. This saves getting bogged down with algebra. The equilibrium constant is very small. Hence, the numerator of the equilibrium expression must be very small—almost all of the $NO(g)$ is consumed at equilibrium. Suppose that *all* of the $NO(g)$ reacts. Then, $2x = 0.22$ and $x = 0.11$. Using this value of x gives an equilibrium pressure of N_2 of $0.41 + 0.11$ or $\boxed{0.52 \text{ atm}}$. The equilibrium pressure of O_2 is, by a similar computation, $\boxed{0.70 \text{ atm}}$. To get the true (non-zero) equilibrium partial pressure of NO, substitute these two pressures back into the original expression

$$4.2 \times 10^{-31} = \frac{P_{NO}^2}{P_{N_2} P_{O_2}} = \frac{(P_{NO})^2}{(0.52)(0.70)}$$

Solving for P_{NO} gives $\boxed{3.9 \times 10^{-16} \text{ atm}}$. The reaction lies *exceedingly* far toward the reactants at equilibrium at 25°.

14.35 Copy the equation for the synthesis of ammonia that is given in the problem and write the corresponding mass-action expression

$$N_2(g) + 3\,H_2(g) \rightleftharpoons 2\,NH_3(g) \qquad \frac{P_{NH_3}^2}{P_{H_2}^3 P_{N_2}} = K_{298} = 6.78 \times 10^5$$

This can be done almost effortlessly. The difficulty starts with translating the other statements in the problem into mathematical terms. If the H to N atom ratio is 3 to 1 then

$$P_{H_2} = 3 P_{N_2}$$

because the third component, NH_3, maintains, within itself, the required 3 to 1 ratio of atoms. The fact that the total pressure is 1.00 atm means

$$P_{N_2} + P_{H_2} + P_{NH_3} = 1.00 \text{ atm}$$

Let x equal the equilibrium partial pressure of N_2. Then $P_{H_2} = 3x$ and $P_{NH_3} = 1.00 - 4x$ at equilibrium. Substitution in the mass-action expression gives

$$6.78 \times 10^5 = \frac{(1.00 - 4x)^2}{(3x)^3 x} = \frac{(1.00 - 4x)^2}{27x^4}$$

This equation is not very hard to solve analytically (see below), but is it necessary even to try? The x is expected to be small because at equilibrium the mixture is mostly NH_3 (K is big). Using this idea simplifies the algebra. Suppose $4x \ll 1.00$. Then

$$6.78 \times 10^5 \approx \frac{1.00}{27x^4} \quad \text{from which} \quad x \approx 0.0153$$

Now, improve this answer by guessing new values for x and computing the right-hand side of the equation for each. Base each new guess on the outcome of the previous computation. Soon, the computed value becomes sufficiently close to 6.78×10^5. The following table maps such a process. It starts with the x from the rough solution

x	$(1.00 - 4x)^2$	$27x^4$	$(1.00 - 4x)^2/27x^4$
0.0153	0.8813	1.480×10^{-6}	5.96×10^5
0.0145	0.8874	1.194×10^{-6}	7.43×10^5
0.0149	0.8843	1.331×10^{-6}	6.64×10^5
0.0147	0.8859	1.261×10^{-6}	7.03×10^5
0.0148	0.8851	1.295×10^{-6}	6.83×10^5
0.01483	0.8849	1.306×10^{-6}	6.776×10^5

Therefore x equals 0.01483 to four significant digits. Then

$$P_{N_2} = \boxed{0.0148 \text{ atm}} \quad P_{H_2} = \boxed{0.0445 \text{ atm}} \quad P_{NH_3} = \boxed{0.941 \text{ atm}}$$

Tip. Obtain the analytical solution of the equation

$$6.78 \times 10^5 = \frac{(1.00 - 4x)^2}{27x^4}$$

by multiplying through by 27 and then taking the square root of both sides. This gives

$$4.278 \times 10^3 = \frac{1.00 - 4x}{x^2} \quad \text{from which comes} \quad (4.278 \times 10^3)x^2 + 4x - 1.00 = 0$$

Substitution into the quadratic formula affords two roots

$$x = \frac{-b \pm \sqrt{b^2 - 4ac}}{2a} = \frac{-4 \pm \sqrt{16 + 17122}}{8556} = +0.0148 \text{ and } -0.0158$$

The first root is the answer obtained by approximation; the second is physically meaningless.

14.37 The reaction and corresponding mass-action expression in terms of partial pressures are

$$CO(g) + Cl_2(g) \rightleftharpoons COCl_2(g) \qquad \frac{P_{COCl_2}}{P_{CO}P_{Cl_2}} = K_{873} = 0.20$$

where the 0.20, the value of the equilibrium constant, comes from problem **14.6** in the text. Write the mass-action expression in terms of concentrations

$$\frac{[COCl_2]}{[CO][Cl_2]} = 0.20 \left(\frac{RT}{P_{ref}} \right)^{-\Delta n_g}$$

where Δn_g equals the change in the number of moles of gas from left to right in the reaction (-1 in this case because there is 1 mol of gas on the right and 2 mol on the left of the equation), and P_{ref} is the reference pressure used in obtaining the given K (1.00 atm in this case). Insert these values along with the gas constant R and the absolute temperature T into the preceding equation

$$\frac{[COCl_2]}{[CO][Cl_2]} = 0.20 \left(\frac{(0.08206 \text{ L atm mol}^{-1}\text{K}^{-1})(873.15 \text{ K})}{1 \text{ atm}} \right)^{-(-1)} = 14.3$$

Finally, substitute the equilibrium concentrations of the CO and Cl_2

$$\frac{[COCl_2]}{[2.3 \times 10^{-4}][1.7 \times 10^{-2}]} = 14.3$$

and solve for $[COCl_2]$. The answer is $\boxed{5.6 \times 10^{-5} \text{ mol L}^{-1}}$.

14.39 Write mass-action expressions for the reduction of iron(III) oxide with hydrogen and the reduction of carbon dioxide with hydrogen. Let these expressions correspond to the form of the equation given in the problem

$$\frac{1}{P_{H_2}^3} = K_1 = 4.0 \times 10^{-6} \qquad \text{and} \qquad \frac{P_{CO}}{P_{CO_2}P_{H_2}} = K_2 = 3.2 \times 10^{-4}$$

If both reactions are simultaneously at equilibrium in the same container, then both equations must be satisfied simultaneously. Compute the partial pressure of the hydrogen from the first equation

$$P_{H_2} = \sqrt[3]{\frac{1}{K_1}} = \sqrt[3]{\frac{1}{4.0 \times 10^{-6}}} = 63 \text{ atm}$$

Insert this partial pressure into the (slightly rearranged) second equation

$$\frac{P_{CO}}{P_{CO_2}} = K_2 P_{H_2} = (3.2 \times 10^{-4})63 = \boxed{0.020}$$

14.41   **a)** The reaction and mass-action expression are

$$NH_3(g) + HCl(g) \rightleftharpoons NH_4Cl(s) \qquad \frac{1}{P_{NH_3} P_{HCl}} = K_{613} = 4.0$$

If P_{NH_3} is 0.80 atm at equilibrium, then, by substitution, the equilibrium partial pressure of $HCl(g)$ is $\boxed{0.31 \text{ atm}}$.

b) The equilibrium cannot occur before addition of $NH_4Cl(s)$ to the container filled with $NH_3(g)$ because the $NH_4Cl(s)$ is the only source of $HCl(g)$. For every mole of $HCl(g)$ that is produced, one mole of $NH_3(g)$ joins the quantity of $NH_3(g)$ that was responsible for the original 1.50 atm. The partial pressures of the $NH_3(g)$ and $HCl(g)$ are directly proportional to their respective chemical amounts (assuming ideal-gas behavior). This means that for every atmosphere of $HCl(g)$ that is produced one additional atmosphere of $NH_3(g)$ is also produced. Let x equal the equilibrium partial pressure of $HCl(g)$. Then the equilibrium partial pressure of $NH_3(g)$ is $1.50 + x$ and

$$P_{NH_3} P_{HCl} = (1.50 + x)x = 0.25 \qquad \text{which gives} \qquad x^2 + 1.50x - 0.25 = 0$$

Solving the quadratic equation gives x equal to 0.151 (the negative root is rejected). Therefore, $P_{HCl} = \boxed{0.15 \text{ atm}}$, and $P_{NH_3} = 1.50 + 0.15 = \boxed{1.65 \text{ atm}}$.

14.43   **a)** The container holds no gas until some of the $NH_4HSe(s)$ decomposes. Breakdown of the solid is the only source of the two gases than eventually fill the container at equilibrium. The stoichiometry of the decomposition reaction requires that the partial pressure of the $H_2Se(g)$ equal the partial pressure of the $NH_3(g)$ (assuming ideal-gas behavior). The total pressure equals the sum of the partial pressures of the two gases. In equation form

$$P_{H_2Se} = P_{NH_3} \qquad \text{and} \qquad P_{H_2Se} + P_{NH_3} = 0.0184 \text{ atm}$$

Clearly both partial pressures equal 0.00920 atm at equilibrium. The equilibrium constant is

$$K = (P_{H_2Se})_{eq}(P_{NH_3})_{eq} = (0.00920)^2 = \boxed{8.46 \times 10^{-5}}$$

b) The size of the container has no effect on the K. The two partial pressures can differ, but their product must equal K at equilibrium

$$8.46 \times 10^{-5} = P_{H_2Se} P_{NH_3} = P_{H_2Se}(0.0252) \quad \text{hence:} \quad P_{H_2Se} = \boxed{0.00336 \text{ atm}}$$

The Direction of Change in Chemical Reactions: Empirical Description

14.45   **a)** The reaction quotient Q has the form of a mass-action expression. Computations of K require the substitution of *equilibrium* partial pressures into this mathematical form. Computations of Q on the other hand may employ whatever partial pressures might temporarily prevail. In the following, the subscript zero means initial P's were used

$$Q_0 = \frac{(P_{Al_3Cl_9})_0^2}{(P_{Al_2Cl_6})_0^3} = \frac{(1.02 \times 10^{-2})^2}{(0.473)^3} = \boxed{9.83 \times 10^{-4}}$$

b) From the solution to problem **14.21**, $K_{454} = 1.04 \times 10^{-4}$. The initial reaction quotient Q_0 exceeds this K so that the reaction approaches equilibrium by "shifting" from the right to the left (generating reactants at the expense of products). The process $\boxed{\text{consumes } Al_3Cl_9}$ and produces Al_2Cl_6.

14.47   The fading of color means that the reaction consumes $Br_2(g)$ as it goes toward equilibrium: change proceeds from left to right in the reaction

$$H_2(g) + Br_2(g) \rightleftharpoons 2\,HBr(g)$$

The initial reaction quotient Q_0 is accordingly less than K_{700}. The data given in the problem allow computation of Q_0

$$Q_0 = \frac{(P_{HBr})_0{}^2}{(P_{H_2})_0(P_{Br_2})_0} = \frac{(0.90 \text{ atm})^2}{(0.40 \text{ atm})(0.40 \text{ atm})} = 5.1$$

Thus $\boxed{K_{700} \text{ must exceed } 5.1}$.

14.49 The "water gas" reaction and its associated mass-action expression are

$$C(s) + H_2O(g) \rightleftharpoons CO(g) + H_2(g) \qquad \frac{P_{CO}P_{H_2}}{P_{H_2O}}$$

If the reaction quotient Q is less than K_{1000}, the reaction tends to proceed (to shift) from left to right at 1000 K; if Q is greater than K_{1000}, the reaction tends to proceed from right to left.

a) Substitute the data into the mass-action expression to obtain a Q

$$Q = \frac{P_{CO}P_{H_2}}{P_{H_2O}} = \frac{(1.525)(0.805)}{0.600} = \boxed{2.05}$$

This Q is less than K_{1000} (which is 2.6) so the reaction shifts from $\boxed{\text{left to right}}$ at 1000 K.

b) All three partial pressures are higher than in part **a)**. Use them in the mass-action expression

$$Q = \frac{P_{CO}P_{H_2}}{P_{H_2O}} = \frac{(1.714)(1.383)}{0.724} = \boxed{3.27}$$

Since Q now exceeds K_{1000}, the reaction shifts from $\boxed{\text{right to left}}$ to reach equilibrium at 1000 K.

14.51 **a)** Let P_{di} stand for the partial pressure of gaseous diphosphorus $P_2(g)$ and P_{tet} for the partial pressure of gaseous tetraphosphorus $P_4(g)$. For the process

$$\frac{P_{di}{}^2}{P_{tet}} = Q$$

Initially, $P_{di} = 2.00$ atm and $P_{tet} = 5.00$ atm making $Q_0 = \boxed{0.800}$. Because $Q_0 > K_{1473}$ the equilibrium shifts to the $\boxed{\text{left}}$ (reducing the numerator and increasing the denominator in the above fraction) at 1200°C (1473 K) until Q equals K.

b) Let x equal the increase in the pressure of $P_4(g)$ during the change. Then

$$K = \frac{(2.00 - 2x)^2}{(5.00 + x)} = 0.612$$

The equation can be solved with the quadratic formula. It is however instructive to get x numerically because progress toward an answer is a lot like the progress of the reaction toward equilibrium. Construct a table:

x	$(2.00 - 2x)^2$	$5.00 + x$	Q
0.000	4.00	5.00	0.800
0.100	3.24	5.10	0.635
0.120	3.10	5.12	0.605
0.115	3.13	5.115	0.612

As x increases Q decreases from 0.800. As the table shows, progress to a good answer is quick. Electronic calculators make it painless. At $x = 0.120$, Q is only slightly less than K. If $x = 0.115$,

$$P_{di} = 2.00 - 2(0.115) = \boxed{1.77 \text{ atm}} \quad \text{and} \quad P_{tet} = 5.00 + 0.115 = \boxed{5.12 \text{ atm}}$$

c) If the volume of the system is increased, then, by LeChâtelier's principle, there will be net $\boxed{\text{dissociation}}$ of P_4. The system responds to its forced rarefaction by producing more molecules to fill the larger volume.

14.53 Chemical systems always tend toward equilibrium.[2] If a stress is applied to a system at equilibrium, the system reacts to minimize the stress and to reach a new equilibrium.

a) The stress is the addition of $N_2O(g)$. The system reacts to decrease the concentration of N_2O. The reaction proceeds from $\boxed{\text{right to left}}$ until a new equilibrium is reached.

b) The stress is the reduction in volume. The partial pressures of all the compounds will momentarily rise. The equilibrium will then shift in such a way as to reduce the number of molecules of gas (chemical amount of gas) in the container and reduce the total pressure. There are three moles of gas on the reactant side of the equation and two moles of gas on the product side. The equilibrium will thus shift from $\boxed{\text{left to right}}$.

c) The reaction is exothermic. Cooling the mixture shifts the equilibrium from $\boxed{\text{left to right}}$ (to favor the products).

d) In order to maintain a constant pressure, the volume of the system must have increased. Thus, the reaction will shift from $\boxed{\text{right to left}}$.

e) The partial pressures of the reacting gases are unchanged by the addition of an inert gas, and the equilibrium law is independent of total pressure. There is $\boxed{\text{no effect}}$ on the position of the equilibrium.

14.55 **a)** The equilibrium constant increases with decreasing temperature. This means that decreasing the temperature favors the products. Because removing heat shifts the reaction to the right, heat must be generated on the right. The reaction is $\boxed{\text{exothermic}}$.

b) Reducing the volume shifts this particular equilibrium to the right. Shrinking favors the side of the reaction with fewer moles of gas. Hence, there is a net $\boxed{\text{decrease}}$ in the number of gas molecules from left to right in the reaction.

14.57 Good design would provide for transferring the heat generated in the chlorination of the ethene (the first reaction) into the dehydrochlorination of the dichloroethane (the second reaction). Removal of the product heat from the first reaction would tend to drive the first reaction toward the right; input of this heat to the second reaction would drive it toward its products. Good design would also arrange for the continuous removal of the gaseous products of both reactions. This would shift them in the desired direction.

The Direction of Change in Chemical Reactions: Thermodynamic Explanation

14.59 The ΔH of the reaction is clearly less than zero. The hydration of ethylene to give ethanol is thus exothermic. Think of the heat as a reaction product

$$C_2H_4(g) + H_2O(g) \longrightarrow C_2H_5OH(g) + \text{heat}$$

By LeChâtelier's principle, the equilibrium production of ethanol is maximized by running the reaction at $\boxed{\text{low temperature}}$ (which allows the "product" heat more readily to escape). There are two moles of gas on the reactant side and only one mol of gas of the product side; $\boxed{\text{high pressure}}$ will force the equilibrium from left to right, increasing the yield of ethanol.

14.61 Use the van't Hoff equation to calculate $\Delta H°$ for the reaction from the two K's and the temperatures at which the K's were measured

$$\ln \frac{K(T_2)}{K(T_1)} = \frac{-\Delta H°}{R} \left(\frac{1}{T_2} - \frac{1}{T_1} \right)$$

$$\ln \left(\frac{0.00121}{6.8} \right) = -8.634 = \frac{-\Delta H°}{8.3145 \text{ J K}^{-1}\text{mol}^{-1}} \left(\frac{1}{473.15 \text{ K}} - \frac{1}{298.15 \text{ K}} \right)$$

$$\Delta H° = -5.8 \times 10^4 \text{ J mol}^{-1}$$

---

[2]A system can sit around "tending" for a thousand years and make no progress. Still, the possibility of spontaneous change is there.

The answer is the (assumedly) constant $\Delta H°$ in the temperature range 25°C to 200°C. The "mol^{-1}" in the answer means per 1 mol of the reaction as it is written in the problem. Thus, the answer is $\boxed{-5.8 \times 10^4 \text{ J}}$.

14.63   **a)** Use the equation $\Delta G° = -RT \ln K$

$$\Delta G°_{298} = -RT \ln K_{298} = (-8.3145 \text{ J K}^{-1}\text{mol}^{-1})(298.15 \text{ K}) \ln(9.3 \times 10^9) = -56.9 \times 10^3 \text{ J mol}^{-1}$$

For one mole of the reaction as written, $\Delta G°_{298}$ is $\boxed{-56.9 \text{ kJ}}$.

b) Use the van't Hoff equation and the values of K_{298} and K_{398} to obtain $\Delta H°$. Then get $\Delta S°_{298}$ from $\Delta G°_{298}$ and the equation $\Delta G° = \Delta H° - T\Delta S°$

$$\ln\left(\frac{3.3 \times 10^7}{9.3 \times 10^9}\right) = \frac{-\Delta H°}{8.3145 \text{ J K}^{-1}\text{mol}^{-1}}\left(\frac{1}{398 \text{ K}} - \frac{1}{298 \text{ K}}\right)$$

$$\Delta H° = -55.6 \text{ kJ mol}^{-1}$$

For one mole of the reaction as written, $\Delta H°$ is $\boxed{-55.6 \text{ kJ}}$. This answer equals *both* $\Delta H°_{298}$ and $\Delta H°_{398}$ because it is assumed in the derivation of the van't Hoff equation that $\Delta H°$ is independent of temperature. Next,

$$\Delta S°_{298} = \frac{\Delta H°_{298} - \Delta G°_{298}}{T} = \frac{(-55.63 \times 10^3 \text{ J}) - (-56.88 \times 10^3 \text{ J})}{298 \text{ K}} = \boxed{4.2 \text{ J K}^{-1}}$$

Tip. Re-do the problem starting with K_{398} instead of K_{298}

$$\Delta G°_{398} = -RT \ln K_{398} = (-8.3145 \text{ J K}^{-1}\text{mol}^{-1})(398.15 \text{ K}) \ln(3.3 \times 10^7) = -57.31 \times 10^3 \text{ J mol}^{-1}$$

$$\Delta S°_{398} = \frac{\Delta H°_{398} - \Delta G°_{398}}{T} = \frac{(-55.63 \times 10^3 \text{ J}) - (-57.31 \times 10^3 \text{ J})}{398 \text{ K}} = 4.2 \text{ J K}^{-1}$$

This is a nice check because $\Delta S°_{398}$ and $\Delta S°_{298}$ come out equal. This has to happen because the derivation of the van't Hoff equation assumes that they are equal.

14.65   Use the van't Hoff equation to calculate K_{600} from K_{298} and the standard enthalpy of the reaction $\Delta H°$, both of which are given

$$\ln\left(\frac{K_{600}}{K_{298}}\right) = \frac{-\Delta H°}{R}\left(\frac{1}{T_2} - \frac{1}{T_1}\right)$$

$$\ln\left(\frac{K_{600}}{5.9 \times 10^5}\right) = \frac{-(-92.2 \times 10^3 \text{ J mol}^{-1})}{8.3145 \text{ J K}^{-1}\text{mol}^{-1}}\left(\frac{1}{600 \text{ K}} - \frac{1}{298 \text{ K}}\right)$$

$$K_{600} = \boxed{4.3 \times 10^{-3}}$$

14.67   **a)** Use the van't Hoff equation to calculate ΔH_{vap}. The equilibrium constants at T_1 and T_2 equal the vapor pressure of the liquid at T_1 and T_2

$$\ln\left(\frac{P_2}{P_1}\right) = -\frac{\Delta H_{\text{vap}}}{R}\left(\frac{1}{T_2} - \frac{1}{T_1}\right)$$

$$\ln\left(\frac{4.2380 \text{ atm}}{0.4034 \text{ atm}}\right) = \frac{-\Delta H_{\text{vap}}}{8.3145 \text{ J K}^{-1}\text{mol}^{-1}}\left(\frac{1}{273.15 \text{ K}} - \frac{1}{223.15 \text{ K}}\right)$$

$$\Delta H_{\text{vap}} = \boxed{23.8 \text{ kJ mol}^{-1}}$$

b) The normal boiling point T_{b} of a liquid is defined as the temperature at which the vapor pressure of the liquid equals 1 atm exactly. Therefore, set P_1 equal to 1.000 atm and T_1 equal to T_{b} in the previous equation. Set $T_2 = 273.15$ K and $P_2 = 4.2380$ atm because 4.2380 atm is the vapor pressure at 273.15 K

$$\ln\left(\frac{4.2380 \text{ atm}}{1.000 \text{ atm}}\right) = \frac{-23.8 \times 10^3 \text{ J}}{8.3145 \text{ J K}^{-1}\text{mol}^{-1}}\left(\frac{1}{273.15} - \frac{1}{T_b}\right)$$

$$T_b = \boxed{240 \text{ K}}$$

Using 0.4034 atm for P_2 along with 223.15 K for T_2 gives the same answer. The observed normal boiling point of liquid ammonia is 239.6 K.

14.69 **a)** The numbers to plot are the second-to-last and last rows of the following table. Put $\ln K$ on the vertical axis and $1/T$ on the horizontal axis.

Temp. / K	276.9	288.5	298.2	308.2	323.4
Equil. Constant K	1160	841	689	533	409
$1/T$ / K^{-1}	0.003611	0.003466	0.003353	0.003245	0.003093
$\ln K$	7.06	6.73	6.54	6.28	6.01

b) A form of the van't Hoff equation is

$$\ln K = \frac{-\Delta H^\circ}{R}\left(\frac{1}{T}\right) + \frac{\Delta S^\circ}{R}$$

Comparing this equation to the equation of a straight line ($y = mx + b$) reveals that the slope m of the straight line that results from plotting $\ln K$ versus $1/T$ equals $-\Delta H^\circ/R$. The slope m of the best straight line in the preceding graph comes out to 2030.7 K. Therefore

$$\Delta H^\circ = -mR = -(2030.7 \text{ K})(8.3145 \text{ J K}^{-1}\text{mol}^{-1}) = \boxed{-16.9 \times 10^3 \text{ J mol}^{-1}}$$

Distribution of a Single Species between Immiscible Phases: Extraction and Separation Processes

14.71 Assume exactly 1 L each of H_2O and CCl_4. Then at equilibrium the water holds 1.30×10^{-4} mol of I_2, and the remaining I_2 must be in the CCl_4 layer

$$n_{I_2 (CCl_4)} = 1.00 \times 10^{-2} - 1.30 \times 10^{-4} = 9.9 \times 10^{-3} \text{ mol}$$

The concentration of I_2 in the CCl_4 layer is this amount divided by the volume

$$[I_2]_{(CCl_4)} = \frac{9.9 \times 10^{-3} \text{ mol}}{1 \text{ L}} = 9.9 \times 10^{-3} \text{ mol L}^{-1}$$

The partition coefficient, which is an equilibrium constant, equals the ratio of the two concentrations

$$K = \frac{[I_2]_{(CCl_4)}}{[I_2]_{(aq)}} = \frac{9.9 \times 10^{-3}}{1.30 \times 10^{-4}} = \boxed{76}$$

14.73 **a)** The mass-action expressions for the dissolution of benzoic acid in water (K_1) and the dissolution of benzoic acid in ether (K_2) are quite simple

$$K_1 = [C_6H_5COOH]_{(aq)} \text{ and } K_2 = [C_6H_5COOH]_{(ether)}$$

The concentrations of C_6H_5COOH at saturation (equilibrium) in water and in ether are

$$c_{water} = \frac{2.00 \text{ g } C_6H_5COOH}{1 \text{ L water solution}} \times \left(\frac{1 \text{ mol } C_6H_5COOH}{122 \text{ g } C_6H_5COOH}\right) = 0.0164 \text{ mol L}^{-1}$$

$$c_{ether} = \frac{660 \text{ g } C_6H_5COOH}{1 \text{ L ether solution}} \times \left(\frac{1 \text{ mol } C_6H_5COOH}{122 \text{ g } C_6H_5COOH}\right) = 5.4 \text{ mol L}^{-1}$$

Hence K_1 is $\boxed{0.0164}$, and K_2 is $\boxed{5.4}$.

b) The partition reaction equals reaction 1 in the previous part subtracted from reaction 2. Subtracting a reaction is the same as reversing it and adding it. The equilibrium constant for the partition reaction (reaction 3) is accordingly

$$K_3 = K_2\left(\frac{1}{K_1}\right) = \frac{5.4}{0.0164} = \boxed{330}$$

ADDITIONAL PROBLEMS

14.75 **a)** The partial pressures of the gases are in direct proportion to their chemical amounts (assuming that Dalton's law and the ideal-gas law apply)

$$P_{gas} = n_{gas}\left(\frac{RT}{V}\right)$$

Suppose that the volume of the container is such that the actual chemical amounts of the four gases are 90, 470, 200, and 45 mol. Then, the partial pressure of the $BCl_3(g)$ is $P_{BCl_3} = 90(RT/V)$. Expressions for the partial pressures of the other three gases are similar. The mass-action expression for reaction 1 is

$$\frac{P_{BFCl_2}^3}{P_{BCl_3}^2 P_{BF_3}} = K_1$$

Substituting the three partial pressures gives

$$\frac{(45RT/V)^3}{(90RT/V)^2(470RT/V)} = K_1 = \frac{(45)^3}{(90)^2(470)} = \boxed{0.024}$$

The RT/V's canceled out! The cancellation means the volume of the container has no effect on the position of this particular equilibrium. This conclusion justifies taking an arbitrary volume for the size of the system. A similar procedure with the mass-action expression for reaction 2 gives $K_2 = \boxed{0.40}$.

b) The given equation (equation 3) equals the sum of equations 1 and 2 divided by three. Obtain the required equilibrium constant by taking the cube root of $(K_1 \times K_2)$. This gives $\boxed{0.21}$.

The same answer is obtained by substituting the original partial pressures into the mass-action expression for reaction 3, but the preceding emphasizes how K_3 derives entirely from K_1 and K_2; K_3 is a new number but adds no new information.

14.77 **a)** Both the *cis* and *trans* isomers of 1-methyl-2-ethylcyclopropane are gaseous at 425.6°C. Assume that the ideal-gas law and Dalton's law apply. Suppose that 1.000 mol of the *cis* form is initially present. Then the equilibrium chemical amounts of the *cis* and *trans* isomers are

	cis $\rightleftharpoons$	*trans*
Initial amount (mol)	1.000	0
Change (mol)	$-x$	$+x$
Equilibrium amount (mol)	$1.000 - x$	x

At equilibrium, 73.6% of the *cis* isomer has been converted to the *trans*. Therefore x is 0.736 mol. This makes the equilibrium chemical amounts of the two isomers

$$n_{cis} = 1.000 - x = 1.000 - 0.736 = 0.264 \text{ mol}$$

$$n_{trans} = x = 0.736 \text{ mol}$$

The equilibrium partial pressures of the two isomers are

$$P_{cis} = \frac{n_{cis}RT}{V} = \frac{(0.264)RT}{V}$$

$$P_{trans} = \frac{n_{trans}RT}{V} = \frac{(0.736)RT}{V}$$

Substitution in the mass-action expression gives K at 698.7 K (425.6°C)

$$K_{699} = \frac{P_{trans}}{P_{cis}} = \frac{(0.736)RT/V}{(0.264)RT/V} = \boxed{2.79}$$

b) From the preceding answer it is clear that at equilibrium

$$P_{trans} = 2.788(P_{cis}) \quad \text{and therefore} \quad n_{trans} = 2.788(n_{cis})$$

The combined chemical amount of the two isomers stays equal to 0.525 mol because one molecule of the *trans* isomer forms for every one molecule of the *cis* isomer that is consumed

$$n_{trans} + n_{cis} = 0.525 \text{ mol}$$

Solving the two equations in two unknowns gives $n_{trans} = 0.386$ mol. Substitution in the ideal-gas equation then gives the answer

$$P_{trans} = n_{trans}\left(\frac{RT}{V}\right) = 0.386 \text{ mol}\left(\frac{(0.08206 \text{ L atm mol}^{-1}\text{K}^{-1})(698.75 \text{ K})}{15.00 \text{ L}}\right) = \boxed{1.48 \text{ atm}}$$

14.79 Imagine for simplicity that the SO_3/SO_2 equilibrium is established in a 1.000 L sealed container. The contents of this container then have a mass of 0.925 g. Assuming ideality, the chemical amount of the gas mixture in the container is

$$n_{gas} = \frac{PV}{RT} = \frac{(1 \text{ atm})(1.000 \text{ L})}{0.082057 \text{ L atm mol}^{-1}\text{K}^{-1}(627 + 273.15) \text{ K}} = 0.01354 \text{ mol}$$

A gas of the same mass that consists of undecomposed SO_3 contains somewhat fewer moles

$$n_{SO_3} = 0.925 \text{ g} \times \left(\frac{1 \text{ mol}}{80.06 \text{ g}} \right) = 0.01155 \text{ mol}$$

The partial decomposition of SO_3

$$SO_3(g) \rightleftharpoons SO_2(g) + \tfrac{1}{2} O_2(g)$$

causes the increase from 0.01155 mol of gas to 0.01354 mol of gas. If x mol of $SO_2(g)$ has formed when the reaction reaches equilibrium, then $x/2$ mole of $O_2(g)$ has also formed and x mol of $SO_3(g)$ has been consumed. The unknown x is readily computed

$$n_{tot} = n_{SO_3} + n_{SO_2} + n_{O_2} = (0.01155 - x) \text{ mol} + x \text{ mol} + x/2 \text{ mol}$$
$$0.01354 \text{ mol} = (0.01155 + x/2) \text{ mol}$$
$$x = 0.00398 \text{ mol}$$

The fractional dissociation or degree of dissociation α of the SO_3 equals the amount converted to products divided by the original amount

$$\alpha = \frac{(0.00398) \text{ mol}}{0.01155 \text{ mol}} = \boxed{0.345}$$

14.81   The set-up of this problem is exactly like the set-up given for problem **14.35** but now K is different, and the total pressure is 100 atm. The set-up yields the equation

$$3.19 \times 10^{-4} = \frac{(100 - 4x)^2}{(3x)^3 x}$$

where x is the equilibrium partial pressure of nitrogen. Factor the numerator and multiply out the denominator to obtain

$$3.19 \times 10^{-4} = \frac{16(25.0 - x)^2}{27x^4} \quad \text{which gives} \quad 5.383 \times 10^{-4} = \frac{(25.0 - x)^2}{x^4}$$

Taking the square root of both sides of this equation leads to a quadratic equation in x that can be solved routinely. It is also quite quick to solve by successive approximations. First, observe that x must be less than 25. If P_{N_2} exceeded 25 atm, the combined pressure of just two (the nitrogen and hydrogen) out of the three gases would exceed the total pressure in the container. Now, guess an x between 0 and 25 and compute the right-hand side of the equation. Revise the guess based on experience until the computed value becomes sufficiently close to 5.383×10^{-4}. The following table shows the process. The first guess is near the middle of the range:

x	$(25.0 - x)^2 / x^4$	Comment
15	1.97×10^{-3}	x is too small
20	1.56×10^{-4}	x is too big
18	4.67×10^{-4}	x is too big
17.5	6.00×10^{-4}	x is too small
17.7	5.43×10^{-4}	x is a bit too small
17.72	5.375×10^{-4}	x is acceptable

The answer is $x = 17.72$. If $x = 17.72$ then

$$P_{N_2} = \boxed{17.7 \text{ atm}} \qquad P_{H_2} = \boxed{53.2 \text{ atm}} \qquad P_{NH_3} = \boxed{29.1 \text{ atm}}$$

Tip. Check the answers by confirming that the sum of the three partial pressures is 100 atm and that

$$\frac{P_{NH_3}^2}{P_{H_2}^3 P_{N_2}} = \frac{(29.1)^2}{(53.2)^3 (17.7)} = 3.18 \times 10^{-4} = K$$

14.83 The first equation and its mass-action expression are

$$Cl_2(g) \rightleftharpoons Cl_2(aq) \qquad \frac{[Cl_2]}{P_{Cl_2}} = K$$

Take the equilibrium concentration of the dissolved chlorine and the partial pressure of the gaseous chlorine from the statement of the problem

$$K_1 = \frac{[Cl_2]}{P_{Cl_2}} = \frac{0.061}{1.00} = 0.061$$

The second reaction and its mass-action expression are

$$Cl_2(aq) + H_2O(l) \rightleftharpoons H^+(aq) + Cl^-(aq) + HOCl(aq) \qquad \frac{[HOCl][H^+][Cl^-]}{[Cl_2]} = K_2$$

K_2 is readily calculated because $[H^+]$ must be equal to $[Cl^-]$ and to $[HOCl]$, based on the 1-to-1-to-1 stoichiometry of the reaction

$$K_2 = \frac{[HOCl][H^+][Cl^-]}{[Cl_2]} = \frac{(0.030)(0.030)(0.030)}{(0.061)} = \boxed{4.4 \times 10^{-4}}$$

14.85 **a)** The decomposition reaction is $NH_4HS(s) \rightleftharpoons NH_3(g) + H_2S(g)$. One mole of gaseous ammonia forms per mole of gaseous hydrogen sulfide. Assuming that the product mixture follows Dalton's law, then at equilibrium the partial pressure of each gas equals half the total pressure

$$P_{NH_3} = P_{H_2S} = \frac{0.659 \text{ atm}}{2}$$

Insert these values into the mass-action expression and compute K

$$K = P_{NH_3} P_{H_2S} = \left(\frac{0.659}{2}\right)^2 = \boxed{0.109}$$

b) Raising the equilibrium partial pressure of $NH_3(g)$ to 0.750 atm will, by LeChâtelier's principle, reduce the partial pressure of $H_2S(g)$. It will not change the equilibrium constant.

$$P_{H_2S} = \frac{K}{P_{NH_3}} = \frac{0.109}{0.750} = \boxed{0.145 \text{ atm}}$$

Tip. In part **a)**, the 2 is used twice: once as a divisor and once as an exponent. This is all right.

14.87 The K for this reduction of nickel oxide to nickel relates the partial pressures of $CO(g)$ and $CO_2(g)$. The nickel and nickel(II) oxide are pure solids and do not enter the mass-action expression

$$\frac{P_{CO_2}}{P_{CO}} = K$$

The total pressure equals the sum of the two partial pressures

$$P_{CO_2} + P_{CO} = 2.50 \text{ atm}$$

Once the system is at equilibrium, the two equations relating the two partial pressures must be satisfied simultaneously. Therefore

$$\frac{P_{CO_2}}{(2.50 - P_{CO_2})} = K$$

At 754°C, K is 255.4. Substitute this value and solve: $P_{CO_2} = \boxed{2.49 \text{ atm}}$. Substitution back into the mass-action equation gives P_{CO} equal to $\boxed{9.75 \times 10^{-3} \text{ atm}}$.

14.89 **a)** The reaction and reaction-quotient expression are

$$PCl_5(g) \rightleftharpoons PCl_3(g) + Cl_2(g) \qquad \frac{P_{Cl_2}P_{PCl_3}}{P_{PCl_5}} = Q$$

The problem gives three initial partial pressures. Substitution gives an initial reaction quotient Q_0 of $\boxed{120}$. Since Q_0 exceeds K at 300°C, the reaction proceeds to the $\boxed{\text{left}}$ at 300°C.

b) Assume that the mixture of gases behaves ideally. As originally mixed, the reaction has $Q = 120$ and $K = 11.5$. Let y equal the partial pressure of $Cl_2(g)$ consumed by the right-to-left reaction. An equal partial pressure of $PCl_3(g)$ is also consumed, and an equal partial pressure of $PCl_5(g)$ is created. This follows from the stoichiometry of the equation. Therefore

$$K = 11.5 = \frac{(6.0 - y)(2.0 - y)}{(0.10 + y)} \qquad \text{from which} \quad y^2 - 19.5y + 10.85 = 0$$

The roots of the quadratic equation are 0.573 and 18.93. Only the first makes physical sense because the second gives negative partial pressures. Complete the computation by finding the various partial pressures

$$P_{PCl_3} = 2.0 - 0.573 = \boxed{1.4 \text{ atm}} \qquad P_{Cl_2} = 6.0 - 0.573 = \boxed{5.4 \text{ atm}}$$
$$P_{PCl_5} = 0.10 + 0.573 = \boxed{0.67 \text{ atm}}$$

c) By LeChâtelier's principle, an increase in volume causes the reaction to shift to the side having *more* moles of gas. The amount of $PCl_5(g)$ will $\boxed{\text{decrease}}$.

14.91 Equilibrium constants depend strongly on temperature. The equation in the problem gives the experimentally determined temperature dependence of the K of the hydrogenation reaction

$$C_5H_5N(g) + 3H_2(g) \rightleftharpoons C_5H_{11}N(g)$$

a) Substitute $T = 500$ K into the expression[3]

$$\log_{10} K = -20.281 + \frac{10,560 \text{ K}}{T} = -20.281 + \frac{10,560}{500} = 0.839$$

Taking the antilog of both sides gives $K = 10^{0.839} = \boxed{6.90}$.

b) Assume that the idea-gas law and Dalton's law apply and that the hydrogenation reaction has reached equilibrium. The fraction f of nitrogen in the form of C_5H_5N (pyridine or "py") equals the chemical amount of pyridine divided by the sum of the chemical amounts of pyridine and $C_5H_{11}N$ (piperidine or "pip"). By Dalton's law these chemical amounts are directly proportional to the partial pressures of the two compounds

$$f = \frac{n_{py}}{n_{py} + n_{pip}} = \frac{P_{py}}{P_{py} + P_{pip}}$$

At equilibrium the partial pressures must satisfy the law of mass action

$$\frac{P_{pip}}{P_{py} P_{H_2}^3} = K \qquad \text{which gives} \qquad \frac{P_{pip}}{P_{py}} = 6.90(1.00)^3$$

after substitution of $K = 6.90$ and $P_{H_2} = 1.00$ atm. Let the partial pressure of the pyridine be y atm. The partial pressure of the piperidine is then, by the preceding equation, $6.90y$ atm. Insert the P's into the equation for f

$$f = \frac{P_{py}}{P_{py} + P_{pip}} = \frac{y}{y + 6.90y} = \frac{1}{1 + 6.90} = \boxed{0.127}$$

---

[3]The equation that follows gives the temperature dependence correctly. In some printings of the text, the comma in "10,281" appears as a decimal point by mistake. If this were correct, K would be almost completely independent of temperature, an unlikely situation.

14.93   Equal chemical amounts of PCB-2 and PCB-11 exist in solution in some volume V of water. Then an equal volume of octanol is added and the mixture is vigorously shaken. Assume that the two solvents are perfectly immiscible, that is, that none of the octanol dissolves in the water and that none of the water dissolves in the octanol. Then the volume of each phase remains equal to V and the total volume of the system equals $2V$. The mixing and shaking do not change the chemical amounts of the two PCB's but do allow both to redistribute themselves between the two phases. Once the redistribution comes to equilibrium

$$n_{\text{PCB-2}} = [\text{PCB-2}]_{(aq)} V + [\text{PCB-2}]_{(oct)} V$$
$$n_{\text{PCB-11}} = [\text{PCB-11}]_{(aq)} V + [\text{PCB-11}]_{(oct)} V$$

where the "(aq)" and "(oct)" refer to the water and octanol phases. Setting the two chemical amounts equal to each other and dividing through by V gives

$$[\text{PCB-2}]_{(aq)} + [\text{PCB-2}]_{(oct)} = [\text{PCB-11}]_{(aq)} + [\text{PCB-11}]_{(oct)}$$

The mass-action expressions for the partition of the two PCB's between the two solvents are

$$K_2 = \frac{[\text{PCB-2}]_{(oct)}}{[\text{PCB-2}]_{(aq)}} \quad \text{and} \quad K_{11} = \frac{[\text{PCB-11}]_{(oct)}}{[\text{PCB-11}]_{(aq)}}$$

from which it follows that

$$[\text{PCB-2}]_{(oct)} = K_2[\text{PCB-2}]_{(aq)} \quad \text{and} \quad [\text{PCB-11}]_{(oct)} = K_{11}[\text{PCB-11}]_{(aq)}$$

Substitution of these expressions into the first equation eliminates all quantities with "oct" subscripts

$$[\text{PCB-2}]_{(aq)} + K_2[\text{PCB-2}]_{(aq)} = [\text{PCB-11}]_{(aq)} + K_{11}[\text{PCB-11}]_{(aq)}$$

Factoring common terms from each side of this equation gives

$$[\text{PCB-2}]_{(aq)} \left(1 + K_2\right) = [\text{PCB-11}]_{(aq)} \left(1 + K_{11}\right)$$

which is readily solved for the ratio of the concentrations of the two PCB's in the water phase

$$\frac{[\text{PCB-2}]_{(aq)}}{[\text{PCB-11}]_{(aq)}} = \frac{1 + K_{11}}{1 + K_2} = \frac{1 + 1.26 \times 10^5}{1 + 3.98 \times 10^4} = \boxed{3.17}$$

Tip. Try figuring out the ratio of the amounts of the two PCB's in the *octanol* phase rather than in the water phase. To start, divide the equation for $[\text{PCB-2}]_{(oct)}$ in terms of K_2 by the similar expression for $[\text{PCB-11}]_{(oct)}$ in terms of K_{11}

$$\frac{[\text{PCB-2}]_{(oct)}}{[\text{PCB-11}]_{(oct)}} = \frac{K_2[\text{PCB-2}]_{(aq)}}{K_{11}[\text{PCB-11}]_{(aq)}} = \frac{K_2}{K_{11}} \left(\frac{[\text{PCB-2}]_{(aq)}}{[\text{PCB-11}]_{(aq)}}\right)$$

The quantity in parentheses was just determined. Substitute for it and insert the numbers

$$\frac{[\text{PCB-2}]_{(oct)}}{[\text{PCB-11}]_{(oct)}} = \frac{K_2}{K_{11}} \left(\frac{1 + K_{11}}{1 + K_2}\right) = \frac{3.98 \times 10^4}{1.26 \times 10^5} \left(\frac{1 + 1.26 \times 10^5}{1 + 3.98 \times 10^4}\right) = 0.99998 \approx 1.00$$

Although the amount of PCB-2 entering the water phase is more than triple the amount of PCB-11, such small amounts of both depart the octanol phase that their relative amounts in that phase remain essentially unchanged.

14.95   **a)** The chemical amount of I_2 initially in the 0.100 L of aqueous solution is

$$n_{I_2} = 0.100 \text{ L} \times \left(\frac{2 \times 10^{-3} \text{ mol } I_2}{1.00 \text{ L}} \right) = 2 \times 10^{-4} \text{ mol } I_2$$

Shaking this solution with 0.025 L of CCl_4 in a separatory funnel allows the I_2 to distribute itself between the two phases. When the partition of the I_2 between the phases comes to equilibrium at 25°C, y mol of I_2 has moved into the CCl_4 phase, and $(2 \times 10^{-4} - y)$ mol remains in the aqueous phase. Assume that the two solvents are perfectly immiscible—that no water dissolves in the CCl_4 and that no CCl_4 dissolves in the water. Then the volumes of the two solvents remain unchanged and the concentrations of the solute in the two phases are

$$[I_2]_{(aq)} = \left(\frac{2 \times 10^{-4} - y}{0.100} \right) \text{ mol L}^{-1} \quad \text{and} \quad [I_2]_{(CCl_4)} = \left(\frac{y}{0.025} \right) \text{ mol L}^{-1}$$

The mass-action expression for this system is

$$\frac{[I_2]_{(CCl_4)}}{[I_2]_{(aq)}} = K = 85 \quad \text{from which} \quad \frac{y/0.025}{(2 \times 10^{-4} - y)/0.100} = 85$$

The last equation is easily solved for y, which equals 1.91×10^{-4} mol. Remember that y is the amount of I_2 that transfers to the CCl_4. and not a concentration. By subtraction, the amount remaining in the water is 0.09×10^{-4} mol. The fraction remaining equals the amount remaining divided by the original amount

$$f = \frac{0.09 \times 10^{-4}}{2 \times 10^{-4}} = 0.045 = \boxed{0.04}$$

b) The first extraction with 0.025 L of CCl_4 leaves only 0.045 (4.5%) of the I_2 in the water. Another extraction with a fresh 0.025 L of CCl_4 will leave only 0.045 of that 0.045. The fraction remaining after these successive treatments is

$$f = 0.045 \times 0.045 = \boxed{0.0020}$$

c) From text Example 14.18, the fraction of I_2 remaining in the water after a single 0.050 L extraction is 0.023, which is substantially larger than 0.0020.

Tip. It is about 11 times more efficient to extract the iodine with two half-sized portions of CCl_4 rather than one large portion. In general, it is more efficient to use several smaller portions of solvent, rather than one or two portions in performing separations by extraction.

CUMULATIVE PROBLEMS

14.97   The reaction is the splitting of tetraphosphorus into diphosphorus

$$P_4(g) \rightleftharpoons 2 P_2(g)$$

a) The pressure of a sample of $P_4(g)$ will always exceed the pressure predicted by the ideal-gas law, because the reaction to give $P_2(g)$ furnishes extra molecules to strike the sides of the container. Consider a P-V experiment performed on a sample of $P_4(g)$ at constant temperature. In the experiment, the pressure is tracked as the volume of the sample is changed, and the results are plotted with P on the vertical axis and $1/V$ on the horizontal axis. Boyle's law predicts a straight-line plot: P rising in direct proportion as $1/V$ rises. A plot of the actual experimental results will be a line that curves toward $\boxed{\text{higher}}$ pressure at low values of $1/V$.

b) Consider a V-T experiment in which the volume of a sample is $P_4(g)$ is tracked as the temperature is changed and the pressure is kept constant. Charles's law predicts that the volume should rise linearly with the temperature. But increasing the temperature shifts the endothermic equilibrium toward the right, generating more molecules in the sample. The "extra" molecules cause the sample to expand an extra amount with temperature. Thus, a plot of the volume of the P_4 sample as a function of T curves toward ┌ higher ┐ volumes than predicted by Charles's law as T increases.

14.99 Even the very strong bond in N_2 breaks at a high enough temperature. Let the partial pressure of $N_2(g)$ before any of its molecules break down equal P_0. Let the fraction that has broken down at equilibrium equal α. The value of α is 0.0065 at 5000 K but rises to 0.116 at 6000 K. Represent the approach to equilibrium at either temperature in the usual way

	$N_2(g)$	$\rightleftharpoons$	$2\,N(g)$
Init. Pressure (atm)	P_0		0
Change in Pressure (atm)	$-\alpha P_0$		$+2\alpha P_0$
Equil. Pressure (atm)	$P_0 - \alpha P_0$		$2\alpha P_0$

The total pressure of the equilibrium mixture equals 1.000 atm at both temperatures

$$P_{N_2} + P_N = (P_0 - \alpha P_0) + 2\alpha P_0 = P_0(1 + \alpha) = 1.000 \text{ atm}$$

Since α is given at both temperatures, the original pressure of N_2 at both temperatures is readily computed from the preceding equation: P_0 was 0.9935 atm at 5000 K and 0.8961 atm at 6000 K. Combine these original pressures with α as indicated above to get the equilibrium partial pressures of N and N_2 at the two temperatures

Temperature	5000 K	6000 K
Equil. P_{N_2} / atm	0.9871	0.7921
Equil. P_N / atm	0.0129	0.2079

The equilibrium partial pressures add up to 1.000 atm at both temperatures, as they must. Substitute these equilibrium partial pressures into the mass-action expression to obtain K's at the two temperatures

$$K_{5000} = \frac{P_N^2}{P_{N_2}} = \frac{(0.0129)^2}{0.9871} = 1.69 \times 10^{-4} \qquad K_{6000} = \frac{P_N^2}{P_{N_2}} = \frac{(0.2079)^2}{0.7921} = 5.46 \times 10^{-2}$$

Next, use the van't Hoff equation to estimate ΔH° for the reaction from the two K's and their temperatures

$$\ln \frac{K_{6000}}{K_{5000}} = \frac{-\Delta H^\circ}{R} \left(\frac{1}{6000 \text{ K}} - \frac{1}{5000 \text{ K}} \right)$$

Substitution gives

$$\ln \left(\frac{5.46 \times 10^{-2}}{1.69 \times 10^{-4}} \right) = 5.78 = \frac{-\Delta H^\circ}{8.3145 \text{ J K}^{-1}\text{mol}^{-1}} \left(\frac{1}{6000 \text{ K}} - \frac{1}{5000 \text{ K}} \right)$$

Solving for ΔH° gives 1440 kJ mol^{-1}. Hence the standard enthalpy of the reaction as written is ┌ 1440 kJ ┐.

Tip. The answer is only a crude estimate because ΔH° and ΔS° do change over the 1000 K range. In fact, 1440 kJ exceeds the N≡N bond enthalpy[4] of 945 kJ mol^{-1}, which is correct at 298 K, by over 50%.

---

[4] Given in text Table 12.5.

14.101 Use ΔG_f°'s from text Appendix D to calculate ΔG_{298}° for the reaction. Then compute the equilibrium constant from ΔG_{298}°

$$\Delta G_{298}^{\circ} = 1\underbrace{(0.00)}_{Ni(s)} + \tfrac{1}{2}\underbrace{(0.00)}_{O_2(g)} - 1\underbrace{(-211.7)}_{NiO(s)} = 211.7 \text{ kJ}$$

$$\ln K_{298} = \frac{-\Delta G_{298}^{\circ}}{RT} = \frac{-(211.7 \times 10^3 \text{ J mol}^{-1})}{(8.3145 \text{ J K}^{-1}\text{mol}^{-1})(298.15 \text{ K})} = -85.40$$

$$K_{298} = 8.2 \times 10^{-38}$$

The partial pressure of oxygen is related to K through the mass-action expression for this reaction. Because two of the three substances involved in the reaction are pure solids, this expression is very simple: $\sqrt{P_{O_2}} = K$. It follows that the equilibrium pressure of oxygen is the square of the equilibrium constant. It equals $\boxed{6.7 \times 10^{-75} \text{ atm}}$. The decomposition of $NiO(s)$ to its elements is *very* slight at room temperature![5]

14.103 The reaction is the sublimation of water, $H_2O(s) \rightleftharpoons H_2O(g)$. The K_{273} for this reaction equals the vapor pressure of ice at 273.15 K, which is 0.0060 atm. Ice at its freezing point sublimes spontaneously whenever the pressure of $H_2O(g)$ is less than 0.0060 atm. Cold ice sublimes less easily. Use the van't Hoff equation to calculate K for sublimation at $-15°C$ (258.15 K)

$$\ln\left(\frac{K_{258}}{K_{273}}\right) = \frac{-\Delta H_{sub}}{R}\left(\frac{1}{258.15 \text{ K}} - \frac{1}{273.15 \text{ K}}\right)$$

$$\ln\left(\frac{K_2}{0.0060}\right) = \frac{-50.0 \times 10^3 \text{ J mol}^{-1}}{8.3145 \text{ J K}^{-1}\text{mol}^{-1}}\left(\frac{1}{258.15 \text{ K}} - \frac{1}{273.15 \text{ K}}\right)$$

$$K_{258} = 0.0017$$

Therefore ice at 258.15 K sublimes if the partial pressure of $H_2O(g)$ is $\boxed{\text{less than 0.0017 atm}}$.

14.105 Water is a better solvent for more polar covalent and ionic substances, and carbon tetrachloride (CCl_4) is a better solvent for non-polar covalent compounds. On this basis

a) The polar covalent compound methanol (CH_3OH) has a larger concentration in $\boxed{H_2O}$.

b) The non-polar covalent compound hexachloroethane (C_2Cl_6) has a larger concentration in $\boxed{CCl_4}$.

c) Bromine (Br_2), which has a non-polar covalent bond, has a larger concentration in $\boxed{CCl_4}$.

d) The ionic compound sodium chloride ($NaCl$) has a larger concentration in $\boxed{H_2O}$.

14.107 a) Represent the dimerization reaction as

$$2\,NO_2(g) \rightleftharpoons N_2O_4(g)$$

Text Appendix D gives the ΔG_f°'s of $NO_2(g)$ and $N_2O_4(g)$ at 298.15 K. Use them to compute ΔG°

$$\Delta G^{\circ} = \Delta G_f^{\circ}(N_2O_4) - 2\,\Delta G_f^{\circ}(NO_2) = 97.82 \text{ kJ mol}^{-1} - 2\,(51.29 \text{ kJ mol}^{-1}) = -4.76 \text{ kJ mol}^{-1}$$

The ΔG_{298}° for the reaction as written above is accordingly $\boxed{-4.76 \text{ kJ}}$.

The logarithm of the equilibrium constant depends on ΔG_T° as follows

$$\ln K_T = \frac{\Delta G_T^{\circ}}{-RT}$$

$$\ln K_{298} = \frac{-4.76 \times 10^3 \text{ J mol}^{-1}}{-(8.3145 \text{ J K}^{-1}\text{mol}^{-1})(298.15 \text{ K})} = 1.92$$

$$K_{298} = \exp(1.92) = \boxed{6.8}$$

---

[5]Effectively no molecules of O_2 form.

b) The ΔG of a reaction at a given temperature depends on both its standard Gibbs energy $\Delta G°$ at that temperature and the relative amounts of the various reactants and products. The reaction quotient Q gives the second effect

$$\Delta G = \Delta G° + RT \ln Q = \Delta G° + RT \ln \left(\frac{P_{N_2O_4}}{(P_{NO_2})^2} \right)$$

$$\Delta G_{298} = -4.76 \times 10^3 \text{ J mol}^{-1} + (8.3145 \text{ J K}^{-1}\text{mol}^{-1})(298.15 \text{ K}) \ln \left(\frac{0.01}{0.01^2} \right)$$

$$= -4.76 \times 10^3 \text{ J mol}^{-1} + 11.4 \times 10^3 \text{ J mol}^{-1} = 6.65 \times 10^3 \text{ J mol}^{-1}$$

Hence the ΔG of this reaction is $\boxed{6.65 \text{ kJ}}$. A positive ΔG means that the forward reaction is non-spontaneous, but that the reverse reaction is; the reaction tends to proceed from $\boxed{\text{right to left}}$.

14.109 The reaction is

$$CO_2(g) + H_2(g) \rightleftharpoons CO(g) + H_2O(g) \qquad K_{3500} = 8.28$$

Use the fundamental relationship between K and $\Delta G°$

$$\Delta G°_{3500} = -RT \ln K_{3500} = -(8.3145 \text{ J K}^{-1}\text{mol}^{-1})(3500 \text{ K}) \ln 8.28 = -61.5 \times 10^3 \text{ J mol}^{-1}$$

This means that $\Delta G°_{3500}$ for 1 mol of reaction is $\boxed{-61.5 \text{ kJ}}$.

The standard Gibbs energy is an exact measure of the driving force of a reaction only in the very special case that all reactants and products are present in standard states. When this is not the case, it is necessary to compute ΔG (without the naught), which is the non-standard (actual) Gibbs energy

$$\Delta G_{3500} = \Delta G°_{3500} + RT \ln Q = \Delta G°_{3500} + RT \ln \left(\frac{P_{CO} P_{H_2O}}{P_{CO_2} P_{H_2}} \right)$$

$$= -61.5 \times 10^3 \text{ J mol}^{-1} + (8.3145 \text{ J K}^{-1}\text{mol}^{-1})(3500 \text{ K}) \ln \left(\frac{(2)(2)}{(0.1)(0.1)} \right)$$

$$= -61.5 \times 10^3 \text{ J mol}^{-1} + 174.4 \times 10^3 \text{ J mol}^{-1} = 112.9 \times 10^3 \text{ J mol}^{-1}$$

The answer is accordingly $\boxed{112.9 \text{ kJ}}$ for the partial pressures of the gases that are specified in the statement of the problem. The reaction runs spontaneously from $\boxed{\text{right to left}}$. This is the reverse of its direction when the reactants and products are present in standard states at 3500 K.

14.111 a) The balanced equation and its mass-action expression are

$$AgCl \cdot NH_3(s) \rightleftharpoons AgCl(s) + NH_3(g) \qquad \boxed{K = P_{NH_3}}$$

b) Addition of $AgCl(s)$ has $\boxed{\text{no effect}}$ on the equilibrium partial pressure of NH_3. The addition of a pure solid does not change its activity in an equilibrium system.

c) Pumping in NH_3 has $\boxed{\text{no effect}}$ on the equilibrium partial pressure of NH_3 as long as some $AgCl(s)$ remains. The added NH_3 simply reacts with $AgCl(s)$ until P_{NH_3} again equals K.

d) Lowering the temperature of an endothermic reaction decreases its equilibrium constant. The P_{NH_3} therefore $\boxed{\text{decreases}}$ in order to stay equal to K.

Chapter 15

Acid-Base Equilibria

Classifications of Acids and Bases

15.1   **a)** The chloride ion Cl^- $\boxed{\text{cannot act}}$ as a Brønsted-Lowry acid because it cannot possibly donate a hydrogen ion.

b) The hydrogen sulfate ion HSO_4^- can act as a Brønsted-Lowry acid; its conjugate base is $\boxed{SO_4^{2-}}$ (the sulfate ion).

c) The ammonium ion NH_4^+ can act as a Brønsted-Lowry acid; its conjugate base is $\boxed{NH_3}$ (ammonia).

d) Ammonia NH_3 can act as a Brønsted-Lowry acid; its conjugate base is $\boxed{NH_2^-}$ (the amide ion). Formation of this ion occurs in non-aqueous solvents.

e) Water H_2O can act as a Brønsted-Lowry acid; its conjugate base is $\boxed{OH^-}$ (the hydroxide ion).

15.3   Citric acid donates hydrogen ion to the hydrogen carbonate ion, $\boxed{HCO_3^-(aq)}$, which serves as a base:

$$H_3C_5H_5O_7(aq) + HCO_3^-(aq) \rightleftharpoons H_2C_5H_5O_7^-(aq) + H_2CO_3(aq)$$

The H_2CO_3 that is produced quickly decomposes to water and gaseous carbon dioxide, which causes the cookies to rise. A preliminary step is the dissolution of the solid sodium hydrogen carbonate in the lemon juice.

15.5   **a)** The slaking of lime: $\boxed{CaO(s) + H_2O(l) \longrightarrow Ca(OH)_2(s)}$.

b) The reaction can be seen as a Lewis acid-base reaction. The $\boxed{CaO}$ is the Lewis base. It donates a pair of electrons (located on the oxide ion) to a hydrogen atom in the H_2O molecule. The result is a new O—H bond.

15.7   **a)** The fluoride ion (F^-) has a negative charge. In the Brønsted-Lowry system, an acid is a donor of a positive entity (the H^+ ion). By plus-minus symmetry then, an acid in this scheme is a $\boxed{\text{fluoride acceptor}}$.

b) In $ClF_3O_2 + BF_3 \longrightarrow ClF_2O_2 \cdot BF_4$, BF_3 accepts a F^- ion from ClF_3O_2,

In $TiF_4 + 2\,KF \longrightarrow K_2[TiF_6]$, TiF_4 accepts an F^- ion from KF, so $\boxed{TiF_4}$ is the acid, and $\boxed{KF}$ is the base.

Tip. TiF_5^- forms from TiF_4 after *one* F^- ion is accepted. This intermediate species acts as an acid when accepting the second F^- ion. If it donated an F^- ion to some other species, then it would be acting as a base.

15.9   Oxides of metals are base anhydrides; oxides of nonmetals are acid anhydrides.

a) MgO is the base anhydride of magnesium hydroxide $Mg(OH)_2$.

b) Cl_2O is the acid anhydride of hypochlorous acid HOCl.

c) SO_3 is the acid anhydride of sulfuric acid H_2SO_4.

d) Cs_2O is the base anhydride of cesium hydroxide CsOH.

15.11   The equations illustrating the amphoteric behavior of SnO(s) are

$$\overset{\text{base}}{SnO}\,(s) + 2\,HCl(aq) \longrightarrow Sn^{2+}(aq) + 2\,Cl^-(aq) + H_2O(l)$$

$$\overset{\text{acid}}{SnO}\,(s) + NaOH(aq) + H_2O(l) \longrightarrow Sn(OH)_3^-(aq) + Na^+(aq)$$

Properties of Acids and Bases in Aqueous Solutions: The Brønsted-Lowry Scheme

15.13   The pH of an aqueous solution equals the negative logarithm of the hydronium-ion concentration
$$pH = -\log[H_3O^+] = -\log(2.0 \times 10^{-4}) = \boxed{3.70}.$$

15.15   The extremes of the pH range for urine each give a H_3O^+ concentration

$$[H_3O^+]_{high} = 10^{-5.5} = \boxed{3 \times 10^{-6} \text{ M}} \qquad [H_3O^+]_{low} = 10^{-6.5} = \boxed{3 \times 10^{-7} \text{ M}}$$

The pOH comes from the pH using the relation $pOH = pK_w - pH$.

$$pOH = 14.0 - 5.5 = 8.5 \qquad pOH = 14.0 - 6.5 = 7.5$$

where the use of $pK_w = 14.0$ assumes a temperature of 25°C. Then

$$[OH^-]_{low} = 10^{-8.5} = \boxed{3 \times 10^{-9} \text{ M}} \qquad [OH^-]_{high} = 10^{-7.5} = \boxed{3 \times 10^{-8} \text{ M}}$$

15.17   The pH of the seawater equals 8.00. Using the definition of pH:

$$[H_3O^+] = 10^{-8.00} = \boxed{1.0 \times 10^{-8} \text{ M}}$$

Use 13.776 instead of 14.00 as pK_w when calculating pOH:

$$pOH = pK_w - pH = 13.776 - 8.00 = 5.78 \qquad [OH^-] = 10^{-5.78} = \boxed{1.7 \times 10^{-6} \text{ M}}$$

15.19   The equation $\boxed{2\,K(s) + 2\,H_2O(l) \longrightarrow 2\,KOH(aq) + H_2(g)}$ is the better representation of what really happens. The reaction starts fast and continues with great vigor. If the equation involving H_3O^+ applied, one would expect a slower process since H_3O^+ is only 10^{-7} mol L^{-1} in pure water. Even if the reaction were vigorous at low concentrations of H_3O^+, the reaction generates $OH^-(aq)$, which lowers the concentration of H_3O^+. Loss of H_3O^+ would quickly cause progress to flag. The other equation represents a direct interaction between K(s) and $H_2O(l)$. The concentration of H_2O remains relatively constant and high (about 56 mol L^{-1}).

Acid and Base Strength

15.21   **a)** A base in the Brønsted-Lowry definition is a hydrogen-ion acceptor. In ephedrine and many other organic bases, the site at which the hydrogen ion attaches is a nitrogen atom

$$\boxed{C_{10}H_{15}ON(aq) + H_2O(l) \rightleftharpoons C_{10}H_{15}ONH^+(aq) + OH^-(aq)}$$

b) Use the equation $K_a K_b = K_w$, which relates the strength of an aqueous acid and its conjugate base. Assume that the given K_b is the value at 25°C. Because $K_w = 1.0 \times 10^{-14}$ at this temperature

$$K_a = \frac{K_w}{K_b} = \frac{1.0 \times 10^{-14}}{1.4 \times 10^{-4}} = \boxed{7.1 \times 10^{-11}}$$

c) According to text Table 15.2 (page 623), the K_a for ammonium ion NH_4^+ equals 5.6×10^{-10} at $25°C$. The K_b for ammonia NH_3 at this temperature is then

$$K_b = \frac{K_w}{K_a} = \frac{1.0 \times 10^{-14}}{5.6 \times 10^{-10}} = 1.8 \times 10^{-5}$$

The larger the K_b the stronger the base. Because K_b for ephedrine (1.4×10^{-4}) exceeds K_b for ammonia (1.8×10^{-5}), ephedrine is a $\boxed{\text{stronger}}$ base than ammonia.

15.23 The equation given in the problem can be obtained by combining the following two equations that have K_a's in text Table 15.2

$$H_2O(l) + HClO_2(aq) \rightleftharpoons H_3O^+(aq) + ClO_2^-(aq) \qquad K_a = 1.1 \times 10^{-2}$$
$$H_2O(l) + HNO_2(aq) \rightleftharpoons H_3O^+(aq) + NO_2^-(aq) \qquad K_a = 4.6 \times 10^{-4}$$

Subtract the second equation from the first. The desired equilibrium constant is then the K_a of the first divided by the K_a of the second

$$HClO_2(aq) + NO_2^-(aq) \rightleftharpoons HNO_2(aq) + ClO_2^-(aq) \qquad K = \frac{1.1 \times 10^{-2}}{4.6 \times 10^{-4}} = \boxed{24}$$

In this case the constant exceeds 1, which means that the products are favored at equilibrium. The relatively small concentration of $HClO_2$ at equilibrium means that $\boxed{HClO_2}$ is a $\boxed{\text{stronger acid}}$ than HNO_2; the comparatively large concentration of HNO_2 at equilibrium means $\boxed{NO_2^-}$ is a $\boxed{\text{stronger base}}$ than ClO_2^-.

15.25 **a)** The color changes of most indicators become complete over a range of 1 to 1.9 pH units.[1] Assume that the center of the range is roughly equal to the pK_a of the indicator. Then

$$\begin{array}{llll}
\text{bromocresol green} & pK_a = 4.6 & K_a = 3 \times 10^{-5} \\
\text{methyl orange} & pK_a = 3.8 & K_a = 2 \times 10^{-4}
\end{array}$$

The acid form of $\boxed{\text{methyl orange}}$ is the stronger acid because it has the larger K_a (smaller pK_a).

b) Text Figure 15.12 shows that bromocresol green is green and methyl orange is orange only in their respective transition ranges. This means that the pH of the solution must simultaneously lie between pH 3.8 and pH 5.4 (bromocresol green) and 3.2 and 4.4 (methyl orange). Therefore, the pH of the solution lies in the range $\boxed{3.8\text{--}4.4}$.

Equilibria Involving Weak Acids and Bases

15.27 Abbreviate the formula for aspirin $HC_9H_7O_4$ and its conjugate base as aspH and asp$^-$ respectively. Compute the concentration of aspH when 0.65 g of it dissolves in 50.0 mL of water

$$n_{aspH} = 0.65 \text{ g aspH} \times \left(\frac{1 \text{ mol aspH}}{180.16 \text{ g aspH}} \right) = 0.00361 \text{ mol}$$

$$[aspH] = \frac{0.00361 \text{ mol}}{0.0500 \text{ L}} = 0.0722 \text{ mol L}^{-1} = 0.0722 \text{ M}$$

This result equals the "initial" concentration of aspH: *after* dissolution but *before* any acid-base reaction with water. Next, consider the reaction between the aspirin and water

$$aspH(aq) + H_2O(l) \rightleftharpoons asp^-(aq) + H_3O^+(aq) \qquad \frac{[H_3O^+][asp^-]}{[aspH]} = K_a = 3.0 \times 10^{-4}$$

---

[1]See text Figure 15.12, page 629

Call the concentration of H_3O^+ at equilibrium x:

	aspH(aq)	$+H_2O(aq)$	$\rightleftharpoons$	$asp^-(aq)$ +	$H_3O^+(aq)$
Init. Conc. (M)	0.0722	—		0	small
Change in Conc. (M)	$-x$	—		$+x$	$+x$
Equil. Conc. (M)	$0.0722 - x$	—		x	x

Substitute the equilibrium concentrations into the mass-action expression

$$\frac{[asp^-][H_3O^+]}{[aspH]} = K_a = \frac{x^2}{0.0722 - x} = 3.0 \times 10^{-4}$$

Solve for x using the quadratic formula or by successive approximation. The answer is $x = 4.5 \times 10^{-3}$. A hydrogen-ion concentration of 4.5×10^{-3} M translates to a pH of $\boxed{2.35}$.

Tip. Neglecting x compared to 0.0722 gives a hydrogen-ion concentration that is over 3% too high and a pH of 2.33, which is close, but incorrect.

15.29   **a)** Benzoic acid is C_6H_5COOH.[2] As the 0.20 M C_6H_5COOH reacts with water, it generates H_3O^+

	$C_6H_5COOH(aq)$	$+H_2O(l)$	$\rightleftharpoons$	$C_6H_5COO^-(aq)$ +	$H_3O^+(aq)$
Init. Conc. (M)	0.20	—		0	small
Change in Conc. (M)	$-x$	—		$+x$	$+x$
Equil. Conc. (M)	$0.20 - x$	—		x	x

The concentration of H_3O^+ arising from the autoionization of water is very small compared to the concentration from the reaction of benzoic acid. Neglecting it means that x equals the concentration of H_3O^+ at equilibrium. Insert x in the mass-action expression

$$K_a = 6.46 \times 10^{-5} = \frac{[C_6H_5COO^-][H_3O^+]}{[C_6H_5COOH]} = \frac{x^2}{0.20 - x}$$

Rearrange to obtain: $x^2 + 6.46 \times 10^{-5}x - 1.29 \times 10^{-5} = 0$. Apply the quadratic formula:

$$x = \frac{-6.46 \times 10^{-5} \pm \sqrt{4.17 \times 10^{-9} + 5.16 \times 10^{-5}}}{2}$$

$$x = \frac{-6.46 \times 10^{-5} \pm 7.18 \times 10^{-3}}{2} = -0.00362 \text{ and } 0.00356$$

The negative root has no physical meaning. Using the positive root gives $[H_3O^+] = 0.00356$ M. The pH is $-\log[H_3O^+] = \boxed{2.45}$.

Tip. If x is neglected in comparison to 0.20 in the mass-action equation, then the very simple equation $x = \sqrt{(0.20)(6.46 \times 10^{-5})}$ results. The positive root of this equation is $x = 0.0036$. This equals, to two significant figures, the answer obtained using the quadratic equation. To get a feel for the way that successful approximation works, carry out a few computations both exactly and approximately and compare the results.

b) The equilibrium concentration of $[H_3O^+]$ must equal 3.56×10^{-3} M. Start as in the preceding part, but now x is known and $[HOAc]_0$, the initial concentration of acetic acid is the goal

$$K_a = \frac{x^2}{[HOAc]_0 - x} \quad \text{hence} \quad 1.76 \times 10^{-5} = \frac{(3.56 \times 10^{-3})^2}{[HOAc]_0 - 3.56 \times 10^{-3}}$$

Solving gives $[HOAc]_0 = 0.72$ M. This means $\boxed{0.72 \text{ mol}}$ of acetic acid must be dissolved per liter of solution.

---

[2] Again, the single H segregated in the formula is the acidic hydrogen.

15.31   Use the approach of problem **15.29a**

$$HIO_3(aq) + H_2O(aq) \rightleftharpoons IO_3^-(aq) + H_3O^+(aq)$$

Init. Conc. (M)	0.100	–	0	small
Change in Conc. (M)	$-x$	–	$+x$	$+x$
Equil. Conc. (M)	$0.100 - x$	–	x	x

$$\frac{[IO_3^-][H_3O^+]}{[HIO_3]} = K_a \quad \text{hence} \quad \frac{x^2}{0.100 - x} = 0.16$$

This can be rearranged to obtain $x^2 + 0.16x - 0.016 = 0$. Substitution in the quadratic formula then gives

$$x = \frac{-0.16 \pm \sqrt{0.0256 - 4(-0.016)}}{2} = 0.070 \text{ and} - 0.23$$

If $[H_3O^+] = 0.070$ M, then pH = $\boxed{1.2}$. The pH is well on the acid side because HIO_3 is a rather strong weak acid.

Tip. The short-cut of neglecting x in comparison to 0.100 gives an H_3O^+ concentration of 0.13 M and a pH of 0.9, which is seriously wrong.

15.33   The $papH^+Cl^-$, a salt, dissolves completely in water to give $papH^+$ ion and Cl^- ion. The Cl^- ion does not react significantly with the water, but the $papH^+$ ion reacts as a weak acid

$$papH^+(aq) + H_2O(l) \rightleftharpoons pap(aq) + H_3O^+(aq)$$

Assume that this reaction is the sole source of H_3O^+. The concentration of H_3O^+ is 4.9×10^{-4} M (calculated from the pH of 3.31), and the concentration of the conjugate base pap is also 4.9×10^{-4} M. The concentration of $papH^+$ equals its original concentration minus the portion converted into pap. This is $(0.205 - 4.9 \times 10^{-4})$ M. Substitute these equilibrium values into the K_a expression

$$K_a = \frac{(4.9 \times 10^{-4})(4.9 \times 10^{-4})}{(0.205 - 4.9 \times 10^{-4})} = \boxed{1.2 \times 10^{-6}}$$

Tip. "Forgetting" to subtract in the denominator has a negligible effect.

15.35   Morphine $C_{17}H_{19}O_3N$ is a potent opiate. Its K_b applies to the reaction

$$C_{17}H_{19}O_3N(aq) + H_2O(l) \rightleftharpoons C_{17}H_{19}O_3NH^+(aq) + OH^-(aq)$$

The initial concentration of morphine (*after* dissolution but *before* reaction with water) is

$$[C_{17}H_{19}O_3N]_0 = \frac{0.0400 \text{ mol}}{0.600 \text{ L}} = 0.0667 \text{ mol L}^{-1} = 0.0667 \text{ M}$$

Let x equal the change in the concentration of $C_{17}H_{19}O_3N$ in coming to equilibrium

$$C_{17}H_{19}O_3N(aq) + H_2O(aq) \rightleftharpoons C_{17}H_{19}O_3NH^+(aq) + OH^-(aq)$$

Init. Conc. (M)	0.0667	–	0	small
Change in Conc. (M)	$-x$	–	$+x$	$+x$
Equil. Conc. (M)	$0.0667 - x$	–	x	x

Substitute the equilibrium concentrations into the mass-action expression

$$\frac{[C_{17}H_{19}O_3NH^+][OH^-]}{[C_{17}H_{19}O_3N]} = K_b = \frac{x^2}{0.0667 - x} = 8 \times 10^{-7}$$

This equation can be solved using the quadratic formula. But, observe that x must be quite small compared to 0.0667. Neglecting the x in the denominator allows the quick conclusion that $x = 2.31 \times 10^{-4}$, corresponding to an equilibrium concentration of OH^- equal to 2.3×10^{-4} M. The pOH is $-\log[OH^-] = 3.64$, and the pH is this number subtracted from 14.0, which is $\boxed{10.4}$. Use of one significant figure (the 4) in the final answer reflects the fact that K_b has only one significant figure.

Tip. It is assumed that the temperature is 25°C. The value of K_w changes with temperature. The more the temperature differs from 25°C the more approximate the equation $pH + pOH = 14.0$ becomes. The value of K_b of course changes with temperature as well.

15.37   Hydrofluoric acid is a weak acid in water

$$HF(aq) + H_2O(l) \rightleftharpoons F^-(aq) + H_3O^+(aq) \quad K_a = 6.6 \times 10^{-4}$$

Because the pH of the HF solution at 25°C is 2.13,

$$[H_3O^+] = \text{antilog}(-2.13) = 10^{-2.13} = 7.4 \times 10^{-3} \text{ M}$$

Since the dissociation of hydrofluoric acid is the only important source of H_3O^+ in the solution, $[F^-]$ is also 7.41×10^{-3} M. The existence of the equilibrium guarantees that

$$\frac{[H_3O^+][F^-]}{[HF]} = K_a = 6.6 \times 10^{-4}$$

All of the concentrations except [HF] are known. Solve for [HF]

$$[HF] = \frac{[H_3O^+][F^-]}{K_a} = \frac{(7.41 \times 10^{-3})^2}{6.6 \times 10^{-4}} = \boxed{0.083 \text{ M}}$$

15.39   Aqueous NaOH contains Na^+ ions and OH^- ions. The latter react with acetic acid molecules to produce water and acetate ions. Acetate ion is a weak base in its own right. Its reaction (as a base) with water causes the solution of sodium acetate that results from treating acetic acid with an equal chemical amount of NaOH (a procedure called neutralization) to be basic, and have $\boxed{\text{pH greater than 7}}$.

Tip. One might guess that "neutralization" gives solutions that are neutral. Not so. Chemical neutralization of an acid by a base in aqueous solution does *not* in most cases give a neutral (pH 7) solution. When a weak acid is reacted with a strong base, the pH at the equivalence point exceeds 7. When a weak base is reacted with a strong acid, the pH at the equivalence point is less than 7. Exact neutralization of a strong acid by a strong base does give a solution of pH 7.

15.41   The ammonium bromide dissolves to give $NH_4^+(aq)$, a weak acid, and $Br^-(aq)$, which is not active as a base. The $NH_4^+(aq)$ ion reacts weakly with water to generate $H_3O^+(aq)$. The NH_4Br solution is therefore somewhat acidic. The hydrogen chloride reacts completely with water to give $Cl^-(aq)$ and $H_3O^+(aq)$, thereby creating a strongly acidic solution. The sodium hydroxide, a strong base, dissolves to $Na^+(aq)$ and $OH^-(aq)$; the latter makes the solution highly basic. The $NaCH_3COO$ gives $Na^+(aq)$ and $CH_3COO^-(aq)$, which reacts weakly as a base. The KI dissolves to $K^+(aq)$ and $I^-(aq)$, neither of which reacts as either acid or base. The order of increasing pH is therefore

$$\boxed{HCl < NH_4Br < KI < NaCH_3COO < NaOH}$$

Buffer Solutions

15.43   Assume that the buffer is prepared at 25°C. Compute the pK_a of the conjugate acid of tris[3] from the pK_b of tris itself

$$pK_a = 14.00 - pK_b = 14.00 - 5.92 = 8.08$$

---

[3]The molecular formula of tris is $(HOCH_2)_3CNH_2$.

The addition of HCl converts some tris to its conjugate acid

$$\text{tris}(aq) + \text{HCl}(aq) \rightleftharpoons \text{trisH}^+(aq) + \text{Cl}^-(aq)$$

The resulting solution is a mixture of a weak acid (trisH^+) and its conjugate base (tris). It is a buffer by virtue of the reaction

$$\text{trisH}^+(aq) + \text{H}_2\text{O}(l) \rightleftharpoons \text{tris}(aq) + \text{H}_3\text{O}^+(aq) \qquad K_a = \frac{[\text{tris}][\text{H}_3\text{O}^+]}{[\text{trisH}^+]}$$

Compute the concentrations of the tris and trisH^+ after complete reaction with the HCl but before the preceding equilibrium is established

$$[\text{tris}]_0 = \frac{(0.050 - 0.025 \text{ mol})}{2.00 \text{ L}} = 0.0125 \text{ M} \qquad \text{and} \qquad [\text{trisH}^+]_0 = \frac{0.025 \text{ mol}}{2.00 \text{ L}} = 0.0125 \text{ M}$$

The equilibrium now reduces the concentration of the trisH^+ as it forms H_3O^+ and tris in equal amounts. If x is the equilibrium concentration of $\text{H}_3\text{O}^+(aq)$ ion, then

	$\text{trisH}^+(aq)$	$+\text{H}_2\text{O}(aq) \rightleftharpoons$	$\text{tris}(aq) +$	$\text{H}_3\text{O}^+(aq)$
Init. Conc. (M)	0.0125	–	0.0125	small
Change in Conc. (M)	$-x$	–	$+x$	$+x$
Equil. Conc. (M)	$0.0125 - x$	–	$0.0125 + x$	x

$$K_a = \frac{[\text{tris}][\text{H}_3\text{O}^+]}{[\text{trisH}^+]} = \frac{(0.0125 + x)x}{(0.0125 - x)}$$

Assume that x is small compared to 0.0125. Then the 0.0125's cancel out and

$$[\text{H}_3\text{O}^+] = K_a \quad \text{so that} \quad \text{pH} = \text{p}K_a = \boxed{8.08}$$

Clearly x is less than 10^{-7}, so the assumption was justified.

Tip. The pH equals the pK_a of the weak acid. This is a general result in buffer solutions in which the concentrations of an acid and its conjugate base are equal (but not extremely low).

15.45 **a)** After mixing and dissolution but before any other chemical change, the solution is 0.10 M in acetic acid and 0.040 M in acetate ion (from the sodium acetate). Then the weak-acid equilibrium comes into play. Let x equal the equilibrium concentration of H_3O^+

	$\text{HOAc}(aq)$	$+\text{H}_2\text{O}(aq) \rightleftharpoons$	$\text{OAc}^-(aq) +$	$\text{H}_3\text{O}^+(aq)$
Init. Conc. (M)	0.10	–	0.040	small
Change in Conc. (M)	$-x$	–	$+x$	$+x$
Equil. Conc. (M)	$0.10 - x$	–	$0.040 + x$	x

$$K_a = 1.76 \times 10^{-5} = \frac{[\text{OAc}^-][\text{H}_3\text{O}^+]}{[\text{HOAc}]} = \frac{(0.040 + x)x}{(0.10 - x)}$$

Assume that x is small compared to 0.10. If so, the x's that are added and subtracted can be omitted in the above expression to obtain

$$x = 1.76 \times 10^{-5} \left(\frac{0.10}{0.040} \right) = 4.4 \times 10^{-5}$$

The result for x clearly justifies the assumption. It is only about 0.1% of 0.040. The pH is $-\log(4.4 \times 10^{-5})$ or $\boxed{4.36}$.

b) The addition of 0.010 mol of $OH^-(aq)$ ion (in the form of NaOH) very quickly converts 0.010 mol of acetic acid (HOAc) to 0.010 mol of acetate ion (OAc^-). The concentrations of HOAc and OAc^- right after the conversion but before the reaction of HOAc as a weak acid are

$$[\text{HOAc}] = \frac{(0.050 - 0.010)\ \text{mol}}{0.500\ \text{L}} = 0.080\ \text{M} \quad \text{and} \quad [\text{OAc}^-] = \frac{(0.020 + 0.010)\ \text{mol}}{0.500\ \text{L}} = 0.060\ \text{M}$$

Now consider the K_a equilibrium. Let y equal the equilibrium concentration of H_3O^+

	$HOAc(aq)$	$+H_2O(aq) \rightleftharpoons$	$OAc^-(aq) +$	$H_3O^+(aq)$
Init. Conc. (M)	0.080	—	0.060	small
Change in Conc. (M)	$-y$	—	$+y$	$+y$
Equil. Conc. (M)	$0.080 - y$	—	$0.060 + y$	y

$$K_a = \frac{[\text{OAc}^-][\text{H}_3\text{O}^+]}{[\text{HOAc}]} = 1.76 \times 10^{-5} = \frac{(0.060 + y)y}{(0.080 - y)}$$

Assume that y is small compared to 0.060 and 0.080. Then

$$y = 1.76 \times 10^{-5} \left(\frac{0.080}{0.060}\right) = 2.34 \times 10^{-5}$$

The assumption is clearly justified. The pH is the negative logarithm of 2.34×10^{-5} or $\boxed{4.63}$.

Tip. The pH rises only from 4.36 to 4.63, despite addition of strong base amounting to 20% of the amount of weak acid present. Such resistance to changes in pH characterizes buffered solutions.

15.47 Buffer solutions are most efficient at resisting changes in pH at their **buffer points**. At the buffer point, the concentrations of the conjugate pair are equal, and the pH of the buffer equals the pK_a of the weak acid. The physician should therefore select a weak acid having a pK_a as close as possible to the desired pH. The best choice on the list is $\boxed{m\text{-chlorobenzoic acid}}$, $pK_a = 3.98$.

15.49 The answer is certainly less than 1000 mL, because adding that much NaOH solution to 500 mL of 0.100 M aqueous formic acid would neutralize all of the acid by the reaction

$$\text{HCOOH}(aq) + \text{NaOH}(aq) \longrightarrow \text{Na}^+(aq) + \text{HCOO}^-(aq) + \text{H}_2\text{O}(l)$$

and create a dilute solution of sodium formate. Such solutions are known to have pH's exceeding 7 (see problem **15.39**). Before any NaOH is added, the solution contains mostly un-ionized formic acid, $\text{HCOOH}(aq)$ and also a small amount of the formate ion, $\text{HCOO}^-(aq)$. Adding the NaOH solution converts $\text{HCOOH}(aq)$ to $\text{HCOO}^-(aq)$ according to the preceding chemical equation. The addition raises the pH and also dilutes the whole system. It does *not* alter the total amount of formate-containing species, which keeps its original value:

$$n_{\text{HCOOH}} + n_{\text{HCOO}^-} = \left(\frac{0.100\ \text{mmol}}{\text{mL}}\right)(500\ \text{mL}) = 50.0\ \text{mmol}$$

The concentrations of formic acid and formate ion are related by the acid-ionization equilibrium

$$\text{HCOOH}(aq) + \text{H}_2\text{O}(l) \rightleftharpoons \text{HCOO}^-(aq) + \text{H}_3\text{O}^+(aq) \qquad \frac{[\text{H}_3\text{O}^+][\text{HCOO}^-]}{[\text{HCOOH}]} = K = 1.77 \times 10^{-4}$$

At pH = 4.00, $[\text{H}_3\text{O}^+] = 1.0 \times 10^{-4}$ M. Inserting this value into the K_a expression gives:

$$1.77 \times 10^{-4} = \frac{(1.0 \times 10^{-4})[\text{HCOO}^-]}{[\text{HCOOH}]} \qquad \text{which gives} \qquad 1.77 = \frac{[\text{HCOO}^-]}{[\text{HCOOH}]}$$

Let the volume of 0.0500 M NaOH that is needed to raise the pH to 4.00 equal V mL. The final volume of the mixture at pH 4.00 then equals $(500 + V)$ mL. Each millimole of NaOH that is added converts one millimole of HCOOH(aq) to HCOO$^-$(aq). Assume that this acid-base reaction is the only significant source of HCOO$^-$(aq). After 0.0500 V mmol of NaOH has been added

$$[\text{HCOO}^-] = \frac{0.0500 V \text{ mmol}}{(500 + V) \text{ mL}}$$

in which the numerator is the chemical amount of HCOO$^-$(aq) produced by the reaction and the denominator is the final volume of the solution. Also

$$[\text{HCOO}^-] + [\text{HCOOH}] = \frac{50.0 \text{ mmol}}{(500 + V) \text{ mL}}$$

Solve the second of the preceding equations for [HCOOH] and insert the expression for [HCOO$^-$]

$$[\text{HCOOH}] = \frac{50.0}{500 + V} - [\text{HCOO}^-] = \frac{50.0}{500 + V} - \frac{0.0500 V}{500 + V}$$
$$= \frac{50.0 - 0.0500 V}{500 + V} \text{ M}$$

Substitute the expressions for [HCOOH] and [HCOO$^-$] into the equation for their ratio

$$1.77 = \frac{[\text{HCOO}^-]}{[\text{HCOOH}]} = \frac{0.0500 V/(500 + V)}{(50.0 - 0.0500 V)/(500 + V)}$$
$$= \frac{0.0500 V}{50.0 - 0.0500 V} = \frac{0.0500 V}{0.0500 (1000 - V)}$$
$$1.77 = \frac{V}{1000 - V}$$

It is now easy to solve for V, which equals $\boxed{639 \text{ mL}}$. The final volume of the solution is 1139 mL and the concentrations of HCOOH(aq) and HCOO$^-$(aq) are 0.0158 and 0.0280 M respectively. Both are large in comparison to the final H$_3$O$^+$ concentration, 1.0×10^{-4} M. This means that the ionization equilibrium affects these two concentrations only negligibly.

Acid-Base Titration Curves

15.51   Assume that the temperature is 25°C. The substance Ba(OH)$_2$ is a strong base in water. It ionizes completely in solution to give one mole of Ba^{2+} ion and two moles of OH$^-$ ion per mole dissolved. Before any acid is added, [OH$^-$] = 2 × 0.3750 = 0.7500 M. The pOH, which equals the negative logarithm of this number, is 0.1249; pH= 14.000− pOH = 14.00 − 0.1249 = $\boxed{13.88}$.

The chemical amount of OH$^-$ ion in the original 100.0 mL of Ba(OH)$_2$ equals its molarity multiplied by its volume (in liters). It is 0.07500 mol. This means that attaining the equivalence point requires 0.07500 mol of HClO$_4$. The volume of 0.4540 M HClO$_4$ that provides this much HClO$_4$ is

$$V = \frac{1 \text{ L}}{0.4540 \text{ mol HClO}_4} \times 0.07500 \text{ mol HClO}_4 = 0.1652 \text{ L} = 165.2 \text{ mL}$$

When the titration is 1.00 mL short of the equivalence point, only 164.2 mL of 0.4540 M HClO$_4$ has been added for a total of 0.07455 mol of HClO$_4$. Some OH$^-$ ion remains unreacted. Its amount equals the difference between the amount of OH$^-$ originally present and the amount reacted. The concentration of OH$^-$ equals this same amount divided by the volume of the solution

$$[\text{OH}^-] = \frac{(0.07500 - 0.07455) \text{ mol}}{(0.1000 + 0.1642) \text{ L}} = 0.0017 \text{ M}$$

Note the (correct) use of two significant figures in the preceding result. The pOH is 2.77, and the pH is therefore $\boxed{11.23}$.

The pH reaches $\boxed{7.00}$ at the equivalence point; this is a titration of a strong base with a strong acid.

When the titration is 1.00 mL past the equivalence point, all of the OH^- from $Ba(OH)_2$ has been reacted away, and excess $HClO_4$ is present. The amount of excess is

$$n_{HClO_4} = \left(\frac{0.4540 \text{ mol}}{1 \text{ L}} \times 0.1662 \text{ L} \right) - 0.07500 \text{ mol} = 4.5 \times 10^{-4} \text{ mol } HClO_4$$

The excess $HClO_4$, a strong acid, is completely ionized to produce 4.5×10^{-4} mol of H_3O^+ (and, of course, an equal amount of ClO_4^- ion). The concentration of this H_3O^+ is

$$[H_3O^+] = \frac{4.5 \times 10^{-4} \text{ mol}}{(0.1000 + 0.1662) \text{ L}} = 0.0017 \text{ M}$$

Hence, $pH = -\log(0.0017) = \boxed{2.77}$. Throughout this analysis, the autoionization of water is ignored. Even very small amounts of strong acid or base completely overshadow water as a source of H_3O^+ or OH^- ion.

Tip. The pH plummets dramatically (from 11.24 to 2.77) upon addition of only 2 mL of acid in the range of the equivalence point.

15.53   Assume that the temperature is 25°C. Hydrazoic acid, a weak acid, reacts with water

$$HN_3(aq) + H_2O(l) \rightleftharpoons N_3^-(aq) + H_3O^+(aq) \qquad \frac{[N_3^-][H_3O^+]}{[HN_3]} = K_a$$

• *Before Addition of Base.* The initial concentration of the HN_3 is 0.1000 M. Let x equal the concentration of H_3O^+ present at equilibrium in the solution. Then

$$\frac{[N_3^-][H_3O^+]}{[HN_3]} = K_a = 1.9 \times 10^{-5} = \frac{x^2}{0.1000 - x}$$

where it has been assumed that hydrazoic acid is the only significant source of H_3O^+. Rearranging gives the quadratic equation

$$x^2 + (1.9 \times 10^{-5})x - 1.9 \times 10^{-6} = 0$$

The positive root of this equation is 1.37×10^{-3}. Therefore

$$[H_3O^+] = 1.37 \times 10^{-3} \text{ M} \quad \text{and} \quad pH = \boxed{2.86}$$

• *After Addition of 25.00 mL of Base.* Sodium hydroxide is a strong base. Each added mole of NaOH converts one mole of $HN_3(aq)$ to one mole of $N_3^-(aq)$. The 25.00 mL of 0.1000 M NaOH furnishes

$$\frac{0.1000 \text{ mol } OH^-}{1 \text{ L}} \times 0.0250 \text{ L} = 2.500 \times 10^{-3} \text{ mol } OH^-$$

Assume that the added OH^- reacts completely with the HN_3. The reaction produces 2.500×10^{-3} mol of N_3^- and leaves 2.500×10^{-3} mol of HN_3. Exactly half of the hydrazoic acid is reacted– this is the **half-equivalence point** of the titration. The total volume of the solution is 0.0750 L, so the "original" concentrations of the weak acid and its conjugate base are both 0.0333 M. "Original" is in quotation marks because these concentrations apply at the state after the mixing of the solutions but before the acid-base equilibrium gets established. As it becomes established, the equilibrium generates H_3O^+ and changes both concentrations slightly. The changes are so slight that the approximate equation

$$\text{pH} \approx \text{p}K_a - \log \frac{[HN_3]_0}{[N_3^-]_0}$$

holds at this point in the titration.[4] Substitution gives

$$\text{pH} = 4.72 - \log \frac{0.0333}{0.0333} = \boxed{4.72}$$

The pH equals the $\text{p}K_a$ of the weak acid being titrated. This is generally true at half-equivalence in practical titrations of weak acids. A titration at half-equivalence is also a buffer at its buffer point. See problem **15.45**.

● *At the Equivalence Point.* The addition of 50.00 mL of the NaOH solution brings the titration to its equivalence point—the number of moles of OH^- added equals the number of moles of HN_3 originally present. If no hydrazoic acid at all is left, then the concentration of N_3^- ion equals

$$[N_3^-] = \frac{0.00500 \text{ mol}}{0.100 \text{ L}} = 0.0500 \text{ M}$$

In fact however, a small proportion of the N_3^- ion reacts with water because N_3^- is a weak base

	$N_3^-(aq)$	$+ H_2O(aq) \rightleftharpoons$	$HN_3(aq) +$	$OH^-(aq)$
Init. Conc. (M)	0.0500	–	0	small
Change in Conc. (M)	$-x$	–	$+x$	$+x$
Equil. Conc. (M)	$0.0500 - x$	–	x	x

The K_b for this reaction is 5.26×10^{-10}, obtained by dividing K_w by the K_a of hydrazoic acid. Use the mass-action expression for the preceding to obtain

$$\frac{[OH^-][HN_3]}{[N_3^-]} = K_b = 5.26 \times 10^{-10} = \frac{x^2}{0.0500 - x}$$

Solving for x gives $[OH^-] = 5.13 \times 10^{-6}$ M. This corresponds to a pOH of 5.29 and therefore a pH of $\boxed{8.71}$.

● *Beyond the Equivalence Point.* A total of 51.00 mL of NaOH(aq) has been added. All of the HN_3 has been reacted, and some OH^- remains in excess. The concentration of leftover OH^- equals the amount of OH^- added minus the amount reacted divided by the total volume of the solution

$$[OH^-] = \frac{(0.05100 \text{ L} \times 0.1000 \text{ M}) - 5.000 \times 10^{-3} \text{ mol}}{(0.05000 + 0.05100) \text{ L}} = 9.901 \times 10^{-4} \text{ M}$$

With this much "other" OH^- in solution, the reaction of N_3^- to HN_3 plus OH^- adds just a pittance to the concentration of OH^- and

$$\text{pOH} = -\log(9.901 \times 10^{-4}) = 3.00 \quad \text{so that} \quad \text{pH} = \boxed{11.00}$$

15.55 The titration of the weak base ethylamine with the strong acid HCl falls into four ranges: *before* the addition of acid; *between* the first addition of acid and the equivalence point; *at* the equivalence point; *beyond* the equivalence point. The pH of the original 40.00 mL of 0.1000 M ethylamine exceeds

---

[4]This is text equation 15.7 written for the particular case of this conjugate acid-base pair.

7 because ethylamine is a base. As 0.1000 M HCl is added, the pH falls. Abbreviate ethylamine and its conjugate acid as $EtNH_2$ and $EtNH_3^+$ respectively and assume that the titration is performed at 25°C.

- *Before Addition of Acid.* Ethylamine raises the pH of pure water by the reaction

$$EtNH_2(aq) + H_2O(l) \rightleftharpoons EtNH_3^+(aq) + OH^-(aq) \qquad \frac{[EtNH_3^+][OH^-]}{[EtNH_2]} = 6.41 \times 10^{-4}$$

Let $[OH^-] = y$, and assume that the concentration of hydroxide ion from the autoionization of water is small. Because no HCl has been added

$$[OH^-] = [EtNH_3^+] = y \quad \text{and} \quad [EtNH_2] = 0.1000 - y$$

$$6.41 \times 10^{-4} = \frac{y^2}{0.1000 - y}$$

This equation has a familiar form. Solving gives $y = [OH^-] = 0.00769$ M. The pOH is 2.11, and the pH is $14.00 - 2.114 = \boxed{11.89}$.

Tip. Formulas very similar to the ones developed in the text for the titration of a weak acid with a strong base work to compute the pH along this titration curve. The only difference is that the natural choice of unknown is $[OH^-]$ rather than $[H_3O^+]$.

- *After First Addition of Acid, Before Equivalence Point.* In this range of the titration:

$$[EtNH_3^+] = \frac{c_t V}{V_0 + V} + y$$

where c_t is the concentration of the titrant, V_0 is the original volume of ethylamine solution, V is the volume of titrant added and y is the concentration of OH^-. The numerator $c_t V$ equals the chemical amount of $EtNH_3^+$ generated by the 1-to-1 reaction between the titrant and the $EtNH_2(aq)$, the denominator $(V_0 + V)$ equals the total volume of the solution, and their quotient $c_t V/(V_0 + V)$ equals the concentration of $EtNH_3^+(aq)$ from the neutralization reaction alone. The reaction of $EtNH_2$ to produce $OH^-(aq)$ gives additional $EtNH_3^+(aq)$ and is responsible for the y on the right-hand side of the above equation. Similarly

$$[EtNH_2] = \frac{c_0 V_0 - c_t V}{V_0 + V} - y$$

where c_0 stands for the original ethylamine concentration. In this titration, c_0 equals 0.1000 M, c_t equals 0.1000 M, and V_0 equals 0.0400 L. If 5.00 mL of HCl has been added, $V = 0.00500$ L. Substitution in the two preceding equations gives

$$[EtNH_3^+] = 0.01111 + y \qquad \text{and} \qquad [EtNH_2] = 0.07778 - y$$

Put the concentrations into the K_b expression to obtain

$$6.41 \times 10^{-4} = \frac{y(0.01111 + y)}{(0.07778 - y)}$$

In this equation, y is not negligible compared to 0.01111 or 0.07777. Omitting it in the addition and subtraction on the right-hand side gives a trial y of 0.004487, which is 40% of 0.01111! Solve the equation by rearranging it and using the quadratic formula. The answer is $[OH^-] = 0.00331$ M from which $pOH = 2.48$ and $\boxed{pH = 11.52}$.

At 20.00 mL, the same formulas give $[OH^-] = 6.18 \times 10^{-4}$ M, and a pH of $\boxed{10.79}$; at 39.90 mL, the same formulas give a pH of $\boxed{8.20}$.

• *At the Equivalence Point.* At equivalence, the reaction mixture consists of 80.00 mL of 0.05000 M ethylammonium chloride, $EtNH_3^+Cl^-$. The cation of this salt reacts as a weak acid

$$EtNH_3^+(aq) + H_2O(l) \rightleftharpoons EtNH_2(aq) + H_3O^+(aq)$$

The equilibrium constant for this reaction is $K_a = K_w/K_b$. Let $x = [H_3O^+]$. Then

$$K_a = 1.56 \times 10^{-11} = \frac{x^2}{(0.05000 - x)}$$

Solving gives $x = 8.83 \times 10^{-7}$ so the pH $= -\log(8.83 \times 10^{-7}) = \boxed{6.05}$

Tip. Using the formulas that work in the range before the equivalence point gives a deceptive result at this point

$$[EtNH_2] = 0 - [OH^-] \quad (?!)$$

This cannot be right since $[OH^-]$ and $[EtNH_2]$ must both be positive numbers.

• *Beyond the Equivalence Point.* In this range, the solution behaves like a simple solution of HCl. Compared to the strong acid HCl, the weakly acidic ethylammonium ion contributes little to the $H_3O^+(aq)$ concentration. When 40.10 mL of HCl has been added, the first 40.00 mL has gone to produce $EtNH_3^+(aq)$ ion by reacting with all the $EtNH_2(aq)$. The remaining 0.10 mL is free to act as a strong acid. The 0.10 mL of 0.1000 M HCl is of course diluted to 80.10 mL. Every HCl generates one H_3O^+ in aqueous solution

$$[H_3O^+] = \frac{0.10}{80.10} \times 0.1000 \text{ M} = 1.25 \times 10^{-4} \text{ M} \qquad pH = \boxed{3.90}$$

After 50.00 mL of titrant has been added $[H_3O^+] = (10.00/90.00) \times 0.1000 = 1.111 \times 10^{-2}$ M; $\boxed{pH = 1.95}$.

15.57 Addition of 46.50 mL of the 0.393 M NaOH(aq) solution to the acidic mixture of hydrochloric acid and sodium benzoate must bring the mixture quite near to an equivalence point because one more drop of base boosts the pH by more than one entire unit. (At and near the equivalence point in titrations, small additions of titrant cause large changes in pH.) Assume that 46.52 mL of the base brings the titration to the equivalence point. The chemical amount of NaOH added at this point is

$$n_{NaOH} = 0.04652 \text{ L} \times \left(\frac{0.393 \text{ mol OH}^-}{1 \text{ L}}\right) = 0.01828 \text{ mol}$$

This amount of strong base completes the neutralization of the HCl, some of which had previously been neutralized by benzoate ion ($C_6H_5COO^-$). The benzoate ion, a base, was present in solution from the ionization of the sodium benzoate in the original sample. Thus

$$n_{HCl} = n_{NaOH} + n_{C_6H_5COO^-}$$

The chemical amount of HCl is

$$n_{HCl} = 0.0500 \text{ L} \times \left(\frac{0.500 \text{ mol HCl}}{1 \text{ L}}\right) = 0.0250 \text{ mol}$$

Substitution gives

$$0.0250 = 0.01828 + n_{C_6H_5COO^-} \quad \text{hence} \quad n_{C_6H_5COO^-} = 6.7 \times 10^{-3} \text{ mol}$$

The mass of C_6H_5COONa in the original sample was

$$m_{C_6H_5COONa} = 6.72 \times 10^{-3} \text{ mol} \times \left(\frac{144.11 \text{ g } C_6H_5COONa}{1 \text{ mol } C_6H_5COONa}\right) = \boxed{0.97 \text{ g}}$$

15.59 Diethylamine and hydrochloric acid react in a 1-to-1 molar ratio. Therefore the chemical amount of HCl to reach the equivalence point equals the chemical amount of diethylamine originally present. This amount equals the volume of the HCl solution used in the titration multiplied by the molarity of that solution

$$n_{\text{HCl}} = (15.90 \text{ mL}) \times \left(\frac{0.0750 \text{ mmol HCl}}{1 \text{ mL}} \right) = 1.1925 \text{ mmol}$$

There was originally 1.1925 mmol of diethylamine. This is readily expressed as a mass

$$m_{(\text{C}_2\text{H}_5)_2\text{NH}} = 1.1925 \times 10^{-3} \text{ mol} \, (\text{C}_2\text{H}_5)_2\text{NH} \times \left(\frac{73.14 \text{ g} \, (\text{C}_2\text{H}_5)_2\text{NH}}{1 \text{ mol} \, (\text{C}_2\text{H}_5)_2\text{NH}} \right) = \boxed{0.0872 \text{ g} \, (\text{C}_2\text{H}_5)_2\text{NH}}$$

Suppose that at the equivalence point, all of the diethylamine is converted to its conjugate acid, the diethylammonium ion $(\text{C}_2\text{H}_5)_2\text{NH}_2^+$. Then the concentration of diethylammonium ion equals its chemical amount, 1.1925 mmol, divided by the volume of the solution (115.90 mL)

$$[(\text{C}_2\text{H}_5)_2\text{NH}_2^+] = \frac{1.1925 \text{ mmol}}{115.90 \text{ mL}} = 0.0103 \text{ mol L}^{-1}$$

In actuality, this concentration is a bit high. Some of the diethylammonium ion reacts away as it donates H^+ ions to increase the H_3O^+ concentration in the solution

$$(\text{C}_2\text{H}_5)_2\text{NH}_2^+(aq) + \text{H}_2\text{O}(l) \rightleftharpoons (\text{C}_2\text{H}_5)_2\text{NH} + \text{H}_3\text{O}^+(aq) \qquad \frac{[\text{H}_3\text{O}^+][(\text{C}_2\text{H}_5)_2\text{NH}]}{[(\text{C}_2\text{H}_5)_2\text{NH}^+]} = K_{\text{a}} = \frac{K_{\text{w}}}{K_{\text{b}}}$$

where the K_{b} is the basicity constant of diethylamine. Let x stand for the concentration of H_3O^+, and assume that all H_3O^+ in the solution comes from the reaction of diethylammonium ion as an acid. Then, at 25°C,

$$K_{\text{a}} = \frac{K_{\text{w}}}{K_{\text{b}}} = \frac{1.0 \times 10^{-14}}{3.09 \times 10^{-4}} = 3.236 \times 10^{-11} = \frac{x^2}{0.0103 - x}$$

Solving give $x = 5.77 \times 10^{-7}$. Then $[\text{H}_3\text{O}^+] = 5.77 \times 10^{-7}$ M for a pH of 6.24.

This concentration of H_3O^+ is only about six times larger than the concentration furnished by autoionization in pure water. How valid then is the assumption that diethylammonium ion furnishes *all* of the H_3O^+? One way to check is to employ the following equation,[5] which takes into account the autoionization of water as a source of H_3O^+ in solutions of weak acids

$$[\text{H}_3\text{O}^+]^3 + K_{\text{a}}[\text{H}_3\text{O}^+]^2 - (K_{\text{w}} + K_{\text{a}}c_{\text{a}})[\text{H}_3\text{O}^+] - K_{\text{a}}K_{\text{w}} = 0$$

Inserting $K_{\text{w}} = 1.0 \times 10^{-14}$, $c_{\text{a}} = 0.0103$, and $K_{\text{a}} = 3.24 \times 10^{-11}$ gives the cubic equation

$$[\text{H}_3\text{O}^+]^3 + (3.24 \times 10^{-11})[\text{H}_3\text{O}^+]^2 - (3.437 \times 10^{-13})[\text{H}_3\text{O}^+] - 3.24 \times 10^{-25} = 0$$

This equation is readily solved using a scientific calculator, but it is instructive to use chemical knowledge to simplify it. The last term is certainly much smaller than any of the other three because $[\text{H}_3\text{O}^+]$ must exceed (or equal) 5.77×10^{-7} M. After all, a second source of H_3O^+ ion can only raise the concentration of that ion, never lower it. Omitting the last term on this basis and dividing through by $[\text{H}_3\text{O}^+]$ gives

$$[\text{H}_3\text{O}^+]^2 + (3.24 \times 10^{-11})[\text{H}_3\text{O}^+] - 3.437 \times 10^{-13} = 0$$

Solution of this quadratic equation is routine even in the absence of a calculator (by means of the quadratic formula). The applicable root gives $[\text{H}_3\text{O}^+] = 5.86 \times 10^{-7}$ M, for a pH of $\boxed{6.23}$.

---

[5]From text Section 15.9.

Tip. The omitted term in the cubic equation equalled $-K_a K_w$. This was the only term reflecting the existence of the K_w equilibrium. The fact that it was dispensable means that the autoionization of water affects the pH of this solution only negligibly.

Tip. Do *not* try to account for the autoionization by adding 1.0×10^{-7} to the answer 5.77×10^{-7} M! Autoionization contributes less H_3O^+ in this solution than in pure water. The reason is that the autoionization equilibrium is shifted to the left by H_3O^+ ion from the diethylammonium ion (LeChâtelier's principle). Of course, this K_a equilibrium is also shifted (slightly) to the left by hydronium ion from the autoionization, which in turn causes an even slighter secondary effect back on the autoionization, which in turn.... The best understanding starts with two points: 1) only one kind of hydronium ion exists in the solution; 2) the equilibrium concentration of H_3O^+ ion comes as a compromise among all the simultaneous competing tendencies to donate or accept it.

A suitable indicator for the titration is $\boxed{\text{bromothymol blue}}$, which changes color in the pH range that includes pH 6.23.[6]

15.61 This buffer solution contains N-ethylmorpholine $C_6H_{13}NO$ (call it "M") and $C_6H_{13}NOH^+$ (MH^+), its conjugate acid, which forms from the reaction of the N-ethylmorpholine with HCl. The two species are in chemical equilibrium

$$M(aq) + H_2O(l) \rightleftharpoons MH^+(aq) + OH^-(aq) \qquad K_b = \frac{[OH^-][MH^+]}{[M]}.$$

Use this mass-action expression to compute K_b. This requires equilibrium values for the concentrations of all three species in the expression. The concentration of OH^- is easy because the pH of the solution is given: $[OH^-] = 1.0 \times 10^{-7}$ (assuming a temperature of 25°C). There was 10.00 mmol of M in the solution before the addition of the HCl, and the added HCl amounts to 8.00 mmol. If the neutralization reaction goes to completion, it forms 8.00 mmol of MH^+ and leaves 2.00 mmol of M unreacted (in excess). The volume of the solution is deliberately brought to 100.0 mL by the addition of water. The equilibrium concentrations of M and MH^+ therefore are

$$[M] = \frac{2.00 \text{ mmol}}{100.0 \text{ mL}} - 1.0 \times 10^{-7} \text{ M} \qquad [MH^+] = \frac{8.00 \text{ mmol}}{100.0 \text{ mL}} + 1.0 \times 10^{-7} \text{ M}$$

The subtraction and addition of the 1.0×10^{-7} account for the slight amount of conversion of M to MH^+ by the action of the equilibrium. Substitution into the K_b expression now gives the answer

$$K_b = \frac{[OH^-][MH^+]}{[M]} = \frac{(1.0 \times 10^{-7})(0.0800 + 1.0 \times 10^{-7})}{(0.0200 - 1.0 \times 10^{-7})} = \boxed{4 \times 10^{-7}}$$

Tip. The 1.0×10^{-7} is so small compared to 0.0200 or 0.0800 that actually subtracting or adding it is not worth the trouble.

Polyprotic Acids

15.63 Aqueous arsenic acid donates H^+ ions in three steps. Each step has a different K_a

$$\begin{aligned}
H_3AsO_4(aq) + H_2O(l) &\rightleftharpoons H_2AsO_4^-(aq) + H_3O^+(aq) & K_{a1} &= 5.0 \times 10^{-3} \\
H_2AsO_4^-(aq) + H_2O(l) &\rightleftharpoons HAsO_4^{2-}(aq) + H_3O^+(aq) & K_{a2} &= 9.3 \times 10^{-8} \\
HAsO_4^{2-}(aq) + H_2O(l) &\rightleftharpoons AsO_4^{3-}(aq) + H_3O^+(aq) & K_{a3} &= 3.0 \times 10^{-12}
\end{aligned}$$

K_{a2} is thousands of times smaller than K_{a1}, and K_{a3} is thousands of times smaller yet. This means that the first reaction will predominate in producing H_3O^+ and that the subsequent reactions will be negligible sources of H_3O^+. Compute the hydronium-ion concentration as if the first step occurred

---

[6]See text Figure 15.12

separately and use the answer in the mass-action expressions for the following steps. Ignoring the interaction of the equilibria avoids complicated systems of simultaneous equations.

When the first step is considered separately, the problem is just like problem **15.29a**. Let x be the equilibrium concentration of H_3O^+, which equals the equilibrium concentration of $H_2AsO_4^-$. The mass-action expression becomes

$$\frac{[H_2AsO_4^-][H_3O^+]}{[H_3AsO_4]} = K_{a1} = 5.0 \times 10^{-3} = \frac{x^2}{(0.1000 - x)}$$

Rearrange and substitute into the quadratic formula to obtain

$$x = \frac{-5.0 \times 10^{-3} \pm \sqrt{2.50 \times 10^{-5} + 2.0 \times 10^{-3}}}{2}$$

The positive root of the equation is 0.0200. Thus

$$[H_3AsO_4] = \boxed{0.080 \text{ M}} \qquad [H_2AsO_4^-] = \boxed{0.020 \text{ M}} \qquad [H_3O^+] = \boxed{0.020 \text{ M}}$$

Now consider the donation of the second hydrogen ion. Let y equal the concentration of $HAsO_4^{2-}$ produced at equilibrium

	$H_2AsO_4^-(aq)$	$+H_2O(l) \rightleftharpoons$	$HAsO_4^{2-}(aq) +$	$H_3O^+(aq)$
Init. Conc. (M)	0.0200	—	0	0.020
Change in Conc. (M)	$-y$	—	$+y$	$+y$
Equil. Conc. (M)	$0.020 - y$	—	y	$0.020 + y$

Use of the mass-action expression for K_{a2} gives the equation

$$\frac{[HAsO_4^{2-}][H_3O^+]}{[H_2AsO_4^-]} = K_{a2} = 9.3 \times 10^{-8} = \frac{y(0.020 + y)}{(0.020 - y)}$$

This equation is easily solved when it is realized that y must be small compared to 0.0200. Then $y = \boxed{9.3 \times 10^{-8} \text{ M}} = [HAsO_4^{2-}]$. Note that $[HAsO_4^{2-}]$ is equal to K_{a2}.

Finally, consider the third of the three reactions. Let z equal the concentration of AsO_4^{3-} produced at equilibrium

	$HAsO_4^{2-}(aq)$	$+H_2O(l) \rightleftharpoons$	$AsO_4^{3-}(aq) +$	$H_3O^+(aq)$
Init. Conc. (M)	9.3×10^{-8}	—	0	0.0200
Change in Conc. (M)	$-z$	—	$+z$	$+z$
Equil. Conc. (M)	$9.8 \times 10^{-8} - z$	—	z	$0.0200 + z$

Use of the mass-action expression for K_{a3} gives the equation

$$\frac{[AsO_4^{3-}][H_3O^+]}{[HAsO_4^{2-}]} = K_{a2} = 3.0 \times 10^{-12} = \frac{z(0.020 + z)}{(9.8 \times 10^{-8} - z)}$$

Solving gives $[AsO_4^{3-}] = \boxed{1.5 \times 10^{-17} \text{ M}}$. This is a very small concentration. One liter of the arsenic acid solution contains fewer than ten million AsO_4^{3-} ions!

15.65 The phosphate ion accepts hydrogen ions from water in three stages

$$PO_4^{3-}(aq) + H_2O(l) \rightleftharpoons HPO_4^{2-}(aq) + OH^-(aq) \qquad K_{b1} = K_w/K_{a3} = 4.55 \times 10^{-2}$$
$$HPO_4^{2-}(aq) + H_2O(l) \rightleftharpoons H_2PO_4^-(aq) + OH^-(aq) \qquad K_{b2} = K_w/K_{a2} = 1.61 \times 10^{-7}$$
$$H_2PO_4^-(aq) + H_2O(l) \rightleftharpoons H_3PO_4(aq) + OH^-(aq) \qquad K_{b3} = K_w/K_{a1} = 1.33 \times 10^{-12}$$

Treat the successive equilibria independently. Set up a three-line table for the reaction between PO_4^{3-} ion and water in the usual way

	$PO_4^{3-}(aq)$	$+H_2O(l) \rightleftharpoons$	$HPO_4^{2-}(aq) +$	$OH^-(aq)$
Init. Conc. (M)	0.050	–	0	small
Change in Conc. (M)	$-x$	–	$+x$	$+x$
Equil. Conc. (M)	$0.050 - x$	–	x	x

Writing the mass-action expression then gives

$$\frac{[HPO_4^{2-}][OH^-]}{[PO_4^{3-}]} = K_{b1} = 4.55 \times 10^{-2} = \frac{x^2}{(0.050 - x)}$$

Rearrange and substitute into the quadratic formula to obtain

$$x^2 + (4.55 \times 10^{-2})x - 2.27 \times 10^{-3} = 0 \quad \text{which gives} \quad x = 0.0302$$

$$[OH^-] = \boxed{0.030 \text{ M}} \quad [HPO_4^{2-}] = \boxed{0.030 \text{ M}} \quad [PO_4^{3-}] = 0.050 - 0.0302 = \boxed{0.020 \text{ M}}$$

Turn to the second stage. Let y equal the concentration of $H_2PO_4^-$ formed at equilibrium, but use the concentration of OH^- established by the first stage of the reaction in the K_{b2} mass-action expression

$$\frac{[H_2PO_4^-][OH^-]}{[HPO_4^{2-}]} = K_{b2} = 1.61 \times 10^{-7} = \frac{y(0.0302 + y)}{0.0302 - y}$$

This equation is easily solved because y must be small compared to 0.0302; $y = \boxed{1.61 \times 10^{-7} \text{ M}} = [H_2PO_4^-]$.

Finally, consider the third reaction. The mass-action expression gives

$$\frac{[H_3PO_4][OH^-]}{[H_2PO_4^-]} = K_{b3} = 1.33 \times 10^{-12} = \frac{z(0.0302 + z)}{1.61 \times 10^{-7} - z}$$

where z equals the equilibrium concentration of phosphoric acid. Solving gives $z = \boxed{7.1 \times 10^{-18} \text{ M}} = [H_3PO_4]$.

15.67 The major natural contributor to the acidity of rainwater is dissolved CO_2, which reacts with water to form carbonic acid $CO_2(g) + H_2O(l) \rightleftharpoons H_2CO_3(aq)$. In recent times, $SO_3(g)$ and $NO_2(g)$, which are air pollutants, have joined $CO_2(g)$ as contributors to the acidity of rain. Carbonic acid donates two hydrogen ions

$$H_2CO_3(aq) + H_2O(l) \rightleftharpoons HCO_3^-(aq) + H_3O^+(aq) \qquad K_{a1} = 4.3 \times 10^{-7}$$
$$HCO_3^-(aq) + H_2O(l) \rightleftharpoons CO_3^{2-}(aq) + H_3O^+(aq) \qquad K_{a2} = 4.8 \times 10^{-11}$$

At equilibrium, the following equations relate the concentrations

$$K_{a1} = \frac{[H_3O^+][HCO_3^-]}{[H_2CO_3]} \qquad K_{a2} = \frac{[H_3O^+][CO_3^{2-}]}{[HCO_3^-]}$$

The pH of the raindrop is 5.60. It follows that $[H_3O^+] = 2.51 \times 10^{-6}$ M. Substituting this value of $[H_3O^+]$ and the two K_a's into the preceding gives

$$0.171 = \frac{[HCO_3^-]}{[H_2CO_3]} \qquad 1.91 \times 10^{-5} = \frac{[CO_3^{2-}]}{[HCO_3^-]}$$

It is convenient to recast these equations so that the concentration of the same species, say HCO_3^-, is in the denominator in both. Take the reciprocal of the first and copy the second

$$5.84 = \frac{[H_2CO_3]}{[HCO_3^-]} \qquad 1.91 \times 10^{-5} = \frac{[CO_3^{2-}]}{[HCO_3^-]}$$

The fraction f of any of the three species present equals its concentration divided by the sum of the concentrations of all three. If the species is H_2CO_3

$$f_{H_2CO_3} = \frac{[H_2CO_3]}{[H_2CO_3] + [HCO_3^-] + [CO_3^{2-}]}$$

This expression can be simplified by dividing both its numerator and denominator by $[HCO_3^-]$ and inserting the ratios just calculated

$$f_{H_2CO_3} = \frac{[H_2CO_3]/[HCO_3^-]}{[H_2CO_3]/[HCO_3^-] + 1 + [CO_3^{2-}]/[HCO_3^-]}$$

$$f_{H_2CO_3} = \frac{5.84}{5.84 + 1 + (1.91 \times 10^{-5})} = \frac{5.84}{6.84} = 0.854$$

Expressions are obtained to compute the fractions of the other three species by changing the numerator as required. The resulting fractions are 0.146 for HCO_3^- and 2.79×10^{-6} for CO_3^{2-}. The total concentration of all three forms of the carbon-containing species is 1.0×10^{-5} M. The answers equal the respective fractions times this total

$$[H_2CO_3] = \boxed{8.5 \times 10^{-6} \text{ M}} \quad [HCO_3^-] = \boxed{1.5 \times 10^{-6} \text{ M}} \quad [CO_3^{2-}] = \boxed{2.8 \times 10^{-11} \text{ M}}$$

Organic Acids and Bases: Structure and Reactivity

15.69   The conjugate base of propane has a single Lewis structure:

The conjugate base of propene has two resonance Lewis structures:

Resonance stabilization makes $\boxed{\text{propene}}$ the stronger acid because it gives it a more stable conjugate base.

A DEEPER LOOK... Exact Treatment of Acid-Base Equilibria

15.71   The molar mass of thiamine hydrochloride is 337.27 g mol^{-1}, as computed from the molecular formula given in the problem. 3.0×10^{-5} g of this substance in 1.00 L of water makes a solution that is 8.89×10^{-8} M in thiH$^+$ ion[7] and of course 8.89×10^{-8} M in Cl$^-$ ion. The thiH$^+(aq)$ cation is a weak acid

$$thiH^+(aq) + H_2O(l) \rightleftharpoons thi(aq) + H_3O^+(aq) \qquad K_a = 3.4 \times 10^{-7} \text{ at } 25°C$$

---

[7]Here "thi" stands for thiamine $C_{12}H_{17}ON_4SCl_2$ and thiH$^+$ stands for $(HC_{12}H_{17}ON_4SCl_2)^+$. The latter is the "thiammonium" cation, the conjugate acid of thiamine.

This equilibrium produces only small amounts of $H_3O^+(aq)$ because K_a is small and the original concentration of $thiH^+(aq)$ is quite small. The simultaneous autoionization of water:

$$2\,H_2O(l) \rightleftharpoons H_3O^+(aq) + OH^-(aq)$$

must be reckoned with as a source of $H_3O^+(aq)$.

The following mathematical relationships always hold in this solution

$$3.4 \times 10^{-7} = K_a = \frac{[thi][H_3O^+]}{[thiH^+]} \qquad 1.0 \times 10^{-14} = K_w = [H_3O^+][OH^-]$$

$$8.89 \times 10^{-8} = c_a = [thi] + [thiH^+] \qquad [H_3O^+] + [thiH^+] = [OH^-] + [Cl^-]$$

The last equation follows from the requirement of electrical neutrality: for every positive charge in the solution there must be a negative charge. The second-to-last equation represents a material balance. Whatever the distribution between its two forms, the *total* concentration of thiamine-material is known. The first two equations are the usual mass-action expressions. The $[Cl^-]$ equals 8.89×10^{-8} M, as stated above, because $Cl^-(aq)$ does not react to any extent with other species. The four simultaneous equations therefore involve four unknowns. It is "merely" a question of algebra to solve for $[H_3O^+]$. The details of the algebra are given in text Section 15.9 for a similar case.[8] The result is the following cubic equation in $[H_3O^+]$

$$[H_3O^+]^3 + K_a[H_3O^+]^2 - (K_w + c_aK_a)[H_3O^+] - K_aK_w = 0$$

Substitute the numbers specific to this case

$$[H_3O^+]^3 + (3.4 \times 10^{-7})[H_3O^+]^2 - (4.02 \times 10^{-14})[H_3O^+] - 3.4 \times 10^{-21} = 0$$

This cubic equation can be solved using a scientific calculator. It has two negative roots and one positive root. A more instructive way to get the positive root, which is the only physically meaningful root, is just to guess a value of $[H_3O^+]$ near 10^{-7} and make successive approximations. The answer is $[H_3O^+] = 1.37 \times 10^{-7}$ M for a pH of $\boxed{6.86}$.

Tip. If the autoionization of water is (mistakenly) neglected, a quadratic equation results

$$[H_3O^+]^2 + (3.4 \times 10^{-7})[H_3O^+] - 3.02 \times 10^{-14} = 0$$

Solving gives $[H_3O^+] = 7.31 \times 10^{-8}$ M for a pH of 7.14. But pH 7.14 is on the basic side of 7, which is impossible when an acid is dissolved in water at 25°C.

15.73 Maleic acid is a diprotic acid for which K_{a1} and K_{a2} differ by about four orders of magnitude. Look at the course of the titration region by region.

• *Before Addition of Base.* Before any 0.1000 M NaOH is added, the predominant source of $H_3O^+(aq)$ in the solution is the equilibrium

$$H_2mal(aq) + H_2O(l) \rightleftharpoons Hmal^-(aq) + H_3O^+(aq)$$

Let x equal $[H_3O^+]$. Then

$$K_{a1} = 1.42 \times 10^{-2} = \frac{[Hmal^-][H_3O^+]}{[H_2mal]} = \frac{x^2}{0.1000 - x}$$

Solving for x using the quadratic formula gives 0.03125. If $[H_3O^+]$ is 0.03125 mol L^{-1}, the pH is $\boxed{1.51}$.[9]

---

[8] In this case of this problem, c_b, the original concentration of thi, the conjugate base of ThiH$^+$, is 0.
[9] In this problem the pH's are computed to the hundredth of a pH unit.

• *After the Addition of Some Base, But Before the Equivalence Point.* The first 5.00 mL of 0.1000 M NaOH reacts with the maleic acid (H_2mal). 5.00 mL of this NaOH solution contains only 5.00×10^{-4} mol of NaOH, which is less than the 50.00×10^{-4} mol of H_2mal that is present. The acid is in excess, and the reaction ends when the base runs out. Assume for the moment that the acid-base reaction goes to completion and that no other reactions take place. The yield of $Hmal^-(aq)$ is 5.00×10^{-4} mol, and 45.00×10^{-4} mol of H_2mal(aq) remains. Adding 5.00 mL of dilute aqueous solution raises the volume of the solution to 55.00 mL so

$$[Hmal^-] = \frac{5.00 \times 10^{-4} \text{ mol}}{0.05500 \text{ L}} = 0.009091 \text{ M} \qquad [H_2\text{mal}] = \frac{45.00 \times 10^{-4} \text{ mol}}{0.05500 \text{ L}} = 0.08182 \text{ M}$$

The K_{a1} equilibrium now acts to alter these concentrations slightly. It generates hydronium ions. For every $H_3O^+(aq)$ produced, one H_2mal(aq) is consumed and one additional $Hmal^-(aq)$ is generated. Let y equal the concentration of H_3O^+ that is generated. Then

$$K_{a1} = [H_3O^+]\frac{[Hmal^-]}{[H_2\text{mal}]} \qquad 1.42 \times 10^{-2} = y\left(\frac{0.009091 + y}{0.08182 - y}\right)$$

Rearranging the last expression reveals it as a quadratic equation. It is readily solved to give $[H_3O^+] = 0.0244$ M.[10] The Ka_2 equilibrium is legitimately neglected as a source of H_3O^+. It proceeds to a far lesser extent, and the "starting" concentration of $Hmal^-$ is about nine times smaller than that of H_2mal. The pH equals $\boxed{1.61}$.

• *Halfway to the First Equivalence Point.* At this point, 25.00 mL of titrant has been added raising the total volume to 75.00 mL. Repeating the reasoning just used gives

$$1.42 \times 10^{-2} = y\frac{0.0333 + y}{0.0333 - y}$$

Again, rearrange and solve for y by use of the quadratic formula. The answer is $[H_3O^+] = 8.45 \times 10^{-3}$ M and pH = $\boxed{2.07}$.

• *At the First Equivalence Point.* 50.00 mL of 0.1000 M NaOH brings the titration to the *first* equivalence point. The solution consists of 100.00 mL of 0.0500 M NaHmal (sodium hydrogen maleate). The $Hmal^-(aq)$ ion is amphoteric. It behaves as an acid

$$Hmal^-(aq) + H_2O(l) \rightleftharpoons H_3O^+(aq) + \text{mal}^{2-}(aq) \qquad K_{a2} = 8.57 \times 10^{-7}$$

And it behaves as a base

$$Hmal^-(aq) + H_2O(l) \rightleftharpoons OH^-(aq) + H_2\text{mal}(aq) \qquad K_{b2} = \frac{K_w}{K_{a1}} = 7.04 \times 10^{-13}$$

$Hmal^-(aq)$ is just like $HCO_3^-(aq)$ in this respect. Copy the analysis presented for $HCO_3^-(aq)$ starting on page 717 in text Section 15.9 to obtain

$$[H_3O^+] \approx \sqrt{\frac{K_{a1}K_{a2}[Hmal^-]_0 + K_{a1}K_w}{K_{a1} + [Hmal^-]_0}}$$

in which $[Hmal^-]_0$ equals the "original" concentration of $Hmal^-$. Substitute 0.0500 M for $[Hmal^-]_0$, insert the other numbers. and evaluate. The result is $[H_3O^+] = 9.73 \times 10^{-5}$ M for a pH of $\boxed{4.01}$. Notice that the approximate formula

$$[H_3O^+] \approx \sqrt{K_{a1}K_{a2}}$$

---

[10]Neglecting y as small compared to 0.00909 and to 0.08182 simplifies the algebra, but y then comes out to equal 0.127, which *exceeds* 0.08182 and is obviously far too large to neglect. Using the quadratic formula (or a calculator) is a must.

gives a wrong pH of 3.96. It fails because K_{a1} of maleic acid is rather large and may not be neglected compared to $[Hmal^-]_0$ in the denominator of the fraction under the square-root sign in the preceding.

• *Halfway to the Second Equivalence Point.* After 75.00 mL of 0.1000 M NaOH has been added, the titration is half-way to the *second* equivalence point. In this range, the main source of $H_3O^+(aq)$ is the reaction

$$Hmal^-(aq) + H_2O(l) \rightleftharpoons mal^{2-}(aq) + H_3O^+(aq) \qquad K_{a2} = 8.57 \times 10^{-7}$$

The concentration of $H_2mal(aq)$ is now very small (because so much base has been added!) and consequently the K_{a1} equilibrium has only a negligible effect on the pH. The solution at this point is equivalent to a solution prepared by addition of 25.00 mL of 0.1000 M NaOH to 100.00 mL of 0.0500 M NaHmal. This amount of NaOH converts half of the $Hmal^-(aq)$ to $mal^{2-}(aq)$. Dilution meanwhile reduces the concentrations of both species by the factor 100/125. After completion of the acid-base neutralization, but before any equilibrium starts

$$[Hmal^-] = 0.025 \times \frac{100}{125} = 0.020 \text{ M} \qquad [mal^{2-}] = 0.025 \times \frac{100}{125} = 0.020 \text{ M}$$

The K_{a2} equilibrium changes these concentrations, but only slightly. It adds z to the concentration of $mal^{2-}(aq)$ and removes z from the concentration of $Hmal^-(aq)$, where z is the concentration of hydronium ion that it produces. The mass-action expression is

$$K_{a2} = 8.57 \times 10^{-7} = \frac{[H_3O^+][mal^{2-}]}{[Hmal^-]} = z\left(\frac{0.020 + z}{0.020 - z}\right)$$

Solving this equation is easy because z can be neglected compared to 0.020. The answer is $z = 8.57 \times 10^{-7}$ so the pH is $\boxed{6.07}$.

• *Just Short of the Second Equivalence Point.* After 99.9 mL of 0.1000 M NaOH has been added, essentially all of the original $H_2mal(aq)$ has been converted to $mal^{2-}(aq)$. Only a small concentration of $Hmal^-(aq)$ ion persists

$$[Hmal^-] = \frac{0.01 \text{ mmol}}{149.9 \text{ mL}} = 6.671 \times 10^{-5} \text{ M} \qquad [mal^{2-}] = \frac{4.99 \text{ mmol}}{149.9 \text{ mL}} = 0.03329 \text{ M}$$

The major source of OH^- ions in solution is hydrolysis of $mal^{2-}(aq)$

$$mal^{2-}(aq) + H_2O(l) \rightleftharpoons Hmal^-(aq) + OH^-(aq) \qquad \frac{[OH^-][Hmal^-]}{[mal^{2-}]} = K_{b1} = \frac{K_w}{K_{a2}} = \frac{1.0 \times 10^{-14}}{8.57 \times 10^{-7}}$$

Let z equal the equilibrium concentration of OH^- from this reaction. Then

$$1.167 \times 10^{-8} = z\left(\frac{6.671 \times 10^{-5} + z}{0.03329 - z}\right)$$

The z in the denominator of the fraction may be neglected (it is small compared to 0.03329). The z in the numerator may not be neglected. Rearrangement then leads to

$$z^2 + (6.671 \times 10^{-5})z - 3.885 \times 10^{-10} = 0$$

Solution gives $[OH^-] = 5.38 \times 10^{-6}$ M for a pOH of 5.27 and so a pH of $\boxed{8.73}$.

• *At the Second Equivalence Point.* The solution consists of 150.00 mL of 0.0333 M Na_2mal. Again, the hydrolysis of $mal^{2-}(aq)$ ion is the major source of OH^- ions. Let x be the equilibrium concentration of $OH^-(aq)$. Then

$$K_{b1} = \frac{1.0 \times 10^{-14}}{8.57 \times 10^{-7}} = \frac{[Hmal^-][OH^-]}{[mal^{2-}]} = \frac{x^2}{0.0333 - x}$$

Solving for x gives 1.97×10^{-5} so the pOH is 4.71 and the pH is $\boxed{9.29}$.

• *Past the Second Equivalence Point.* Once excess NaOH has been added, the hydrolysis of mal^{2-} ion is completely overshadowed by NaOH as a source of OH$^{-}(aq)$. The first 100.00 mL of NaOH was used up neutralizing the H$_2$mal. The next 5.00 mL of 0.100 M NaOH makes a solution that is $(5.00/155.00) \times 0.100$ or 3.23×10^{-3} M in OH$^{-}(aq)$. This corresponds to pOH 2.49 or pH $\boxed{11.51}$.

ADDITIONAL PROBLEMS

15.75 The equation for the dissociation of HOAc (acetic acid) in water is

$$\text{HOAc}(aq) + \text{H}_2\text{O}(l) \rightleftharpoons \text{OAc}^{-}(aq) + \text{H}_3\text{O}^{+}(aq)$$

LeChâtelier's principle predicts that this equilibrium shifts to the right if the volume of the system is increased by the addition of the solvent H$_2$O(l). This shift increases the degree of dissociation of the HOAc, but *not* the equilibrium constant of the reaction. The point can be understood mathematically. Suppose that some concentration c of acetic acid is dissolved in water and comes to equilibrium as indicated in the preceding equation. Then

$$K = \frac{(c\alpha)(c\alpha)}{c - c\alpha} = \frac{c\alpha^2}{1 - \alpha}$$

where α is the fraction of the acetic acid that dissociates. If c is decreased (by adding water to the solution) then α must increase to maintain the equality.

15.77 The equilibrium constants are for the reaction

$$\text{ClCH}_2\text{COOH}(aq) + \text{H}_2\text{O}(l) \rightleftharpoons \text{ClCH}_2\text{COO}^{-}(aq) + \text{H}_3\text{O}^{+}(aq)$$

Call the first constant (observed at 273.15 K) K_{273} and the second (observed at 313.15 K) K_{313}. Now use the van't Hoff equation (text Section 14.7, equation 14.12), which gives the temperature dependence of the equilibrium constant. For this case

$$\ln \frac{K_{313}}{K_{273}} = \ln \left(\frac{1.230 \times 10^{-3}}{1.528 \times 10^{-3}} \right) = \frac{-\Delta H^{\circ}}{R} \left(\frac{1}{313.15 \text{ K}} - \frac{1}{273.15 \text{ K}} \right)$$

Substituting $R = 8.3145$ J K^{-1}mol^{-1} and doing the arithmetic gives $\Delta H^{\circ} = -3.86 \times 10^3$ J mol^{-1}. The ΔH° of the reaction as it is written above therefore equals $\boxed{-3.86 \text{ kJ}}$.

Tip. This answer equals both ΔH°_{313} and ΔH°_{273} because use of the van't Hoff equation has a built-in assumption: that ΔH° is independent of temperature. The truth is that ΔH° *does* change with the temperature, but relatively little.

15.79 Use the relations: pOH = 14.00 − pH; [H$_3$O^{+}] = $10^{-\text{pH}}$; [OH^{-}] = $10^{-\text{pOH}}$.

	Material	[H$_3$O^{+}]	pOH	[OH^{-}]
a)	Orange Juice (pH 2.8)	2×10^{-3} M	11.2	6×10^{-12} M
b)	Tomato Juice (pH 3.9)	1×10^{-4} M	10.1	8×10^{-11} M
c)	Milk (pH 4.1)	8×10^{-5} M	9.9	1×10^{-10} M
d)	Borax Solution (pH 8.5)	3×10^{-9} M	5.5	3×10^{-6} M
e)	Household Ammonia (pH 11.9)	1×10^{-12} M	2.1	8×10^{-3} M

15.81 The pH should be $\boxed{\text{low}}$. LeChâtelier's principle indicates that increasing [H$_3$O^{+}] shifts the equilibrium in the problem to the left, favoring Cl$_2(aq)$ at the expense of Cl$^{-}(aq)$.

Tip. The reaction in this problem is a disproportionation. See text Section 11.4.3, on text page 457.

15.83 Oxoacids donate H^+ ions from one or more —OH groups that are attached to a central atom of another element. The acid ionization constant of H_3PO_2 equals $K_a = 8 \times 10^{-2}$. This makes it about ten times stronger that H_3PO_4, for which $K_a = 7.52 \times 10^{-3}$ (according to Table 15.2). This unexpected strength indicates the presence of a P=O bond in addition to the required P—OH bond. A double-bonded O would withdraw electron density from the P and so weaken the O—H bond, favoring loss of an H^+ ion. The only way to have —OH and =O bond to the central P is to have the other two H atoms also bond directly to the central P. The molecular structure is $\boxed{H_2P=O(OH)}$.

The acid would be $\boxed{\text{monoprotic}}$ because the two H's bound to the central P would not readily donated.

Tip. H_3PO_2 is called phosphinic acid or sometimes hypophosphorous acid. It is a strong reducing agent. When it acts as a reducing agent it is ususaly oxidized to phosphorous acid $P(OH)_3$ or phosphoric acid $O=P(OH)_3$. This removes the P—H bonds.

15.85 **a)** The conjugate acid of urea is $\boxed{NH_2CONH_3^+}$, called the urea acidium ion.

b) The following equilibrium generates urea from its conjugate acid

$$NH_2CONH_3^+(aq) + H_2O(l) \rightleftharpoons NH_2CONH_2(aq) + H_3O^+(aq) \qquad K_a = 10^{(13.8-14.0)} = 1.58$$

The original concentration of the conjugate acid of urea is 0.15 M. Its large K_a indicates that it is nearly a strong acid. If it were a strong acid, the equilibrium concentration of the urea would be 0.15 M. As it is

$$1.58 = \frac{[H_3O^+][NH_2CONH_2]}{[NH_2CONH_3^+]}$$

Let the concentration of H_3O^+ and $[NH_2CONH_3^+]$ equal x. Then

$$1.58 = \frac{x^2}{0.15 - x}$$

The positive root of this equation is $x = 0.140$. The equilibrium concentration of $[NH_2CONH_2]$ is therefore $0.150 - 0.140 = \boxed{0.010 \text{ M}}$.

15.87 The K_a of acetic acid tells its strength, which is its ability to donate H^+ ions. The equation for this donation is

$$HC_2H_3O_2(aq) \rightleftharpoons C_2H_3O_2^-(aq) + H^+(aq)$$

Increasing the temperature from 25 to 50°C lowers the K_a of acetic acid to about 0.93 of its original value while simultaneously increasing K_w, which measures the ability of water of donate H^+ ions, by a factor of 5.47. This latter increase cannot possible compensate for the original extreme weakness of water compared to acetic acid. The contribution of water to the concentration of hydrogen ion in the solution was negligible before the rise in temperature and remains negligible.

The rise in temperature causes the whole solution to expand without changing the amount of dissolved acetic acid. It causes the concentration of acetic acid to drop from 0.10 M to .099 M. The acetic acid solution is less concentrated at 50°C, and its K_a of the acid is less. The equilibrium concentration of H^+ ion therefore drops slightly. Consequently $\boxed{\text{the pH of the solution rises slightly.}}$

15.89 The compounds are a strong acid (HCl), a salt of a strong acid and a weak base (NH_4Cl), a salt of a weak acid and a strong base (Na_3PO_4), a salt of a weak acid and a strong base ($NaC_2H_3O_2$), and a salt of a strong acid and a strong base (KNO_3). The 0.100 M solutions of HCl and NH_4Cl are acidic. The 0.100 M solutions of Na_3PO_4 and $NaCH_3COO$ are basic. The 0.100 M solution of KNO_3 is neutral.

15.91 The single quantity that measures the strength of an acid in a given solvent is its K_a. The K_a's of NH_4^+ and HCN in water at room temperature are 5.6×10^{-10} and 6.17×10^{-10} respectively. Therefore HCN is only about 10% stronger than NH_4^+ ion at room temperature.

15.93   The initial concentrations of C_6F_5COOH and $C_6F_5COO^-$ are

$$[C_6F_5COOH]_0 = \frac{0.050 \text{ mol}}{2.00 \text{ L}} = 0.025 \text{ M} \quad [C_6F_5COO^-]_0 = \frac{0.060 \text{ mol}}{2.00 \text{ L}} = 0.030 \text{ M}$$

These are the concentrations *after* the complete mixing of the two solutions, but *before* any reactions involving the pentafluorobenzoic acid and its conjugate base the pentafluorobenzoate ion have a chance to occur. Then let the following reaction go to equilibrium

$$C_6F_5COOH(aq) + H_2)(l) \rightleftharpoons H_3O^+(aq) + C_6F_5COO^-$$

Let y equal the equilibrium concentration of hydronium ion. Then

$$K_a = \frac{[H_3O^+][C_6F_5COO^-]}{[C_6F_5COOH]} = \frac{y(0.030 + y)}{(0.025 - y)} = 0.033$$

If y is neglected compared to 0.025 and 0.030, then the equation is easy to solve, and y equals 0.0275. This is obviously *wrong* because 0.0275 is *larger* than 0.025 rather than being a great deal smaller. When such things happen, start over and neglect nothing. The original equation rearranges to

$$y^2 + 0.063y - 0.000825 = 0$$

Use of the quadratic formula gives $y = 0.01113$. It follows that the pH of the buffer is $\boxed{1.95}$.

15.95 $\boxed{\text{Procedure d)}}$ would not make an effective buffer. A good buffer results when substantial amounts of a weak acid and its conjugate base are mixed in solution. Procedure d) yields a solution that is 1.77×10^{-5} M in HCl mixed with some NaCl. The solution would change pH greatly upon addition of either strong acid or strong base. All the other solutions are acetic acid-acetate buffers.

15.97   Assume that the inadvertent heating of the original sample of $Na_2CO_3 \cdot 10H_2O$ drove off only water (and not, for example, some CO_2). Then, the ability of the sample on a molar basis to neutralize acids remained unchanged, despite its loss of mass. The sample was mixed with the base NaOH and then with the acid HCl. It neutralized some of the HCl; the NaOH neutralized the rest of the HCl. The chemical amounts of the HCl and NaOH were

$$n_{HCl} = 30.0 \text{ mL} \times \frac{0.100 \text{ mmol HCl}}{\text{mL}} = 3.00 \text{ mmol HCl}$$

$$n_{NaOH} = 6.4 \text{ mL} \times \frac{0.200 \text{ mmol NaOH}}{\text{mL}} = 1.28 \text{ mmol NaOH}$$

Because HCl and NaOH neutralize each other in a 1 : 1 molar ratio, 3.00 mmol − 1.28 mmol = 1.72 mmol of acid were neutralized by the partially dehydrated sample. Formulate the compound in the sample as $Na_2CO_3 \cdot xH_2O$ (where $0 \le x \le 10$). The equation for the reaction between the sample and the HCl is

$$Na_2CO_3 \cdot xH_2O(s) + 2 \, HCl(aq) \longrightarrow 2 \, NaCl(aq) + CO_2(g) + (1 + x) \, H_2O(l)$$

The chemical amount of $Na_2CO_3 \cdot xH_2O$ was then

$$n_{Na_2CO_3 \cdot xH_2O} = 1.72 \text{ mmol HCl} \times \frac{1 \text{ mmol Na}_2\text{CO}_3}{2 \text{ mmol HCl}} = 0.860 \text{ mmol}$$

The original chemical amount of the decahydrate ($Na_2CO_3 \cdot 10H_2O$) was also 0.860 mmol because the inadvertent dehydration reaction

$$Na_2CO_3 \cdot 10H_2O(s) \longrightarrow Na_2CO_3 \cdot xH_2O(s) + (10 - x)H_2O(g)$$

keeps the decahydrate and the partially dehydrated sodium carbonate in a 1 : 1 molar ratio. The original mass of $Na_2CO_3 \cdot 10H_2O$ was

$$m_{Na_2CO_3} = 0.860 \text{ mmol } Na_2CO_3 \cdot 10H_2O \times \frac{286.14 \text{ mg } Na_2CO_3 \cdot 10H_2O}{1 \text{ mmol } Na_2CO_3 \cdot 10H_2O} = 246.1 \text{ mg}$$

and the chemical amount of water incorporated in the original decahydrate was $10 \times 0.860 = 8.60$ mmol H_2O. Heating the sample reduced its mass from the original 246.1 mg to 200 mg. The fraction of the water that was lost was

$$\text{fraction lost} = \frac{(246.1 - 200) \text{ mg } H_2O}{8.60 \text{ mmol } H_2O(18.015 \text{ mg } H_2O \text{ /mmol } H_2O)} = \boxed{0.30}$$

Tip. The x in the formula is 7, because 3/10 of the original 10 H_2O's were lost, but 7/10 remain. $Na_2CO_3 \cdot 7H_2O$, sodium carbonate heptahydrate, is a known substance but is thermodynamically stable relative to other hydrates of sodium carbonate only within a narrow range of temperature, from about 32°C to 35°C.

15.99 The aqueous solutions in the three flasks contain different acids (HNO_3, $HCOOH$, and $C_6H_5NH_3^+$ ion) at various concentrations but have the same pH.

a) To determine which flask contains which solution: take equal volumes of solution (say 10 mL) from flasks A, B, and C and titrate them (separately) with a strong base (NaOH would work) to the same endpoint (the phenolphthalein endpoint would work). The 10 mL of $C_6H_5NH_3Cl$ solution would require the most titrant because it would contain the largest amount of acid. The 10 mL of HNO_3 solution would require the least titrant because it would contain the smallest amount of acid. These three acids all react with NaOH in a 1 : 1 molar ratio. This is important. If one were a diprotic or triprotic acid, its neutralizing power would be doubled or tripled.

b) The relative strengths of the three acids are

$$\boxed{\text{strongest } HNO_3 > HCOOH > C_6H_5NH_3^+ \text{ weakest}}$$

It requires a larger concentration of a weaker acid to lower the pH to a given value than it does of a stronger acid. A weaker acid donates H^+ ion less effectively.

c) Nitric acid (HNO_3) is a strong acid in water (it is necessary to know this). It dissociates essentially completely. Effectively no molecules of HNO_3 remain in solution after its acid dissociation, and the hydrogen-ion concentration in the solution equals 1.0×10^{-3} M. This is also the hydrogen-ion concentration in the other two flasks (equal pH implies equal H_3O^+ concentration).

Now, write acid-dissociation equations and K_a expressions for the two weak acids:

$$C_6H_5NH_3^+(aq) + H_2O(l) \rightleftharpoons H_3O^+(aq) + C_6H_5NH_2(aq) \quad K_{C_6H_5NH_3^+} = \frac{[H_3O^+][C_6H_5NH_2]}{[C_6H_5NH_3^+]}$$

$$HCOOH(aq) + H_2O(l) \rightleftharpoons H_3O^+(aq) + HCOO^-(aq) \quad K_{HCOOH} = \frac{[H_3O^+][HCOO^-]}{[HCOOH]}$$

The sole source of H_3O^+ ion in either of the two solutions, apart from negligible contributions from the autoionization of water, is the dissociation of the weak acid. The acid-dissociation reactions reduce the concentration of the weak acid by 1.0×10^{-3} M and increase the concentration of their conjugate bases by the same amount

$$[C_6H_5NH_3^+] = (4 \times 10^{-2} - 1.0 \times 10^{-3}) \text{ M} \quad [C_6H_5NH_2] = (0 + 1.0 \times 10^{-3}) \text{ M}$$
$$[HCOOH] = (6 \times 10^{-3} - 1.0 \times 10^{-3}) \text{ M} \quad [HCOO^-] = (0 + 1.0 \times 10^{-3}) \text{ M}$$

Substitution into the K_a expressions gives the two K_a's

$$K_{C_6H_5NH_3^+} = \frac{(1.0 \times 10^{-3})(1.0 \times 10^{-3})}{3.9 \times 10^{-2}} = 2.6 \times 10^{-4}$$

$$K_{HCOOH} = \frac{(1.0 \times 10^{-3})(1.0 \times 10^{-3})}{5 \times 10^{-3}} = \boxed{2 \times 10^{-4}}$$

The K_b for aniline, which is the conjugate base of $C_6H_5NH_3^+$ ion is computed using the fact that $K_a K_b = K_w$ for an acid-base conjugate pair

$$K_{C_6H_5NH_2} = \frac{1.0 \times 10^{-14}}{2.6 \times 10^{-4}} = \boxed{4 \times 10^{-11}}$$

Tip. Check the answers against tabulated values in reliable sources. For example, Text table 15.2 gives 1.77×10^{-4} for the K_a of formic acid.

15.101 a) The 0.1000 M solution of weak acid HA has a volume of 50.00 mL because 50.00 mL of the 0.1000 M base titrates it to equivalence. The addition of 40.00 mL of the base converts 40.00/50.00 of the acid to its conjugate base and creates a solution with a volume of 90.00 mL. According to the problem, the pH of this 90.00 mL of solution is 4.50. This means:

$$[H_3O^+] = 10^{-4.50} = 3.16 \times 10^{-5} \text{ M}$$

The solution is at equilibrium. The concentration of HA is the amount of unreacted HA divided by the volume of the solution minus the small concentration lost by the donation of H^+ to water:

$$[HA] = \frac{10.00 \text{ mL}}{90.00 \text{ mL}}(0.100 \ m) - [H_3O^+] = (0.01111 - 3.16 \times 10^{-5}) \text{ M}$$

Similarly, the equilibrium concentration of A^- is the amount of A^- formed divided by the volume of the solution plus the small concentration added by the HA/A^- equilibrium

$$[A^-] = \frac{40.00 \text{ mL}}{90.00 \text{ mL}}(0.100 \ m) + [H_3O^+] = (0.04444 + 3.16 \times 10^{-5}) \text{ M}$$

Substitute these values into the K_a expression

$$K_a = \frac{[H_3O^+][A^-]}{[HA]} = \frac{(3.16 \times 10^{-5})(0.044476)}{0.011079} = \boxed{1.3 \times 10^{-4}}$$

Skipping the subtraction and addition of the $[H_3O^+]$ gives $K_a = 1.3 \times 10^{-4}$, which is the same answer.

b) The solution at the equivalence point could have been prepared by dissolving 5.000 mmol of NaA in 100.00 mL of water. In such a solution, the initial concentration (before any acid-base equilibria) of A^- ion is 0.05000 M. To determine the pH, consider the reaction of A^- ion with water:

$$A^-(aq) + H_2O(l) \rightleftharpoons HA(aq) + OH^-(aq) \qquad \frac{[HA][OH^-]}{[A^-]} = K_b$$

Let x equal the equilibrium concentration of OH^- ion. Then

$$\frac{x^2}{0.05000 - x} = \frac{K_w}{K_a} = 7.9 \times 10^{-11}$$

Solving for x gives a $[OH^-]$ of 2.0×10^{-6} M. The pOH is accordingly 5.70, and the pH is $\boxed{8.30}$.

Tip. A solution of the salt of a strong base (NaOH in this case) and a weak acid (HA in this case) is always basic (pH > 7).

15.103 Phosphonocarboxylic acid ("H$_3$Pho") donates H$^+$ in three steps

$$H_3Pho(aq) + H_2O(l) \rightleftharpoons H_2Pho^-(aq) + H_3O^+(aq) \quad K_{a1} = 1.0 \times 10^{-2}$$
$$H_2Pho^-(aq) + H_2O(l) \rightleftharpoons HPho^{2-}(aq) + H_3O^+(aq) \quad K_{a2} = 7.8 \times 10^{-6}$$
$$HPho^{2-}(aq) + H_2O(l) \rightleftharpoons Pho^{3-}(aq) + H_3O^+(aq) \quad K_{a3} = 2.0 \times 10^{-9}$$

The law of mass action gives the following three equations

$$K_{a1} = \frac{[H_3O^+][H_2Pho^-]}{[H_3Pho]} \qquad K_{a2} = \frac{[H_3O^+][HPho^{2-}]}{[H_2Pho^-]} \qquad K_{a3} = \frac{[H_3O^+][Pho^{3-}]}{[HPho^{2-}]}$$

The pH of the blood is 7.40, and the buffering action of the blood maintains this pH despite the addition of the drug. It follows that $[H_3O^+] = 3.98 \times 10^{-8}$ M. Substituting the known value of $[H_3O^+]$ and the three K's gives

$$2.513 \times 10^5 = \frac{[H_2Pho^-]}{[H_3Pho]} \qquad 195.98 = \frac{[HPho^{2-}]}{[H_2Pho^-]} \qquad 0.05025 = \frac{[Pho^{3-}]}{[HPho^{2-}]}$$

Recast these equations so that the concentration of the same species, say H$_2$Pho$^-$, is in the denominator

$$3.980 \times 10^{-6} = \frac{[H_3Pho]}{[H_2Pho^-]} \qquad 195.98 = \frac{[HPho^{2-}]}{[H_2Pho^-]} \qquad 9.8480 = \frac{[Pho^{3-}]}{[H_2Pho^-]}$$

The first new equation is the reciprocal of the first of the preceding group. The third comes by multiplying the second and third equations in the preceding group. The fraction f of any of the four Pho-containing species equals its concentration divided by the sum of the concentrations of all four. For example,

$$f_{H_3Pho} = \frac{[H_3Pho]}{[H_3Pho] + [H_2Pho^-] + [HPho^{2-}] + [Pho^{3-}]}$$

This expression can be evaluated by dividing numerator and denominator by $[H_2Pho^-]$ and inserting the ratios just calculated

$$f_{H_3Pho} = \frac{[H_3Pho]/[H_2Pho^-]}{[H_3Pho]/[H_2Pho^-] + 1 + [HPho^{2-}]/[H_2Pho^-] + [Pho^{3-}]/[H_2Pho^-]}$$

$$f_{H_3Pho} = \frac{3.98 \times 10^{-6}}{3.98 \times 10^{-6} + 1 + 195.98 + 9.8480} = \frac{3.98 \times 10^{-6}}{206.83} = 1.92 \times 10^{-8}$$

The same approach gives the fractions of the other three species. Simply change the numerator as required. The resulting fractions are 0.004835 for H$_2$Pho$^-$, 0.94754 for HPho^{2-}, and 0.0476 for Pho^{3-}. The total concentration of all four forms of the drug is 1.0×10^{-5} M. The desired concentrations equal this total multiplied by the respective fractions

$$[H_3Pho] = \boxed{1.9 \times 10^{-13} \text{ M}} \qquad [H_2Pho^-] = \boxed{4.8 \times 10^{-8} \text{ M}}$$

$$[HPho^{2-}] = \boxed{9.5 \times 10^{-6} \text{ M}} \qquad [Pho^{3-}] = \boxed{4.8 \times 10^{-7} \text{ M}}$$

15.105 As explained in the problem, hyperventilating causes the loss of CO$_2$ dissolved in the blood. Carbon dioxide is a weak acid in water. Loss of the weak acid causes the blood to $\boxed{\text{rise in pH}}$.

15.107 Represent the amino acid glycine as glyH. Its conjugate acid is then glyH$_2^+$, and its conjugate base is gly$^-$. Re-write the two equilibria given in the problem together with the water autoionization equilibrium

$$glyH(aq) + H_2O(l) \rightleftharpoons gly^-(aq) + H_3O^+(aq) \qquad K_a = 1.7 \times 10^{-10}$$
$$glyH(aq) + H_2O(l) \rightleftharpoons glyH_2^+(aq) + OH^-(aq) \qquad K_b = 2.2 \times 10^{-12}$$
$$2H_2O(l) \rightleftharpoons H_3O^+(aq) + OH^-(aq) \qquad K_w = 1.0 \times 10^{-14}$$

The following equations convey the *same* information

$$glyH_2^+(aq) + H_2O(l) \rightleftharpoons glyH(aq) + H_3O^+(aq) \qquad K_1 = K_w/K_b$$
$$glyH(aq) + H_2O(l) \rightleftharpoons gly^-(aq) + H_3O^+(aq) \qquad K_2 = K_a$$

In this pair of equilibria, the second equation quoted in the problem has been *reversed* and *added* to the water autoionization equation. The change puts the emphasis on $glyH_2^+$ (^+H_3N—CH_2—COOH) as a diprotic acid rather than on self-neutralization by the glycine. The new equations reveal that solutions of glycine are just like solutions of other amphoteric species—0.10 M aqueous glycine is like 0.10 M $HCO_3^-(aq)$, for example. Spotting the similarity is nice because the case of the pH of an aqueous solution of $HCO_3^-(aq)$ is extensively treated in the text Section 15.9. The text develops the approximate formula $[H_3O^+] \approx \sqrt{K_1 K_2}$. According to this formula, the $[H_3O^+]$ in the solution is independent of the concentration of the glycine! Substitution into the formula gives

$$[H_3O^+] \approx \sqrt{K_1 K_2} = \sqrt{\left(\frac{K_w}{K_b}\right) K_a} = \sqrt{\left(\frac{1.0 \times 10^{-14}}{2.2 \times 10^{-12}}\right)(1.7 \times 10^{-10})} = 8.8 \times 10^{-7} \text{ M}$$

The pH is approximately 6.06. A more nearly exact analysis, also in text Section 15-9, gives

$$[H_3O^+] \approx \sqrt{\frac{K_1 K_2 c_0 + K_1 K_w}{K_1 + c_0}}$$

The treatment leading to this formula includes only one approximation: that the equilibrium concentration of glycine is close to c_0, its original concentration. Substitution of $c_0 = 0.10$ M and the three constants gives $[H_3O^+] = 8.6 \times 10^{-7}$ M, and the pH equals $\boxed{6.07}$.

Tip. If c_0 were 0.01 M (one-tenth of the c_0 given in the problem) the pH of the glycine solution would be 6.14. This deviates significantly from 6.06, the approximate answer. Thus, as c_0 gets smaller it becomes *more* important in determining the pH, at least at first. As c_0 becomes quite small, the autoionization of water washes out the glycine as a source of H_3O^+. That is, the solution approximates pure water, which has a pH of 7.0 at this temperature.

15.109 The *p*-nitrophenol should be the stronger acid because its conjugate base can form a resonance structure with the negative charge on the electronegative —NO_2 group

No such structure is possible for the conjugate base of *m*-nitrophenol.

CUMULATIVE PROBLEMS

15.111 a) The neutralization reaction that ensues when sour milk and baking soda are mixed is

$$HC_3H_5O_3(aq) + NaHCO_3(aq) \longrightarrow NaC_3H_5O_3(aq) + H_2O(l) + CO_2(g)$$

b) Compute the chemical amount of sodium bicarbonate that is neutralized

$$n_{NaHCO_3} = \tfrac{1}{2} \text{ tsp} \times \left(\frac{236.6 \text{ mL}}{48 \text{ tsp}}\right)\left(\frac{2.16 \text{ g NaHCO}_3}{1 \text{ mL}}\right)\left(\frac{1 \text{ mol NaHCO}_3}{84.01 \text{ g NaHCO}_3}\right) = 0.0634 \text{ mol}$$

Lactic acid and sodium bicarbonate react in a 1-to-1 molar ratio. Hence the one cup of sour milk contains 0.0634 mol of lactic acid. The concentration of lactic acid is 0.0634 mol/0.2366 L = $\boxed{0.268 \text{ mol L}^{-1}}$.

c) The same chemical amount (0.0634 mol) of $CO_2(g)$ will be produced. Compute its volume using the ideal-gas law

$$V_{CO_2} = \frac{nRT}{P} = \frac{(0.0633 \text{ mol})(0.08206 \text{ L atm mol}^{-1}\text{K}^{-1})(177 + 273.15) \text{ K}}{1 \text{ atm}} = \boxed{2.34 \text{ L}}$$

15.113 Label the gaseous monoprotic acid HY. The data on the temperature, pressure, and density of HY provide its molar mass. It is easy to show (if ideal-gas behavior is assumed) that the density of of a gas is

$$\rho = \frac{m}{V} = \frac{P}{RT}\mathcal{M}$$

Solve this equation for $\mathcal{M}$ and insert the numbers for HY in this case

$$\mathcal{M}_{HY} = \rho\left(\frac{RT}{P}\right) = \frac{1.05 \text{ g}}{L}\left(\frac{0.08206 \text{ L atm mol}^{-1}\text{K}^{-1}(313.15 \text{ K})}{1.00 \text{ atm}}\right) = 26.98 \text{ g mol}^{-1}$$

The concentration of HY when 1.85 g of it is dissolved in 450 mL of water is

$$c_{HY} = \frac{n_{HY}}{V} = \frac{m_{HY}}{\mathcal{M}_{HY}V} = \frac{1.85 \text{ g}}{(26.98 \text{ g mol}^{-1})0.450 \text{ L}} = 0.1524 \text{ mol L}^{-1}$$

As the HY reacts with water, it generates H_3O^+. The final concentration of H_3O^+ equals $10^{-5.01} = 9.77 \times 10^{-6}$ M. For HY as an acid

	HY(aq)	+ H_2O(l) ⇌	Y^-(aq) +	H_3O^+(aq)
Init. Conc. (M)	0.1524	–	0	small
Change in Conc. (M)	-9.77×10^{-6}	–	$+9.77 \times 10^{-6}$	$+9.77 \times 10^{-6}$
Equil. Conc. (M)	0.1524	–	9.77×10^{-6}	9.77×10^{-6}

Now that the equilibrium concentrations of all of the products and reactants in the acid ionization of HY are known, it is easy to compute K_a

$$K_a = \frac{[H_3O^+][Y^-]}{[HY]} = \frac{(9.77 \times 10^{-6})^2}{0.1524} = 6.27 \times 10^{-10}$$

Inspection of text Table 15.2 reveals that HY is very likely $\boxed{HCN}$, which has a K_a of 6.17×10^{-10} at 25°C. This identification is strongly supported by a comparison of the molar mass of HCN (27.03 g mol^{-1}) to the molar mass of HY from the gas-density data (27.0 g mol^{-1}).

Chapter 16

Solubility and Precipitation Equilibria

The Nature of Solubility Equilibria

16.1 The chemical equation for the setting of plaster of paris (calcium sulfate hemihydrate) is

$$\tfrac{3}{2}\,H_2O + CaSO_4 \cdot \tfrac{1}{2}H_2O \longrightarrow CaSO_4 \cdot 2H_2O$$

Use molar ratios from this equation in a series of unit-factors

$$V_{H_2O} = 25.0 \text{ kg } CaSO_4 \cdot \tfrac{1}{2}H_2O \times \left(\frac{1 \text{ kmol } CaSO_4 \cdot \tfrac{1}{2}H_2O}{145.15 \text{ kg } CaSO_4 \cdot \tfrac{1}{2}H_2O} \right) \left(\frac{3 \text{ kmol } H_2O}{2 \text{ kmol } CaSO_4 \cdot \tfrac{1}{2}H_2O} \right)$$
$$\times \left(\frac{18.02 \text{ kg } H_2O}{1 \text{ kmol } H_2O} \right) \left(\frac{1 \text{ L } H_2O}{1 \text{ kg } H_2O} \right) = \boxed{4.65 \text{ L } H_2O}$$

16.3 On the graph, read across from "80" on the vertical axis until the solubility curve is reached. Then drop down to the horizontal axis and read the temperature. A solubility of 80 g KBr per 100 g H_2O is reached at a temperature of approximately 48°C. The last of the KBr will dissolve at $\boxed{\text{about } 48°C}$.

Tip. The solubility says nothing about the rate at which the KBr dissolves in the water. Solids often dissolve slowly.

Ionic Equilibria between Solids and Solutions

16.5 Solubility-product constant expressions[1] follow the general rules for heterogeneous equilibria. Pure solids and the solvent do not appear in these expressions. For the dissolution of iron(III) sulfate in water

$$\boxed{Fe_2(SO_4)_3(s) \rightleftharpoons 2\,Fe^{3+}(aq) + 3\,SO_4^{2-}(aq)} \qquad \boxed{K_{sp} = [Fe^{3+}]^2[SO_4^{2-}]^3}$$

16.7 The dissolution of thallium(I) iodate is represented

$$TlIO_3(s) \rightleftharpoons Tl^+(aq) + IO_3^-(aq) \qquad [Tl^+][IO_3^-] = K_{sp} = 3.07 \times 10^{-6}$$

If S mol per liter of $TlIO_3$ dissolves, then S mol per liter of Tl^+ and also S mol per liter of IO_3^- are present at equilibrium, as long as neither ion reacts further in solution.

$$K_{sp} = 3.07 \times 10^{-6} = [Tl^+][IO_3^-] = S^2 \quad \text{which gives} \quad S = 1.752 \times 10^{-3} \text{ mol L}^{-1}$$

This means that 1.75×10^{-3} mol of thallium(I) iodate saturates a liter of solution at 25°C. This equals the solubility of the thallium(I) iodate in 1000 mL of water because the small amount of

---

[1]Also called solubility-product expressions and K_{sp} expressions.

solute does not alter the volume of the solution measurably from the volume of the pure solvent. This chemical amount of thallium(I) iodate has a mass of 0.665 g (obtained by multiplying by $\mathcal{M}_{TlIO_3} = 379.3$ g mol^{-1}). The mass of $TlIO_3$ that saturates 100.0 mL of water at 25°C is 1/10 as much: $\boxed{0.0665 \text{ g } TlIO_3/100.0 \text{ mL}}$.

16.9 The equilibrium equation and K_{sp} for the dissolution of potassium perchlorate are

$$KClO_4(s) \rightleftharpoons K^+(aq) + ClO_4^-(aq) \qquad [K^+][ClO_4^-] = K_{sp} = 1.07 \times 10^{-2}$$

If S mol L^{-1} of $KClO_4$ dissolves, then S mol L^{-1} of K^+ and S mol L^{-1} of ClO_4^- are present at equilibrium, as long as neither ion reacts further. Substitute S into the K_{sp} expression

$$K_{sp} = 1.07 \times 10^{-2} = [K^+][ClO_4^-] = S^2$$

Solving gives $S = 0.103$ mol L^{-1}. Then

$$S = \frac{0.103 \text{ mol } KClO_4}{L} \times \left(\frac{138.55 \text{ g } KClO_4}{1 \text{ mol } KClO_4} \right) = \boxed{\frac{14.3 \text{ g } KClO_4}{L}} \text{ at } 25°C$$

16.11 The equilibrium and K_{sp} expressions are

$$Hg_2I_2(s) \rightleftharpoons Hg_2^{2+}(aq) + 2I^-(aq) \qquad [Hg_2^{2+}][I^-]^2 = K_{sp}$$

If S mol L^{-1} of Hg_2I_2 dissolves, then S mol L^{-1} of Hg_2^{2+} and $2S$ mol L^{-1} of I^- are produced. Assume that neither ion reacts further (with the solvent, for example). Substitute into the K_{sp} mass-action expression

$$[Hg_2^{2+}][I^-]^2 = S(2S)^2 = 1.2 \times 10^{-28} \text{ at } 25°C$$

from which:

$$4S^3 = 1.2 \times 10^{-28} \qquad \text{which gives} \qquad S = 3.1 \times 10^{-10} \text{ mol L}^{-1}$$

$$[Hg_2^{2+}] = S = \boxed{3.1 \times 10^{-10} \text{ mol L}^{-1}} \qquad [I^-] = 2S = \boxed{6.2 \times 10^{-10} \text{ mol L}^{-1}}$$

16.13 As seen in problems **16.9** and **16.11**, a solubility differs from a solubility-product constant, although there is often a simple relationship between the two. The dissolution equilibrium of interest and its associated K are

$$Ag_2CrO_4(s) \rightleftharpoons 2Ag^+(aq) + CrO_4^{2-}(aq) \qquad K_{sp} = [Ag^+]^2[CrO_4^{2-}]$$

The problem states that 0.0129 g of silver chromate dissolves in 500 mL (0.500 L) of water at 25°C. Assume that neither Ag^+ nor CrO_4^{2-} reacts with other species once in the water. Then the chemical amounts of the two ions in solution at equilibrium are

$$n_{Ag^+} = 0.0129 \text{ g} \times \left(\frac{1 \text{ mol } Ag_2CrO_4}{331.7 \text{ g } Ag_2CrO_4} \right) \left(\frac{2 \text{ mol } Ag^+}{1 \text{ mol } Ag_2CrO_4} \right) = 7.78 \times 10^{-5} \text{ mol}$$

$$n_{CrO_4^{2-}} = \tfrac{1}{2} n_{Ag^+} = 3.89 \times 10^{-5} \text{ mol}$$

The equilibrium concentrations of the two ions equal their respective chemical amounts divided by 0.500 L, which is the volume of the solution

$$[Ag^+] = 15.6 \times 10^{-5} \text{ mol L}^{-1} \qquad [CrO_4^{2-}] = 7.78 \times 10^{-5} \text{ mol L}^{-1}$$

Substitute into the K_{sp} expression to obtain a numerical K_{sp}

$$K_{sp} = [Ag^+]^2[CrO_4^{2-}] = (15.6 \times 10^{-5})^2(7.78 \times 10^{-5}) = \boxed{1.9 \times 10^{-12}} \text{ at } 25°C$$

Tip. This equals the value in text Table 16.2.

16.15   The dissolution reaction and its K_{sp}-expression are

$$AgCl(s) \rightleftharpoons Ag^+(aq) + Cl^-(aq) \qquad [Ag^+][Cl^-] = K_{sp}$$

At equilibrium in 1.00 L of the solution at 100°C

$$n_{Ag^+} = 0.018 \text{ g} \times \left(\frac{1 \text{ mol AgCl}}{143.3 \text{ g AgCl}} \right) \left(\frac{1 \text{ mol Ag}^+}{1 \text{ mol AgCl}} \right) = 1.26 \times 10^{-4} \text{ mol}$$

Since the volume of the solution is 1.00 L, the concentration of Ag^+ ion is 1.26×10^{-4} M. The concentration of Cl^- ion is the same—one mole per liter of chloride ion is produced in solution for every mole per liter of silver ion. Substitute the concentrations into the mass-action expression

$$K_{sp} = [Ag^+][Cl^-] = (1.26 \times 10^{-4})(1.26 \times 10^{-4}) = \boxed{1.6 \times 10^{-8}} \text{ at } 100°C$$

Tip. This K_{sp} exceeds the K_{sp} in text Table 16.2 (by a factor of 100!) because the temperature is higher.

Precipitation and the Solubility Product

16.17   Get the initial concentrations of Ba^{2+} ion and CrO_4^{2-} ion and use them to calculate the initial reaction quotient Q_0 for the reaction

$$BaCrO_4(s) \rightleftharpoons Ba^{2+}(aq) + CrO_4^{2-}(aq)$$

"Initial" in this case means after the solution cools to 25°C, but before any reaction occurs. Compare Q_0 with K_{sp}. If Q_0 exceeds K_{sp}, then a precipitate will eventually form as the reaction proceeds from right to left. If Q_0 is less than K_{sp}, then there can be no precipitate. The initial amount of dissolved Ba^{2+} is

$$n_{Ba^{2+}} = 0.0063 \text{ g} \times \left(\frac{1 \text{ mol BaCrO}_4}{253 \text{ g BaCrO}_4} \right) \left(\frac{1 \text{ mol Ba}^{2+}}{1 \text{ mol BaCrO}_4} \right) = 2.49 \times 10^{-5} \text{ mol}$$

The initial concentration of Ba^{2+} ion equals 2.49×10^{-5} M, if the volume of the cooled solution is taken as 1.00 L. The initial concentration of the CrO_4^{2-} ion is, by a similar calculation, also equal to 2.49×10^{-5} M. Then

$$Q_0 = [Ba^{2+}]_0[CrO_4^{2-}]_0 = (2.49 \times 10^{-5})(2.49 \times 10^{-5}) = 6.2 \times 10^{-10}$$

From text Table 16.2, K_{sp} for $BaCrO_4$ equals 2.1×10^{-10} at 25°C. $\boxed{BaCrO_4 \text{ precipitates}}$ until Q is lowered to equal K_{sp}.

Tip. The answer can be refined somewhat. Hot water contracts as it cools. If "hot" is 100°C this contraction in cooling to 25°C amounts to about 4%. This *raises* the "initial" concentrations of the Ba^{2+} and CrO_4^{2-} ions by about 4% each, which in turns raises Q_0 by 1.04^2, which is about 8%.

Tip. In practice in the laboratory, the predicted precipitate might not form right away or even anytime soon because of supersaturation.[2] Only about 3 mg of solid $BaCrO_4$ needs to form before equilibrium is reached. It might be hard to see this amount of solid, especially if it stays suspended in the water.

16.19   Calculate the reaction quotient for the reaction

$$Ce(IO_3)_3(s) \rightleftharpoons Ce^{3+}(aq) + 3\,IO_3^-(aq) \qquad K_{sp} = 1.9 \times 10^{-10} \text{ at } 25°C$$

just after the solutions are mixed and compare it to K_{sp}. The desired reaction quotient is $Q_0 = [Ce^{3+}]_0[IO_3^-]_0^3$. where the subscript zero refers to the concentrations prevailing just after mixing.

---

[2]Text Section 16.1.

After mixing, the volume of the solution equals 400.0 mL, the sum of the two starting volumes. Mixing dilutes both solutions. Before mixing, the concentration of the Ce^{3+} ion was 0.0020 M. After mixing

$$[Ce^{3+}]_0 = 0.0020 \text{ M} \times \left(\frac{250.0 \text{ mL}}{400.0 \text{ mL}}\right) = 1.25 \times 10^{-3} \text{ M}$$

Before mixing, the concentration of the IO_3^- ion was 0.10 M. After mixing

$$[IO_3^-]_0 = 0.010 \text{ M} \times \left(\frac{150.0 \text{ mL}}{400.0 \text{ mL}}\right) = 3.75 \times 10^{-2} \text{ M}$$

$$Q_0 = [Ce^{3+}]_0[IO_3^-]_0^3 = (1.25 \times 10^{-3})(3.75 \times 10^{-2})^3 = 0.66 \times 10^{-8}$$

Q_0 is greater than K_{sp}; $\boxed{\text{a precipitate tends to form}}$.

16.21 The 50.0 mL of 0.0500 M $Pb(NO_3)_2$ contains 2.50 mmol of $Pb^{2+}(aq)$ ion; the 40.0 mL of 0.200 $NaIO_3$ contains 8.00 mmol of $IO_3^-(aq)$ ion. The two ions react

$$Pb^{2+}(aq) + 2\,IO_3^-(aq) \rightleftharpoons Pb(IO_3)_2(s)$$

Assume that this reaction goes to completion. Then zero Pb^{2+} ion remains in solution (it is the limiting reactant), but $8.00 - 2(2.50) = 3.00$ mmol of IO_3^- remains in solution. The concentration of the excess IO_3^- is 3.00 mmol/90.00 mL = 0.0333 M. The reaction in fact does not go to completion, but stops short in the equilibrium state. And even if it did go to completion, a small amount of the solid product would soon redissolve anyway

$$Pb(IO_3^-)_2(s) \rightleftharpoons Pb^{2+}(aq) + 2\,IO_3^-(aq) \qquad K_{sp} = [Pb^{2+}][IO_3^-]^2 = 2.6 \times 10^{-13} \text{ at } 25°C$$

The equilibrium state is of course the same regardless of how it is attained. Let S equal the concentration of Pb^{2+} furnished by back-reaction

	$Pb(IO_3)_2(s) \rightleftharpoons$	$Pb^{2+}(aq) +$	$2\,IO_3^-(aq)$
Init. Conc. (mol L^{-1})	–	0.0	0.0333
Change in Conc. (mol L^{-1})	–	$+S$	$+2S$
Equil. Conc. (mol L^{-1})	–	S	$0.0333 + 2S$

Substitution from the third line of the table into the K_{sp} expression gives

$$(S)(0.0333 + 2S)^2 = 2.6 \times 10^{-13}$$

This cubic equation is simplified by assuming that $2S << 0.0333$. Then

$$(0.0333)^2(S) = 2.6 \times 10^{-13} \qquad \text{which gives} \qquad S = 2.3 \times 10^{-10}$$

Hence: $[Pb^{2+}] = \boxed{2.3 \times 10^{-10} \text{ M}}$ and $[IO_3^-] = \boxed{0.033 \text{ M}}$. The assumption that $2S$ was much smaller than 0.0333 was obviously justified.

16.23 Follow the procedure used in problem **16.21**. The solution contains 5.00 mmol of $AgNO_3$ and 1.8 mmol of Na_2CrO_4 after mixing but before reaction. Assume that the precipitation reaction

$$2\,Ag^+(aq) + CrO_4^{2-}(aq) \rightleftharpoons Ag_2CrO_4(s)$$

goes to completion. As it does, it consumes 2 mmol of Ag^+ for every 1 mmol of CrO_4^{2-} ion. Because the chemical amount of Ag^+ ion is more than twice that of CrO_4^{2-} ion, CrO_4^{2-} ion is the limiting reactant. The chemical amount of Ag^+ ion left in excess is $5.00 - 2(1.80) = 1.40$ mmol. The concentration of Ag^+ ion in the solution equals this amount divided by the volume of the solution

$$[Ag^+] = \frac{1.40 \text{ mmol}}{(50.0 + 30.0) \text{ mL}} = 0.01750 \text{ M}$$

Now let some of the precipitate redissolve

$$Ag_2CrO_4(s) \rightleftharpoons 2\,Ag^+(aq) + CrO_4^{2-}(aq) \qquad K_{sp} = 1.9 \times 10^{-12} = [Ag^+]^2[CrO_4^{2-}]$$

Let S equal the concentration of CrO_4^{2-} in solution after this equilibrium is attained. The concentration of Ag^+ ion is $0.01750 + 2S$ and

$$K_{sp} = (0.01750 + 2S)^2 S$$

Assume that $2S$ is much smaller than 0.0175. Then

$$(0.01750)^2 S = 1.9 \times 10^{-12} \quad \text{which gives} \quad S = 6.2 \times 10^{-9}$$

The final concentrations of the two ions are

$$[CrO_4^{2-}] = \boxed{6.2 \times 10^{-9} \text{ M}} \qquad [Ag^+] = 0.01750 + 2(6.2 \times 10^{-9}) = \boxed{0.018 \text{ M}}$$

16.25 The $F^-(aq)$ ion from the dissolved NaF depresses the solubility of CaF_2. This is the common-ion effect. CaF_2 dissolves according to the equation

	$CaF_2(s) \rightleftharpoons$	$Ca^{2+}(aq) +$	$2\,F^-(aq)$
Init. Conc. (mol L^{-1})	–	0.0	0.040
Change in Conc. (mol L^{-1})	–	$+S$	$+2S$
Equil. Conc. (mol L^{-1})	–	S	$0.040 + 2S$

where S is the solubility of the CaF_2. For this dissolution reaction

$$K_{sp} = [Ca^{2+}][F^-]^2 = 3.9 \times 10^{-11}$$

Substitute the equilibrium concentrations from the table, assuming that $2S \ll 0.040$

$$S(0.040)^2 = 3.9 \times 10^{-11} \quad \text{which gives} \quad S = 2.4 \times 10^{-8} \text{ mol L}^{-1}$$

The assumption is obviously justified; the solubility of CaF_2 is $\boxed{2.4 \times 10^{-8} \text{ mol L}^{-1}}$ at 25°C.

16.27 **a)** For every y mol L^{-1} of $Ni(OH)_2$ that is dissolved at equilibrium, y mol L^{-1} of Ni^{2+} and $2y$ mol L^{-1} of OH^- have formed according to the equation

$$Ni(OH)_2(s) \rightleftharpoons Ni^{2+}(aq) + 2\,OH^-(aq) \qquad [Ni^{2+}][OH^-]^2 = K_{sp} = 1.6 \times 10^{-16}$$

Substituting into the K_{sp} expression gives the equation $(y)(2y)^2 = 1.6 \times 10^{-16}$. Solving gives $y = 3.4 \times 10^{-6}$. The solubility of $Ni(OH)_2$ is $\boxed{3.4 \times 10^{-6} \text{ mol L}^{-1}}$ at 25°C.

b) The presence of a common ion (the OH^- ion) reduces the solubility of the nickel(II) hydroxide. Set up the usual three-line table:

	$Ni(OH)_2(s) \rightleftharpoons$	$Ni^{2+}(aq) +$	$2\,OH^-(aq)$
Init. Conc. (mol L^{-1})	–	0	0.100
Change in Conc. (mol L^{-1})	–	$+z$	$+2z$
Equil. Conc. (mol L^{-1})	–	z	$0.100 + 2z$

Substitute the equilibrium concentrations into the K_{sp} expression

$$K_{sp} = [\text{Ni}^{2+}][\text{OH}^-]^2 = z(0.100 + 2z)^2 = 1.6 \times 10^{-16}$$

Assume that $2z$ is much smaller than 0.100 mol L^{-1}. Then

$$(0.100)^2(z) = 1.6 \times 10^{-16} \quad \text{so that} \quad z = 1.6 \times 10^{-14} \text{ mol L}^{-1}$$

The assumption is obviously justified. The solubility is $\boxed{1.6 \times 10^{-14} \text{ mol L}^{-1}}$ at $25°\text{C}$.

16.29 As long as the solution is in equilibrium with $\text{Mg(OH)}_2(s)$ at room temperature, the following equation holds

$$K_{sp} = 1.2 \times 10^{-11} = [\text{Mg}^{2+}][\text{OH}^-]^2$$

Let S represent the solubility of the Mg(OH)_2 *before* any NaOH is added. At this stage

$$[\text{Mg}^{2+}] = S \text{ and } [\text{OH}^-] = 2S \quad \text{so that} \quad 1.2 \times 10^{-11} = [\text{Mg}^{2+}][\text{OH}^-]^2 = 4S^3$$

Solving gives $S = 1.44 \times 10^{-4}$ mol L^{-1}. Sodium hydroxide dissociates completely to $\text{Na}^+(aq)$ and $\text{OH}^-(aq)$ ions. The concentration of OH^- goes up with the addition of NaOH, and the concentration of Mg^{2+} must diminish to keep the mass-action expression equal to the constant K_{sp}. Additional magnesium hydroxide precipitates as NaOH is added. This is the common-ion effect in action. The problem states that the solubility of Mg(OH)_2 is reduced to 0.0010 of its original value. This means that after the addition

$$[\text{Mg}^{2+}] = 0.0010(1.44 \times 10^{-4}) = 1.44 \times 10^{-7} \text{ mol L}^{-1}$$

Then

$$[\text{OH}^-] = \sqrt{\frac{K_{sp}}{[\text{Mg}^{2+}]}} = \sqrt{\frac{1.2 \times 10^{-11}}{1.44 \times 10^{-7}}} = \boxed{9.1 \times 10^{-3} \text{ mol L}^{-1}}$$

The Effects of pH on Solubility

16.31 Assume that the concentration of OH^- ion from the dissolution greatly exceeds the concentration of OH^- ion from the autoionization of water. Then, when y mol L^{-1} of the AgOH(s) dissolves, y mol L^{-1} of Ag^+ ion and y mol L^{-1} of OH^- ion are produced.

	AgOH(s) $\rightleftharpoons$	$\text{Ag}^+(aq)$ +	$\text{OH}^-(aq)$
Init. Conc. (mol L^{-1})	–	0.0	small
Change in Conc. (mol L^{-1})	–	$+y$	$+y$
Equil. Conc. (mol L^{-1})	–	y	y

At equilibrium at $25°\text{C}$

$$K_{sp} = [\text{Ag}^+][\text{OH}^-] = y^2 = 1.5 \times 10^{-8} \quad \text{so that} \quad y = 1.2 \times 10^{-4} \text{ mol L}^{-1}$$

The molar solubility of AgOH in water equals $\boxed{0.00012 \text{ mol L}^{-1}}$ at $25°\text{C}$.

If the solution is buffered at pH 7.00, then the pH stays at 7.00 even though the dissociation of AgOH produces OH^-. A pH of 7.00 means $[\text{OH}^-] = 1.0 \times 10^{-7}$ mol L^{-1}. Put this concentration into the K_{sp}-expression and solve for the equilibrium concentration of $\text{Ag}^+(aq)$

$$[\text{Ag}^+] = \frac{K_{sp}}{[\text{OH}^-]} = \frac{1.5 \times 10^{-8}}{1.0 \times 10^{-7}} = 0.15 \text{ mol L}^{-1}$$

The solubility of the AgOH(s) equals the chemical amount of Ag^+ ion in solution per liter; it is therefore $\boxed{0.15 \text{ mol L}^{-1}}$ in water buffered at pH 7. This solubility is 1250 times larger than the solubility of AgOH in pure water at this temperature.

16.33   **a)** The solubility of PbI_2 will remain ┃unchanged┃ as the pH of its solution is lowered. The anion is an exceedingly weak base; it has little interaction with H_3O^+ even at a high concentration of H_3O^+.

b) The solubility of AgOH will ┃increase┃ as the pH of its solution is lowered. The additional H_3O^+ drives additional dissolution by removing OH^- ion.

c) The solubility of $Ca_3(PO_4)_2$ will ┃increase┃ as the pH of its solution is lowered from 7. A higher concentration of H_3O^+ drives the dissolution by removing product PO_4^{3-} ion as HPO_4^{2-} ion.

Complex Ions and Solubility

16.35   The copper(II) nitrate dissolves readily to give $Cu^{2+}(aq)$ and $NO_3^-(aq)$ ions. Imagine that the $Cu^{2+}(aq)$ reacts to completion with the $NH_3(aq)$ (which is in excess) to form $Cu(NH_3)_4^{2+}(aq)$

$$Cu^{2+}(aq) + 4\,NH_3(aq) \longrightarrow Cu(NH_3)_4^{2+}(aq)$$

Then
$$[Cu(NH_3)_4^{2+}] = 0.10 \text{ M} \quad \text{and} \quad [NH_3] = 1.50 - 4(0.10) = 1.10 \text{ M}$$

Now, reason that any free Cu^{2+} ion forms from back-reaction (dissociation of $Cu(NH_3)_4^{2+}$). The back-reaction would proceed in four steps

$$Cu(NH_3)_4^{2+}(aq) \rightleftharpoons Cu(NH_3)_3^{2+}(aq) + NH_3(aq) \qquad \frac{[Cu(NH_3)_3^{2+}][NH_3]}{[Cu(NH_3)_4^{2+}]} = 1.1 \times 10^{-2}$$

$$Cu(NH_3)_3^{2+}(aq) \rightleftharpoons Cu(NH_3)_2^{2+}(aq) + NH_3(aq) \qquad \frac{[Cu(NH_3)_2^{2+}][NH_3]}{[Cu(NH_3)_3^{2+}]} = 2 \times 10^{-3}$$

$$Cu(NH_3)_2^{2+}(aq) \rightleftharpoons Cu(NH_3)^{2+}(aq) + NH_3(aq) \qquad \frac{[Cu(NH_3)^{2+}][NH_3]}{[Cu(NH_3)_2^{2+}]} = 5 \times 10^{-4}$$

$$Cu(NH_3)^{2+}(aq) \rightleftharpoons Cu^{2+}(aq) + NH_3(aq) \qquad \frac{[Cu^{2+}][NH_3]}{[Cu(NH_3)^{2+}]} = 1.0 \times 10^{-4}$$

where the four constants equal the reciprocals of K_4 through K_1 in text Table 16.3.[3] The product of K_1 through K_4 equals 0.9×10^{12}, but K_f equals 1.1×10^{12} in the table. This inconsistency arises from rounding off the step-wise K's and is unimportant. Label the concentrations of the four Cu-containing products x, y, z, and w respectively, and calculate them in turn. Treat the steps as if they occurred independently. That is, neglect the amount of Cu-containing product reacted away by later steps and assume that NH_3 from later steps adds only negligibly to the 1.10 M NH_3 present when dissociation starts. For the first step

	$Cu(NH_3)_4^{2+}(aq)$	$\rightleftharpoons$	$Cu(NH_3)_3^{2+}(aq)$	$+\,NH_3(aq)$
Init. Conc. (mol L^{-1})	0.10		0	1.1
Change in Conc. (mol L^{-1})	$-x$		$+x$	$+x$
Equil. Conc. (mol L^{-1})	$0.10 - x$		x	$1.1 + x$

$$\frac{[Cu(NH_3)_3^{2+}][NH_3]}{[Cu(NH_3)_4^{2+}]} = 1.1 \times 10^{-2} = \frac{x(1.1 + x)}{0.10 - x} \quad \text{which leads to} \quad x = 9.9 \times 10^{-4}$$

The set-up for the second step is similar. It gives

$$\frac{[Cu(NH_3)_2^{2+}][NH_3]}{[Cu(NH_3)_3^{2+}]} = 2 \times 10^{-3} = \frac{y(1.1 + y)}{9.9 \times 10^{-4} - y} \quad \text{which leads to} \quad y = 1.8 \times 10^{-6}$$

---

[3]The equilibrium constants in text Table 16.3 were determined at 25°C.

For the third step

$$\frac{[\text{Cu(NH}_3)^{2+}][\text{NH}_3]}{[\text{Cu(NH}_3)_2^{2+}]} = 5 \times 10^{-4} = \frac{z(1.1+z)}{1.8 \times 10^{-6} - z} \quad \text{which leads to} \quad z = 8.2 \times 10^{-10}$$

For the fourth step

$$\frac{[\text{Cu}^{2+}][\text{NH}_3]}{[\text{Cu(NH}_3)^{2+}]} = 1.0 \times 10^{-4} = \frac{w(1.1+w)}{8.2 \times 10^{-10} - w} \quad \text{which leads to} \quad w = 7.4 \times 10^{-14}$$

This w equals the concentration of free Cu^{2+}. The concentrations of the partially dissociated complexes are low. Indeed, all dissociation reduces the concentration of $\text{Cu(NH}_3)_4^{2+}$ by less than 1%. Therefore

$$[\text{Cu}^{2+}] = \boxed{7 \times 10^{-14} \text{ mol L}^{-1}} \qquad\qquad [\text{Cu(NH}_3)_4^{2+}] = \boxed{0.10 \text{ mol L}^{-1}}$$

16.37 Imagine that the 1-to-1 reaction between $\text{K}^+(aq)$ and 18-crown-6(aq) goes 100% to completion and all free $\text{K}^+(aq)$ then comes from a back-dissociation of the product $(\text{K-crown})^+(aq)$. The equation for this dissociation is the reverse of the equation in the problem

	(Kcrown)$^+$(aq) $\rightleftharpoons$	K$^+$(aq) +	crown(aq)
Init. Conc. (mol L^{-1})	0.0080	0	0
Change in Conc. (mol L^{-1})	$-x$	$+x$	$+x$
Equil. Conc. (mol L^{-1})	$0.0080 - x$	x	x

This reaction has the following mass-action expression and K

$$\frac{[\text{K}^+][\text{crown}]}{[(\text{Kcrown})^+]} = \frac{1}{111.6} \qquad \text{hence at equilibrium} \qquad \frac{1}{111.6} = \frac{x^2}{0.0080 - x}$$

Rearranging gives the quadratic equation

$$x^2 + (8.961 \times 10^{-3})x - 7.169 \times 10^{-5} = 0$$

Solving gives $x = 0.0051$ or $-.0145$, but only the positive root makes physical sense. The equilibrium concentration of $\text{K}^+(aq)$ is $\boxed{0.0051 \text{ M}}$.

Calculation of the concentration of free Na^+ proceeds in the same fashion. The quadratic equation

$$y^2 + 0.152y - 0.00121 = 0$$

(where y is the concentration of free Na^+ ion) arises when $K = 1/6.6$. Solution of this quadratic equation gives a Na^+ ion concentration of $\boxed{0.0076 \text{ M}}$.

16.39 The question can be answered with certainty only by calculating the solubility of AgCl(s) in a solution that is 1.00 M in Cl$^-$ ion. Let S equal this solubility. Assume that for every mole of AgCl that dissolves, either an Ag$^+$ ion or an AgCl$_2^-$ ion[4] forms. In mathematical form

$$S = [\text{Ag}^+] + [\text{AgCl}_2^-]$$

The solubility equilibrium for AgCl assures that

$$K_{\text{sp}} = [\text{Ag}^+][\text{Cl}^-]$$

---

[4]Dichloroargenate(I) ion.

as long as solid silver chloride is present. The $AgCl_2^-$ complex ion is generated by the reaction

$$Ag^+(aq) + 2\,Cl^-(aq) \rightleftharpoons AgCl_2^-(aq) \qquad K_f = \frac{[AgCl_2^-]}{[Ag^+][Cl^-]^2}$$

Solve the K_{sp} and K_f equations for $[Ag^+]$ and $[AgCl_2^-]$ respectively and substitute the results into the first equation

$$\begin{aligned}
S &= \frac{K_{sp}}{[Cl^-]} + K_f[Ag^+][Cl^-]^2 \\
&= \frac{K_{sp}}{[Cl^-]} + K_f K_{sp}[Cl^-] \\
&= \frac{1.6 \times 10^{-10}}{[Cl^-]} + (1.8 \times 10^5)(1.6 \times 10^{-10})[Cl^-]
\end{aligned}$$

The numerical values of K_{sp} and K_f come from text Tables 16.2 and 16.3 respectively The assumption in using them is that the temperature is 25°C. Now, reason that the concentration of Cl^- ion is so large at 1.00 M that it is not substantially reduced by reaction with Ag^+ ion. If $[Cl^-] = 1.00$ M, then

$$S = \frac{1.6 \times 10^{-10}}{1.00} + (1.8 \times 10^5)(1.6 \times 10^{-10})[1.00] = 2.9 \times 10^{-5}\ \text{M}$$

The notion that $[Cl^-]$ is only negligibly reduced from its original value is vindicated by this low solubility—no more than about 6×10^{-5} mol L^{-1} of Cl^- ion can be tied up in the complex. The solubility of AgCl in this solution is more than double the solubility of AgCl in pure water, which is 1.3×10^{-5} M by a computation like the one in problem **16.7**. Hence, AgCl(s) is $\boxed{\text{more soluble}}$ in 1.00 M NaCl than in pure water.

In 0.100 M NaCl, in which $[Cl^-]$ is 0.100 M, the equation for S becomes

$$S = \frac{1.6 \times 10^{-10}}{0.100} + (1.8 \times 10^5)(1.6 \times 10^{-10})[0.100] = 2.9 \times 10^{-6}\ \text{M}$$

AgCl is $\boxed{\text{less soluble}}$ in dilute NaCl than in pure water.

Tip. The reversal is remarkable. Complexation plays a potent role in determining solubilities.

16.41 $CuSO_4$ dissolves to give $Cu^{2+}(aq)$ and $SO_4^{2-}(aq)$. The aquated Cu^{2+} ion exists as the $Cu(H_2O)_4^{2+}$ complex ion. This ion acts as a Brønsted-Lowry acid according to the equation

$$\boxed{Cu(H_2O)_4^{2+}(aq) + H_2O(l) \rightleftharpoons H_3O^+(aq) + Cu(H_2O)_3OH^+(aq)}$$

The solution is acidic because the K_a for the complex exceeds K_b for the SO_4^{2-} ion. An equivalent answer is the equation

$$Cu^{2+}(aq) + 2\,H_2O(l) \rightleftharpoons H_3O^+(aq) + (CuOH)^+(aq)$$

16.43 The computation is like other computations of the pH of solutions of weak acids.[5] The coordinated cobalt(II) ion is acidic

	$Co(H_2O)_6^{2+}$	$+H_2O \rightleftharpoons$	$H_3O^+(aq) +$	$Co(H_2O)_5OH^+$
Init. Conc. (mol L^{-1})	0.10		small	0
Change in Conc. (mol L^{-1})	$-x$		$+x$	$+x$
Equil. Conc. (mol L^{-1})	$0.10 - x$		x	x

---

[5]Such as problem **15.29a**.

The equilibrium expression is

$$\frac{[H_3O^+][Co(H_2O)_5OH^+]}{[Co(H_2O)_6^{2+}]} = K_a = 3 \times 10^{-10} = \frac{x^2}{0.10 - x}$$

Solve the equation to obtain $x = 5.5 \times 10^{-6}$. The concentration of H_3O^+ is 5.5×10^{-6} M, and the pH is therefore $\boxed{5.3}$.

16.45 The reaction $Pt(NH_3)_4^{2+}(aq) + H_2O(l) \rightleftharpoons H_3O^+(aq) + Pt(NH_3)_3NH_2^+(aq)$ causes the solution to be acidic. The mass-action expression for this reaction is

$$\frac{[Pt(NH_3)_3NH_2^+][H_3O^+]}{[Pt(NH_3)_4^{2+}]} = K_a$$

The concentration of the $Pt(NH_3)_3NH_2^+$ ion equals the concentration of the H_3O^+ ion in solution as long as this reaction is the only source of either ion. The concentration of the $Pt(NH_3)_4^{2+}$ ion equals 0.15 M minus the concentration of H_3O^+ ion. The H_3O^+ concentration can be calculated from the pH given in the problem:

$$[H_3O^+] = 10^{-4.92} = 1.20 \times 10^{-5} \text{ M}$$

Substitute this and the other concentrations in the K_a-expression

$$K_a = \frac{(1.20 \times 10^{-5})(1.20 \times 10^{-5})}{0.15 - 1.20 \times 10^{-5}} = \boxed{9.6 \times 10^{-10}}$$

16.47 The problem concerns the fate of $Pb^{2+}(aq)$ in a solution adjusted to pH 13.0 by the addition of NaOH. The precipitation reaction

$$Pb^{2+}(aq) + 2\,OH^-(aq) \rightleftharpoons Pb(OH)_2(s)$$

tends to reduce the concentration of $Pb^{2+}(aq)$. The K for this reaction is the reciprocal of the K_{sp} of $Pb(OH)_2(s)$. It is quite large (K_{sp} is 4.2×10^{-15} so $1/K_{sp}$ is 2.38×10^{14}). This seems to mean that 1.00 M Pb^{2+} ion gives a precipitate of $Pb(OH)_2(s)$ at pH 13.0. But there is a complication. The equilibrium

$$Pb^{2+}(aq) + 3\,OH^-(aq) \rightleftharpoons Pb(OH)_3^-(aq)$$

ties up Pb^{2+} ion in a *soluble* form, thereby opposing the precipitation of $Pb(OH)_2(s)$. The K for this reaction is the K_f for $Pb(OH)_3^-(aq)$ and is large (given as 4×10^{14} in the problem). Whether $Pb(OH)_2$ precipitates depends how the competition between these reactions plays out.

Suppose $Pb(OH)_2(s)$ *does* precipitate. Then, at pH 13.0, where $[OH^-] = 0.10$ M, the concentration of Pb^{2+} must fulfill the equation

$$K_{sp} = 4.2 \times 10^{-15} = [Pb^{2+}][OH^-]^2 = [Pb^{2+}](0.10)^2$$

This means $[Pb^{2+}]$ is locked at 4.2×10^{-13} M if solid $Pb(OH)_2$ is present. The mass-action expression for the complexation equilibrium is

$$K_f = 4 \times 10^{14} = \frac{[Pb(OH)_3^-]}{[Pb^{2+}][OH^-]^3}$$

Substitute $[OH^-] = 0.10$ and $[Pb^{2+}] = 4.2 \times 10^{-13}$ M, and solve for $[Pb(OH)_3^-]$. The answer is 0.17 M. Any concentration of Pb^{2+} ion that exceeds this threshold value causes precipitation. Because 1.00 M exceeds 0.17 M, $Pb(OH)_2(s)$ precipitates in the case defined in this problem. At equilibrium, $[Pb^{2+}]$ equals $\boxed{4.2 \times 10^{-13} \text{ M}}$, and the concentration of $Pb(OH)_3^-$ equals $\boxed{0.17 \text{ M}}$.

The problem also asks whether $Pb(OH)_2(s)$ precipitates if the initial concentration of Pb^{2+} is 0.050 M. This value is less than the precipitation threshold of 0.17 M so no precipitate of $Pb(OH)_2(s)$ can form. The K_{sp} equilibrium is *not* in effect. Essentially all of the Pb^{2+} ion is tied up in the complex, making the concentration of $Pb(OH)_3^-$ equal $\boxed{0.050\ M}$. Put this value into the K_f mass-action expression

$$K_f = 4 \times 10^{14} = \frac{[Pb(OH)_3^-]}{[Pb^{2+}][OH^-]^3} = \frac{0.050}{[Pb^{2+}](0.10)^3}$$

Solving gives the concentration of $Pb^{2+}(aq)$ (also called "free" Pb^{2+} ion) as $\boxed{1 \times 10^{-13}\ M}$.

A DEEPER LOOK... Selective Precipitation of Ions

16.49 **a)** Imagine slowly adding oxalate ions to the mixture containing Mg^{2+} and Pb^{2+} ions. Both oxalate salts will stay in solution until their respective reaction quotients exceed their K_{sp}'s. The applicable equilibria and K_{sp} expressions are

$$PbC_2O_4(s) \rightleftharpoons Pb^{2+}(aq) + C_2O_4^{2-}(aq) \qquad [Pb^{2+}][C_2O_4^{2-}] = K_{sp} = 2.7 \times 10^{-11}$$
$$MgC_2O_4(s) \rightleftharpoons Mg^{2+}(aq) + C_2O_4^{2-}(aq) \qquad [Mg^{2+}][C_2O_4^{2-}] = K_{sp} = 8.6 \times 10^{-5}$$

The concentration of oxalate ion needed to precipitate magnesium oxalate from the 0.10 M Mg^{2+} solution is

$$[C_2O_4^{2-}] = \frac{K_{sp}}{[Mg^{2+}]} = \frac{8.6 \times 10^{-5}}{0.10} = 8.6 \times 10^{-4}\ M$$

If the $[C_2O_4^{2-}]$ is kept at or below $\boxed{8.6 \times 10^{-4}\ mol\ L^{-1}}$, then magnesium oxalate cannot precipitate. Only lead oxalate can precipitate. The lead oxalate precipitates first because its solubility is smaller than the solubility of magnesium oxalate.

b) If the $[C_2O_4^{2-}]$ is held at 8.6×10^{-4} M, then

$$[Pb^{2+}] = \frac{K_{sp}}{[C_2O_4^{2-}]} = \frac{2.7 \times 10^{-11}}{8.6 \times 10^{-4}} = 3.1 \times 10^{-8}\ M$$

$$\text{fraction } Pb^{2+}\text{ remaining} = \frac{3.1 \times 10^{-8}\ M}{0.10\ M} = \boxed{3.1 \times 10^{-7}}$$

16.51 Calculate the $[I^-]$ that just suffices to bring about precipitation of each metal ion. The dissolution equations and their K_{sp}'s are

$$Hg_2I_2(s) \rightleftharpoons Hg_2^{2+}(aq) + 2\,I^-(aq) \qquad K_{sp} = [Hg_2^{2+}][I^-]^2 = 1.2 \times 10^{-28}$$
$$PbI_2(s) \rightleftharpoons Pb^{2+}(aq) + 2\,I^-(aq) \qquad K_{sp} = [Pb^{2+}][I^-]^2 = 8.5 \times 10^{-9}$$

The required concentration is quite small in the case of $Hg_2I_2(s)$ because the K_{sp} is very small. In the case of $PbI_2(s)$ the required concentration is

$$[I^-] = \sqrt{\frac{8.5 \times 10^{-9}}{0.0500}} = 4.1 \times 10^{-4}\ M$$

The optimum $[I^-]$ would be just below $\boxed{4.1 \times 10^{-4}\ M}$. If the concentration of I^- is set to this value, then $PbI_2(s)$ cannot precipitate, but almost all of the $Hg_2^{2+}(aq)$ can precipitate as $Hg_2I_2(s)$.

16.53 Metal ions form sulfides of greatly varying but generally low solubility. Careful control of the pH, which strongly affects these solubilities, allows the separation of the metal ions by differential precipitation of the sulfides. The dissolution of zinc sulfide is represented

$$ZnS(s) + H_2O(l) \rightleftharpoons Zn^{2+}(aq) + HS^-(aq) + OH^-(aq) \qquad K = [Zn^{2+}][HS^-][OH^-]$$

The equilibrium expression is a triple product, but is otherwise not exceptional. Both the OH^- and HS^- concentrations depend strongly on the concentration of H_3O^+ according to the equations

$$2\,H_2O(l) \rightleftharpoons H_3O^+(aq) + OH^-(aq) \qquad K_w = [H_3O^+][OH^-]$$

$$H_2S(aq) + H_2O(aq) \rightleftharpoons H_3O^+(aq) + HS^-(aq) \qquad K_{a1} = \frac{[HS^-][H_3O^+]}{[H_2S]}$$

Solve these equations for $[OH^-]$ and $[HS^-]$, and substitute into the triple-product K expression

$$[Zn^{2+}]\left(\frac{K_{a1}[H_2S]}{[H_3O^+]}\right)\left(\frac{K_w}{[H_3O^+]}\right) = K$$

Assume that the temperature is 25°C. Then K_{a1} of H_2S equals 9.1×10^{-8}, K_w equals 1.0×10^{-14}, and K equals 2×10^{-25}.[6] Also, the concentration of H_2S equals 0.10 M, and the concentration of H_3O^+ equals 1.0×10^{-5} M. Solve the preceding equation for the concentration of Zn^{2+}, and insert the three equilibrium constants and two concentrations

$$[Zn^{2+}] = K\left(\frac{[H_3O^+]}{K_{a1}[H_2S]}\right)\left(\frac{[H_3O^+]}{K_w}\right)$$

$$= 2 \times 10^{-25}\left(\frac{1.0 \times 10^{-5}}{(9.1 \times 10^{-8})(0.10)}\right)\left(\frac{1.0 \times 10^{-5}}{1.0 \times 10^{-14}}\right) = \boxed{2 \times 10^{-13} \text{ M}}$$

16.55 Precipitation of $FeS(s)$ can begin only if the pH is high enough to make the reaction quotient Q for the reaction

$$FeS(s) + H_2O(l) \rightleftharpoons Fe^{2+}(aq) + HS^-(aq) + OH^-(aq) \quad [Fe^{2+}][HS^-][OH^-] = 5 \times 10^{-19}$$

exceed the K.[7] Compute the $[H_3O^+]$ that barely causes $FeS(s)$ to precipitate. This concentration is attained when the following equation (obtained as in problem **16.53**) is satisfied

$$[Fe^{2+}]\left(\frac{K_{a1}[H_2S]}{[H_3O^+]}\right)\left(\frac{K_w}{[H_3O^+]}\right) = 5 \times 10^{-19}$$

Solve for $[H_3O^+]$ and substitute the various numbers. The K_{a1} of H_2S equals 9.1×10^{-8}; K_w equals 1.0×10^{-14}; the concentration of H_2S is 0.10 M; the concentration of Fe^{2+} is 0.10 M:

$$[H_3O^+] = \sqrt{\frac{[Fe^{2+}]K_{a1}[H_2S]K_w}{5 \times 10^{-19}}}$$

$$= \sqrt{\frac{(0.10)(9.1 \times 10^{-8})(0.10)(1.0 \times 10^{-14})}{5 \times 10^{-19}}} = 4.3 \times 10^{-3} \text{ M}$$

This is the minimum concentration of H_3O^+ that keeps FeS in solution. The maximum pH is therefore $\boxed{2.4}$. Higher pH, (implying lower $[H_3O^+]$ and higher $[OH^-]$) shifts the dissolution reaction to the left, causing a precipitate.

The equation for the dissolution of $PbS(s)$ is similar to that for the dissolution of $FeS(s)$

$$PbS(s) + H_2O(l) \rightleftharpoons Pb^{2+}(aq) + HS^-(aq) + OH^-(aq) \quad [Pb^{2+}][HS^-][OH^-] = 3 \times 10^{-28}$$

and the expression

$$[Pb^{2+}]\left(\frac{K_{a1}[H_2S]}{[H_3O^+]}\right)\left(\frac{K_w}{[H_3O^+]}\right) = 3 \times 10^{-28}$$

---

[6]See text Tables 15.2 and 16.3.
[7]The K's in this problem come from tables in the text.

is therefore easily written. Substitution of 0.10 for $[H_2S]$, 4.3×10^{-3} for $[H_3O^+]$, and 9.1×10^{-8} for K_{a1} gives $[Pb^{2+}]$ equal to $\boxed{6 \times 10^{-11} \text{ M}}$, which is very small.

Tip. The point is that at pH 2.4, all the Fe^{2+} but (essentially) none of the Pb^{2+} stays in solution.

ADDITIONAL PROBLEMS

16.57 According to text Table 16.2 and problem **16.11**, the mercury(I) ion exists as a dimer, Hg_2^{2+}, in aqueous solution. The chemical equation and associated equilibrium law for dissolution of its chloride are therefore

$$\boxed{Hg_2Cl_2(s) \rightleftharpoons Hg_2^{2+}(aq) + 2\,Cl^-(aq)} \qquad \boxed{[Hg_2^{2+}][Cl^-]^2 = K_{sp}}$$

16.59 The dissolution of barium sulfate is represented by the equation and equilibrium law

$$BaSO_4(s) \rightleftharpoons Ba^{2+}(aq) + SO_4^{2-}(aq) \qquad [Ba^{2+}][SO_4^{2-}] = K_{sp} = 1.1 \times 10^{-10}$$

If the equilibrium concentration of Ba^{2+} equals x, then the concentration of SO_4^{2-} is also x, as long as there are no additional sources (or sinks) of either ion. Then $x^2 = 1.049 \times 10^{-10}$ so that $x = [Ba^{2+}] = \boxed{1.0 \times 10^{-5} \text{ M}}$. This concentration is too low for any bad effects on patients who drink the suspension of $BaSO_4(s)$.

16.61 It takes 860 mL of 0.0050 M NaF to make 1.00 L of solution when mixed with 140 mL of 0.0010 M $Sr(NO_3)_2$. Just after mixing, but before any precipitation reactions occur, the concentrations of Sr^{2+} and F^- ions are

$$[Sr^{2+}] = \left(\frac{140}{1000}\right) 0.0010 = 1.4 \times 10^{-4} \text{ M} \qquad [F^-] = \left(\frac{860}{1000}\right) 0.0050 = 4.3 \times 10^{-3} \text{ M}$$

The dissolution reaction and associated K_{sp} expression are

$$SrF_2(s) \rightleftharpoons Sr^{2+}(aq) + 2\,F^-(aq) \qquad [Sr^{2+}][F^-]^2 = K_{sp} = 2.8 \times 10^{-9}$$

The mixture will tend to precipitate SrF_2 only if the "initial" reaction quotient Q_0 (after mixing but before reaction starts) exceeds K_{sp}. Compute Q_0 by substitution of the initial concentrations into the mass-action expression

$$Q_0 = [Sr^{2+}]_0[F^-]_0^2 = (1.4 \times 10^{-4})(4.3 \times 10^{-3})^2 = 2.6 \times 10^{-9}$$

Because Q_0 is less than K_{sp}, $\boxed{\text{no precipitate of } SrF_2}$ occurs.

Tip. The volumes of differing aqueous solutions are not in general additive (that is, mixing 500 mL of solution A with 500 mL of solution B does *not* always give 1000 mL of solution). However, the solutions in this problem are so dilute that figuring the volume of the NaF(*aq*) solution by subtracting 140 mL from 1000 mL is certainly acceptable.

16.63 The K_2CO_3 and $Ca(NO_3)_2$ are both strong electrolytes. They dissolve in water by dissociation

$$K_2CO_3(s) \longrightarrow 2\,K^+(aq) + CO_3^{2-}(aq)$$
$$Ca(NO_3)_2(s) \longrightarrow Ca^{2+}(aq) + 2\,NO_3^-(aq)$$

When the two solutions are mixed, insoluble $CaCO_3(s)$ precipitates according to the net ionic equation

$$Ca^{2+}(aq) + CO_3^{2-}(aq) \longrightarrow CaCO_3(s)$$

Compute the chemical amounts of the four ions in the two solutions before the two are mixed

$$n_{K^+} = 0.150 \text{ L solution} \times \left(\frac{0.200 \text{ mol } K_2CO_3}{\text{L solution}}\right)\left(\frac{2 \text{ mol } K^+}{1 \text{ mol } K_2CO_3}\right) = 0.0600 \text{ mol } K^+$$

$$n_{CO_3^{2-}} = 0.150 \text{ L solution} \times \left(\frac{0.200 \text{ mol } K_2CO_3}{\text{L solution}}\right)\left(\frac{1 \text{ mol } CO_3^{2-}}{1 \text{ mol } K_2CO_3}\right) = 0.0300 \text{ mol } CO_3^{2-}$$

$$n_{Ca^{2+}} = 0.100 \text{ L solution} \times \left(\frac{0.400 \text{ mol } Ca(NO_3)_2}{\text{L solution}}\right)\left(\frac{1 \text{ mol } Ca^{2+}}{1 \text{ mol } Ca(NO_3)_2}\right) = 0.0400 \text{ mol } Ca^{2+}$$

$$n_{NO_3^-} = 0.100 \text{ L solution} \times \left(\frac{0.400 \text{ mol } Ca(NO_3)_2}{\text{L solution}}\right)\left(\frac{2 \text{ mol } NO_3^-}{1 \text{ mol } Ca(NO_3)_2}\right) = 0.0800 \text{ mol } NO_3^-$$

The Ca^{2+} ion reacts with CO_3^{2-} ion in a $1:1$ molar ratio. If the resulting $CaCO_3(s)$ is completely insoluble, this precipitation reaction continues until the entire 0.0300 mol of CO_3^{2-} ion, which is the limiting reactant, is consumed. At that point, the amount of $CaCO_3(s)$ sitting on the bottom of the container is 0.0300 mol. The mass of this $CaCO_3$ is

$$m_{CaCO_3} = 0.0300 \text{ mol } CaCO_3 \times \left(\frac{100.1 \text{ g } CaCO_3}{\text{mol } CaCO_3}\right) = \boxed{3.00 \text{ g } CaCO_3}$$

Obtain the concentrations of other three ions in the solution by dividing their chemical amounts by the total volume of the solution

$$c_{K^+} = \frac{0.0600 \text{ mol}}{(0.150 + 0.100) \text{ L}} = \boxed{0.240 \text{ mol L}^{-1}}$$

$$c_{Ca^{2+}} = \frac{(0.0400 - 0.0300) \text{ mol}}{(0.150 + 0.100) \text{ L}} = \boxed{0.0400 \text{ mol L}^{-1}}$$

$$c_{NO_3^-} = \frac{0.0800 \text{ mol}}{(0.150 + 0.100) \text{ L}} = \boxed{0.320 \text{ mol L}^{-1}}$$

The concentration of the carbonate ion is $\boxed{\text{zero mol L}^{-1}}$ under the assumption that $CaCO_3$ is completely insoluble.

Tip. The actual residual concentration of the carbonate ion can be estimated as follows

$$c_{CO_3^{2-}} = [CO_3^{2-}] = \frac{K_{sp}}{[Ca^{2+}]} = \frac{8.7 \times 10^{-9}}{0.0400} = 2.2 \times 10^{-7} \text{ mol L}^{-1}$$

The K_{sp} used here comes from text Table 16.2, which lists K_{sp}'s at 25°C.

16.65 The problem asks what would happen to the aqueous solubility of $CaCO_3$ if CO_2 were extremely soluble in water but the rest of its chemistry were unchanged. Represent the dissolution of calcium carbonate in an aqueous solution of a strong acid as the sum of these four steps

$$CaCO_3(s) \rightleftharpoons Ca^{2+}(aq) + CO_3^{2-}(aq) \qquad K_1 = K_{sp}$$

$$CO_3^{2-}(aq) + 2 H_3O^+ \rightleftharpoons H_2CO_3(aq) + 2 H_2O(l) \qquad K_2 = \frac{1}{K_{a1}K_{a2}}$$

$$H_2CO_3(aq) \rightleftharpoons CO_2(aq) + H_2O(l) \qquad K_3$$

$$CO_2(aq) \rightleftharpoons CO_2(g) \qquad K_4$$

The question amounts to asking what would happen to the position of the first of these four equilibrium if K_4 became very small, K_1, K_2, and K_3 stayed the same. Decreasing K_4 would not affect the way the second equilibrium removes CO_3^{2-} from among the products of the first equilibrium. Therefore calcium carbonate $\boxed{\text{would still dissolve in strong acids}}$.

16.67   Two solubility equilibria are in simultaneous effect

$$AgBr(s) \rightleftharpoons Ag^+(aq) + Br^-(aq) \qquad K_{sp} = [Ag^+][Br^-] = 7.7 \times 10^{-13}$$
$$CuBr(s) \rightleftharpoons Cu^+(aq) + Br^-(aq) \qquad K_{sp} = [Cu^+][Br^-] = 4.2 \times 10^{-8}$$

Divide the K_{sp} expression for the second by the K_{sp} expression for the first

$$\frac{[Cu^+][Br^-]}{[Ag^+][Br^-]} = \frac{4.2 \times 10^{-8}}{7.7 \times 10^{-13}} = 5.45 \times 10^4$$

A single solution at equilibrium can have only one concentration of Br^- ion (or any other ion). Hence the $[Br^-]$ in the numerator equals the $[Br^-]$ in the denominator. Cancellation then gives

$$\frac{[Cu^+]}{[Ag^+]} = \boxed{5.5 \times 10^4}$$

16.69   The Mohr method takes advantage of the differing solubilities of $AgCl(s)$ and $Ag_2CrO_4(s)$. Calculate the concentration of $Ag^+(aq)$ that just suffices to precipitate each salt. The applicable equilibria and K_{sp}-expressions are

$$AgCl(s) \rightleftharpoons Ag^+(aq) + Cl^-(aq) \qquad\qquad [Ag^+][Cl^-] = 1.6 \times 10^{-10}$$
$$Ag_2CrO_4(s) \rightleftharpoons 2\,Ag^+(aq) + CrO_4^{2-}(aq) \qquad [Ag^+]^2[CrO_4^{2-}] = 1.9 \times 10^{-12}$$

Inserting the given concentrations of Cl^- ion and CrO_4^{2-} ion and solving for the concentration of Ag^+ ion establishes that precipitation of $AgCl(s)$ requires only 1.6×10^{-9} M $Ag^+(aq)$ but that precipitation of $AgCrO_4(s)$ requires 2.8×10^{-5} M $Ag^+(aq)$. Thus, $\boxed{AgCl(s)}$ will precipitate first.

Ag_2CrO_4 starts to come down only after $[Ag^+]$ reaches 2.8×10^{-5} M. Use the K_{sp} expression to calculate $[Cl^-]$ when $[Ag^+] = 2.8 \times 10^{-5}$ M

$$[Cl^-] = \frac{K_{sp}}{[Ag^+]} = \frac{1.6 \times 10^{-10}}{2.8 \times 10^{-5}} = 5.7 \times 10^{-6} \text{ M}$$

The fraction of Cl^- remaining at this point is

$$f_{Cl^-} = \frac{[Cl^-]}{[Cl^-]_0} = \frac{5.7 \times 10^{-6} \text{ M}}{0.100 \text{ M}} = \boxed{5.7 \times 10^{-5}}$$

16.71   Magnesia dissolves in water and raises the pH by releasing $OH^-(aq)$ ion

$$MgO(s) + H_2O(l) \rightleftharpoons Mg^{2+}(aq) + 2\,OH^-(aq) \qquad [Mg^{2+}][OH^-]^2 = K_{sp}$$

The pH of the solution at 25°C equals 10.16. At this temperature, the sum of the pH and pOH of an aqueous solution equals 14.00. This means that the pOH of the saturated solution of magnesia equals 3.84. By the definition of pOH, $[OH^-] = 10^{-3.84} = 1.44 \times 10^{-4}$ M. The pH of the solution greatly exceeds the pH of water. This means that dissolution of magnesia far surpasses water as a source of OH^- ion and that the autoionization of water can accordingly be ignored. Then

$$2\,[Mg^{2+}] = [OH^-]$$

Substitute into the K_{sp} expression

$$K_{sp} = [Mg^{2+}][OH^-]^2 = \frac{[OH^-]}{2}[OH^-]^2 = \frac{(1.44 \times 10^{-4})^3}{2} = 1.5 \times 10^{-12}$$

The solubility of the $MgO(s)$ equals the final concentration of $Mg^{2+}(aq)$

$$S_{MgO} = [Mg^{2+}] = \frac{[OH^-]}{2} = \boxed{7.2 \times 10^{-5} \text{ mol L}^{-1}}$$

16.73   Silver ion is complexed strongly by ammonia

$$\mathrm{Ag^+}(aq) + 2\,\mathrm{NH_3}(aq) \rightleftharpoons \mathrm{Ag(NH_3)_2^{2+}}(aq) \qquad K_\mathrm{f} = 1.7 \times 10^7$$

When $\mathrm{AgBr}(s)$ is placed in aqueous ammonia, a new dissolution reaction

$$\mathrm{AgBr}(s) + 2\,\mathrm{NH_3}(aq) \rightleftharpoons \mathrm{Ag(NH_3)_2^{2+}}(aq) + \mathrm{Br^-}(aq) \qquad K = K_\mathrm{f} K_\mathrm{sp}$$

replaces the "regular" dissolution reaction. The new dissolution reaction equals the sum of the regular K_sp reaction and the complexation reaction. Its K is 1.7×10^7 times larger than the K_sp, which explains the increase in solubility of AgBr when ammonia is present.

16.75   **a)** The aqueous HCl reacts with $\mathrm{CdS}(s)$ to give both $\mathrm{CdCl_4^{2-}}(aq)$ and $\mathrm{H_2S}(aq)$

$$\boxed{\mathrm{CdS}(s) + 2\,\mathrm{H_3O^+}(aq) + 4\,\mathrm{Cl^-}(aq) \rightleftharpoons \mathrm{CdCl_4^{2-}}(aq) + \mathrm{H_2S}(aq) + 2\,\mathrm{H_2O}(l)}$$

The $\mathrm{HCl}(aq)$ acts on the $\mathrm{CdS}(s)$ by removing both sulfide ion (as $\mathrm{H_2S}$) and cadmium ion (as the tetrachloro complex). The problem states that some $\mathrm{CdS}(s)$ remains, so the above equation is an accurate description of the final equilibrium.

b) The equilibrium in the preceding part can be constructed as the sum of four simpler equilibria: one representing the dissolution of $\mathrm{CdS}(s)$ in pure water; a second and third representing the acid-base reactions of the $\mathrm{HS^-}$ and $\mathrm{OH^-}$ ions (both are bases) with water; a fourth representing the complexation of $\mathrm{Cd^{2+}}$ by $\mathrm{Cl^-}$ ions

$$\mathrm{CdS}(s) + \mathrm{H_2O}(l) \rightleftharpoons \mathrm{Cd^{2+}}(aq) + \mathrm{OH^-}(aq) + \mathrm{HS^-}(aq) \qquad K_1$$
$$\mathrm{HS^-}(aq) + \mathrm{H_2O}(l) \rightleftharpoons \mathrm{H_2S}(aq) + \mathrm{OH^-}(aq) \qquad K_2 = K_\mathrm{w}/K_\mathrm{a1}$$
$$2\,\mathrm{OH^-}(aq) + 2\,\mathrm{H_3O^+}(aq) \rightleftharpoons 4\,\mathrm{H_2O}(l) \qquad K_3 = (1/K_\mathrm{w})^2$$
$$\mathrm{Cd^{2+}}(aq) + 4\,\mathrm{Cl^-}(aq) \rightleftharpoons \mathrm{CdCl_4^{2-}}(aq) \qquad K_4 = K_\mathrm{f}$$

where K_a1 is the K_a for the first stage of the acid ionization of $\mathrm{H_2S}(aq)$. The equilibrium constants of the four reactions are numbered for identification. The desired constant is the product of the four constants because the equation of interest is the sum of the four equations

$$K = K_1 K_2 K_3 K_4 = K_1 \left(\frac{K_\mathrm{w}}{K_\mathrm{a1}}\right) \left(\frac{1}{K_\mathrm{w}}\right)^2 (K_\mathrm{f})$$

Text Table 16.5 lists K_1 as 7×10^{-28} and text Table 15.2 lists K_a1 as 9.1×10^{-8} (both at 25°C). The problem says that K_f is 800. Assume that this value was also measured at 25°C. Insert these numbers and the well-known value of K_w at 25°C into the preceding

$$K = (7 \times 10^{-28}) \left(\frac{1.0 \times 10^{-14}}{9.1 \times 10^{-8}}\right) \left(\frac{1}{1.0 \times 10^{-14}}\right)^2 (8 \times 10^2) = 6.2 \times 10^{-4}$$

Thus, at equilibrium at 25°C

$$\frac{[\mathrm{CdCl_4^{2-}}][\mathrm{H_2S}]}{[\mathrm{H_3O^+}]^2[\mathrm{Cl^-}]^4} = K = \boxed{6 \times 10^{-4}}$$

c) Let S equal the molar solubility of the $\mathrm{CdS}(s)$ in 6 M HCl. Assume that at equilibrium the cadmium in solution is all in the form of $\mathrm{CdCl_4^{2-}}(aq)$, and that the sulfur in solution is all in the form of $\mathrm{H_2S}(aq)$. Then

$$S = [\mathrm{CdCl_4^{2-}}] = [\mathrm{H_2S}]$$

Now, every Cd^{2+} ion that goes into solution consumes 4 Cl^- ions, and every S^{2-} ion that goes into solution consumes 2 H_3O^+ ions. This means that at equilibrium

$$[H_3O^+] = 6 - 2S \quad\text{and}\quad [Cl^-] = 6 - 4S$$

The 6 comes from the original concentration of HCl, which was 6 M. Substitute in the mass-action expression derived in the preceding parts

$$\frac{[CdCl_4^{2-}][H_2S]}{[H_3O^+]^2[Cl^-]^4} = \frac{S^2}{(6-2S)^2(6-4S)^4} = 6.2 \times 10^{-4}$$

Assuming that S is negligible compared to 6 M in this equation gives an unacceptable solution.[8] However, S must lie between 0 and 6 mol L^{-1}. Guess that S equals 1.0 mol L^{-1}. Then the left side equals 3.9×10^{-3}, which exceeds 6.2×10^{-4}. Guess a smaller S, such as 0.6 mol L^{-1}. The left side then equals 9.3×10^{-5}, which is less than 6.2×10^{-4}. Further adjustments in the range between 0.6 and 1.0 give improved fits. The value $S = 0.8$ fits the equation fairly well. The solubility of CdS(s) in 6 M HCl is thus *apparently* 0.8 mol L^{-1}. Is this possible? It was assumed that all of the sulfur stays in solution in the form of $H_2S(aq)$. If this is correct then the concentration of $H_2S(aq)$ also equals 0.8 M. But 0.8 M exceeds the solubility of H_2S at room temperature, which is only 0.1 M.[9] Adding 6 M HCl to CdS(s) therefore causes gaseous H_2S to bubble out until the concentration of $H_2S(aq)$ falls to 0.1 M. The previous equation is replaced by:

$$\frac{S(0.1)}{(6-2S)^2(6-4S)^4} = 6.2 \times 10^{-4}$$

Solving this new equation (by successive approximation) gives $S = 1.046$ mol L^{-1}. The solubility of CdS(s) is $\boxed{1 \text{ mol L}^{-1}}$, after rounding off.

16.77 The solution is prepared by mixing 0.020 mol of $CuCl_2$ and 0.100 mol of NaCN in 1.0 L of water. The original concentration of Cu^{2+} ion is therefore 0.020 M, and the original concentration of CN^- is 0.100 M. These concentrations do not last long. The $Cu^{2+}(aq)$ ion and $CN^-(aq)$ soon combine to form a complex ion

$$Cu^{2+}(aq) + 4\,CN^-(aq) \rightleftharpoons Cu(CN)_4^{2-}(aq) \qquad \frac{[Cu(CN)_4^{2-}]}{[Cu^{2+}][CN^-]^4} = K_f = 2.0 \times 10^{30}$$

The very large K_f means that nearly all of the Cu^{2+} ion is complexed. At equilibrium then, $[Cu(CN)_4^{2-}] = 0.020$ M. Complexation reduces the concentration of CN^- ion, which is in excess, from its original 0.100 M to $0.100 - 4(0.020) = 0.020$ M because 1 mol of Cu^{2+} accounts for 4 mol of $CN^-(aq)$. The values

$$[CN^-] = 0.020 \text{ M} \quad\text{and}\quad [Cu(CN)_4^{2-}] = 0.020 \text{ M}$$

might now be substituted into the K_f mass-action expression and used to compute an equilibrium concentration of Cu^{2+} ion. However, $CN^-(aq)$ also hydrolyzes (reacts with H_2O) to give HCN(aq)

$$CN^-(aq) + H_2O(l) \rightleftharpoons HCN(aq) + OH^-(aq) \qquad \frac{[HCN][OH^-]}{[CN^-]} = \frac{K_w}{K_a} = \frac{1.0 \times 10^{-14}}{6.17 \times 10^{-10}} = 1.62 \times 10^{-5}$$

This reaction lowers the concentration of $CN^-(aq)$ from 0.020 M. Let x equal the concentration of CN^- that reacts in this way. Then

$$\frac{[HCN][OH^-]}{[CN^-]} = 1.62 \times 10^{-5} = \frac{x^2}{0.020 - x}$$

---

[8]S comes out to equal 5.38 M, which is far too large to neglect relative to 6 M.
[9]Aqueous H_2S is saturated at a concentration of 0.1 M. See text Example 16.10.

Solving gives x equal to 5.61×10^{-4} M. A better value for the equilibrium concentration of CN^- is therefore $0.020 - 5.61 \times 10^{-4} = 0.0194$ M. Put this value into the K_f expression and solve for $[Cu^{2+}]$

$$\frac{[Cu(CN)_4^{2-}]}{[Cu^{2+}][CN^-]^4} = 2.0 \times 10^{30} = \frac{(0.020)}{[Cu^{2+}](0.0194)^4} \qquad [Cu^{2+}] = \boxed{7.0 \times 10^{-26} \text{ M}}$$

Tip. If the hydrolysis of CN^- ion is ignored, the answer is 6.3×10^{-26} M—barely a significant difference.

16.79 **a)** Text Example 16.9 gives K_{a1} for $Fe(H_2O)_6^{3+}$ as 7.7×10^{-3}. The problem gives K_{a2} for this ion as 2.0×10^{-5}. Because K_{a2} is about 400 times smaller than K_{a1}, the second stage of the acid ionization of $Fe(H_2O)_6^{3+}$ has essentially $\boxed{\text{no effect}}$ on the pH. The pH of the 0.100 M solution of $Fe(NO_3)_3$ is 1.62, based solely on the first H^+-donation. This corresponds to $[H_3O^+] = 2.4 \times 10^{-2}$ M. Thus, considering only the first stage

$$[H_3O^+] = [Fe(H_2O)_5(OH)^{2+}] = 2.4 \times 10^{-2} \text{ M}$$

The K_a- expression for the *second* stage is

$$K_{a2} = 2.0 \times 10^{-5} = \frac{[H_3O^+][Fe(H_2O)_4(OH)_2^+]}{[Fe(H_2O)_5OH^{2+}]}$$

Substitute the concentrations obtained from the first-stage-only calculation into this expression and solve for $[Fe(H_2O)_4(OH)_2^+]$, which is very easy. The answer is $\boxed{2.0 \times 10^{-5} \text{ M}}$. This concentration is negligible compared to 0.024 M, the concentration of $Fe(H_2O)_5OH^{2+}(aq)$. The $[H_3O^+]$ that arises in the second stage is similarly negligible compared to 0.024 M.

b) The question refers to the dissociation of the $Fe(OH)_2^+$ complex ion. Dissociation proceeds through steps that are the reverse of the steps for formation

$$Fe(OH)_2^+(aq) \rightleftharpoons Fe(OH)^{2+}(aq) + OH^-(aq) \qquad \frac{[Fe(OH)^{2+}][OH^-]}{[Fe(OH)_2^+]} = K_{d1}$$

$$Fe(OH)^{2+}(aq) \rightleftharpoons Fe^{3+}(aq) + OH^-(aq) \qquad \frac{[Fe^{3+}][OH^-]}{[Fe(OH)^{2+}]} = K_{d2}$$

The d's in the subscripts on the K's in the preceding stand for dissociation. Divide the first of the two mass-action equations into the equation $K_w = [H_3O^+][OH^-]$. The result is

$$\frac{K_w}{K_{d1}} = \frac{[H_3O^+][Fe(OH)_2^+]}{[Fe(OH)^{2+}]}$$

The right-hand side of this equation is identical to the mass-action expression for K_{a2} in the previous part except that the chemical formulas in the K_{a2} expression show associated H_2O molecules explicitly. That is, $Fe(OH)_2^+$ replaces its equivalent $Fe(H_2O)_4(OH)_2^+$, and $Fe(OH)^{2+}$ replaces *its* equivalent $Fe(H_2O)_5OH^{2+}$. Hence, $K_{a2} = K_w/K_{d1}$. Similarly, $K_{a1} = K_w/K_{d2}$. Numerical values for K_{a1} and K_{a2} are given in the problem and in text Example 16.9. Substitution gives

$$K_{d1} = \frac{K_w}{K_{a2}} = \frac{1.0 \times 10^{-14}}{2.0 \times 10^{-5}} = 5.0 \times 10^{-10} \qquad K_{d2} = \frac{K_w}{K_{a1}} = \frac{1.0 \times 10^{-14}}{7.7 \times 10^{-3}} = 1.3 \times 10^{-12}$$

Because the two-step dissociation of the complex exactly reverses the two-step formation of the complex, K_f of the complex equals the reciprocal of the product of K_{d1} and K_{d2}.

$$K_f = \frac{1}{K_{d1}K_{d2}} = \frac{1}{(5.0 \times 10^{-10})(1.3 \times 10^{-12})} = \boxed{1.5 \times 10^{21}}$$

CUMULATIVE PROBLEMS

16.81 The molar mass of codeine ($C_{18}H_{21}NO_3$) equals 299.370 g mol^{-1}. At room temperature, the molal solubility of this substances is

$$\frac{1.00 \text{ g codeine}}{120 \text{ mL water}} \times \left(\frac{1 \text{ mL water}}{1 \text{ cm}^3 \text{ water}}\right) \times \left(\frac{1 \text{ cm}^3 \text{ water}}{1.00 \text{ g water}}\right) \times \left(\frac{1000 \text{ g water}}{1 \text{ kg water}}\right)$$

$$\times \left(\frac{1 \text{ mol codeine}}{299.37 \text{ g codeine}}\right) = \boxed{0.0278 \text{ mol kg}^{-1}}$$

At 80°, the same mass of codeine dissolves in half the amount of water, so the molal solubility is double what it is at room temperature: $\boxed{0.0557 \text{ mol kg}^{-1}}$. The dissolution reaction is driven to the right (favoring the dissolved codeine) by the increase in temperature. The reaction is therefore $\boxed{\text{endothermic}}$; the "heat term" for the dissolution reaction is on the left.

16.83 The problem combines a calculation on a formic acid/formate buffer (Chapter 15) with a solubility calculation for the slightly soluble salt CaF_2 (Chapter 16). Start by treating the two calculations separately. Compute the pH of the buffer and assume that the pH remains unchanged as the CaF_2 dissolves. Recognize that this approach neglects the effect of the reaction of F^- ion, a weak base, on the pH and also neglects the autoionization of water. Plan to check these points.

The addition of 50.0 mL of 0.15 M HNO_3 to 100.0 mL of 0.12 M $NaHCOO$ creates the buffer. The nitric acid converts 0.0075 mol of $HCOO^-$ ion to $HCOOH$ and leaves 0.0045 mol of $HCOO^-$ ion unreacted. The concentrations of these two species immediately after the conversion, but before either interacts further are

$$[HCOOH] = \frac{0.0075 \text{ mol}}{0.150 \text{ L}} = 0.050 \text{ M} \qquad [HCOO^-] = \frac{0.0045 \text{ mol}}{0.150 \text{ L}} = 0.030 \text{ M}$$

Both concentrations change slightly as the formic acid/formate equilibrium takes effect. This equilibrium generates x mol L^{-1} of H_3O^+ as shown in the following:

	$HCOOH(aq)$	$+H_2O(aq) \rightleftharpoons$	$HCOO^-(aq) +$	$H_3O^+(aq)$
Init. Conc. (M)	0.050	—	0.030	small
Change in Conc. (M)	$-x$	—	$+x$	$+x$
Equil. Conc. (M)	$0.050 - x$	—	$0.030 + x$	x

Take K_a from text Table 15.2 and substitute it and the equilibrium concentrations into the mass-action expression

$$K_a = 1.77 \times 10^{-4} = \frac{[HCOO^-][H_3O^+]}{[HCOOH]} = \frac{(0.030 + x)x}{(0.050 - x)}$$

Solving for x is routine. The $[H_3O^+]$ equals 2.90×10^{-4} M.

Dissolution of $CaF_2(s)$ generates $F^-(aq)$ ion. Some of this ion reacts with water to give $HF(aq)$

$$F^-(aq) + H_2O(l) \rightleftharpoons HF(aq) + OH^-(aq) \qquad \frac{[HF][OH^-]}{[F^-]} = K_b$$

Once equilibrium is established, the K_b equation holds. The following equation, which is obtained by dividing the K_b equation into the usual K_w equation, therefore also holds

$$K_a = \frac{[H_3O^+][F^-]}{[HF]} = 6.6 \times 10^{-4}$$

By previous assumption, the concentration of hydronium ion is fixed at 2.90×10^{-4} M by the formic acid/formate buffer. Therefore

$$\frac{2.90 \times 10^{-4}[\text{F}^-]}{[\text{HF}]} = 6.6 \times 10^{-4} \quad \text{which means} \quad \frac{[\text{F}^-]}{[\text{HF}]} = 2.276$$

Now, CaF_2 dissolves:

$$CaF_2(s) \rightleftharpoons Ca^{2+}(aq) + 2\,F^-(aq) \qquad K_{sp} = [Ca^{2+}][F^-]^2 = 3.9 \times 10^{-11}$$

Obviously, the dissolution reaction generates F^- ion and Ca^{2+} ion in a 2 : 1 ratio. But it is *not* true at equilibrium that the concentration of F^- ion equals twice that of Ca^{2+} ion. Some F^- ion reacts with water to give HF. What *is* true is

$$2\,[Ca^{2+}] = [F^-] + [HF]$$

Eliminating [HF] between this and the equation for the ratio $[\text{F}^-]/[\text{HF}]$ gives

$$[F^-] = 2\,[Ca^{2+}] - \frac{[\text{F}^-]}{2.276} \quad \text{which rearranges to} \quad [F^-] = 1.389\,[Ca^{2+}]$$

Insertion of the preceding equation for $[F^-]$ in terms of $[Ca^{2+}]$ into the K_{sp} expression gives

$$[Ca^{2+}](1.389)^2[Ca^{2+}]^2 = 3.9 \times 10^{-11}$$
$$[Ca^{2+}] = 2.72 \times 10^{-4} \text{ M}$$

One mole of Ca^{2+} is present in solution for every one mole of CaF_2 that dissolved. Therefore, the molar solubility of CaF_2 in the buffer equals $\boxed{2.7 \times 10^{-4} \text{ mol L}^{-1}}$.

Now, check on the assumptions. A side calculation gives $[F^-]$ as 3.8×10^{-4} M, and [HF] as 1.7×10^{-4} M. These concentrations are both small compared to the concentrations of $HCOO^-$ and HCOOH. Therefore the HF/F^- equilibrium can affect the pH only negligibly. The water autoionization is similarly drowned out as a source of H_3O^+ by the large concentration of HCOOH.

Tip. It is instructive to set up the equations for an exact treatment[10] to solve this problem. First, list all the species present in the solution. In this case, nine different ions or molecules exist at equilibrium (in addition to H_2O itself):

$$Na^+, H_3O^+, Ca^{2+}, NO_3^-, OH^-, F^-, HF, HCOO^-, \text{and HCOOH}$$

Next, establish all possible relationships among the concentrations of the species. The electrical neutrality of the solution requires that

$$[Na^+] + [H_3O^+] + 2[Ca^{2+}] = [NO_3^-] + [OH^-] + [F^-] + [HCOO^-]$$

The Na^+ and NO_3^- ions do not react with other species in the solution, and it is easy to obtain their final concentrations

$$[Na^+] = 0.080 \text{ M} \qquad [NO_3^-] = 0.050 \text{ M}$$

Four mass-action relationships are satisfied at equilibrium

$$1.77 \times 10^{-4} = \frac{[HCOO^-][H_3O^+]}{[HCOOH]} \qquad 6.6 \times 10^{-4} = \frac{[F^-][H_3O^+]}{[HF]}$$
$$3.9 \times 10^{-11} = [Ca^{2+}][F^-]^2 \qquad 1.0 \times 10^{-14} = [H_3O^+][OH^-]$$

[10]Text Section 15.9.

Finally, two material-balance relationships exist

$$[HF] + [F^-] = 2[Ca^{2+}] \qquad \text{and} \qquad [HCOOH] + [HCOO^-] = 0.080$$

There are now seven equations in seven unknowns, and the problem becomes "merely" to solve the simultaneous equations. Computer programs can be used for this, but mechanization ignores the chemical insights provided by the non-exact method. Resort to the exact method to check assumptions or when confusion threatens to sink all efforts.

Chapter 17

Electrochemistry

Electrochemical Cells

17.1 Electrons flow from the Pt electrode (left) through the connecting wire to the Cu electrode (right). The electrons come from the oxidation of Cr^{2+} ions to Cr^{3+} ions. Cu^{2+} ions accept the electrons at the surface of the right electrode, and metallic copper plates out there. In the salt bridge, negative ions move from right to left and positive ions from left to right so that the contents of both half-cells (the beakers) remain neutral.

The overall reaction is represented $\boxed{2\,Cr^{2+}(aq) + Cu^{2+}(s) \longrightarrow 2\,Cr^{3+}(aq) + Cu(s)}$.

17.3 Formation of 1 mol of $Sn(s)$ from $Sn^{4+}(aq)$ ions requires 4 mol of electrons: $Sn^{4+}(aq) + 4\,e^- \longrightarrow Sn(s)$. Hence

$$n_{Sn} = 6.95 \times 10^4 \text{ C} \times \left(\frac{1 \text{ mol } e^-}{96\,485.3 \text{ C}}\right) \times \left(\frac{1 \text{ mol Sn}}{4 \text{ mol } e^-}\right) = \boxed{0.180 \text{ mol Sn}}$$

Tip. Note the careful use of the word "maximum" in the statement of the problem. No $Sn(s)$ is necessarily formed at all! Some other electrochemical reaction might be partially or even entirely responsible for the flow of the current. For example, the reduction might produce Sn^{2+} ion.

17.5 At the anode, $Zn(s)$ is being oxidized to $Zn^{2+}(aq)$; at the cathode, $Cl_2(g)$ is being reduced to $Cl^-(aq)$.

a) $\boxed{Zn(s) + Cl_2(g) \longrightarrow Zn^{2+}(aq) + 2\,Cl^-(aq)}$.

b) An ampere is a coulomb per second. Hence

$$Q = \left(\frac{0.800 \text{ C}}{1 \text{ s}}\right) \times 25.0 \text{ min} \left(\frac{60 \text{ s}}{1 \text{ min}}\right) = \boxed{1.20 \times 10^3 \text{ C}}$$

In moles

$$n_{e^-} = \left(\frac{0.800 \text{ C}}{1 \text{ s}}\right) \times 25.0 \text{ min} \left(\frac{60 \text{ s}}{1 \text{ min}}\right) \left(\frac{1 \text{ mol } e^-}{96\,485.3 \text{ C}}\right) = \boxed{0.0124 \text{ mol } e^-}$$

c) The oxidation half-reaction is $Zn(s) \longrightarrow Zn^{2+}(aq) + 2\,e^-$. This half-reaction removes metallic zinc from the anode. Therefore

$$\Delta m_{\text{anode}} = 0.0124 \text{ mol } e^- \times \left(\frac{1 \text{ mol Zn}}{2 \text{ mol } e^-}\right)\left(\frac{65.39 \text{ g Zn}}{1 \text{ mol Zn}}\right) = \boxed{-0.407 \text{ g}}$$

The calculation assumes that no side-reactions occur that divert electrons.

d) The reduction half-reaction is $Cl_2(g) + 2\,e^- \longrightarrow 2\,Cl^-(aq)$. Therefore the passage of 0.0124 mol of electrons reduces 0.00622 mol of $Cl_2(g)$ (if no side-reactions occur). Use the ideal-gas law to calculate the volume of this amount of chlorine at 25°C (298 K) and a pressure of 1 atm

$$V_{Cl_2} = \frac{n_{Cl_2}RT}{P} = \frac{(0.00622 \text{ mol})(0.08206 \text{ L atm mol}^{-1}\text{K}^{-1})(298 \text{ K})}{1 \text{ atm}} = \boxed{0.152 \text{ L}}$$

Tip. It would be hard to build a practical cell employing the chlorine-zinc reaction and producing a steady current of 0.800 A for 25 minutes. The point of the problem however is the computation of the quantities, not the engineering of the cell.

17.7 Calculate the ratio of the chemical amounts of oxygen and copper generated by the operation of the cell. The molar mass of oxygen is 32.0 g mol^{-1}, and the molar mass of copper is 63.54 g mol^{-1}. The cell therefore forms 0.500 mol of O_2 as it forms 1.00 mol of Cu. A balanced half-equation for the oxidation of water to gaseous oxygen is $3\,H_2O(l) \longrightarrow 1/2\,O_2(g) + 2\,H_3O^+(aq) + 2\,e^-$. This equation states that the production of 1/2 mol of O_2 releases 2 mol of electrons. Hence

$$\frac{0.500 \text{ mol } O_2}{1.00 \text{ mol Cu}} \times \left(\frac{2 \text{ mol } e^-}{1/2 \text{ mol } O_2}\right) = \frac{2.00 \text{ mol } e^-}{1 \text{ mol Cu}}$$

The copper starts in the $\boxed{+2 \text{ oxidation state}}$ and is reduced as follows: $Cu^{2+}(aq) + 2e^- \longrightarrow Cu(s)$.

17.9 **a)** In the electrolysis of molten KCl, the half-reactions are

$$K^+(melt) + e^- \longrightarrow K(l) \qquad \text{reduction} \quad \text{cathode}$$
$$Cl^-(melt) \longrightarrow \tfrac{1}{2}\,Cl_2(g) + e^- \qquad \text{oxidation} \quad \text{anode}$$

The sum of these two half-equations represents the overall cell reaction

$$K^+(melt) + Cl^-(melt) \longrightarrow K(l) + \tfrac{1}{2}\,Cl_2(g)$$

b) Use a series of unit-factors in each case

$$m_K = 5.00 \text{ h} \left(\frac{3600 \text{ s}}{1 \text{ h}}\right)\left(\frac{2.00 \text{ C}}{1 \text{ s}}\right)\left(\frac{1 \text{ mol } e^-}{96\,485.3 \text{ C}}\right)\left(\frac{1 \text{ mol K}}{1 \text{ mol } e^-}\right)\left(\frac{39.098 \text{ g K}}{1 \text{ mol K}}\right) = \boxed{14.6 \text{ g K}}$$

$$m_{Cl_2} = 5.00 \text{ h} \left(\frac{3600 \text{ s}}{1 \text{ h}}\right)\left(\frac{2.00 \text{ C}}{1 \text{ s}}\right)\left(\frac{1 \text{ mol } e^-}{96\,485.3 \text{ C}}\right)\left(\frac{1 \text{ mol } Cl_2}{2 \text{ mol } e^-}\right)\left(\frac{70.906 \text{ g } Cl_2}{1 \text{ mol } Cl_2}\right) = \boxed{13.2 \text{ g } Cl_2}$$

Tip. A tacit assumption is that the cell has at least $14.6 + 13.2 = 27.8$ g of KCl in it.

Cell Potentials and the Gibbs Free Energy

17.11 Represent the reduction half-reaction as $Ag^+(aq) + e^- \longrightarrow Ag(s)$. Use the molar mass of Ag and molar ratios from the balanced half-equation to compute the chemical amount of electrons transferred in the production of 1.00 g of metallic silver

$$n_{e^-} = 1.00 \text{ g Ag} \times \left(\frac{1 \text{ mol Ag}}{107.9 \text{ g Ag}}\right)\left(\frac{1 \text{ mol } e^-}{1 \text{ mol Ag}}\right) = 0.00927 \text{ mol } e^-$$

The free energy change of the cell during the production of the 1.00 g of Ag at 298 K equals the standard free energy change of the reaction at that temperature. This is the case because all of the reactants and products stay in their standard states

$$\Delta G = \Delta G^\circ = -n\mathcal{F}\Delta E^\circ_{cell} = -(0.00927 \text{ mol})(96\,485 \text{ C mol}^{-1})(1.03 \text{ V}) = -921 \text{ J}$$

The negative sign in the answer means that the reaction is spontaneous. The maximum electrical work *produced* by the cell equals $-w_{elec,max}$ because positive work is work that is absorbed.

$$-w_{elec,max} = -w_{elec,rev} = -\Delta G = \boxed{921 \text{ J}}$$

This maximum is attained if the cell is operated reversibly.

17.13 **a)** A succinct answer to the question is the representation $Co|Co^{2+}\|Br_2|Br_2$. The balanced half-equations and the overall equation are

$$\text{anode:} \qquad Co(s) \longrightarrow Co^{2+}(aq) + 2\,e^-$$
$$\text{cathode:} \qquad Br_2(l) + 2\,e^- \longrightarrow 2\,Br^-(aq)$$
$$\text{overall:} \qquad Co(s) + Br_2(l) \longrightarrow Co^{2+}(aq) + 2\,Br^-(aq)$$

The standard potential of the cell ΔE°_{cell} equals the standard reduction potential at the cathode minus the standard reduction potential at the anode

$$\Delta E^\circ_{cell} = E^\circ_{cathode} - E^\circ_{anode} = 1.065 - (-0.28) = \boxed{1.34 \text{ V}}$$

It is assumed that the temperature is 25°C.

Tip. Standard reduction potentials *do* change with temperature. The values compiled in text Appendix E are correct only at 25°C (298.15 K). The amount and the direction in which E°'s vary with temperature depend on the specific half-reaction.

17.15 **a)** The $In^{3+}|In$ half-reaction must be the reduction because metallic indium is observed to plate out as the cell runs. The $Zn^{2+}|Zn$ half-reaction accordingly proceeds as an oxidation

$$\text{anode:} \quad Zn(s) \longrightarrow Zn^{2+}(aq) + 2\,e^- \qquad\qquad \text{cathode:} \quad In^{3+}(aq) + 3\,e^- \longrightarrow In(s)$$

b) The reactants and products in the cell reaction are in their standard states at 298.15 K. The experimental potential difference of the cell (0.425 V) is therefore a *standard* potential difference. According to text Appendix E, the standard reduction potential at the zinc anode is −0.763 V. The standard reduction potential at the indium cathode is then

$$E^\circ_{cathode} - E^\circ_{anode} = \Delta E^\circ_{cell}$$
$$E^\circ_{cathode} = \Delta E^\circ_{cell} + E^\circ_{anode}$$
$$= 0.425 \text{ V} + (-0.763 \text{ V})$$
$$E^\circ_{cathode} = \boxed{-0.338 \text{ V}}$$

17.17 Powdered metallic aluminum should act as a $\boxed{\text{reducing agent}}$. A reducing agent is itself oxidized as it acts. The +3 oxidation state of Al is well-known. It is the obvious product when aluminum gives up electrons. There are no common negative oxidation states of Al; such states would necessarily result if Al served as an oxidizing agent. Finally, according to text Appendix E, the reduction of Al^{3+} to $Al(s)$ has $E^\circ = -1.706$ V. This large negative E° means that Al^{3+} is hard to reduce and, consequently, that $Al(s)$ is easy to oxidize. Metallic aluminum is a powerful reducing agent.

Tip. Massive metallic aluminum (as in aluminum pots and pans) has the same reducing power as powered aluminum. Reactions with powered aluminum however usually happen faster because of the greater surface area that a powder presents.

17.19   The stronger an oxidizing agent is, the easier it is to reduce (see problem **17.17**). The more powerful oxidizing agent will have the algebraically larger reduction potential. For these two elements, the standard reduction potentials at 298 K are

$$Cl_2(g) + 2\,e^- \longrightarrow 2\,Cl^-(aq) \qquad E° = +1.3583\ V$$
$$Br_2(l) + 2\,e^- \longrightarrow 2\,Br^-(aq) \qquad E° = +1.065\ V$$

Since the $Cl_2(g)|Cl^-$ couple has a larger $E°$ than the $Br_2(l)|Br^-$ couple, $Cl_2(g)$ is the stronger oxidizing agent, and $\boxed{Br_2(l)\ \text{is worse than}\ Cl_2(g)}$ as a disinfectant, all other factors (such as environmental toxicity) being equal. Note the assumption that the temperature is 298 K or close to it.

17.21   **a)** The problem is to find the strongest oxidizing agent under standard acidic conditions in this group of six species: $Co(s)$, $Ag^+(aq)$, $Cl^-(aq)$, $Cr(s)$, $BrO_3^-(aq)$ and $I_2(s)$. Compare the ease of reduction of the six by examining their standard reduction potentials. The metals $Co(s)$ and $Cr(s)$ and the $Cl^-(aq)$ ion are *hard* to reduce: negative oxidation states of the metals are rare, and the $Cl^{2-}(aq)$ ion is unknown. These three are immediately eliminated. Text Appendix E gives these reduction half-equations and standard potentials for the other three

$$BrO_3^-(aq) + 6\,H^+(aq) + 5\,e^- \longrightarrow \tfrac{1}{2}\,Br_2(l) + 3\,H_2O(l) \qquad E° = 1.52\ V$$
$$Ag^+(aq) + e^- \longrightarrow Ag(s) \qquad E° = 0.7996\ V$$
$$I_2(s) + 2\,e^- \longrightarrow I^-(aq) \qquad E° = 0.535\ V$$

$\boxed{BrO_3^-\ \text{ion}}$ is the strongest oxidizing agent (most easily reduced, largest $E°$, top of the list).

b) To find the strongest reducing agent, compare the ease of oxidation of the species. These half-equations and standard reduction potentials come from text Appendix E

$$Cl_2(g) + 2\,e^- \longrightarrow 2\,Cl^-(aq) \qquad E° = 1.358\ V$$
$$I_2(s) + 2\,e^- \longrightarrow 2\,I^-(aq) \qquad E° = 0.535\ V$$
$$Co^{2+}(aq) + 2\,e^- \longrightarrow Co(s) \qquad E° = -0.28\ V$$
$$Cr^{3+}(aq) + 3\,e^- \longrightarrow Cr(s) \qquad E° = -0.74\ V$$
$$Cr^{2+}(aq) + 2\,e^- \longrightarrow Cr(s) \qquad E° = -0.905\ V$$

The species under comparison are on the *right* sides of these half-equations. $\boxed{Cr(s)}$ is the strongest reducing agent (most easily oxidized, smallest $E°$, bottom of the list). Oxidation of the very poor reducing agent $Ag^+(aq)$ would give Ag^{2+}; oxidation of the very poor reducing agent $BrO_3^-(aq)$ would give $BrO_4^-(aq)$. No half-equations involving these species are given in Appendix E.

c) The standard reduction potential of $Co^{2+}(aq)$ ion (-0.28 V) is algebraically less than that of $Pb^{2+}(aq)$ ion (-0.1263 V), but more than that of $Cd^{2+}(aq)$ ion (-0.4026 V). Therefore $\boxed{Co(s)}$ will reduce $Pb^{2+}(aq)$ ion, but not reduce $Cd^{2+}(aq)$ under standard acidic conditions.

17.23   **a)** The standard potential for the half-reaction: $Mn^{3+}(aq) + 3\,e^- \longrightarrow Mn(s)$ is *not* equal to the simple sum of the half-cell potentials of the half-reactions

$$Mn^{2+}(aq) + 2\,e^- \longrightarrow Mn(s) \qquad E° = -1.029\ V$$
$$Mn^{3+}(aq) + e^- \longrightarrow Mn^{2+}(aq) \qquad E° = 1.51\ V$$

Instead the standard potential is a *weighted average*

$$E_3° = \frac{n_1 E_1° + n_2 E_2°}{n_3}$$

where the three subscripted n's are the numbers of electrons transferred in the two half-reactions being combined and in the half-reaction that results. Substitution gives

$$E° = \frac{2(-1.029 \text{ V}) + 1(1.51 \text{ V})}{3} = \boxed{-0.183 \text{ V}}$$

b) The disproportionation $3 \text{ Mn}^{2+}(aq) \longrightarrow \text{Mn}(s) + 2 \text{ Mn}^{3+}(aq)$ combines the reduction of $\text{Mn}^{2+}(aq)$ to $\text{Mn}(s)$ and the oxidation of $\text{Mn}^{2+}(aq)$ to $\text{Mn}^{3+}(aq)$. It is represented by the second of the following half-equations subtracted from the first

$$2\,e^- + \text{Mn}^{2+}(aq) \longrightarrow \text{Mn}(s) \qquad E° = -1.029 \text{ V}$$
$$2\,\text{Mn}^{3+}(aq) + 2\,e^- \longrightarrow 2\,\text{Mn}^{2+}(aq) \qquad E° = 1.51 \text{ V}$$

The coefficients in the second half-equation are all double the coefficients in text Appendix E. Because the final disproportionation reaction is a whole reaction, not a half-reaction, this doubling can be ignored in the computation of $\Delta E°_{\text{cell}}$

$$\Delta E°_{\text{cell}} = E°_{\text{reduction}} - E°_{\text{oxidation}} = -1.029 - 1.51 = -2.539 \text{ V}$$

1 M $\text{Mn}^{2+}(aq)$ ion $\boxed{\text{does not disproportionate}}$ to $\text{Mn}(s)$ and $\text{Mn}^{3+}(aq)$; the standard potential difference for the process is negative.

Tip. To see why this calculation succeeds, write a weighted-average formula like the one in part **a)**, and use it to compute $\Delta E°_{\text{cell}}$. Watch what happens when the values of n_1, n_2, and n_3 are substituted

$$\Delta E°_{\text{cell}} = \frac{n_1 E°_1 - n_2 E°_2}{n_3} = \frac{2(-1.029) - 2(1.51)}{2} = -1.029 - 1.51 = -2.539 \text{ V}$$

In combining half-reactions to give a whole reaction n_1, n_2, and n_3 are *always* equal to each other and so cancel out. The weighted-average formula is always correct. Simply adding or subtracting half-cell potentials works only in an special case, but it is a very important special case.

17.25 **a)** The disproportionation of $\text{Br}_2(l)$ in acid is represented

$$6\,\text{Br}_2(l) + 18\,\text{H}_2\text{O}(l) \longrightarrow 2\,\text{BrO}_3^-(aq) + 10\,\text{Br}^-(aq) + 12\,\text{H}_3\text{O}^+(aq)$$

for which the standard potential difference is

$$\Delta E°_{\text{cell}} = E°_{\text{reduction}} - E°_{\text{oxidation}}$$
$$= E°(\text{Br}_2|\text{Br}^-) - E°(\text{BrO}_3^-|\text{Br}_2) = 1.065 - 1.52 = -0.46 \text{ V}$$

The negative $\Delta E°_{\text{cell}}$ means $\text{Br}_2(l)$ $\boxed{\text{will not disproportionate}}$ to $\text{Br}^-(aq)$ and $\text{BrO}_3^-(aq)$ in water under standard acidic conditions.

b) The reduction giving Br_2 has a larger reduction potential than the one giving Br^- ion. Hence $\boxed{\text{Br}^-}$ is more easily oxidized than Br_2 and must be a stronger reducing agent under standard acidic conditions

Concentration Effects and the Nernst Equation

17.27 The overall reaction in this galvanic cell is

$$2\,\text{Cr}^{2+}(aq) + \text{Pb}^{2+}(aq) \longrightarrow \text{Pb}(s) + 2\,\text{Cr}^{3+}(aq)$$

and its standard potential difference is

$$\Delta E°_{\text{cell}} = E°_{\text{reduction}} - E°_{\text{oxidation}} = E°_{\text{cathode}} - E°_{\text{anode}}$$
$$= E°(\text{Pb}^{2+}|\text{Pb}) - E°(\text{Cr}^{3+}|\text{Cr}^{2+}) = -0.1263 - (-0.424) = 0.298 \text{ V}$$

A standard potential difference is observed in an actual cell only if one has constructed it in such a way that all reactants and products are in their standard states. In this problem *none* of the solute concentrations equals 1 M. The Nernst equation provides the means to account for non-standard concentrations. It relates an actual cell voltage E_{cell} to a standard cell voltage E°_{cell}

$$E_{cell} = E^{\circ}_{cell} - \frac{RT}{n\mathcal{F}} \ln Q \quad \text{from which} \quad E_{cell} = E^{\circ}_{cell} - \frac{0.0592 \text{ V}}{n} \log Q \quad \text{(at } 25°C\text{)}$$

where Q is the reaction quotient. For this reaction and initial set of conditions

$$Q = \frac{[Cr^{3+}]^2}{[Cr^{2+}]^2[Pb^{2+}]} = \frac{(0.0030)^2}{(0.15)(0.20)^2}$$

and n equals 2. Substitution of these values gives

$$E_{cell} = 0.2977 \text{ V} - \frac{0.0592 \text{ V}}{2} \log \left(\frac{(0.0030)^2}{(0.15)(0.20)^2} \right) = \boxed{0.381 \text{ V}}$$

Tip. Confirm that "0.0592 V" is correct:

$$\frac{RT}{n\mathcal{F}} \ln Q = \frac{RT}{n\mathcal{F}} 2.3026 \log Q = \frac{(8.3145 \text{ J K}^{-1}\text{mol}^{-1})(298.15 \text{ K})}{n(96485 \text{ C mol}^{-1})} 2.3026 \log Q$$
$$= \frac{0.059160 \text{ V}}{n} \log Q$$

17.29 The Nernst equation works for half-reactions as well as for whole reactions. The half-reaction occurring in the half-cell in this problem is

$$Cr^{3+}(aq) + e^- \longrightarrow Cr^{2+}(aq) \qquad E^{\circ} = -0.424 \text{ V}$$

for which

$$E_{hc} = E^{\circ}_{hc} - \frac{RT}{n\mathcal{F}} \ln Q = -0.424 \text{ V} - \left(\frac{0.0592 \text{ V}}{1} \right) \log \left(\frac{0.0019}{0.15} \right) = \boxed{-0.312 \text{ V}}$$

Tip. By using "0.0592 V", one implicitly assumes that the temperature is 25°C. Platinum is included in the notation for the half-cell because Pt is an electrode. It conducts electrons in or out but is not changed chemically.

17.31 The $I_2|I^-$ half-reaction is at the cathode, the site of reduction. Hence the $H_3O^+|H_2$ half-reaction is an oxidation (at the anode). The overall reaction is

$$2 \text{ H}_2O(l) + I_2(s) + H_2(g) \longrightarrow 2 \text{ H}_3O^+(aq) + 2 I^-(aq)$$

Assume that the temperature is 25°C and use the data in text Appendix E to obtain ΔE°_{cell} of the cell. The appropriate standard reduction potentials are combined as follows

$$E^{\circ}_{cell} = E^{\circ}_{cathode} - E^{\circ}_{anode} = 0.535 \, (I_2|I^-) - (0.000) \, (H_3O^+|H_2) = 0.535 \text{ V}$$

The measured cell voltage depends on the concentrations and partial pressures of reactants and products according to the Nernst equation. At 25°C

$$0.841 \text{ V} = 0.535 \text{ V} - \left(\frac{0.0592 \text{ V}}{2} \right) \log \left(\frac{[H_3O^+]^2[I^-]^2}{P_{H_2}} \right)$$

The concentration of the iodide ion $[\text{I}^-]$ equals 1.00 M; the partial pressure of hydrogen P_{H_2} equals 1 atm. Substitution gives

$$(0.841 - 0.535) \ \text{V} = -\frac{0.0592 \ \text{V}}{2} \log\left(\frac{[\text{H}_3\text{O}^+]^2(1.00)^2}{1.00}\right)$$

The unit V (for volts) cancels out. Solving for $\log[\text{H}_3\text{O}^+]$ gives -5.17. The pH equals $-\log[\text{H}_3\text{O}^+]$, so it is $\boxed{5.17}$.

17.33 **a)** Assume that the temperature is 25°C. In the reaction

$$3\,\text{HClO}_2(aq) + 2\,\text{Cr}^{3+}(aq) + 12\,\text{H}_2\text{O}(l) \longrightarrow 3\,\text{HClO}(aq) + \text{Cr}_2\text{O}_7^{2-}(aq) + 8\,\text{H}_3\text{O}^+(aq)$$

Cr^{3+} is oxidized to $\text{Cr}_2\text{O}_7^{2-}$, and HClO_2 is reduced to HClO. Text Appendix E gives the half-equations and standard reduction potentials for these changes at 25°C

$$\text{Cr}_2\text{O}_7^{2-}(aq) + 14\,\text{H}_3\text{O}^+(aq) + 6\,e^- \longrightarrow 2\,\text{Cr}^{3+}(aq) + 21\,\text{H}_2\text{O}(l) \qquad E° = 1.33 \ \text{V}$$
$$3\,\text{HClO}_2(aq) + 6\,\text{H}_3\text{O}^+(aq) + 6\,e^- \longrightarrow 3\,\text{HClO}(aq) + 9\,\text{H}_2\text{O}(l) \qquad E° = 1.64 \ \text{V}$$

The first half-reaction, the oxidation, occurs at the anode; the second, the reduction, occurs at the cathode. The difference between the standard reduction potentials is the standard cell potential

$$\Delta E°_{\text{cell}} = E°_{\text{cathode}} - E°_{\text{anode}} = 1.64 \ \text{V} - 1.33 \ \text{V} = \boxed{0.31 \ \text{V}}$$

b) The concentration of Cr^{3+} ion is related to the measured cell potential by the Nernst equation. For the above overall reaction, the Nernst equation at 25°C is

$$0.15 \ \text{V} = 0.31 \ \text{V} - \frac{0.0592 \ \text{V}}{6} \log\left(\frac{[\text{HClO}]^3[\text{Cr}_2\text{O}_7^{2-}][\text{H}_3\text{O}^+]^8}{[\text{HClO}_2]^3[\text{Cr}^{3+}]^2}\right)$$

Insert the various concentrations, which are given in the problem (recall a pH of 0 means a $[\text{H}_3\text{O}^+]$ of 1.00 M)

$$0.15 \ \text{V} = 0.31 \ \text{V} - \frac{0.0592 \ \text{V}}{6} \log\left(\frac{(0.20)^3(0.80)(1.00)^8}{(0.15)^3[\text{Cr}^{3+}]^2}\right)$$

Solving for the concentration of Cr^{3+} is straightforward

$$\frac{6(0.15 - 0.31) \ \text{V}}{0.0592 \ \text{V}} = -\log\left(\frac{1.896}{[\text{Cr}^{3+}]^2}\right)$$

$$10^{+16.216} = \frac{1.896}{[\text{Cr}^{3+}]^2}$$

$$[\text{Cr}^{3+}] = \boxed{1 \times 10^{-8} \ \text{M}}$$

17.35 Compute the equilibrium constant K of the reaction

$$3\,\text{HClO}_2(aq) + 2\,\text{Cr}^{3+}(aq) + 12\,\text{H}_2\text{O}(l) \longrightarrow 3\,\text{HClO}(aq) + \text{Cr}_2\text{O}_7^{2-}(aq) + 8\,\text{H}_3\text{O}^+(aq)$$

It is known (from problem **17.33a**) that $E°_{\text{cell}}$ equals 0.31 V. The equilibrium constant K and the standard cell potential $E°_{\text{cell}}$ are related as follows

$$\ln K = \frac{n\mathcal{F}E°_{\text{cell}}}{RT} = \frac{6(9.6485 \times 10^4 \ \text{C mol}^{-1})(0.31 \ \text{V})}{(8.3145 \ \text{J K}^{-1}\text{mol}^{-1})(298.15 \ \text{K})} = 72.4 \qquad K = \boxed{3 \times 10^{31}}$$

This answer is so very large that for any practical purpose this reaction goes to completion. In other words, at 298 K HClO_2 and Cr^{3+} tend to react until one or the other is completely consumed.

After the 2.00 L of 1.00 $mol\,L^{-1}$ $HClO_2$ and 2.00 L of 0.5 $mol\,L^{-1}$ Cr^{3+} are mixed but before they start to react, the solution contains 2.00 mol of $HClO_2$ and 1.0 mol of Cr^{3+} ion. These amounts are in a 4 : 2 ratio, but the two substances react in a 3 : 2 ratio. Once the 1.0 mol of green Cr^{3+} ion is entirely converted to orange $Cr_2O_7^{2-}$ ion the reaction stops, and the solution is $\boxed{\text{orange}}$.

Tip. The species in this reaction that do *not* contain chromium are colorless. This means that the only possible answers are green and orange. How does one know that $HClO_2(aq)$ (for example) is colorless? One must have seen a solution of the species or have read a description written by someone who has. Either way, it's nice to know some descriptive facts. It saves work and doubt.

17.37 In a disproportionation reaction, a single species is simultaneously oxidized and reduced. Here, some In^+ ion is oxidized to In^{3+} ion while some is reduced to elemental In. Calculate E°_{cell}, the standard potential that would be observed if the reaction took place in an electrochemical cell. Use standard reduction potentials from text Appendix E. They apply exactly because the temperature is 298 K

cathode:	$2\,In^+(aq) + 2\,e^- \longrightarrow 2\,In(s)$	$E^\circ = -0.21$ V
anode:	$In^{3+}(aq) + 2\,e^- \longrightarrow In^+(aq)$	$E^\circ = -0.40$ V
cell reaction:	$3\,In^+(aq) \longrightarrow In^{3+}(aq) + 2\,In(s)$	$\Delta E^\circ_{\text{cell}} = 0.19$ V

Now get the value of K

$$\ln K = \frac{n\mathcal{F}\Delta E^\circ_{\text{cell}}}{RT} = \frac{2(9.6485 \times 10^4\text{ C mol}^{-1})(0.19\text{ V})}{(8.3145\text{ J K}^{-1}\text{mol}^{-1})(298.15\text{ K})} = 15 \qquad K = e^{15} = \boxed{3 \times 10^6}$$

Tip. Remember: a coulomb-volt equals a joule (1 C V = 1 J).

17.39 At the anode, H_2 is oxidized to H_3O^+; at the cathode, H_3O^+ to reduced to H_2. The sum of the oxidation and reduction half-equations is

$$H_2(g, \text{anode}) + 2\,H_3O^+(aq, \text{cathode}) \longrightarrow H_2(g, \text{cathode}) + 2\,H_3O^+(aq, \text{anode})$$

The two sides of this equation are identical with respect to their chemistry, so $\Delta E^\circ_{\text{cell}}$, the *standard* potential difference of the cell, equals zero. The observed non-zero potential is caused by the unequal concentrations of H_3O^+ on the two sides. Write the Nernst equation for this reaction, which transfers 2 mol of electrons for every 1 mol of H_2

$$0.150\text{ V} = 0.00\text{ V} - \frac{0.0592\text{ V}}{2}\log\left(\frac{P_{H_2,\,\text{cathode}}[H_3O^+]^2_{\text{anode}}}{P_{H_2,\,\text{anode}}[H_3O^+]^2_{\text{cathode}}}\right)$$

$$0.150\text{ V} = -\frac{0.0592\text{ V}}{2}\log\left(\frac{(1.00)[H_3O^+]^2_{\text{anode}}}{(1.00)(1.00)^2_{\text{cathode}}}\right)$$

Solve for the H_3O^+ concentration at the anode. The answer is 0.00293 M. The pH at the anode is therefore $\boxed{2.53}$.

Buffer action in the solution surrounding the anode depends on the reaction

$$HA(aq) + H_2O(l) \rightleftharpoons H_3O^+(aq) + A^-(aq) \qquad K_a = \frac{[H_3O^+][A^-]}{[HA]}$$

According to the problem, the concentrations of A^- and HA at equilibrium in the buffer solution are equal to 0.10 M at the same time that the pH is 2.53. Substitution in the K_a expression gives

$$K_a = \frac{[H_3O^+][A^-]}{[HA]} = \frac{(10^{-2.53})(0.10)}{(0.10)} = \boxed{0.0029}$$

17.41 **a)** The equation for the cell reaction is

$$\mathrm{Br}_2(l) + \mathrm{H}_2(g) + 2\,\mathrm{H}_2\mathrm{O}(l) \longrightarrow 2\,\mathrm{Br}^-(aq) + 2\,\mathrm{H}_3\mathrm{O}^+(aq)$$

Break this equation down into two half-equations

$$\text{reduction:} \qquad \mathrm{Br}_2(l) + 2\,e^- \longrightarrow 2\,\mathrm{Br}^-(aq)$$

$$\text{oxidation:} \qquad \mathrm{H}_2(g) + 2\,\mathrm{H}_2\mathrm{O}(l) \longrightarrow 2\,\mathrm{H}_3\mathrm{O}^+(aq) + 2\,e^-$$

Reduction *always* takes place at the cathode and oxidation *always* takes place at the anode. The standard potential difference is then the standard *reduction* potential for the oxidation subtracted from the standard reduction potential for the reduction. In this case

$$\Delta E^\circ_{\text{cell}} = E^\circ_{\text{cathode}} - E^\circ_{\text{anode}} = 1.065\ \mathrm{V} - 0.000\ \mathrm{V} = \boxed{1.065\ \mathrm{V}}$$

where the numbers come from the listing in text Appendix E.

b) The concentration of Br^- in the cell is related to the measured cell potential and other concentrations (and partial pressures) by the Nernst equation. For this cell (at 25°C) the Nernst equation takes this form

$$1.710\ \mathrm{V} = 1.065\ \mathrm{V} - \left(\frac{0.0592}{2}\right)\log\left(\frac{[\mathrm{Br}^{2-}]^2[\mathrm{H}_3\mathrm{O}^+]^2}{P_{\mathrm{H}_2}}\right)$$

At pH 0 the concentration of $\mathrm{H}_3\mathrm{O}^+$ is 1.00 M. The partial pressure of H_2 is 1.0 atm. Substitution of these values gives

$$1.710\ \mathrm{V} = 1.065\ \mathrm{V} - \left(\frac{0.0592\ \mathrm{V}}{2}\right)\log\left(\frac{[\mathrm{Br}^-]^2(1.00)^2}{(1.0)}\right)$$

Solve for the concentration of bromide ion. The result, to two significant figures, is $\boxed{1.3 \times 10^{-11}\ \mathrm{M}}$.

c) The dissolution equilibrium of $\mathrm{AgBr}(s)$ is governed by a K_{sp} expression

$$\mathrm{AgBr}(s) \rightleftharpoons \mathrm{Ag}^+(aq) + \mathrm{Br}^-(aq) \qquad K_{\mathrm{sp}} = [\mathrm{Ag}^+][\mathrm{Br}^-]$$

The $\mathrm{Br}^-(aq)$ ion in the cell is at equilibrium with $\mathrm{AgBr}(s)$ and $\mathrm{Ag}^+(aq)$ ion. The concentrations of both ions are known: $[\mathrm{Br}^-]$ was computed in the previous part, and $[\mathrm{Ag}^+]$ is 0.060 M. Hence

$$K_{\mathrm{sp}} = (0.060)(1.27 \times 10^{-11}) = \boxed{7.6 \times 10^{-13}}$$

Molecular Electrochemistry

17.43 A cyclic voltammetry experiment is carried out on the hexaammineruthenium(III) ion. The half-reaction at the *working* electrode, which is the half-reaction that is under study, is

$$\boxed{[\mathrm{Ru(NH_3)_6}]^{3+} + e^- \longrightarrow [\mathrm{Ru(NH_3)_6}]^{2+}} \quad E^\circ = -0.215\ \mathrm{V} \text{ vs. saturated calomel electrode (SCE)}$$

The experiment is however performed using an $\mathrm{Ag}|\mathrm{AgCl}$, KCl(sat.) electrode as the *reference* electrode. The half-reaction at this electrode and its standard reduction potential are

$$\mathrm{AgCl}(s) + e^- \longrightarrow 2\,\mathrm{Ag}(s) + 2\,\mathrm{Cl}^-(\text{sat}) \quad E^\circ = 0.197\ \mathrm{V} \text{ vs. standard hydrogen electrode (SHE)}$$

The standard reduction potential of the saturated calomel electrode is 0.045 V more positive

$$\mathrm{Hg}_2\mathrm{Cl}_2(s) + 2\,e^- \longrightarrow 2\,\mathrm{Hg}(l) + 2\,\mathrm{Cl}^-(\text{sat}) \quad E^\circ = 0.242\ \mathrm{V} \text{ vs. SHE (standard hydrogen electrode)}$$

Using a Ag|AgCl, KCl(sat) electrode as a reference instead of a standard calomel electrode shrinks the potential difference between the working electrode and reference electrode by 0.045 V

$$[Ru(NH_3)_6]^{3+} + e^- \longrightarrow [Ru(NH_3)_6]^{2-} \quad E^\circ = -0.170 \text{ V vs. Ag|AgCl, KCl(sat)}$$

The cyclic voltammogram closely resembles the one sketched in text Example 17.10. The big difference is that the midpoint between the waves occurs at a voltage of -0.170 V (assuming that the experiment is carried out at 298 K) instead of at 0.43 V.

The rising wave, which corresponds to the onset of reduction of the $[Ru(NH_3)_6]^{3+}$ ion, should set in at about -0.1 V as the imposed potential is scanned toward more negative values (from left to right in the diagram in text Example 17.10). When the direction of scan is reversed, a reverse current should start to flow at about -0.25 V as the half-reaction starts to occur in reverse.

Tip. The potential of the Ag|AgCl, KCl(sat) reference electrode is *not* 0.222 V, the potential listed in text Appendix E for the half reaction $AgCl(s) + e^- \longrightarrow Ag(s) + Cl^-(aq)$. This is because the concentration of Cl^- ion in saturated $KCl(aq)$ exceeds 1 M, the standard state concentration.

Tip. Here is a formula that gives the potential of a working or observed (obs.) half-reaction relative to reference electrode 2 when its potential is known relative to reference electrode 1 and the potential difference between reference electrode 2 and reference electrode 1 is also known

$$E_{\text{obs,ref2}} = E_{\text{obs,ref1}} - \Delta E_{\text{ref2-ref1}}$$

In this problem, $E_{\text{obs,ref2}} = -0.215 \text{ V} - (0.197 - 0.242) \text{ V} = -0.170 \text{ V}$.

17.45 The difference in potential between the highest occupied molecular orbital and the lowest unoccupied molecular orbital in the material employed in the electrogenerated chemiluminescence experiment is

$$\Delta E = E_{\text{HOMO}} - E_{\text{LUMO}} = 1.5 \text{ V} - (-0.9 \text{ V}) = 2.4 \text{ V}$$

Transferring electrons across this difference in potential releases quanta having energies of 2.4 eV and wavelength

$$\lambda = \frac{hc}{E} = \frac{(6.626 \times 10^{-34} \text{ J s})(3.00 \times 10^8 \text{ m s}^{-1})}{2.4 \text{ eV}} \left(\frac{1 \text{ eV}}{1.602 \times 10^{-19} \text{ J}} \right) = 5.2 \times 10^{-7} \text{ m} = \boxed{520 \text{ nm}}$$

This light is $\boxed{\text{green.}}$

17.47 Refer to text Figure 17.13. The potential of the conduction band (labelled "LUMO CB") must be more negative than the potential of the standard hydrogen electrode[1], which is 0 V, if it is successfully to provide electrons to reduce H^+ ions to H_2. Also, the potential of the valence band (labelled "HOMO VB") must be more positive than 1.229 V successfully to accept electrons from the oxidation of H_2O to O_2. The two potentials quoted in the problem, -0.65 V and $+1.75$ V, satisfy these criteria respectively. Therefore the answer is $\boxed{\text{yes}}$.

17.49 The redox potential of the HOMO must be less than the potential of the conduction band of strontium titanate in order to inject electrons. This means it must be $\boxed{\text{less than } -0.2 \text{ V}}$. The redox potential of the LUMO must be more positive than the potential for the oxidation of water. This is $\boxed{\text{greater than 1.229 V}}$. The minimum HOMO LUMO gap must be around $\boxed{1.5 \text{ eV}}$.

Batteries and Fuel Cells

17.51 The half-reactions in a lead-acid cell are

$$\text{cathode:} \quad PbO_2(s) + SO_4^{2-}(aq) + 4 H_3O^+(aq) + 2 e^- \longrightarrow PbSO_4(s) + 6 H_2O(l)$$

$$\text{anode:} \quad Pb(s) + SO_4^{2-}(aq) \longrightarrow PbSO_4(s) + 2 e^-$$

---

[1]A standard hydrogen electrode (SHE) is the same as a normal hydrogen electrode (NHE).

The standard reduction potentials are 1.685 V (cathode) and -0.356 V (anode). The standard potential difference is $1.685 - (-0.356) = \boxed{2.041 \text{ V}}$. When electrochemical cells are connected in series, their voltages add. The voltage generated by six lead-acid cells connected in series equals $6(2.041 \text{ V}) = \boxed{12.25 \text{ V}}$.

Tip. The 12-volt batteries used in cars consist of six 2-volt lead-acid cells connected in series.

17.53 **a)** Compute the quantity of charge furnished through an outside circuit as the battery oxidizes 10 kg (10 000 g) of Pb. Oxidation of 1 mol of Pb requires passage of 2 mol of electrons (see problem **17.43**). Hence

$$Q = 10\,000 \text{ g Pb} \times \left(\frac{1 \text{ mol Pb}}{207 \text{ g Pb}}\right)\left(\frac{2 \text{ mol } e^-}{1 \text{ mol Pb}}\right)\left(\frac{9.65 \times 10^4 \text{ C}}{1 \text{ mol } e^-}\right) = \boxed{9.3 \times 10^6 \text{ C}}$$

b) The maximum amount of electrical work that a cell absorbs during its operation is

$$w_{\text{elec,max}} \text{ (absorbed)} = -Q\,E_{\text{cell}} \quad \text{which implies} \quad w_{\text{elec,max}} \text{ (released)} = +Q\,E_{\text{cell}}$$

The standard potential of a lead-acid cell is 2.041 V, as computed in problem **17.51**. This battery consists of six such cells connected in series. Generally the voltage of a lead-acid cell drops as it is discharged, as shown by the Nernst equation

$$\Delta E = \Delta E^\circ - \frac{0.0592}{2} \log \frac{1}{[\text{SO}_4^{2-}]^2[\text{H}_3\text{O}^+]^4}$$

Assume that this battery is so large that the concentrations of SO_4^{2-} ion and H_3O^+ ion in it change only negligibly during the oxidation of the 10 kg of Pb(s). Also assume that the $\text{SO}_4^{2-}(aq)$ and $\text{H}_3\text{O}^+(aq)$ start out in their standard states and that the temperature is 25°C. Then

$$w_{\text{elec,max}} \text{ (released)} = +Q\,\Delta E^\circ = (9.3 \times 10^6 \text{ C})(6)(2.041 \text{ V}) = 1.1 \times 10^8 \text{ V C} = \boxed{1.1 \times 10^8 \text{ J}}$$

Tip. This amount of work is unattainable, quite apart from the fact that the concentrations of SO_4^{2-} ion and H_3O^+ ion in normal-sized batteries *do* diminish significantly during discharge. The more fundamental limitation is that truly reversible discharge of a cell is impossible.

17.55 The amounts of Pb and PbO_2 in the battery diminish during discharge. Without reactants, the battery cannot function. Therefore, simply replacing the dilute H_2SO_4 with concentrated H_2SO_4 is not enough to recharge the battery. The PbSO_4 that accumulates must be removed and fresh Pb and PbO_2 added. The answer is $\boxed{\text{no}}$.

17.57 The maximum amount of electrical work absorbed by an electrochemical reaction at constant T and P equals the change in free energy for that reaction. This quantity is in turn related to the potential difference and the amount of electricity passing the cell

$$w_{\text{elec,max}} = \Delta G = -Q\,\Delta E^\circ_{\text{cell}}$$

If all of the gases in this fuel cell are kept at a pressure of 1 atm and the temperature is 298.15 K, then all reactants and products of the cell process are in standard states. The preceding equation becomes

$$w_{\text{elec,max}} = \Delta G^\circ = -Q\,\Delta E^\circ_{\text{cell}}$$

Break the equation for the reaction in the fuel cell down into half-equations and look up their standard reduction potentials in text Appendix E

$$\text{anode:} \quad \text{H}_2(g) + 2\,\text{H}_2\text{O}(l) \longrightarrow 2\,\text{H}_3\text{O}^+(aq) + 2e^- \qquad E^\circ = 0.00 \text{ V}$$
$$\text{cathode:} \quad \tfrac{1}{2}\,\text{O}_2(g) + 2\,\text{H}_3\text{O}^+(aq) + 2\,e^- \longrightarrow 3\,\text{H}_2\text{O}(l) \qquad E^\circ = 1.229 \text{ V}$$

The standard potential of the cell E°_{cell} is clearly 1.229 V. Now for Q. It equals the quantity of charge transferred as the fuel cell generates one gram of water

$$Q = 1\text{ g H}_2\text{O} \times \left(\frac{1\text{ mol H}_2\text{O}}{18.015\text{ g H}_2\text{O}}\right)\left(\frac{2\text{ mol }e^-}{1\text{ mol H}_2\text{O}}\right)\left(\frac{9.6485\times10^4\text{ C}}{1\text{ mol }e^-}\right) = 1.0711\times10^4\text{ C}$$

The maximum electrical work produced in the surroundings per gram of water equals the negative of the maximum electrical work absorbed by the fuel cell

$$-w_{elec,max} = Q\,\Delta E^\circ = (1.0711\times10^4\text{ C})(1.229\text{ V}) = 1.316\times10^4\text{ J}$$

At 60% efficiency only six-tenths of this maximum work is realized. This is $\boxed{+7900\text{ J g}^{-1}}$.

Corrosion and Corrosion Prevention

17.59 Iron is oxidized, and water is reduced

$$\text{Fe}(s) + 2\text{H}_2\text{O}(l) \longrightarrow \text{Fe}^{2+}(aq) + 2\text{OH}^-(aq) + \text{H}_2(g)$$

At 298 K the reaction has a standard cell potential of

$$\Delta E^\circ_{cell} = E^\circ_{reduction} - E^\circ_{oxidation} = -0.8277 - (-0.409) = \boxed{-0.419\text{ V}}$$

where the E°'s come from text Appendix E. A negative E°_{cell} means the reaction is not spontaneous with the products and reactants in standard states. But recall LeChâtelier's principle. Low concentrations of $\text{Fe}^{2+}(aq)$ and $\text{OH}^-(aq)$ and a low partial pressure of $\text{H}_2(g)$ would tend to drive the reaction to the right, the product side. Under practical circumstances, all three of these species would have far lower than their standard concentrations (or pressures). For example, the concentration of OH^- ion would be much closer to 10^{-7} M, its value at pH 7 than to 1 M, its concentration in its standard state. Corrosion of iron might well tend to occur by this reaction.

Tip. The *rate* of corrosion (which is a crucial aspect of all practical situations) is not considered.

17.61 In a remote theoretical sense, metallic sodium could be used as a sacrificial anode—it is far more easily oxidized than iron according to the standard reduction potentials (see text Appendix E). In practice, however, the ocean water would very rapidly oxidize it. Metallic sodium in fact reacts violently with H_2O to form $\text{NaOH}(aq)$ and $\text{H}_2(g)$. Sodium would be a $\boxed{\text{bad sacrificial anode}}$.

Electrometallurgy

17.63 The half-equations for the Downs process for sodium are

$$\text{Cl}^- \longrightarrow \tfrac{1}{2}\text{Cl}_2(g) + e^- \quad \text{anode} \qquad \text{Na}^+ + e^- \longrightarrow \text{Na}(l) \quad \text{cathode}$$

17.65 A steady current of 55,000 A for 24 h means that 4.75×10^9 C passes through a single cell. Dividing by the Faraday constant gives the chemical amount of electricity passing through the cell. It is 4.93×10^4 mol. It takes 3 mol of electrons to deposit 1 mol of Al. The theoretical yield of Al is therefore 1.64×10^4 mol, which is 4.43×10^5 g of Al per cell. There are 100 cells, so the total theoretical yield of Al is 100 times larger than for a single cell. It is $\boxed{4.4\times10^7\text{ g}}$.

17.67 The Kroll process uses the reaction

$$\boxed{\text{TiCl}_4(l) + 2\text{Mg}(s) \longrightarrow \text{Ti}(s) + 2\text{MgCl}_2(s)}$$

The minimum mass of magnesium to produce 100 kg of titanium by this process is

$$m_{Mg} = 100\text{ kg Ti} \times \left(\frac{1\text{ kmol Ti}}{47.88\text{ kg Ti}}\right)\left(\frac{2\text{ kmol Mg}}{1\text{ kmol Ti}}\right)\left(\frac{24.305\text{ kg Mg}}{1\text{ kmol Mg}}\right) = \boxed{102\text{ kg Mg}}$$

Tip. The word "minimum" is important. More would undoubtedly be required in a practical operation.

17.69 7.32 g of zinc is to be coated onto the steel garbage can. This is 0.112 mol of Zn. Each mole of Zn requires 2 mol of electrons to plate it out, and a mole of electrons is 96 485 C. The total charge passed through the cell is therefore 2.161×10^4 C. A current of 8.50 A means that 8.50 C passes through the cell every second. The time required to pass the required charge is

$$t = \frac{Q}{I} = \frac{2.161 \times 10^4 \text{ C}}{8.50 \text{ C s}^{-1}} = 2.54 \times 10^3 \text{ s} = \boxed{42.4 \text{ min}}$$

A Deeper Look... Electrolysis of Water and Aqueous Solutions

17.71 a) The product at the cathode could be either gaseous hydrogen from the reduction of 1.0×10^{-5} M $H_3O^+(aq)$ or metallic nickel from the reduction of 1.00 M $Ni^{2+}(aq)$. Direct observation of an operating cell could settle the issue at a glance.[2] In this cell, at 25°C, the potential for the reduction of $Ni^{2+}(aq)$ to $Ni(s)$ is −0.23 V, as tabulated in text Appendix E, because the concentration of the Ni^{2+} ion is 1 M. The reduction potential for $H_3O^+(aq)$ to $H_2(g)$ is not equal to its tabulated value of 0.00 V because the concentration of H_3O^+ ion is 1.0×10^{-5} M and not the standard 1 M. Use the Nernst equation to figure the $H_3O^+(aq) \,|\, H_2(g)$ reduction potential under this circumstance. Assume that the cell operates at 25°C

$$E_{H_3O^+(aq)\,|\,H_2(g)} = E^\circ - \frac{0.0592 \text{ V}}{1} \log\left(\frac{P_{H_2}^{1/2}}{[H_3O^+]}\right) = 0.0 - (0.0592 \text{ V}) \log\left(\frac{1}{10^{-5}}\right) = -0.296 \text{ V}$$

This result is algebraically less than the −0.23 V for $Ni^{2+}(aq)|Ni(s)$. Therefore $\boxed{\text{nickel}}$ forms first.

b) A current of 2.00 amperes for 10 hours is a current of 2.00 C s^{-1} for 36000 s. Therefore 7.20×10^4 C passes through the cell. The mass of Ni deposited is

$$m_{Ni} = 7.20 \times 10^4 \text{ C} \times \left(\frac{1 \text{ mol } e^-}{96\,485 \text{ C}}\right)\left(\frac{1 \text{ mol Ni}(s)}{2 \text{ mol } e^-}\right)\left(\frac{58.69 \text{ g Ni}}{1 \text{ mol Ni}}\right) = \boxed{21.9 \text{ g Ni}}$$

The volume of the electrolyte has to be so large that removal of 21.9 g of nickel does not lower the concentration of $Ni^{2+}(aq)$ ion to the point that H_3O^+ ion starts to be reduced.

c) If the pH is 1.0, then $[H_3O^+]$ is 0.10 M. The Nernst equation for the reduction of H_3O^+ to $H_2(g)$ at 1 atm and 25°C becomes

$$E = 0.00 - 0.0592 \text{ V} \log\left(\frac{1.00}{1 \times 10^{-1}}\right) = -0.0592 \text{ V}$$

At this pH, $\boxed{H_2(g)}$ rather than $Ni(s)$ tends to form at the cathode.

Tip. The results make sense with respect to LeChâtelier's principle. In **a)**, the highly dilute $H_3O^+(aq)$ is harder to reduce than the $Ni^{2+}(aq)$. In **b)**, the higher concentration of $H_3O^+(aq)$ raises its reduction potential above than of the $Ni^{2+}(aq)$ ion, the concentration of which is unchanged.

ADDITIONAL PROBLEMS

17.73 When Drano is mixed with water, the NaOH dissolves to give $OH^-(aq)$ ion. The solution attacks the metallic aluminum to oxidize it as water is reduced to hydrogen

$$\boxed{2\,Al(s) + 6\,H_2O(l) + 2\,OH^-(aq) \longrightarrow 2\,Al(OH)_4^-(aq) + 3\,H_2(g)}$$

17.75 The desired reaction equals the reverse of the non-spontaneous reaction that is forced to occur by the application of the outside voltage. It is

$$\boxed{NaClO_3(aq) + 3\,H_2(g) \longrightarrow NaCl(aq) + 3\,H_2O(l)}$$

---

[2] The candidates for reduction at the *cathode* are both *cations*. In electrolytic cells, positively charged cations migrate toward the negatively charged cathode.

17.77 **a)** Compute the chemical amounts of zinc and nickel(II) available for the reaction

$$n_{\text{Zn}} = 32.68 \text{ g Zn} \times \left(\frac{1 \text{ mol Zn}}{65.39 \text{ g Zn}} \right) = 0.4998 \text{ mol Zn}$$

$$n_{\text{Ni}^{2+}} = 0.575 \text{ L solution} \times \left(\frac{1.00 \text{ mol Ni}^{2+}}{1 \text{ L solution}} \right) = 0.575 \text{ mol Ni}^{2+}$$

The balanced equation $\text{Zn}(s) + \text{Ni}^{2+}(aq) \longrightarrow \text{Zn}^{2+}(aq) + \text{Ni}(s)$ has Zn and Ni^{2+} reacting in a 1-to-1 ratio, so $\boxed{\text{zinc}}$ is the limiting reactant.

b) The cell is discharged (stops generating a potential difference) when the zinc is all gone

$$t = 0.4998 \text{ mol Zn} \times \left(\frac{2 \text{ mol } e^-}{1 \text{ mol Zn}} \right) \left(\frac{9.6485 \times 10^4 \text{ C}}{1 \text{ mol } e^-} \right) \left(\frac{1 \text{ s}}{0.0715 \text{ C}} \right) = \boxed{1.35 \times 10^6 \text{ s}}$$

c) The cell produces 1 mol of Ni(s) for every 1 mol of Zn(s) that it consumes

$$m_{\text{Ni}} = 0.4998 \text{ mol Zn} \times \left(\frac{1 \text{ mol Ni}}{1 \text{ mol Zn}} \right) \left(\frac{58.69 \text{ g Ni}}{1 \text{ mol Ni}} \right) = \boxed{29.33 \text{ g Ni}}$$

d) The reaction has reduced 0.4998 mol of Ni^{2+} ion when it comes to a stop for want of zinc. There remains $0.575 - 0.4998 = 0.075$ mol of $\text{Ni}^{2+}(aq)$. This remaining nickel ion is still dissolved in the original 575 mL of solution. Its concentration is 0.075 mol/0.575 L = $\boxed{0.13 \text{ mol L}^{-1}}$.

17.79 Set up a string of unit factors

$$\mathcal{Q} = 1.83 \text{ g Zn} \times \left(\frac{1 \text{ mol Zn}}{65.39 \text{ g Zn}} \right) \left(\frac{2 \text{ mol } e^-}{1 \text{ mol Zn}} \right) \left(\frac{9.6485 \times 10^4 \text{ C}}{1 \text{ mol } e^-} \right) \left(\frac{100 \text{ C total}}{0.25 \text{ C in meter}} \right)$$

$$= \boxed{2.16 \times 10^6 \text{ C total}}$$

Tip. The unit-factors from the molar mass of Zn and the stoichiometry of the half-reaction are routine. The factor derived from the Faraday constant is new in this chapter, but of the same type as the others. The fourth unit factor is marginally creative in defining total coulombs as different from coulombs passing through the electric meter.

17.81 In the conversion $\text{Al}_2\text{O}_3 \longrightarrow \text{Al}$, aluminum passes from the +3 oxidation state to the 0 oxidation state; 3 mol of electrons is transferred for every 1 mol of aluminum formed. The total charge $\mathcal{Q}$ transferred for the year's supply of Al is

$$\mathcal{Q} = 1.5 \times 10^{10} \text{ kg} \times \left(\frac{1 \text{ mol Al}}{0.02698 \text{ kg}} \right) \left(\frac{3 \text{ mol } e^-}{1 \text{ mol Al}} \right) \left(\frac{9.65 \times 10^4 \text{ C}}{1 \text{ mol } e^-} \right) = 1.61 \times 10^{17} \text{ C}$$

The work required to transfer charge through this electrolysis cell depends on the amount of charge and its difference in potential

$$w_{\text{elec}} = -\mathcal{Q} \Delta E_{\text{cell}} = -(1.61 \times 10^{17} \text{ C})(-5.0 \text{ V}) = 8.05 \times 10^{17} \text{ J}$$

where the negative potential difference reflects the fact that this is an electrolytic cell. Then

$$\text{cost} = 8.05 \times 10^{17} \text{ J} \times \left(\frac{1 \text{ kWh}}{3.6 \times 10^6 \text{ J}} \right) \left(\frac{\$0.10}{1 \text{ kWh}} \right) = \$2.2 \times 10^{10} = \boxed{\$22 \text{ billion}}$$

17.83 **a)** The problem describes in detail the identity and state of the reacting species in two half-cells connected by a salt bridge and a wire to make a galvanic cell. Translate the descriptions into balanced reduction half-equations

$$\text{O}_2(g) + 2\,\text{H}_3\text{O}^+(aq) + 2e^- \longrightarrow \text{H}_2\text{O}_2(aq) + 2\,\text{H}_2\text{O}(l) \qquad E^\circ = 0.682 \text{ V}$$

$$\text{MnO}_2(s) + 4\,\text{H}_3\text{O}^+(aq) + 2e^- \longrightarrow \text{Mn}^{2+}(aq) + 6\,\text{H}_2\text{O}(l) \qquad E^\circ = 1.208 \text{ V}$$

All reactants and products in the cell are in standard states. The quoted standard reduction potentials apply without correction to this cell. In a galvanic cell the two half-cells must combine to give a potential difference greater than 0.00 V. This means that reduction occurs in the half-cell with the algebraically larger $E°$. Hence, the first half-reaction is the oxidation, and the second is the reduction. The overall reaction is the second half-reaction minus the first

$$\boxed{H_2O_2(aq) + MnO_2(s) + 2\,H_3O^+(aq) \longrightarrow O_2(g) + Mn^{2+}(aq) + 4\,H_2O(l)}$$

b) Since all participating species are in standard states

$$\Delta E_{cell} = \Delta E°_{cell} = E°_{cathode} - E°_{anode} = 1.208 - 0.682 = \boxed{+0.526\text{ V}}$$

Tip. The schematic representation of this galvanic cell is

$$Pt(s)\,|\,H_2O_2(aq), H_3O^+(aq)\,|\,O_2(g)\ ||\ MnO_2(s)\,|\,H_3O^+(aq), Mn^{2+}(aq)\,|\,C(graphite)$$

By convention, the anode half-cell appears on the left of the double vertical lines in the preceding and the cathode half-cell on the right, as explained in text Section 17.1

17.85 **a)** A piece of metallic iron at the bottom of a solution of $Fe^{2+}(aq)$ removes unwanted $Fe^{3+}(aq)$ by reacting with it

$$\boxed{Fe(s) + 2\,Fe^{3+}(aq) \longrightarrow 3\,Fe^{2+}(aq)}$$

This redox reaction is a "comproportionation," the reverse of a disproportionation. It combines the oxidation of $Fe(s)$ to $Fe^{2+}(aq)$ with the reduction of $Fe^{3+}(aq)$ to $Fe^{2+}(aq)$. It is represented by the first of the following half-equations subtracted from the second

$$2\,e^- + Fe^{2+}(aq) \longrightarrow Fe(s) \qquad \mathcal{E}° = -0.409\text{ V}$$
$$2\,Fe^{3+}(aq) + 2\,e^- \longrightarrow 2\,Fe^{2+}(aq) \qquad \mathcal{E}° = 0.770\text{ V}$$

The standard cell potential is

$$\Delta E°_{cell} = 0.770\text{ V} - (-0.409\text{ V}) = \boxed{1.179\text{ V}}$$

b) The $\Delta E°_{cell}$ for $Mn(s)$ reacting with $Mn^{3+}(aq)$ to give $Mn^{2+}(aq)$ is positive. Problem **17.23** establishes this. Therefore, $\boxed{\text{put } Mn(s) \text{ in}}$ a solution of $Mn^{2+}(aq)$ to minimize the concentration of $Mn^{3+}(aq)$.

17.87 The $Fe^{3+}|Fe^{2+}$ half-reaction has a standard reduction potential of 0.770 V. Model the insoluble Mn(III) and Mn(IV) compounds by MnO_2, which has a reduction potential of 1.208 V at pH 0. A reducing agent such as $\boxed{Br^-(aq)}$ would reduce MnO_2 at pH 0 without reducing the Fe^{3+} because the 1.065 V reduction potential for $Br_2|Br^-$ is less than 1.208 V but more than 0.770 V. Deviations from standard conditions of course affect potentials. This fact and the different rates at which reactions might proceed in a restoration operation make the suggested use of Br^- very tentative.

17.89 The cell reaction is $2\,Ag^+(aq) + Zn(s) \longrightarrow 2\,Ag(s) + Zn^{2+}(aq)$. The standard potential difference for the reaction is

$$\Delta E°_{cell} = \underbrace{(0.7996)}_{Ag^+|Ag} - \underbrace{(-0.7628)}_{Zn^{2+}|Zn} = 1.5624\text{ V}$$

since Ag^+ is clearly reduced (at the cathode) and Zn oxidized (at the anode). The required voltage of 1.50 V is somewhat less than 1.5624 V. Even if 1 M solutions of $Ag^+(aq)$ and $Zn^{2+}(aq)$ were available, using them would not give the required voltage. The concentrations of one or both of the available solutions must be adjusted (by dilution) so that the cell generates 1.50 V. The Nernst equation for this cell relates its actual to its standard potential difference

$$1.50 = 1.5624 - \frac{0.0592}{2} \log \left(\frac{[Zn^{2+}]}{[Ag^+]^2} \right) \quad \text{which gives} \quad \frac{[Zn^{2+}]}{[Ag^+]^2} = 10^{2.108} = 128$$

This last equation must be satisfied to generate the required voltage. If $[Ag^+]$ equals 0.010 M, then $[Zn^{2+}]$ must equal 0.0128 M. There are many other combinations of concentrations that generate the required potential difference, but this one is probably the simplest to prepare

Its voltage will equal 1.50 volt with the zinc anode electrically negative and the silver cathode positive.

Tip. No electrons are allowed to flow in a high-quality measurement of the potential difference (voltage) between the electrodes. The passage of a current would change the concentration of the ions in the two half-cell compartments and so change the voltage of the cell. Common voltmeters contain a large electrical resistance to minimize the passage of current. Specialized instruments that measure the *potential* for a flow of electrons while preventing any actual flow are called *potentiometers*.

17.91 A gas confined at high pressure tends to expand spontaneously to a lower pressure. It does so quickly if a good path is available. In a pressure cell, the free-energy decrease accompanying such a spontaneous expansion can appear as electrical work. In this case, the gas is Cl_2. At the cathode, $Cl_2(g)$ is reduced to 1 M $Cl^-(aq)$. At the anode, 1 M $Cl^-(aq)$ is oxidized to $Cl_2(g)$

$$1/2\,Cl_2(g) + e^- \longrightarrow Cl^-(aq) \qquad \text{cathode, reduction}$$
$$Cl^-(aq) \longrightarrow 1/2\,Cl_2(g) + e^- \qquad \text{anode, oxidation}$$

The anode reaction *generates* $Cl_2(g)$ so the oxidation must occur in the half-cell held at the *lower* pressure of $Cl_2(g)$, the $\boxed{0.010 \text{ atm half-cell}}$. The overall reaction is

$$Cl_2(g) \text{ (0.50 atm)} \longrightarrow Cl_2(g) \text{ (0.01 atm)}$$

The ΔE°_{cell}, the standard potential difference in this cell, equals 0.000 V. The Nernst equation at 25°C gives the actual potential difference

$$E_{cell} = E^\circ_{cell} - \frac{0.0592 \text{ V}}{2} \log \left(\frac{0.010}{0.50} \right) = 0.000 - (-0.050) = \boxed{0.050 \text{ V}}$$

17.93 The half-reactions in the cell are

$$2\,MnO_2(s) + 2\,NH_4^+(aq) + 2\,e^- \longrightarrow Mn_2O_3(s) + 2\,NH_3(aq) + H_2O(l) \qquad \text{cathode, reduction}$$
$$Zn(s) \longrightarrow Zn^{2+}(aq) + 2\,e^- \qquad \text{anode, oxidation}$$

Electrons are released at the anode and taken up at the cathode. Thus, electrons flow from the Zn electrode to the graphite electrode, where they reduce MnO_2. The stud on the top of a common flashlight battery is positive, the flat surface at the bottom is negative.

17.95   **a)** Oxygen is reduced to water at the cathode while ethanol is oxidized to CO_2 at the anode

$$O_2(g) + 4\,H_3O^+ + 4\,e^- \longrightarrow 6\,H_2O(l) \qquad \text{cathode, reduction}$$
$$15\,H_2O(l) + C_2H_5OH(l) \longrightarrow 2\,CO_2(g) + 12\,H_3O^+ + 12\,e^- \qquad \text{anode, oxidation}$$

The alloy and nickel electrodes do not react. They conduct electrons away from and back to the fuel cell (and, in the case of the alloy, speed up the desired reaction).

b) The overall reaction is the sum of the two half-reactions given above (after the reduction half-reaction is multiplied by 3 so that the electrons cancel out)

$$C_2H_5OH(l) + 3\,O_2(g) \longrightarrow 2\,CO_2(g) + 3\,H_2O(l)$$

This reaction transfers 12 mol of e^- when it is run as written. Compute its ΔG°_{298} and use it to get E°_{cell} at 298 K. The necessary ΔG°_f's come from text Appendix D

$$\Delta G^\circ_{298} = 2\,\underbrace{(-394.36)}_{CO_2(g)} + 3\,\underbrace{(-237.18)}_{H_2O(l)} - 1\,\underbrace{(-174.89)}_{C_2H_5OH(l)} = -1325.37 \text{ kJ}$$

$$\Delta E^\circ_{\text{cell}} \text{ at 298 K} = \frac{-\Delta G^\circ_{298}}{n\mathcal{F}} = \frac{-(-1325.37 \times 10^3 \text{ J})}{(12 \text{ mol})(96\,485 \text{ C mol}^{-1})} = 1.1447 \text{ J C}^{-1} = \boxed{1.1447 \text{ V}}$$

c) The standard potential difference equals the standard reduction potential of the half-reaction at the cathode minus the standard reduction potential of the half-reaction at the anode. The E°_{cathode} is 1.229 V (see Appendix E). Use it to compute E°_{anode}

$$\Delta E^\circ_{\text{cell}} = E^\circ_{\text{cathode}} - E^\circ_{\text{anode}}$$
$$= E^\circ(H_3O^+, O_2|H_2O) - E^\circ(CO_2, H_3O^+|C_2H_5OH)$$
$$1.1447 \text{ V} = 1.229 \text{ V} - E^\circ(CO_2, H_3O^+|C_2H_5OH)$$
$$E^\circ(CO_2, H_3O^+|C_2H_5OH) = \boxed{+0.084 \text{ V}}$$

17.97   **a)** Water is oxidized at the anode of the first cell: $\boxed{6\,H_2O(l) \longrightarrow O_2(g) + 4\,H_3O^+ + 4\,e^-}$. Water is far more easily oxidized than $Pt(s)$, the other possible candidate. In the second cell, metallic nickel is oxidized: $\boxed{Ni(s) \longrightarrow Ni^{2+}(aq) + 2\,e^-}$. Nickel is more easily oxidized than water (by comparison of their standard reduction potentials).

b) The total electrical charge Q passing through the cells equals the current, assuming that it is steady multiplied by the elapsed time. Compute this in coulombs and use the Faraday constant to convert to moles

$$n_{e^-} = 0.10 \text{ C s}^{-1} \times 10 \text{ h} \left(\frac{3600 \text{ s}}{1 \text{ hr}}\right) \left(\frac{1 \text{ mol } e^-}{9.6485 \times 10^4 \text{ C}}\right) = 0.0373 \text{ mol } e^-$$

Next, set up unit-factors from the half-equations. Treat the electron like any other product or reactant. At the anode in the first cell

$$m_{O_2} = 0.0373 \text{ mol } e^- \times \left(\frac{1 \text{ mol } O_2}{4 \text{ mol } e^-}\right) \left(\frac{32.00 \text{ g } O_2}{1 \text{ mol } O_2}\right) = \boxed{0.30 \text{ g } O_2}$$

At the anode in the second cell

$$m_{Ni^{2+}} = 0.0373 \text{ mol } e^- \times \left(\frac{1 \text{ mol Ni}}{2 \text{ mol } e^-}\right) \left(\frac{58.69 \text{ g Ni}}{1 \text{ mol Ni}}\right) = \boxed{1.1 \text{ g Ni}}$$

17.99   **a)** Text Appendix E gives these standard reduction potentials:

$$Co^{2+}(aq) \longrightarrow Co(s) \quad E^\circ = -0.28 \text{ V} \qquad \text{and} \qquad Sn^{2+}(aq) \longrightarrow Sn(s) \quad E^\circ = -0.1364 \text{ V}$$

$Sn^{2+}(aq)$ is clearly easier to reduce under standard conditions. Both ions have non-standard concentrations, but because their concentrations are equal and both are $+2$ ions, the effect on the reduction potential is the same for both. Therefore $\boxed{Sn(s) \text{ appears first}}$ if a mixture of 0.10 M $CoCl_2$ and 0.10 M $SnCl_2$ is electrolyzed.

b) The decomposition potential is the minimum potential that suffices to drive a non-spontaneous electrochemical reaction. Start by establishing the overall reaction that is expected to take place when the outside potential is applied. At the cathode, $Sn^{2+}(aq)$ ion should be reduced. At the anode, water should be oxidized to $O_2(g)$.[3] The overall reaction then should be

$$Sn^{2+}(aq) + 3\,H_2O(l) \longrightarrow Sn(s) + \tfrac{1}{2}\,O_2(g) + 2\,H_3O^+(aq)$$

The standard cell potential of this reaction at 298 K is

$$\Delta E^\circ_{\text{cell}} = E^\circ(Sn^{2+}|Sn) - E^\circ(O_2|H_2O) = -0.1364 - 1.229 = -1.3654 \text{ V}$$

None of the reactants or products (except water) is in a standard state in this cell. Use the Nernst equation to obtain E_{cell} for the concentrations that prevail

$$\Delta E_{\text{cell}} = \Delta E^\circ_{\text{cell}} - \frac{RT}{n\mathcal{F}} \ln Q = -1.3654 \text{ V} - \frac{0.0592 \text{ V}}{2} \log \left(\frac{P_{O_2}^{1/2}[H_3O^+]^2}{[Sn^{2+}]} \right)$$

Assume that the $O_2(g)$ is generated at a partial pressure of 1.00 atm and that the concentration of H_3O^+ is 1.0×10^{-7} (pH 7.00). Then

$$\Delta E_{\text{cell}} = -1.3654 \text{ V} - \frac{0.0592 \text{ V}}{2} \log \left(\frac{(1.00)^{1/2}[1.0 \times 10^{-7}]^2}{[0.10]} \right) = -0.9806 \text{ V}$$

The decomposition potential is therefore $\boxed{+0.981 \text{ V}}$.

Tip. If the $O_2(g)$ is generated at a partial pressure of 0.20 atm, which is the approximate partial pressure of $O_2(g)$ in the atmosphere, the decomposition potential is slightly less, 0.970 V.

c) The balanced equation for the electrolytic reduction of $Co^{2+}(aq)$ ion is

$$Co^{2+} + 3\,H_2O(l) \longrightarrow Co(s) + 1/2\,O_2(g) + 2\,H_3O^+(aq)$$

The $\Delta E^\circ_{\text{cell}}$ for this reaction is at 298 K is

$$E^\circ_{\text{cell}} = E^\circ(Co^{2+}|Co) - E^\circ(O_2|H_2O) = -0.28 - 1.229 = -1.509 \text{ V}$$

The reduction of Co^{2+} cannot start until the potential for the reduction of Sn^{2+} has risen high enough to make the E°_{cell}'s for the two reactions equal. Write the Nernst equations for the two reactions and set their E°_{cell}'s equal. Doing this gives

$$-1.509 \text{ V} - \frac{0.0592 \text{ V}}{2} \log \left(\frac{P_{O_2}^{1/2}[H_3O^+]^2}{[Co^{2+}]} \right) = -1.3654 \text{ V} - \frac{0.0592 \text{ V}}{2} \log \left(\frac{P_{O_2}^{1/2}[H_3O^+]^2}{[Sn^{2+}]} \right)$$

---

[3]As shown in text Section 17.8, $Cl^-(aq)$ has a lesser tendency to be oxidized than $H_2O(l)$.

Simplify as follows[4]

$$-0.1436 \text{ V} = \frac{0.0592 \text{ V}}{2} \log\left(\frac{P_{O_2}^{1/2}[\text{H}_3\text{O}^+]^2}{[\text{Co}^{2+}]}\right) - \frac{0.0592 \text{ V}}{2} \log\left(\frac{P_{O_2}^{1/2}[\text{H}_3\text{O}^+]^2}{[\text{Sn}^{2+}]}\right)$$

$$-4.851 = \log\left(\frac{P_{O_2}^{1/2}[\text{H}_3\text{O}^+]^2}{[\text{Co}^{2+}]}\right) - \log\left(\frac{P_{O_2}^{1/2}[\text{H}_3\text{O}^+]^2}{[\text{Sn}^{2+}]}\right)$$

$$-4.851 = \log\left(\frac{[\text{Sn}^{2+}]}{[\text{Co}^{2+}]}\right) \quad \text{which gives} \quad 1.408 \times 10^{-5} = \frac{[\text{Sn}^{2+}]}{[\text{Co}^{2+}]}$$

The concentrations of the two metal ions start out equal, so only $\boxed{0.000014}$ of the Sn^{2+} remains when the Co^{2+} finally can plate out.

17.101 Compute the volume of zinc to be coated onto the steel, then the mass of that volume of zinc and then the chemical amount in that mass. Take the density of zinc from text Appendix F

$$V_{\text{Zn}} = \frac{(0.250 \times 10^{-3} \text{ m})(1.00 \text{ m})(100 \text{ m})}{1 \text{ side}} \times 2 \text{ sides} = 0.0500 \text{ m}^3$$

$$m_{\text{Zn}} = 0.0500 \text{ m}^3 \times \left(\frac{10^6 \text{ cm}^3}{\text{m}^3}\right) \times \left(\frac{7.133 \text{ g Zn}}{\text{cm}^3}\right) = 3.566 \times 10^5 \text{ g}$$

$$n_{\text{Zn}} = 3.566 \times 10^5 \text{ g Zn} \times \left(\frac{1 \text{ mol Zn}}{65.39 \text{ g Zn}}\right) = 5.454 \times 10^3 \text{ mol Zn}$$

Each mole of Zn requires 2 mol of electrons to plate it out. A a mole of electrons is 96485 C. Hence

$$Q = 5.454 \times 10^3 \text{ mol Zn} \times \left(\frac{2 \text{ mol } e^-}{1 \text{ mol Zn}}\right) \times \left(\frac{96485 \text{ C}}{\text{mol } e^-}\right) = 1.052 \times 10^9 \text{ C}$$

The energy used in the plating operation is the amount of electrical charge passed through the circuit multiplied by the voltage that pushes it through. The voltage is 3.5 V in this case so the energy is 3.68×10^9 V C which is 3.68×10^9 J. Divide this amount of energy by 0.9 because the galvanizing is only 90% efficient. This raises the energy consumption to 4.09×10^9 J, equivalent to 1.14×10^3 kW hr, which costs (to two significant figures) $\boxed{\$110}$.

Tip. The problem draws on ideas covered in problems **17.69** and **17.5**.

CUMULATIVE PROBLEMS

17.103 Eliminate ΔG° between the following two equations

$$\Delta G^\circ = \Delta H^\circ - T\Delta S^\circ \quad \text{and} \quad -n\mathcal{F}\Delta E_{\text{cell}}^\circ = \Delta G^\circ$$

Solving the resulting equation for $\Delta E_{\text{cell}}^\circ$ gives

$$\Delta E_{\text{cell}}^\circ = \frac{-\Delta H^\circ}{n\mathcal{F}} + \frac{T\Delta S^\circ}{n\mathcal{F}}$$

This equation gives the temperaturedependence of the standard potential difference of *any* cell. If ΔS° of the cell reaction is positive, then raising the constant temperature T at which the cell operates raises the cell's voltage; if ΔS° is negative, then raising the T at which the cell operates lowers the cell's voltage.

Now, calculate ΔS_{298}° for the reaction $\text{Cu}(s) + 2\,\text{Ag}^+(aq) \longrightarrow \text{Cu}^{2+}(aq) + 2\,\text{Ag}(s)$ taking the needed thermodynamic data from text Appendix D

$$\Delta S_{298}^\circ = \underbrace{(-99.6)}_{\text{Cu}^{2+}(aq)} + 2\underbrace{(42.55)}_{\text{Ag}(s)} - \underbrace{(33.15)}_{\text{Cu}(s)} - 2\underbrace{(72.68)}_{\text{Ag}^+(aq)} = -193.01 \text{ J K}^{-1}$$

---

[4]Recall that $\log a - \log b = \log(a/b)$.

The $\Delta S°$ of the cell reaction is negative,[5] so the voltage $\boxed{\text{decreases with increasing } T}$.

Tip. Exactly how strongly does the cell voltage depend on the temperature? Take the derivative of $Delta E°_{cell}$ with respect to T in the equation for $\Delta E°_{cell}$, and substitute the various values

$$\frac{d(E°_{cell})}{dT} = \frac{\Delta S°}{n\mathcal{F}} = \frac{-193.01 \text{ J K}^{-1}}{2(9.6485 \times 10^4 \text{ C})} = -1.00 \times 10^{-3} \text{ V K}^{-1}$$

The voltage drops by a millivolt for every kelvin that the temperature increases.

17.105 a) Combine the two given half-equations to obtain this equation for dissolution

$$\text{AgBr}(s) \longrightarrow \text{Ag}^+(aq) + \text{Br}^-(aq)$$

Since AgBr(s) must end up on the left side, the first half-equation was *subtracted* from the second. Assume that the temperature is 25°C (298 K). The standard potential for the reaction at 298 K is

$$\Delta E°_{cell} = E°_{cathode} - E°_{anode} = (0.0713 - 0.7996) \text{ V} = -0.7283 \text{ V}$$

Write the Nernst equation for this reaction at 298 K

$$\Delta E_{cell} = -0.7283 \text{ V} - \frac{0.0592 \text{ V}}{1} \log\left([\text{Ag}^+][\text{Br}^-]\right)$$

When the reaction reaches equilibrium the argument of the logarithm equals K_{sp} and E_{cell} equals zero. Hence

$$0 = -0.7283 \text{ V} - 0.0592 \text{ V} \log K_{sp}$$
$$\log K_{sp} = \frac{0.7283 \text{ V}}{-0.0592 \text{ V}}$$
$$K_{sp} = 10^{-12.30} = \boxed{5 \times 10^{-13}}$$

b) Set up the Nernst equation just as in the preceding, but put in 0.10 M as the concentration of Br$^-$ ion

$$0 = -0.7283 \text{ V} - (0.0592 \text{ V}) \log\left([\text{Ag}^+][0.10]\right) \qquad \text{from which} \qquad [\text{Ag}^+] = 5 \times 10^{-12} \text{ M}$$

The solubility of AgBr(s) equals the equilibrium concentration of Ag$^+$(aq) *if* AgBr(s) is the only source of Ag$^+$(aq) and *if* all of the dissolved silver is in the form of Ag$^+$(aq). The estimated solubility is therefore $\boxed{5 \times 10^{-12} \text{ mol L}^{-1}}$.

Tip. Text Table 16.2 gives 7.7×10^{-13} as the K_{sp} of AgBr. This is about 50% larger than the answer in part **a)** and leads to an answer that is about 50% larger in part **b)**. The K_{sp} in text Table 16.2 is a "true" or "thermodynamic" equilibrium constant. It includes correction factors to account for various deviations from ideal-solution behavior. The K_{sp} obtained in this problem is a "classical" or "concentration" equilibrium constant.

---

[5]$\Delta S°$ (and $H°$) depend rather weakly on temperature and so are fairly well approximated by $\Delta S°_{298}$ and $\Delta H°_{298}$ at temperatures near 298 K.

Chapter 18

Chemical Kinetics

Rates of Chemical Reactions

18.1 Text Figure 18.3 shows the concentration of NO during the progress of the reaction

$$NO_2(g) + CO(g) \longrightarrow NO(g) + CO_2(g)$$

The line bends to the right as it ascends, indicating that the rate of production of NO slows with time. The instantaneous rate of production of NO at a point in time t equals the slope of the tangent to the curving line at that point. Pencil in a line tangent to the curve at $t = 200$ s and estimate its slope from the graph (as shown in the Figure for $t = 150$ s). A reasonable result is $\boxed{5.3 \times 10^{-5} \text{ mol L}^{-1}\text{s}^{-1}}$.

18.3 The rate of a reaction is expressed in terms of the rate of disappearance of a reactant, or formation of a product

$$\text{rate} = \boxed{-\frac{1}{1}\frac{d[N_2]}{dt} = -\frac{1}{3}\frac{d[H_2]}{dt} = +\frac{1}{2}\frac{d[NH_3]}{dt}}$$

Rate Laws

18.5 **a)** The way in which a change in the concentration of a reactant affects the rate of a reaction under otherwise constant conditions gives the order of the reaction with respect to that reactant. The reaction

$$NO_2(g) + 2\,H_2(g) \longrightarrow N_2 + 2\,H_2O$$

is first order in H_2 because the rate varies as to the first power of the concentration of H_2. It is second order in NO because the rate varies as to the second power of the concentration of NO. The rate expression is $\boxed{\text{rate} = k[H_2][NO]^2}$. The units of the rate constant k are $\boxed{\text{L}^2\,\text{mol}^{-2}\,\text{s}^{-1}}$ This can also be written as $\text{M}^{-2}\text{s}^{-1}$. The reaction has an overall order of 3.

b) Multiplying $[H_2]$ by 2 would double the rate; multiplying [NO] by 3 would increase the rate by a factor of 9. The combined effect would be to increase the rate by a $\boxed{\text{factor of 18}}$.

Tip. Notice that the order of this reaction with respect to the reactant NO_2 is *not* equal to the coefficient of NO_2 in the balanced chemical equation. This same goes for the reactant H_2.

18.7 **a)** Compare the second and third lines of data in the table, assuming that the differences shown there are the *only* differences between the two runs. When $[C_5H_5N]$ is held constant and $[CH_3I]$ is doubled, the rate doubles. The reaction is first order in CH_3I. Next, compare the first and second lines in the table. When both concentrations are doubled, the rate is increased by a factor of 4. Half of this is due to the change in the concentration of CH_3I, so the other half is due to the change in the concentration of C_5H_5N, and

$$\text{rate} = k[C_5H_5N][CH_3I]$$

255

b) Calculate k by substituting the data from any one of the three data points into the rate equation. Using the first line of the table

$$7.5 \times 10^{-7} \text{ mol L}^{-1}\text{s}^{-1} = k(1.00 \times 10^{-4} \text{ mol L}^{-1})(1.00 \times 10^{-4} \text{ mol L}^{-1})$$

Doing the arithmetic gives $\boxed{k = 75 \text{ L mol}^{-1}\text{s}^{-1}}$.

c) Substitute the calculated k and the two given concentrations into the rate equation. Remember that "M" stands for mol L^{-1}.

$$\text{rate} = (75 \text{ L mol}^{-1}\text{s}^{-1})(5.0 \times 10^{-5} \text{ mol L}^{-1})(2.0 \times 10^{-5} \text{ mol L}^{-1}) = \boxed{7.5 \times 10^{-8} \text{ mol L}^{-1}\text{s}^{-1}}$$

18.9 The problem asks for the time it will take for the partial pressure of $SO_2Cl_2(g)$ to fall to one-half of its initial value in the first-order reaction

$$SO_2Cl_2(g) \longrightarrow SO_2(g) + Cl_2(g)$$

given the rate constant. Assume ideal-gas behavior (apart from the existence of the reaction itself). Then the answer equals the half-life of the reaction. In a first-order process the half-life depends solely upon the rate constant. In this case,

$$t_{1/2} = \frac{\ln 2}{k} = \frac{0.69315}{2.2 \times 10^{-5} \text{ s}^{-1}} = \boxed{3.2 \times 10^4 \text{ s}}$$

Tip. Recall that partial pressures of gases are directly proportional to their concentrations (assuming ideality). Also, the *total* pressure in this vessel *increases* as a consequence of the reaction.

18.11 Decomposition of benzene diazonium chloride at 20°C is first-order. Write the integrated rate law

$$P_{C_6H_5N_2Cl} = (P_{C_6H_5N_2Cl})_0 \, e^{-kt} = (P_{C_6H_5N_2Cl})_0 \exp(-kt)$$

Convert the elapsed time (10.0 h) to seconds and substitute it, together with the initial partial pressure and the rate constant, into the preceding equation

$$P_{C_6H_5N_2Cl} = (0.0088 \text{ atm}) \exp\left((-4.3 \times 10^{-5} \text{ s}^{-1})(3.60 \times 10^4 \text{ s})\right) = \boxed{0.0019 \text{ atm}}$$

18.13 The decomposition of chloroethane follows this integrated rate law

$$[C_2H_5Cl] = [C_2H_5Cl]_0 \, e^{-kt}$$

Divide through by $[C_2H_5Cl]_0$, and take the natural logarithm of both sides

$$\ln \frac{[C_2H_5Cl]}{[C_2H_5Cl]_0} = -k\,t$$

Substitution of the two concentrations and the time gives

$$\ln \frac{0.0016 \text{ M}}{0.0098 \text{ M}} = -k(340 \text{ s}) \quad \text{which is easily solved:} \quad k = \boxed{5.3 \times 10^{-3} \text{ s}^{-1}}$$

18.15 The integrated rate law for this very fast second-order association of iodine atoms is

$$\frac{1}{[I]} = \frac{1}{[I]_0} + 2kt$$

Substitute the given concentration, rate constant, and time

$$\frac{1}{[I]} = \frac{1}{1.00 \times 10^{-4} \text{ mol L}^{-1}} + 2(8.2 \times 10^9 \text{ L mol}^{-1} \text{ s}^{-1})(2.0 \times 10^{-6} \text{ s})$$

$$= 1.00 \times 10^4 \text{ L mol}^{-1} + 3.28 \times 10^4 \text{ L mol}^{-1} = 4.28 \times 10^4 \text{ L mol}^{-1}$$

$$[I] = \boxed{2.3 \times 10^{-5} \text{ mol L}^{-1}}$$

18.17   The reaction is the neutralization of $OH^-(aq)$ with $NH_4^+(aq)$. Aqueous acid-base reactions are generally fast. This reaction is no exception, as shown by the huge room-temperature rate constant of $k = 3.4 \times 10^{10}$ M^{-1} s^{-1}. The answer will be a very short time. If 1.00 L of 0.0010 M NaOH and 1.00 L of 0.0010 M NH_4Cl are mixed, then *after* the mixing, but *before* the reaction can start each reactant has a concentration of 5.0×10^{-4} M. The kinetics are second-order overall

$$\text{rate} = \frac{-d[OH^-]}{dt} = k[OH^-][NH_4^+]$$

Throughout the reaction $[OH^-] = [NH_4^+]$. Let this concentration be represented by c. Then

$$\frac{-dc}{dt} = kc^2$$

Integrating this equation and inserting the initial condition gives

$$\frac{1}{c} = \frac{1}{c_0} + kt$$

This equation does not include the factor of 2 that appears in text equation 18.4 because the stoichiometry of the reaction lacks that factor. For c equal to 1.0×10^{-5} M and k equal to 3.4×10^{10} $M^{-1}s^{-1}$, the equation becomes

$$\frac{1}{1.0 \times 10^{-5} \text{ M}} = \frac{1}{5.0 \times 10^{-4} \text{ M}} + (3.4 \times 10^{10} \text{ M}^{-1}\text{s}^{-1})\, t$$
$$9.8 \times 10^4 \text{ M}^{-1} = (3.4 \times 10^{10} \text{ M}^{-1}\text{s}^{-1})\, t$$
$$t = \boxed{2.9 \times 10^{-6} \text{ s}}$$

Reaction Mechanisms

18.19   **a)** Two particles collide in an elementary reaction. The reaction is therefore $\boxed{\text{bimolecular}}$. Its rate law is $\boxed{\text{rate} = k[HCO][O_2]}$.

   b) Three particles collide in an elementary reaction; the reaction is therefore $\boxed{\text{termolecular}}$. Its rate law is $\boxed{\text{rate} = k[CH_3][O_2][N_2]}$.

   c) A particle decomposes spontaneously in an elementary reaction. The reaction is $\boxed{\text{unimolecular}}$. Its rate law is $\boxed{\text{rate} = k[HO_2NO_2]}$.

18.21   **a)** The first step is $\boxed{\text{unimolecular}}$; the three subsequent steps are $\boxed{\text{bimolecular}}$. This is determined simply by counting the number of interacting particles on the left sides of the four equations. Molecularity has meaning only in reference to elementary reactions.

   b) The overall reaction is the sum of the steps: $\boxed{H_2O_2 + O_3 \longrightarrow H_2O + 2\,O_2}$.

   c) The intermediates are $\boxed{O,\ ClO,\ CF_2Cl,\ \text{and}\ Cl}$. These species are produced in the course of the reaction and later consumed.

   Tip. The CF_2Cl_2 is a catalyst. It reacts in an early stage in the mechanism but is later regenerated. The overall reaction could not proceed according to this mechanism without the presence and participation of CF_2Cl_2, but the compound is neither consumed nor produced.

18.23   The equilibrium constant of this elementary reaction equals the ratio of the rate constant of the forward reaction to the rate constant of the reverse reaction

$$K = \frac{k_f}{k_r} = 5.0 \times 10^{10} = \frac{1.3 \times 10^{10} \text{ L mol}^{-1}\text{s}^{-1}}{k_r}$$

Solving for k_r gives 0.26 L mol^{-1}s^{-1}. The reaction in question is in fact just the reverse of the original elementary reaction, so the answer is $\boxed{0.26 \text{ L mol}^{-1}\text{s}^{-1}}$.

Reaction Mechanisms and Rate

18.25 **a)** The rate-limiting elementary step in a mechanism determines the overall reaction rate. In this case, the slow step is $C + E \longrightarrow F$. A preliminary version of the rate law is

$$\text{rate} = k_2[C][E]$$

Unfortunately, the expression involves the concentration of C, an intermediate. This is unacceptable. To eliminate [C] in the rate law, consider how C is formed. It arises in the first step of the mechanism, a fast equilibrium. For that first step

$$k_1[A][B] = k_{-1}[C][D] \quad \text{which is equivalent to} \quad \frac{k_1}{k_{-1}} = \frac{[C][D]}{[A][B]}$$

Solve either of these equations for the concentration of C and substitute into the preliminary rate law

$$\boxed{\text{rate} = \frac{k_1 k_2}{k_{-1}} \frac{[A][B][E]}{[D]}}$$

The overall reaction is the sum of the two steps $\boxed{A + B + E \longrightarrow D + F}$.

Tip. In this reaction system, the accumulation of one product (D) slows down the reaction, but the accumulation of the other product (F) does not.

b) The overall reaction in this case is $\boxed{A + D \longrightarrow B + F}$. For the two fast equilibria these relationships hold

$$k_1[A] = k_{-1}[B][C] \quad \text{and} \quad k_2[C][D] = k_{-2}[E]$$

The last, slow step is the rate-determining step

$$\text{rate} = k_3[E]$$

But E is an intermediate and its concentration may not appear in the final rate expression. To eliminate [E], solve the second of the preceding pair of equations for [E] and substitute

$$\text{rate} = \frac{k_2 k_3}{k_{-2}}[C][D]$$

This expression *still* contains the concentration of an intermediate (C now). Eliminate [C] by solving the first of the pair for [C] and substituting

$$\boxed{\text{rate} = \frac{k_1 k_2 k_3}{k_{-1} k_{-2}} \frac{[A][D]}{[B]}} = k_{\text{expt}} \frac{[A][D]}{[B]}$$

where the experimental k is the algebraic composite of the several step-wise rate constants. The reaction is first order in both A and D, is -1 order in B and first order overall.

18.27 The reaction between hydrogen chloride (HCl) and propene (CH_3CHCH_2) is first order in propene and third order in HCl. This is an experimental fact.

Mechanism (a) proposes the rapid formation of the intermediate H from 2 HCl's followed by slow combination of the H with CH_3CHCH_2. The H is required in the rate-determining step and getting it depends on the collision of 2 HCl's. In this way this mechanism predicts second-order kinetics in HCl which is incorrect.

Mechanism (b) proposes *two* fast equilibria. The first involves HCl only and the second involves HCl and propene. The slow step is the chemical combination of the two intermediates that these equilibria produce. The rate is accordingly proportional to the concentrations of all the reactants in the two fast equilibria. HCl occurs three times among these reactants and propane occurs once so mechanism (b) fits with the observed rate law.

Mechanism (c) involves HCl in the fast production of two different intermediates which then slowly combine to give the product (and to regenerate some HCl). It predicts second-order kinetics in HCl (2 HCl's consumed to furnish reactants for the slow step). It is therefore not consistent with observation.

Tip. A correct mechanism for a reaction must predict the observed rate law; a mechanism that correctly predicts the observed rate law may *still* be wrong. In other words, predicting the rate law is necessary but not sufficient for correctness in a mechanism.

18.29   Write rate expressions for the slow elementary step in each mechanism. Then eliminate from the expressions the concentrations of any intermediates. The results for the three mechanisms (a), (b) and (c) are the three rate laws

$$\text{rate (a)} = k_1[NO_2Cl] \qquad \text{rate (b)} = \frac{k_1 k_2}{k_{-1}} \frac{[NO_2Cl]^2}{[Cl_2]} \qquad \text{rate (c)} = \frac{k_1 k_2 k_3}{k_{-1} k_{-2}} \frac{[NO_2Cl]^2}{[NO_2]}$$

The clusters of constants in the second and third rate expressions can be regarded as composite rate constants. Only mechanism (a) predicts the reaction to be first-order in NO_2Cl. Only mechanism (a) is consistent with experiment.

Tip. Nothing allows one to conclude that mechanism (a) is the true mechanism.

18.31   The overall reaction is $A + B + E \longrightarrow D + F$. It proceeds first by an equilibrium between A and B giving D and the intermediate C, and then by the consumption of C in reaction with E to give F. The rate of appearance of C equals $k_1[A][B]$, and the rate of *disappearance* of C equals $k_{-1}[C][D] + k_2[C][E]$. This latter is a sum because C disappears both by back reaction (with D to give A plus B) and by *further* reaction (with E to give F). If the concentration of intermediate C is constant (the steady state approximation), the rate of disappearance of C must equal its rate of appearance

$$k_1[A][B] = k_{-1}[C][D] + k_2[C][E]$$

The rate of the reaction can be expressed in terms of the rate of appearance of a product, for example,

$$\text{rate} = \frac{d[F]}{dt} = k_2[E][C]$$

Solve the steady-state equation for [C] and substitute the answer in the preceding

$$\text{rate} = \boxed{\frac{k_1 k_2 [A][B][E]}{k_2[E] + k_{-1}[D]}}$$

If $\boxed{k_2[E] \text{ is much smaller than } k_{-1}[D]}$ then this expression becomes

$$\text{rate} \approx \left(\frac{k_1 k_2}{k_{-1}}\right) \frac{[A][B][E]}{[D]}$$

The second-step rate constant k_2 is much smaller than k_{-1} when the first step of the reaction is a fast equilibrium.

18.33 The reaction is the decomposition of nitryl chloride $2\,NO_2Cl \longrightarrow 2\,NO_2 + Cl_2$. The mechanism involves an equilibrium breakdown of the reactant to NO_2 plus Cl followed by reaction of the atomic chlorine with a second molecule of NO_2Cl to generate the products. The change in the concentration of Cl with time is

$$\frac{d[Cl]}{dt} = k_1[NO_2Cl] - k_{-1}[Cl][NO_2] - k_2[Cl][NO_2Cl]$$

because the rate of change of the concentration of Cl equals the rate of its production minus the rate of its consumption. The steady-state approximation is that $d[Cl]/dt = 0$. If so, then

$$k_1[NO_2Cl] - k_{-1}[Cl][NO_2] - k_2[Cl][NO_2Cl] = 0$$

and

$$[Cl] = \frac{k_1[NO_2Cl]}{k_{-1}[NO_2] + k_2[NO_2Cl]}$$

The rate of the overall reaction equals the rate of the final elementary step, which generates the two products

$$\text{rate} = \frac{d[Cl_2]}{dt} = k_2[NO_2Cl][Cl]$$

Substitute the expression for the concentration of the Cl into this equation:

$$\boxed{\text{rate} = \frac{d[Cl_2]}{dt} = \frac{k_1 k_2 [NO_2Cl]^2}{k_{-1}[NO_2] + k_2[NO_2Cl]}}$$

Tip. The answer differs from the rate expression given for the same reaction in problem **18.29**. It becomes equal to that expression as the second term in the denominator becomes large compared to the first term in the denominator. This happens if k_2 is large compared to k_{-1} or if $[NO_2]$ is small, as it would be in the initial stages of the reaction.

Tip. Lewis structures for NO_2Cl (nitryl chloride) are given in the answer to problem **3.93**.

Effect of Temperature on Reaction Rates

18.35 **a)** The rate constant of an elementary reaction depends on the absolute temperature and activation energy E_a according to the Arrhenius equation

$$k = A e^{-E_a/RT} \qquad \text{from which} \qquad \ln k = \ln A - \frac{E_a}{RT}$$

This means that a plot of $\ln k$ versus the reciprocal of T should be a straight line with a slope of $-E_a/R$ and an intercept (when $1/T = 0$) of $\ln A$. Two points determine a line. A quick way to estimate E_a is to select any two of the four data points in the problem (such as the first two) and insert the values in the above equation

$$\ln(5.49 \times 10^6) = \ln A - \frac{E_a}{(5\,000\ K)R} \qquad \text{and} \qquad \ln(9.86 \times 10^8) = \ln A - \frac{E_a}{(10\,000\ K)R}$$

Then solve for E_a (by eliminating $\ln A$ between the equations). This gives $E_a = 4.3 \times 10^5$ J mol^{-1}.[1] Selecting *another* pair of points (for example the second two) and doing the same thing gives a somewhat different answer: $E_a = 3.9 \times 10^5$ J mol^{-1}. The discrepancy means that the experimental data do not fall exactly on a straight line. The best way to use all data is to perform a *least-squares fit*, mathematically determining the slope of the straight line that comes closest to all four data points.

---

[1]Note that E_a has the same units as RT.

Many electronic calculators are equipped to complete the necessary calculations almost without effort. Based on the minimization of the sum of the squares of the deviations, E_a is $\boxed{425 \text{ kJ mol}^{-1}}$.

b) As $1/T$ goes to zero, $\ln k$ approaches $\ln A$. Recall that using just the first two data points gave $E_a = 432$ kJ mol^{-1}. Substituting this value and the k and T values of the first data point into the Arrhenius equation gives $\ln A$ equal to 25.9. Using an E_a of 392 kJ mol^{-1} and the third or fourth (k, T) pair makes $\ln A$ equal 25.5. The least-squares fitting gives $\ln A = 25.76$ so that $A = \boxed{1.54 \times 10^{11} \text{ L mol}^{-1}\text{s}^{-1}}$. This is the best answer. The units of A are always the same as the units of k.

18.37 **a)** The reaction involves the tetrahydridoborate ion and the ammonium ion. Calculate $\ln A$ of the reaction using the Arrhenius equation. The T is equal to 303.2 K (30.0°C); the rate constant k and activation energy E_a are also given in the problem

$$\ln A = \ln k + \frac{E_a}{RT} = \ln(1.94 \times 10^{-4}) + \left(\frac{1.61 \times 10^5 \text{ J mol}^{-1}}{(8.3145 \text{ J K}^{-1}\text{mol}^{-1})(303.2 \text{ K})} \right) = 55.31$$

The value of A might be calculated at this point, but it is not needed. Simply put $\ln A$ back into the Arrhenius equation with T equal to 313.2 (40.0°C)

$$\ln k = \ln A - \frac{E_a}{RT} = 55.31 - \left(\frac{1.61 \times 10^5 \text{ J mol}^{-1}}{(8.3145 \text{ J K}^{-1}\text{mol}^{-1})(313.2 \text{ K})} \right) = -6.51$$

Taking the antilogarithm of -6.51 gives k equal to $\boxed{1.49 \times 10^{-3} \text{ L mol}^{-1}\text{s}^{-1}}$. A 10 K increase in temperature (fairly small) increases the rate constant of the reaction nearly eight-fold.

b) This reaction is second order, but the following reasoning applies to reactions of any order. The larger the rate constant, the more rapid is the reaction. Faster reactions require less time to reach any designated point in their progress. Increasing the temperature of this reaction from 30.0 to 40.0°C increases k from 1.94×10^{-4} to 14.9×10^{-4} L mol^{-1}s^{-1}, which is a factor of 7.68. The time to reach the half-way mark in the reaction is therefore reduced by a factor of 7.68. The 50% conversion requires only $\boxed{1.30 \times 10^3 \text{ s}}$ at 40.0°C instead of the 10 000 s it requires at 30.0°C.

18.39 **a)** Solve the Arrhenius equation for A and insert the quantities given in the problem

$$A = \frac{k}{e^{-E_a/RT}} = \frac{0.41 \text{ s}^{-1}}{\exp\left(-1.61 \times 10^5 \text{ J mol}^{-1}/(8.3145 \text{ J K}^{-1}\text{mol}^{-1})(600 \text{ K}) \right)} = \boxed{4.3 \times 10^{13} \text{ s}^{-1}}$$

b) Assume that neither the activation energy E_a nor the Arrhenius A changes with temperature. Solve the Arrhenius equation for k, set T equal to 1000 K, and insert these values

$$k = 4.25 \times 10^{13} \text{ s}^{-1} \exp\left(\frac{-1.61 \times 10^5 \text{ J mol}^{-1}}{(8.3145 \text{ J K}^{-1}\text{mol}^{-1})(1000 \text{ K})} \right) = \boxed{1.7 \times 10^5 \text{ s}^{-1}}$$

18.41 The activation energy is the difference in energy between the initial state and the activated complex. The activated complex is 3.5 kJ mol^{-1} higher in energy than the reactants, which are OH(g) plus HCl(g). The products, H$_2$O(g) plus Cl(g), are themselves 66.8 kJ mol^{-1} lower in energy than the reactants. It follows that the activated complex is 70.3 kJ mol^{-1} higher in energy than the products. To pass from the products to the activated complex requires $\boxed{70.3 \text{ kJ mol}^{-1}}$.

Molecular Theories of Elementary Reactions

18.43 Text equation 18.19 gives the rate constant k_r of a gas-phase reaction in terms of three molecular parameters: the reduced mass μ of the reacting pair of molecules, their reaction cross-section σ_c, and their relative kinetic energy ϵ_B. The other quantities are the Boltzmann constant k_B and the temperature T

$$k_r = \sigma_c \left(\frac{8k_B T}{\pi \mu} \right)^{1/2} \exp(-\epsilon_B/k_B T)$$

For the purposes of this problem the collection of factors in front of the exponential can be treated as a constant even though it include the temperature (see text page 787). This is what is meant by "fitting well to the Arrhenius expression."

The problem provides numbers for both the pre-exponential factor and the exponential factor for the reactions of the hydroxyl radical with ethane and with ethene:

for $C_2H_6 + OH$: $k_r = 1.1 \times 10^{-10}$ cm^3 mol^{-1} s^{-1} $\times \exp(-2537 \text{ K}/T)$

for $C_2H_4 + OH$: $k_r = \quad 6 \times 10^{-12}$ cm^3 mol^{-1} s^{-1} $\times \exp(0 \text{ K}/T)$

The zero in the exponential factor for the ethene reaction means that the whole exponential factor equals 1. It appears because the problem states that the rate of the ethene/hydroxyl reaction is independent of the temperature. Text equation 18.19 indicates that this can happen only in the barrier height for the ethene reaction is zero: $\boxed{\text{for the ethene reaction: } \epsilon_B = 0 \text{ kJ mol}^{-1}}$.

A similar analysis of the information for the reaction of ethane establishes that

$$\epsilon_B = (2537 \text{ K})(1.3806 \times 10^{-23} \text{ J K}^{-1}) = 3.503 \times 10^{-20} \text{ J molec}^{-1}$$

Multiplying by Avogadro's number puts the result on a per-mole basis: $\boxed{\epsilon_B = 21.09 \times \text{ kJ mol}^{-1}}$.

For the second part of the problem, use equation 9.20 (text page 392) to compute the average speeds of OH and C_2H_6 molecules at 1000 K

$$\bar{v}_{C_2H_6} = \sqrt{\frac{8\,(8.314 \text{ J mol}^{-1} \text{ K}^{-1})(1000 \text{ K})}{\pi\,(0.030 \text{ kg mol}^{-1})}} = 840 \text{ m s}^{-1}$$

$$\bar{v}_{OH} = \sqrt{\frac{8\,(8.314 \text{ J mol}^{-1} \text{ K}^{-1})(1000 \text{ K})}{\pi\,(0.017 \text{ kg mol}^{-1})}} = 1116 \text{ m s}^{-1}$$

The problem asks about a situation in which the average velocities of the molecules are double this: 1680 m s^{-1} for C_2H_6 and 2232 m s^{-1} for OH. The relative kinetic energy in a head-on encounter between an OH and C_2H_6 moving at these speeds is

$$\epsilon_{rel} = \frac{1}{2}\mu(v_{rel})^2 = \frac{1}{2}\frac{(0.030)(0.017)}{(0.030 + 0.017)} \text{ kg mol}^{-1} \cdot (1680 + 2232 \text{ m s}^{-1})^2$$

$$= 83.3 \text{ kg m}^2 \text{ s}^{-2} \text{ mol}^{-1} = 83.3 \text{ kJ mol}^{-1}$$

This exceeds 21 kJ mol^{-1}, the barrier energy for the reaction. $\boxed{\text{The } C_2H_6 \text{ reaction can occur.}}$

Tip In a "catch-up" encounter, the vector sum of the average velocities is only 552 m s^{-1}. This makes the relative energy of the encounter only 1.65 kJ mol^{-1}. Such an encounter does not lead to reaction because it does not provide enough energy to overcome the barrier.

18.45 The reaction $NOCl(g) + NOCl(g) \longrightarrow 2\,NO(g) + Cl_2(g)$ is a bimolecular elementary reaction. The Arrhenius equation (text equation 18.5) gives its rate constant k as the product of an exponential factor in the temperature T and a pre-exponential factor A. Write the Arrhenius equation using molecular rather than molar quantities. This entails replacing the gas constant R with the Boltzmann constant k_B and the molar energy of activation E_a with the molecular energy of activation ϵ_a. The outcome is

$$k = A \exp(-\epsilon_a/k_B T)$$

Also write text equation 18.14, which results from the application of collision theory to the calculation of the k's of bimolecular reactions

$$k = \sigma_c \sqrt{\frac{8k_B T}{\pi \mu}} \exp\left(-\epsilon_a / k_B T\right)$$

A term-by-term comparison of these two equations reveals that the pre-exponential factor A depends on the square root of the temperature and on two molecular quantities as well: the collision cross-section σ_c, and the reduced mass μ of the colliding particles

$$A \propto \sigma_c \sqrt{\frac{8k_B T}{\pi \mu}}$$

Collisions of molecules in the wrong orientation are non-productive. Insert a steric factor P (discussed on text page 787) to account for this. The result is

$$A \propto P \sigma_c \sqrt{\frac{8k_B T}{\pi \mu}}$$

The cross-section and reduced mass in the collision of a molecule A with molecule B are defined as

$$\sigma_c = \pi \left(\frac{d_A + d_B}{2}\right)^2 \quad \text{and} \quad \mu = \frac{m_A m_B}{m_A + m_B}$$

The notion that σ_c and μ are joint properties of the colliding molecules makes sense. A collision is a joint event. In this particular reaction, A and B are both nitrosyl chloride, NOCl. Consequently

$$\sigma_c = \pi (d_{NOCl})^2 \quad \text{and} \quad \mu = \frac{m_{NOCl}}{2}$$

Substitute these equations into the expression for A. At the same times insert a factor of $\frac{1}{2}$ to account for the fact that the collision of a pair of NOCl molecules counts as only one collision and not as one collision for the first NOCl molecule hitting the second plus another collision for the second NOCl molecule hitting the first. The result is

$$A = \frac{1}{2} P \pi (d_{NOCl})^2 \sqrt{\frac{8k_B T}{\pi} \frac{2}{m_{NOCl}}}$$

Put in the numerical values and complete the arithmetic

$$A = (0.5)(0.16)\, \pi \, (3.0 \times 10^{-10}\ \text{m})^2 \sqrt{\frac{8(1.38 \times 10^{-23}\ \text{J K}^{-1})(298\ \text{K})}{\pi} \frac{2}{1.0870 \times 10^{-25}\ \text{kg}}}$$

$$= (0.5)(0.16)\,(28.27 \times 10^{-20}\ \text{m}^2)\,(439.0\ \text{m s}^{-1}) = 1.0 \times 10^{-17}\ \text{m}^3\ \text{s}^{-1}$$

This result is on a per molecule basis. Multiply by Avogadro's number to put it on a molar basis

$$A = \frac{1.0 \times 10^{-17}\ \text{m}^3\ \text{s}^{-1}}{\text{molecule}} \left(\frac{6.022 \times 10^{23}\ \text{molecules}}{\text{mol}}\right) = 6.0 \times 10^6\ \text{m}^3\ \text{mol}^{-1}\text{s}^{-1}$$

The cubic meter is an unaccustomed unit for volume in laboratory-scale operations. It equals 1000 L. Converting cubic meters to liters gives $\boxed{k = 6.0 \times 10^9\ \text{L mol}^{-1}\text{s}^{-1}}$.

Reactions in Solution

18.47   The rate constant for the reaction

$$CH_3Br + Cl^- \longrightarrow CH_3Cl + Br^- \quad \text{in acetone}$$

is 5.9×10^{-3} L mol^{-1}s^{-1}. Diffusion-controlled reactions have their rate constants limited only by the frequency at which reacting species encounter each other ("to collide is to react"). Even at the freezing point of acetone ($-94.7°$ C), collisions would be much more frequent than this k suggests. Therefore successful reaction must require $\boxed{\text{considerable activation energy}}$.

Catalysis

18.49   **a)** In the two-step reaction

$$E + S \rightleftharpoons ES \qquad \text{followed by} \qquad ES \longrightarrow E + \text{product}$$

S is penicillin, the substrate, and E is penicillinase,[2] the enzyme that accelerates its destruction. Write the Michaelis-Menten equation (text equation 18.20) and insert the constants K_m and k_2 that the problem furnishes

$$\text{rate} = \frac{k_2[E]_0[S]}{[S] + K_m} = \frac{(2 \times 10^3 \text{ s}^{-1})(6 \times 10^{-7} \text{ mol L}^{-1})[S]}{[S] + 5 \times 10^{-5} \text{ mol L}^{-1}}$$

The rate of the reaction increases as $[S]$, the concentration of penicillin, increases. The maximum rate will be reached when $[S]$ is large compared to 5×10^{-5} mol L^{-1}. At this point, the denominator of the fraction essentially equals $[S]$ and cancels out the $[S]$ in the numerator

$$(\text{rate})_{\max} = \frac{(2 \times 10^3 \text{ s}^{-1})(6 \times 10^{-7} \text{ mol L}^{-1})}{1} = \boxed{1 \times 10^{-3} \text{ mol L}^{-1}\text{s}^{-1}}$$

b) Rewrite the Michaelis-Menten equation inserting the desired rate (which equals half the maximum rate), the given concentration of enzyme, and the two constants

$$\text{rate} = \frac{(\text{rate})_{\max}}{2} = \frac{1.2 \times 10^{-3} \text{ mol L}^{-1}\text{s}^{-1}}{2} = \frac{(2 \times 10^3 \text{ s}^{-1})(6 \times 10^{-7} \text{ mol L}^{-1})[S]}{[S] + 5 \times 10^{-5} \text{ mol L}^{-1}}$$

Solving for $[S]$ gives the concentration of penicillin required: $\boxed{5 \times 10^{-5} \text{ mol L}^{-1}}$.

ADDITIONAL PROBLEMS

18.51   The rate expression for the process is rate $= k$ [hemoglobin][O_2]. Substitute the given data

$$\text{rate} = (4 \times 10^7 \text{ L mol}^{-1}\text{s}^{-1})(2 \times 10^{-9} \text{ mol L}^{-1})(5 \times 10^{-5} \text{ mol L}^{-1}) = \boxed{4 \times 10^{-6} \text{ mol L}^{-1}\text{s}^{-1}}$$

Tip. Review the discussion of the equilibria involved in the update and release of O_2 by hemoglobin in text Section 14.6.

18.53   **a)** For this first-order process, $\ln[\text{In}^+] = -kt + \ln[\text{In}^+]_0$. Prepare a table of the natural logarithm of the concentration of In$^+$ versus time and plot the data:

t / s	0	240	480	720	1000	1200	10000
$\ln[\text{In}^+]$	-4.80	-5.05	-5.30	-5.55	-5.80	-5.80	-5.80

---

[2]Names ending in "-ase" signify enzymes.

A common error is to make the values on the y axis go up the axis as they become more negative. They should go down the axis as shown above. Then the plot consists of a straight line that slopes from northwest to southeast and levels off beyond 1000 s. The initial slope of the line is $-k$ and the intercept is $\ln[\,\text{In}^+]_0$. The slope can be read off of the graph using the gridlines, but a least-squares analysis gives a slightly better answer. From a least-square fit, the slope is -1.01×10^{-3} s^{-1}. The rate constant is therefore $\boxed{1.01 \times 10^{-3} \text{ s}^{-1}}$.

b) The half-life can be determined from the rate constant,

$$t_{1/2} = \frac{\ln 2}{k} = \frac{0.6931}{1.01 \times 10^{-3} \text{ s}^{-1}} = \boxed{686 \text{ s}}$$

c) At 10000 seconds, the concentration of In$^+$ has persisted unchanged for 9000 seconds. The reaction has clearly reached equilibrium. The concentration of In$^+$ is 3.03×10^{-3} M, and the amount of In$^+$ in the 1.00 L solution is accordingly 3.03×10^{-3} mol. The equilibrium concentration of In^{3+} is calculated by subtracting this amount from the original amount of In$^+$, and using the stoichiometric relationship between the amounts of In$^+$ and In^{3+} as follows

$$n_{\text{In}^{3+}} = \frac{8.23 \times 10^{-3} - 3.03 \times 10^{-3}}{3} = 1.73 \times 10^{-3} \text{ mol}$$

The equilibrium concentration of In^{3+} therefore equals 1.73×10^{-3} M. Substitute into the usual mass-action expression

$$K = \frac{[\text{In}^{3+}]}{[\text{In}^+]^3} = \frac{1.73 \times 10^{-3}}{(3.03 \times 10^{-3})^3} = \boxed{6.22 \times 10^4}$$

18.55 The reaction is $\text{OH}^-(aq) + \text{HCN}(aq) \longrightarrow \text{H}_2\text{O}(l) + \text{CN}^-(aq)$. This reaction is first-order in both OH$^-$ and HCN and second-order overall

$$\text{rate (forward)} = k_\text{f}[\text{OH}^-][\text{HCN}]$$

At equilibrium the forward rate is exactly balanced by the reverse rate, which, because the concentration of water is essentially constant, depends solely on the concentration of CN$^-$

$$\text{rate (reverse)} = k_\text{r}[\text{CN}^-]$$

It follows that

$$k_f[OH^-][HCN] = k_r[CN^-] \quad \text{from which:} \quad \frac{k_r}{k_f} = \frac{[OH^-][HCN]}{[CN^-]}$$

Note that the units of the quantity on the right-hand side of the second equation are mol L^{-1}. Numerically, the right-hand side of the preceding equation equals the equilibrium constant K_b of the reaction

$$CN^-(aq) + H_2O(l) \rightleftharpoons HCN(aq) + OH^-(aq)$$

The K_b of CN^- ion is related to K_a of its conjugate acid HCN[3]

$$K_b = \frac{K_w}{K_a} = \frac{1.0 \times 10^{-14}}{6.17 \times 10^{-10}} = 1.62 \times 10^{-5}$$

Therefore

$$\frac{k_r}{k_f} = 1.62 \times 10^{-5} \text{ mol } L^{-1}$$

Substitution of 3.7×10^{-9} L mol^{-1}s^{-1} for k_f gives $\boxed{k_r = 6.0 \times 10^{-14} \text{ s}^{-1}}$.

18.57   Both reactions are third-order, but [M], the concentration of M, can be treated as a constant inasmuch as it is much larger than the original concentrations of the I and Br and is only very slightly changed by the progress of the reaction. The integrated rate laws for the two reactions are then

$$\frac{1}{[I]} - \frac{1}{[I]_0} = 2k_I[M]t \quad \text{and} \quad \frac{1}{[Br]} - \frac{1}{[Br]_0} = 2k_{Br}[M]t$$

After one half-life, $[I] = 1/2\,[I]_0$ and $[Br] = 1/2\,[Br]_0$ so:

$$\frac{1}{[I]_0} = 2k_I[M]t_{1/2,\,I} \quad \text{and} \quad \frac{1}{[Br]_0} = 2k_{Br}[M]t_{1/2,\,Br}$$

Dividing the first equation by the second gives

$$\frac{[Br]_0}{[I]_0} = \left(\frac{k_I}{k_{Br}}\right)\left(\frac{t_{1/2,\,I}}{t_{1/2,\,Br}}\right)$$

Substitute $[I]_0 = 2\,[Br]_0$ and $k_I = 3.0\,k_{Br}$

$$\frac{1}{2} = 3.0\left(\frac{t_{1/2,\,I}}{t_{1/2,\,Br}}\right) \quad \text{from which} \quad \left(\frac{t_{1/2,\,I}}{t_{1/2,\,Br}}\right) = \frac{1}{6.0} = \boxed{0.17}$$

18.59   Both reactions involves the combination of hydrogen with a halogen to form a hydrohalic acid. They must proceed by different mechanisms because their experimental rate laws are different

$$\text{rate (bromine)} = k[H_2][Br_2]^{1/2} \quad \text{rate (iodine)} = k[H_2][I_2]$$

The currently accepted mechanisms for both reactions involve a fast equilibrium to split the halogen molecule into its two atoms followed by reaction of one atom with H_2. In the case of Br, this elementary process is rate-limiting, but in the case of I, it is fast and *not* rate-limiting. The third steps in the mechanisms differ

$$H + Br_2 \longrightarrow HBr + Br \quad \text{fast} \qquad H_2I + I \longrightarrow 2\,HI \quad \text{rate-limiting}$$

Also, in the iodine reaction, the intermediate H_2I is thought to exist, whereas the analogous H_2Br does not appear in the proposed mechanism for the bromine reaction.

---

[3]Taken from text Table 15.2.

18.61 The reaction is the decomposition of ozone by light: $2\,O_3 + \text{light} \longrightarrow 3\,O_2$. The mechanism involves the production of the intermediate O atom from O_3 (in the first step), and its consumption either to regenerate O_3 (the second step) or to make $2\,O_2$ in an encounter with an O_3 molecule (the third step). The change in the concentration of O with time is

$$\frac{d[O]}{dt} = k_1[O_3] - k_2[O][O_2][M] - k_3[O][O_3]$$

This equation states that the rate of change of the concentration of O equals its rate of production minus its rate of consumption. The steady-state approximation is that the concentration of O [O] comes to a *steady* value. If [O] is steady it is unchanging, and $d[O]/dt = 0$. Then

$$k_1[O_3] - k_2[O][O_2][M] - k_3[O][O_3] = 0$$

so that

$$[O] = \frac{k_1[O_3]}{k_2[O_2][M] + k_3[O_3]}$$

All of this concerns the intermediate. The rate of the overall reaction is

$$\text{rate} = \frac{1}{3}\frac{d[O_2]}{dt} = k_3[O][O_3]$$

Insert the expression for the steady-state concentration of intermediate O into the preceding equation

$$\boxed{\text{rate} = \frac{k_3 k_1[O_3]^2}{k_2[O_2][M] + k_3[O_3]}}$$

Divide the top and bottom of the fraction by k_3

$$\text{rate} = \frac{k_1[O_3]^2}{(k_2/k_3)\,[O_2][M] + [O_3]}$$

This form of the rate equation shows that only the ratio k_2/k_3 affects the rate, not the actual values of k_2 and k_3. This ratio tells how much of the intermediate O cycles back to O_3 relative to how much goes on to give the product.

18.63 The chemical reaction is $A + B + C \longrightarrow D + E$. The species F is an intermediate. In the mechanism, F forms rapidly from A and B and then interacts more slowly with C. It interacts in two different ways, forming D in the first way and forming E in the second.

a) Let K_1 equal the equilibrium constant for the first step.

$$K_1 = \frac{k_1}{k_{-1}} = \frac{[F]}{[A][B]}$$

From the second and third steps in the mechanism, the rates of formation of products D and E are

$$\frac{d[D]}{dt} = k_2[C][F] = k_2 K_1[C][A][B]$$

$$\frac{d[E]}{dt} = k_3[C][F] = k_3 K_1[C][A][B]$$

The rate of formation of D is first-order in the concentrations of A, B, and C; the rate of formation of E is *also* first-order in the concentrations of A, B, and C.

b) Product D forms 10 times faster than product E because a 10 times larger amount of it has formed at the end of the reaction. Therefore $\boxed{k_2 \approx 10\,k_3}$.

c) The fact that a trace of acid greatly alters the relative amounts of D and E suggests that the acid is a catalyst. Quite possibly, E is thermodynamically favored in comparison to D, and the acid provides a rapid route for the reaction $D \longrightarrow E$.

Tip. Reactions that give products other than the thermodynamically most stable products are said to be under kinetic control. The favored product does not form because there is not time enough for it to form. Changing the conditions (or, in principle, waiting long enough) allows expression of the underlying tendency to give the thermodynamically favored product. Waiting is often not a realistic option. It may be that no perceptible progress toward the favored product occurs in a lifetime.

18.65   The reaction of interest is an equilibrium

$$NH_3(aq) + H_2O(l) \rightleftharpoons NH_4^+(aq) + OH^-(aq) \qquad K_b = \frac{k_1}{k_{-1}}$$

The problem gives k_1 and K_b. The latter can also be computed from K_a for NH_4^+ in text Table 15.2. Therefore

$$k_{-1} = \frac{k_1}{K_b} = \frac{2 \times 10^5}{1.8 \times 10^{-5}} = \boxed{1 \times 10^{10} \text{ L mol}^{-1}\text{s}^{-1}}$$

18.67   The initiation step is $CH_3CHO \longrightarrow CH_3 + CHO$. The propagation steps involve the intermediates CH_3 (the methyl radical) and CH_2CHO (the acetyl radical). The propagation step is the combination of the second and third elementary reactions in the problem. The CH_3 is consumed in the second reaction and regenerated in the third, which forms the product CO. This chain continues until all available CH_3CHO is consumed unless interrupted by the termination reaction, which forms the by-product CH_3CH_3 when two methyl radicals encounter each other and react.

18.69   **a)** Assume that the rate constants of the reactions that occur in cooking vary with the temperature according to the Arrhenius equation. Since the food cooks two times faster at 112°C (385 K) than at 100°C (373 K), the rate constant must be twice as large at 385 K: $k_{385} = 2\,k_{373}$. Write an Arrhenius equation for each of the two temperatures, and then eliminate the k's between the two equations to obtain

$$Ae^{-E_a/385R} = 2\,Ae^{-E_a/373R}$$

Cancelling out the A's and taking the natural logarithm of both sides gives

$$\frac{-E_a}{385 \text{ K}} - \frac{-E_a}{373 \text{ K}} = R\ln 2 = (8.3145 \text{ J K}^{-1}\text{mol}^{-1})(0.6931)$$

Solving for E_a gives $\boxed{69.0 \text{ kJ mol}^{-1}}$.

b) The rates of the cooking reactions at 94.4°C (367.6 K) depend on the rate constant at that temperature. The following equation involving the rate constant for cooking at 367.6 K comes from dividing the Arrhenius equation at 367.6 K by the Arrhenius equation at 373 K

$$\frac{k_{367.6}}{k_{373}} = \frac{e^{-E_a/367.6R}}{e^{-E_a/373R}}$$

Taking the natural logarithm of both sides gives

$$\ln\left(\frac{k_{367.6}}{k_{373}}\right) = \frac{-E_a}{367.6R} - \frac{-E_a}{373R} = \frac{-69.0 \times 10^3 \text{ J mol}^{-1}}{8.3145 \text{ J K}^{-1}\text{mol}^{-1}}\left(\frac{1}{367.6} - \frac{1}{373}\right) = -0.327$$

where the value of E_a has been taken from part a). Take the antilogarithm of both sides

$$\frac{k_{367.6}}{k_{373}} = e^{-0.327} = 0.721$$

The rate constant at 94.4°C is smaller than the rate constant at 100°C by the factor 0.721. Consequently the food takes $1/0.732 = 1.387$ times longer to cook. Instead of 10 minutes, the cooking requires $\boxed{14 \text{ minutes}}$.

18.71   The ratio of the rate constants of the forward and reverse reactions in the equilibrium equals the equilibrium constant

$$H_2(g) + I_2(g) \rightleftharpoons 2\,HI(g) \qquad K = \frac{k_f}{k_r}$$

Write Arrhenius equations for both the forward reaction and the reverse reaction

$$k_f = Ae^{-E_{a,f}/RT} \qquad \text{and} \qquad k_r = Ae^{-E_{a,r}/RT}$$

Divide the first equation by the second and take the logarithm of both sides

$$\ln\left(\frac{k_f}{k_r}\right) = \frac{-E_{a,f}}{RT} - \frac{-E_{a,r}}{RT} = \frac{1}{RT}(E_{a,r} - E_{a,f})$$

The left side of this equation equals $\ln K$, the logarithm of the equilibrium constant. But $\ln K$ of the reaction at the temperature T is related to the standard free energy at that temperature by the thermodynamic equation $\Delta G_T^\circ = -RT\ln K$. Substitution gives

$$-\frac{\Delta G_T^\circ}{RT} = \frac{1}{RT}(E_{a,r} - E_{a,f})$$

Multiply both sides by $-RT$

$$\Delta G_T^\circ = -(E_{a,r} - E_{a,f}) = +(E_{a,f} - E_{a,r})$$

Assume that ΔH° and ΔS° do not change very much going from 298 to 1000 K.[4] Then the standard free energy of this reaction at 1000 K is

$$\Delta G_{1000}^\circ = \Delta H_{298}^\circ - 1000\,\Delta S_{298}^\circ$$

Use the data in Appendix D to compute ΔH_{298}° and ΔS_{298}°

$$\Delta H_{298}^\circ = 2\underbrace{(26.48)}_{HI(g)} - 1\underbrace{(62.44)}_{I_2(g)} - 1\underbrace{(0.00)}_{H_2(g)} = -9.48\ \text{kJ}$$

$$\Delta S_{298}^\circ = 2\underbrace{(206.48)}_{HI(g)} - 1\underbrace{(260.58)}_{I_2(g)} - 1\underbrace{(130.57)}_{H_2(g)} = 21.81\ \text{J K}^{-1}$$

$$\Delta G_{1000}^\circ = -9.48 \times 10^3\ \text{J} - (1000\ \text{K})(21.81\ \text{J K}^{-1}) = -31.29 \times 10^3\ \text{J}$$

Now, two of the three quantities in the relationship $\Delta G_{1000}^\circ = (E_{a,f} - E_{a,r})$ are known. Substitute them

$$-31.29\ \text{kJ mol}^{-1} = (165\ \text{kJ mol}^{-1} - E_{a,r})$$

Solving gives $\boxed{196\ \text{kJ mol}^{-1}}$ as the activation energy of the reverse reaction.

A quick way to obtain k_r is to calculate the equilibrium constant K_{1000} and use the known k_f in the equation $K = k_f/k_r$. Thus

$$\ln K_{1000} = \frac{-\Delta G_{1000}^\circ}{RT} = \frac{-(-31.29 \times 10^3\ \text{J mol}^{-1})}{8.3145\ \text{J K}^{-1}\text{mol}^{-1}(1000\ \text{K})} = 3.763 \quad \text{so that} \quad K = 43.08$$

$$k_r = \frac{k_f}{K} = \frac{240\ \text{L mol}^{-1}\text{s}^{-1}}{43.08} = \boxed{5.6\ \text{L mol}^{-1}\text{s}^{-1}}$$

---

[4]This is a common assumption.

18.73 The CF_2Cl_2 enters into the chemical reaction and presumably speeds it up, but is not itself consumed in the reaction. It does not appear in the balanced overall equation representing the reaction because it is consumed in one step of the mechanism but regenerated later. It is a $\boxed{\text{catalyst}}$.

18.75 The mechanism of enzyme catalysis, as modified to allow for the action of an inhibitor, is

$$E + S \rightleftharpoons ES \qquad \text{fast equilibrium} \qquad k_1 \text{ and } k_{-1}$$
$$ES \longrightarrow E + P \qquad \text{slow} \qquad k_2$$
$$E + I \rightleftharpoons EI \qquad \text{fast equilibrium} \qquad k_3 \text{ and } k_{-3}$$

This is a case of competitive inhibition. The inhibitor (symbolized I) competes with the substrate S in binding to the enzyme E. The generation of product P is thereby slowed because P forms only through the complex ES. Follow the pattern of the derivation in text Section 18.8, but allow for the complication of the inhibitor. Label the total concentration of enzyme $[E]_0$. The enzyme is present in one of three states: free, bound to the inhibitor, or bound to the substrate

$$[E]_0 = [E] + [EI] + [ES]$$

Write the equilibrium expression for the third step of the mechanism, letting k_3/k_{-3} equal K_3

$$K_3 = \frac{[EI]}{[E][I]}$$

Solve this expression for $[EI]$, substitute into the first equation, and then solve for $[E]$

$$[E]_0 = [E] + K_3[E][I] + [ES]$$
$$[E]_0 = [E]\left(1 + K_3[I]\right) + [ES]$$
$$[E] = \frac{[E]_0 - [ES]}{1 + K_3[I]}$$

Now, make the steady-state approximation for ES, the intermediate. This approximation is that the ES is generated as fast as it is consumed—its concentration does not change with time. The first step in the mechanism forms it; the reverse of the first step removes it; the second step also removes it

$$0 = \frac{d[ES]}{dt} = k_1[E][S] - k_{-1}[ES] - k_2[ES]$$

Solve for $[ES]$, insert the previous expression for $[E]$, and simplify. The result is

$$[ES] = \frac{k_1[E]_0[S]}{k_1[S] + \left(1 + K_3[I]\right)(k_{-1} + k_2)}$$

The rate of the reaction equals the rate of the slow step, which equals $k_2[ES]$. Therefore

$$\boxed{\text{rate} = k_2[ES] = \frac{k_2 k_1[E]_0[S]}{k_1[S] + \left(1 + K_3[I]\right)(k_{-1} + k_2)} = \frac{k_2[E]_0[S]}{[S] + K_m(1 + K_3[I])}}$$

where K_m is defined as $(k_{-1} + k_2)/k_1$. This K_m is the Michaelis-Menten constant. Any concentration of the inhibitor increases the denominator of this expression and lowers the initial rate of the reaction.

Tip. Imagine that the inhibitor I is left out. This corresponds to letting $[I] = 0$ in the preceding, which then becomes

$$\text{rate} = \frac{k_2[E]_0[S]}{[S] + K_m}$$

which is the Michaelis-Menten equation (text Equation 18.20). This does not prove that the answer is correct, but is nevertheless very reassuring.

CUMULATIVE PROBLEMS

18.77   The reaction is the gas-phase combination of H_2 and I_2 to give HI. The $\Delta H°$ of this change equals -9.48 kJ when two moles of HI are produced. See problem **18.71**. Obtain the initial rate of the reaction by substituting the rate constant and the initial concentrations of H_2 and I_2 into the rate equation

$$\text{rate} = k[H_2][I_2] = 0.0242 \text{ L mol}^{-1} \text{ s}^{-1}[0.081 \text{ mol L}^{-1}][0.036 \text{ mol L}^{-1}] = 7.06 \times 10^{-5} \text{ mol L}^{-1} \text{ s}^{-1}$$

The rate of formation of HI is twice this rate (because HI has a coefficient of 2 in the balanced reaction). The rate of absorption of heat equals the rate at which HI forms times the heat absorbed per mole of HI

$$\text{rate} = \frac{2(7.06 \times 10^{-5} \text{ mol HI})}{\text{L s}} \times \left(\frac{-9.48 \text{ kJ}}{2 \text{ mol HI}} \right) = \boxed{-6.7 \times 10^{-4} \text{ kJ L}^{-1}\text{s}^{-1}}$$

18.79   Substitution of the initial concentrations of NO and O_3 into the second-order rate law gives the desired initial rate of the reaction. Use the ideal-gas law to obtain these concentrations from the mole fractions of the two gases

$$\left(\frac{n}{V} \right)_{NO} = \frac{P_{NO}}{RT} = \frac{0.00057(3.26 \text{ atm})}{(0.08206 \text{ L atm mol}^{-1}\text{K}^{-1})(500 \text{ K})} = 4.53 \times 10^{-5} \text{ mol L}^{-1}$$

$$\left(\frac{n}{V} \right)_{O_3} = \frac{P_{O_3}}{RT} = \frac{0.00026(3.26 \text{ atm})}{(0.08206 \text{ L atm mol}^{-1}\text{K}^{-1})(500 \text{ K})} = 2.07 \times 10^{-5} \text{ mol L}^{-1}$$

Then proceed with the substitution

$$\text{rate} = k[NO][O_3] = (7.6 \times 10^7 \text{ L mol}^{-1}\text{s}^{-1})(4.53 \times 10^{-5} \text{ mol L}^{-1})(2.07 \times 10^{-5} \text{ mol L}^{-1})$$

$$= \boxed{0.071 \text{ mol L}^{-1}\text{s}^{-1}}$$

Chapter 19

Nuclear Chemistry

Mass-Energy Relationships in Nuclei

19.1 Use these tests for balance in nuclear equations: the sums of the left superscripts on the two sides of the equation must be equal; the sums of the left subscripts on the two sides of the equation must be equal.

a) $2\,{}^{12}_{6}\mathrm{C} \longrightarrow {}^{23}_{12}\mathrm{Mg} + {}^{1}_{0}n$ **b)** ${}^{15}_{7}\mathrm{N} + {}^{1}_{1}\mathrm{H} \longrightarrow {}^{12}_{6}\mathrm{C} + {}^{4}_{2}\mathrm{He}$ **c)** $2\,{}^{3}_{2}\mathrm{He} \longrightarrow {}^{4}_{2}\mathrm{He} + 2\,{}^{1}_{1}\mathrm{H}$

19.3 The binding energy of a nuclide depends on the change in mass that occurs during formation of the nuclide from its component particles according to the Einstein equation: $E_{\mathrm{B}} = -c^2 \Delta m$.

a) The nuclear equation is $20\,{}^{1}_{1}\mathrm{H} + 20\,{}^{1}_{0}n \longrightarrow {}^{40}_{20}\mathrm{Ca}$.

$$\Delta m = m[{}^{40}_{20}\mathrm{Ca}] - 20\,m[{}^{1}_{1}\mathrm{H}] - 20\,m[{}^{1}_{0}n]$$
$$= 39.9625912 - 20(1.007825032) - 20(1.0086649158) = -0.3672078 \text{ u}$$
$$E_{\mathrm{B}} = -c^2 \Delta m$$
$$= -(2.9979246 \times 10^8 \text{ m s}^{-1})^2(-0.3672078 \text{ u})\left(\frac{1.660540 \times 10^{-27} \text{ kg}}{1 \text{ u}}\right)$$
$$= 5.480279 \times 10^{-11} \text{ J}$$

E_{B} for a mole of atoms equals E_{B} for one atom multiplied by Avogadro's number

$$E_{\mathrm{B}} = (5.480279 \times 10^{-11} \text{ J})(6.022142 \times 10^{23} \text{ mol}^{-1}) = \boxed{3.300301 \times 10^{10} \text{ kJ mol}^{-1}}$$

The energy equivalent of 1 u is 931.494 MeV. Hence

$$E_{\mathrm{B}} = -(-0.3672078 \text{ u})\left(\frac{931.494 \text{ MeV}}{1 \text{ u}}\right) = \boxed{342.052 \text{ MeV atom}^{-1}}$$

${}^{40}_{20}\mathrm{Ca}$ has 40 nucleons so its E_{B} per nucleon is $342.052 \text{ MeV}/40 = \boxed{8.55130 \text{ MeV nucleon}^{-1}}$.

b) The nuclear equation is $50\,{}^{1}_{0}n + 37\,{}^{1}_{1}\mathrm{H} \longrightarrow {}^{87}_{37}\mathrm{Rb}$. The calculations proceed exactly as in part **a)**. The Δm for the change is -0.813589 u; the binding energy per atom is $\boxed{757.853 \text{ MeV atom}^{-1}}$; the binding energy per mole is $\boxed{7.31218 \times 10^{10} \text{ kJ mol}^{-1}}$; the binding energy per nucleon is E_{B} divided by 87 nucleons or $\boxed{8.7110 \text{ MeV nucleon}^{-1}}$.

c) Uranium-238 has 92 protons and 146 neutrons. Compute Δm for the making of the atom by adding up the mass of 92 hydrogen atoms (not protons) and 146 neutrons and subtracting the result from the mass of the U-238 atom. The answer is -1.934198 u. Continue the approach of part **a)**.

272

E_B is $\boxed{1801.69 \text{ MeV atom}^{-1}}$ or $\boxed{17.3837 \times 10^{10} \text{ kJ mol}^{-1}}$. The binding energy per nucleon is the total binding energy divided by 238 nucleons or $\boxed{7.57013 \text{ MeV nucleon}^{-1}}$.

Tip. To save effort in calculations observe that according to the Einstein equation

$$\Delta m = 1 \text{ u per atom} \quad \text{implies} \quad \Delta E = 8.98755179 \times 10^{10} \text{ kJ mol}^{-1}$$

19.5   According to text Table 19.1, the mass of a single ^{8}Be atom equals 8.00530509 u whereas the mass of two ^{4}He atoms equals twice 4.0026033 u, which is 8.0052066 u. Because the particles have a larger mass when organized as a ^{8}Be atom, the ^{8}Be atom is less stable than the pair of ^{4}He atoms. In other words, the Δm for the nuclear reaction: $^8_4\text{Be} \longrightarrow 2\,^4_2\text{He}$ is negative. The Δm is $\boxed{-9.85 \times 10^{-5} \text{ u}}$.

Nuclear Decay Processes

19.7   Represent the decay as

$$^8_5\text{B} \longrightarrow\, ^8_4\text{Be} + ^0_1e^+ + \nu$$

The difference in mass between the two sides of the reaction appears from this equation to be

$$\Delta m = \underbrace{8.00530509}_{^8\text{Be}} + \underbrace{0.00054858}_{^0_1e^+} - \underbrace{8.024607}_{^8\text{B}} = -0.018753 \text{ u} \quad ?$$

using the numbers found in text Table 19.1 for the boron atom, beryllium atom, and positron. But is it? The tabulated masses include mass of the electrons that surround the nuclei of the atoms. Beryllium has only four electrons, but boron has five. The computation at this stage has wrongly allowed one electron to vanish. This is put right by including another electron mass with the mass of the products. The true change in mass is

$$\Delta m = 8.00530509 + 2(0.00054858) - 8.024607 = -0.018205 \text{ u}$$

This Δm is equivalent to a ΔE of -16.9576 MeV. Therefore $\boxed{16.958 \text{ MeV}}$ is released by the reaction of 1 atom of ^{8}B.

Tip. The difficulty here is avoided by taking care to balance the original equation as to charge. The positive charge on the positron cannot appear without compensation. Therefore Be must be a negative ion

$$^8_5\text{B} \longrightarrow\, ^8_4\text{Be}^- + ^0_1e^+ + \nu$$

19.9   Positron emission (loss of $^0_1e^+$) and electron capture by a nucleus always lower the atomic number by one; electron emission (loss of $^{\ 0}_{-1}e^-$) raises the atomic number by one.

a) $^{39}_{17}\text{Cl} \longrightarrow\, ^{39}_{18}\text{Ar} + ^{\ 0}_{-1}e^- + \bar\nu$ b) $^{22}_{11}\text{Na} \longrightarrow\, ^{22}_{10}\text{Ne} + ^{\ 0}_{-1}e^+ + \nu$

c) $^{224}_{88}\text{Ra} \longrightarrow\, ^{220}_{86}\text{Rn} + ^4_2\text{He}$ d) $^{82}_{38}\text{Sr} + ^{\ 0}_{-1}e^- \longrightarrow\, ^{82}_{37}\text{Rb} + \nu$

19.11   The ^{19}Ne nucleus is neutron deficient and decays by converting a proton to a neutron. The ^{23}Ne nucleus is too rich in neutrons and decays by converting a neutron to a proton. Thus, the decay of ^{19}Ne proceeds by $\boxed{\text{positron emission}}$ to give ^{19}F, and the decay of ^{23}Ne proceeds by $\boxed{\text{electron emission (beta decay)}}$ to give ^{23}Na.

19.13   The electrically neutral neutron decays to a positive proton, a negative electron ($^{\ 0}_{-1}e^-$), and an antineutrino. Determine the change in mass and find the equivalent change in energy

$$\Delta E = (1.0072764669 + 0.00054857991 - 1.0086649158) \text{ u} \left(\frac{931.494 \text{ MeV}}{1 \text{ u}} \right) = -0.782333 \text{ MeV}$$

The electron has a maximum kinetic energy of $\boxed{0.7823 \text{ MeV}}$.

Tip. This is a maximum because the antineutrino and proton carry off some kinetic energy in most decay events.

19.15 $^{30}_{14}\text{Si} + ^1_0n \longrightarrow ^{31}_{14}\text{Si}$ $^{31}_{14}\text{Si} \longrightarrow ^{31}_{15}\text{P} + ^0_{-1}e^- + \bar{\nu}.$

19.17 $^{210}_{84}\text{Po} \longrightarrow ^{206}_{82}\text{Pb} + ^4_2\text{He}$ $^4_2\text{He} + ^9_4\text{Be} \longrightarrow ^{12}_6\text{C} + ^1_0n$

Kinetics of Radioactive Decay

19.19   Compute how many atoms are present in 0.0010 g of ^{209}Po

$$N_{\text{Po}} = 0.0010 \text{ g} \times \left(\frac{1 \text{ mol Po}}{209 \text{ g Po}}\right)\left(\frac{6.02214 \times 10^{23} \text{ atom Po}}{1 \text{ mol Po}}\right) = 2.88 \times 10^{18} \text{ atom } ^{209}\text{Po}$$

The activity A_{Po}, which equals the instantaneous rate of disintegration in the polonium, depends on the number of atoms present

$$A_{\text{Po}} = -\frac{dN}{dt} = kN_{\text{Po}}$$

The k in this equation is a first-order rate constant. It equals $\ln 2/t_{1/2}$ where $t_{1/2}$ is the half-life of the polonium isotope. Substituting gives

$$A_{\text{Po}} = \left(\frac{\ln 2}{t_{1/2}}\right) N_{\text{Po}} = \left(\frac{0.6931}{103 \text{ yr}}\right) 2.88 \times 10^{18} = 1.94 \times 10^{16} \text{ yr}^{-1}$$

Convert to a per-minute basis

$$A_{\text{Po}} = 1.94 \times 10^{16} \text{ yr}^{-1} \times \left(\frac{1 \text{ yr}}{365.2 \text{ d}}\right)\left(\frac{1 \text{ d}}{1440 \text{ min}}\right) = 3.7 \times 10^{10} \text{ min}^{-1}$$

That is, $\boxed{3.7 \times 10^{10} \text{ atoms of } ^{209}\text{Po}}$ decay per minute.

19.21   **a)** At any moment the sample has activity (rate of decay) $A_{^{19}\text{O}}$. This activity is directly proportional to $N_{^{19}\text{O}}$, the number of atoms of ^{19}O that is present

$$A_{^{19}\text{O}} = kN_{^{19}\text{O}}$$

The rate constant k is ($\ln 2/29$ s), which equals 0.0239 s^{-1}. Then

$$N_{^{19}\text{O}} = \frac{A_{^{19}\text{O}}}{k} = \frac{2.5 \times 10^4 \text{ s}^{-1}}{0.0239 \text{ s}^{-1}} = 1.046 \times 10^6 \text{ atoms} = \boxed{1.0 \times 10^6 \text{ atoms } ^{19}\text{O}}$$

b) The answer to the previous part is the number of atoms of the radionuclide present when the sample is fresh. Call it N_i. The number of atoms that remains at any time t after this is

$$N = N_i e^{-kt}$$

In this case t is 2.00 min or 120 s. Substitution of k and N_i gives

$$N = (1.046 \times 10^6) \exp\left(-(0.0239 \text{ s}^{-1})(120 \text{ s})\right) = \boxed{5.9 \times 10^4 \text{ atoms}}$$

19.23   First calculate the number of atoms in 44 mg of ^{219}At

$$N = 44 \times 10^{-3} \text{ g} \times \left(\frac{1 \text{ mol}}{219.01 \text{ g}}\right)\left(\frac{6.022 \times 10^{23} \text{ atoms}}{1 \text{ mol}}\right) = 1.21 \times 10^{20} \text{ atoms}$$

Then compute the activity of this amount of astatine using the given half-life

$$A = kN = \left(\frac{\ln 2}{t_{1/2}}\right) N = \left(\frac{0.6931}{54 \text{ s}}\right)(1.21 \times 10^{20}) = \boxed{1.6 \times 10^{18} \text{ s}^{-1}}$$

19.25 The activity A of any sample of radionuclide (including ^{14}C) decays exponentially

$$A = A_i e^{-kt} \qquad \text{which gives} \qquad -kt = \ln\left(\frac{A}{A_i}\right)$$

For the papyrus, A is 0.153 Bq g^{-1} and A_i is 0.255 Bq g^{-1}. Also, k is $\ln 2/t_{1/2}$ or 1.21×10^{-4} yr^{-1}. Substitution gives

$$-(1.21 \times 10^{-4} \text{ yr}^{-1})\, t = \ln\left(\frac{0.153 \text{ Bq g}^{-1}}{0.255 \text{ Bq g}^{-1}}\right) \qquad \text{so that} \qquad t = \boxed{4.2 \times 10^3 \text{ yr}}$$

19.27 Use the ideal-gas law to compute the number of moles of He per gram of rock

$$n_{He} = \frac{PV}{RT} = \frac{(1 \text{ atm})(9.0 \times 10^{-8} \text{ L})}{(0.08206 \text{ L atm mol}^{-1}\text{K}^{-1})(273.15 \text{ K})} = 4.015 \times 10^{-9} \text{ mol}$$

For every atom of U-238 that decays, eight atoms of He are produced. The number of atoms of U-238 that has decayed since creation of the gram of rock is therefore

$$n_{238U} \text{ decayed} = (4.015 \times 10^{-9} \text{ mol He})\left(\frac{6.022 \times 10^{23} \text{ atom}}{1 \text{ mol}}\right)\left(\frac{1 \text{ atom } ^{238}U}{8 \text{ atom He}}\right) = 3.02 \times 10^{14} \text{ atom}$$

Now compute the number of atoms of ^{238}U not yet decayed in the gram of rock

$$n_{238U} = 2.0 \times 10^{-7} \text{ g } ^{238}U \times \left(\frac{1 \text{ mol}}{238.0 \text{ g}}\right)\left(\frac{6.022 \times 10^{23} \text{ atom}}{1 \text{ mol}}\right) = 5.06 \times 10^{14} \text{ atom}$$

The initial number of atoms of U-238 equals the number decayed plus the number remaining. It is 8.08×10^{14} atoms per gram of rock. The decay kinetics are described by the equation

$$\ln\left(\frac{N}{N_i}\right) = -kt = -\left(\frac{\ln 2}{t_{1/2}}\right) t$$

Substitution gives

$$\ln\left(\frac{5.06 \times 10^{14} \text{ atoms}}{8.08 \times 10^{14} \text{ atoms}}\right) = -\left(\frac{\ln 2}{4.47 \times 10^9 \text{ yr}}\right) t$$

Solving for t gives the approximate age of the rock: $\boxed{3.0 \text{ billion years}}$.

19.29 First-order kinetics govern the radioactive decay of both ^{235}U and ^{238}U

$$N(^{235}U) = N_i(^{235}U)e^{-kt} \qquad \text{and} \qquad N(^{238}U) = N_i(^{238}U)e^{-kt}$$

At the time of the supernova, the two isotopes were equally abundant, but now U-238 is 137.7 times more prevalent. In equation form

$$\text{Then:} \quad N_i(^{238}U) = N_i(^{235}U) \qquad \text{Now:} \quad N(^{238}U) = 137.7 N(^{235}U)$$

Assume that the change was caused entirely by the faster decay of U-235. Then

$$\frac{137.7}{1} = \frac{N_i(^{238}U)\, e^{-k_{238}t}}{N_i(^{235}U)\, e^{-k_{235}t}} = \frac{e^{-k_{238}t}}{e^{-k_{235}t}}$$

where k_{238} is the rate constant for the decay of ^{238}U and k_{235} is the rate constant for the decay of ^{235}U. Take the natural logarithm of both sides of the equation

$$\ln 137.7 = (-k_{238}t) - (-k_{235}t) = t(k_{235} - k_{238})$$

For each isotope $k = \ln 2/t_{1/2}$ so

$$\ln 137.7 = t \left(\frac{\ln 2}{t_{1/2,235}} - \frac{\ln 2}{t_{1/2,238}} \right)$$

The half-lives of the isotopes are 7.04×10^8 yr and 4.47×10^9 yr respectively. Inserting these numbers in the equation and solving for t gives $\boxed{5.9 \times 10^9 \text{ yr}}$. The supposed supernova occurred about 1.4 billion years before the estimated time of the formation of the solar system.

Radiation in Biology and Medicine

19.31 Positron emission is accompanied by emission of a neutrino
$$^{11}_{6}\text{C} \longrightarrow {}^{11}_{5}\text{B} + {}^{0}_{1}e^+ + \nu \qquad {}^{15}_{8}\text{O} \longrightarrow {}^{15}_{7}\text{N} + {}^{0}_{1}e^+ + \nu$$

19.33 Assume decay of all of the ^{11}C and ^{15}O atoms before any are excreted (or else assume that equal chemical amounts of the two radioactive nuclides are excreted). O-15 deposits 1.74 times more energy per kilogram of body mass than the C-11 because its positrons, which are emitted in equal number, are on the average more energetic by the factor $1.72/0.99 = \boxed{1.74}$.

19.35 a) Determine the number of atoms of I-131 ingested

$$N_{\text{I}} = 5.0 \times 10^{-6} \text{ g } {}^{131}\text{I} \left(\frac{1 \text{ mol } {}^{131}\text{I}}{131 \text{ g } {}^{131}\text{I}} \right) \left(\frac{6.022 \times 10^{23} \text{ atoms } {}^{131}\text{I}}{1 \text{ mol } {}^{131}\text{I}} \right) = 2.3 \times 10^{16} \text{ atoms } {}^{131}\text{I}$$

Express the half-life of I-131 in seconds: $8.041 \text{ d} \times 86\,400 \text{ s d}^{-1} = 6.947 \times 10^5$ s. The ingested radioactive iodine has an initial activity of

$$A_{\text{I}} = k N_{\text{I}} = \left(\frac{\ln 2}{t_{1/2}} \right) N_{\text{I}} = \left(\frac{0.6931}{6.947 \times 10^5 \text{ s}} \right) (2.3 \times 10^{16}) = 2.3 \times 10^{10} \text{ s}^{-1} = \boxed{2.3 \times 10^{10} \text{ Bq}}$$

b) Compute the initial rate r_i at which energy is evolved from the decay of I-131. Use the definition of a becquerel (Bq): 1 decay event per second

$$r_i = \left(\frac{2.3 \times 10^{10} \text{ events}}{1 \text{ s}} \right) \left(\frac{0.40 \text{ MeV}}{1 \text{ event}} \right) \left(\frac{1.602 \times 10^{-13} \text{ J}}{1 \text{ MeV}} \right) = 1.474 \times 10^{-3} \text{ J s}^{-1}$$

The victim has a mass of 60 kg and all of the energy is deposited internally. The initial rate of deposition on a body-mass basis is

$$r_i = \frac{1.474 \text{ mJ s}^{-1}}{60 \text{ kg}} = 0.0246 \text{ mJ s}^{-1} \text{ kg}^{-1}$$

This rate drops off over time, but the change in the first second is negligible. A radiation absorbed dose of 1 mJ kg^{-1} equals 1 mGy (1 milligray). The victim gets a dose equal to $\boxed{0.025 \text{ mGy}}$ in the first second.

c) After one half-life (8.04 days) half of the original 2.3×10^{16} atoms of radioactive iodine has decayed inside the body of the victim. This releases a lot of energy

$$E = 1.15 \times 10^{16} \text{ atoms decayed} \times \left(\frac{0.40 \text{ MeV}}{1 \text{ atom}} \right) \left(\frac{1.602 \times 10^{-13} \text{ J}}{1 \text{ MeV}} \right) = 737 \text{ J}$$

Dividing by 60 kg gives a dose of 12 J kg^{-1} = 12 Gy in the first 8.04 days. Dosage continues at a significant rate after that. Without immediate aggressive treatment to flush the radioactive iodine from the victim's body, this case of radiation poisoning will $\boxed{\text{surely end in death}}$.

Tip. Compare closely to problem **19.36**. The inhaled plutonium in that problem is equal in mass to the radioactive iodine ingested here, but is probably not a lethal dose. Make sure to understand why.

Nuclear Fission

19.37 **a)** The balanced nuclear reaction is $^{90}_{38}\text{Sr} \longrightarrow {}^{90}_{40}\text{Zr} + 2\,{}^{0}_{-1}e^- + 2\,\bar{\nu}$. The version $^{90}_{38}\text{Sr} \longrightarrow {}^{90}_{40}\text{Zr}^{2+} + 2\,{}^{0}_{-1}e^- + 2\,\bar{\nu}$ shows an exact charge balance between the two sides (in the right superscripts), in addition to balance as to Z (left subscripts) and A (left superscripts).

b) The overall nuclear reaction is two consecutive beta decays. The change in mass in beta decay is

$$\Delta m = m\,(\text{daughter atom}) - m\,(\text{parent atom})$$

To get the Δm of the overall reaction, simply subtract the mass of an atom of ^{90}Sr from the mass of an atom of ^{90}Zr. The masses of the beta particles (the electrons) are automatically accounted for when this is done. The required isotopic masses are given in the problem. The result is a Δm of -0.0030 u. The corresponding energy (taking 1 u as equivalent to 931.494 MeV) is $\boxed{2.8 \text{ MeV}}$.

c) This part is similar to problem **19.19**. Compute the number of atoms in 1.00 g of ^{90}Sr

$$N_{\text{Sr-90}} = 1.00 \text{ g} \times \left(\frac{1 \text{ mol}}{89.9073 \text{ g}}\right)\left(\frac{6.02214 \times 10^{23} \text{ atom}}{1 \text{ mol}}\right) = 6.698 \times 10^{21} \text{ atoms } {}^{90}\text{Sr}$$

The activity A, which is the instantaneous rate of disintegration of the Sr-90 depends on the number of Sr-90 atoms present

$$A = -\frac{dN}{dt} = kN$$

The k in this equation equals $\ln 2/t_{1/2}$ where $t_{1/2}$ is the half-life. Substituting gives

$$A = \left(\frac{\ln 2}{t_{1/2}}\right) N = \left(\frac{0.6931}{28.1 \text{ yr}}\right)(6.698 \times 10^{21}) = 1.65 \times 10^{20} \text{ yr}^{-1}$$

This is the number of disintegrations per year at the moment that the ^{90}Sr is released. The problem asks for the activity on a per-second basis

$$A = 1.65 \times 10^{20} \text{ yr}^{-1} \times \left(\frac{1 \text{ yr}}{365.2 \text{ d}}\right)\left(\frac{1 \text{ d}}{86\,400 \text{ s}}\right) = \boxed{5.23 \times 10^{12} \text{ s}^{-1}}$$

d) The activity of the Sr-90 falls off with time as the number of Sr-90 atoms persisting in the 1.00 g sample diminishes. The relationship is

$$A = A_{\text{i}}e^{-kt} = A_{\text{i}}\exp\left(-\frac{\ln 2\,t}{t_{1/2}}\right)$$

Substitute the initial activity from part **c)**, the specified interval of 100 yr, and the half-life of 28.1 yr

$$A = (5.23 \times 10^{12} \text{ s}^{-1})\exp\left(\frac{-0.6931(100 \text{ yr})}{28.1 \text{ yr}}\right) = \boxed{4.44 \times 10^{11} \text{ s}^{-1}}$$

Tip. The activity of the isotope falls off to about 8.5% of its original value in 100 years. The problem could have asked for activities in becquerels (Bq) rather than disintegrations per second (s^{-1}). The conversion from s^{-1} to Bq is particularly simple: the answers are 5.23×10^{12} Bq and 4.44×10^{11} Bq.

19.39 The lighter isotopes of uranium happen to decay faster than the heavier isotopes. The quicker breakdown of the light isotopes leaves heavy isotopes behind, causing the average atomic mass of the uranium to $\boxed{\text{increase}}$ with time. This assumes that the lighter isotopes are not important products of decay of the heavier isotopes.

19.41   The change in mass when one atom of U-235 gains a neutron and then undergoes fission as specified in the problem is the mass of the products minus the mass of the reactants

$$\Delta m = 1 \underbrace{(93.919)}_{^{94}Kr} + 1 \underbrace{(138.909)}_{^{139}Ba} + 3 \underbrace{(1.0086649)}_{^{1}_{0}n} - 1 \underbrace{(235.043925)}_{^{235}U} - 1 \underbrace{(1.0086649)}_{^{1}_{0}n} = -0.1986 \text{ u}$$

Convert to kJ mol^{-1} using the mass/energy equivalence established in problem **19.3c**. The result is -1.785×10^{10} kJ mol^{-1}. The problem asks for the energy change per gram of ^{235}U

$$-1.785 \times 10^{10} \text{ kJ mol}^{-1} \times \left(\frac{1 \text{ mol } ^{235}U}{235.04 \text{ g } ^{235}U} \right) = -7.59 \times 10^{7} \text{ kJ g}^{-1}$$

The energy released is the negative of the energy change of the system: $\boxed{+7.59 \times 10^{7} \text{ kJ g}^{-1}}$.

ADDITIONAL PROBLEMS

19.43   When a positron and electron meet, they annihilate each other to generate gamma radiation

$$^{0}_{1}e^{+} + ^{0}_{-1}e^{-} \longrightarrow 2\gamma$$

The mass of the electron and positron both equal 0.00054858 u. Therefore, Δm for the annihilation reaction is -0.00109716 u and

$$\Delta E = -0.00109716 \text{ u} \times \left(\frac{931.494 \text{ MeV}}{1 \text{ u}} \right) = -1.02200 \text{ MeV}$$

Because neither particle had any kinetic energy, only this amount of energy appears in the surroundings, borne by two gamma rays of energy $\boxed{0.51100 \text{ MeV}}$.

19.45   **a)** The beta decay (electron emission decay) of ^{64}Cu is

$$^{64}_{29}Cu \longrightarrow ^{64}_{30}Zn + ^{0}_{-1}e^{-} + \bar{\nu} \qquad \text{for which} \qquad \Delta m = m(^{64}_{30}Zn) - m(^{64}_{29}Cu)$$

The ΔE for the process is given as -0.58 MeV. Convert this to atomic mass units and use it in the previous equation

$$-0.58 \text{ Mev} \times \left(\frac{1 \text{ u}}{931.494 \text{ MeV}} \right) = m(^{64}_{30}Zn) - m(^{64}_{29}Cu)$$

$$-0.000623 \text{ u} = m(^{64}_{30}Zn) - 63.92976 \text{ u}$$

Hence the daughter ^{64}Zn weighs $\boxed{63.92914 \text{ u}}$.

b) This nuclear reaction produces ^{64}Ni from ^{64}Cu

$$^{64}_{29}Cu \longrightarrow ^{64}_{28}Ni + ^{0}_{1}e^{+} + \nu$$

Its Δm is

$$\Delta m = \left(m(^{64}_{28}Ni) - m(^{64}_{29}Cu) \right) + 2 \left(m(^{0}_{1}e^{+}) \right)$$

The last term must be included because a positron is lost from the daughter-parent atom pair and the neutral daughter atom has one fewer electrons than the parent atom. The masses of a positron and an electron are equal. The value of ΔE is given as -0.65 MeV. Convert this energy to atomic mass units and set it equal to Δm in the previous equation

$$-0.65 \text{ Mev} \times \left(\frac{1 \text{ u}}{931.494 \text{ MeV}} \right) = \left(m(^{64}_{28}Ni) - m(^{64}_{29}Cu) \right) + 2 \,(0.00054858 \text{ u})$$

Solving for the difference in mass between daughter and parent gives -0.00179 u. The parent atom ^{64}Cu weighs 63.92976 u, so the daughter atom ^{64}Ni weighs $\boxed{63.92797 \text{ u}}$.

19.47   Only an element having Z larger by 2 can decay directly to Ac by alpha emission. Since Ac is element 89, this would be element 91, | protactinium |. Element 91 is in fact named as the parent of actinium ("proto-actinium"). Only an element with Z less by 1 can decay directly to Ac by emission of an electron (beta emission). This would be element 88, which is | radium |. The fact that compounds of radium contain no actinium rules out beta emission by radium as a significant source of actinium.

19.49   **a)** The formation of a ^{30}P atom is represented $15\,^1_1\text{H} + 15\,^1_0 n \longrightarrow \,^{30}_{15}\text{P}$. Note that the mass of the electrons is included in the mass of the ^1_1H atoms. The mass of the product is 29.9783138 u, and the mass of the reactants is $15(1.007825032 + 1.0086649158)$ or 30.24734922 u. The difference between these masses is -0.2690354 u, which corresponds to a difference in energy of -250.605 MeV. The binding energy equals the negative of this figure; the binding energy per nucleon is then $+250.605/30$ or | 8.35350 MeV per nucleon |.

b) The equation for position emission by ^{30}P is $^{30}_{15}\text{P} \longrightarrow \,^{30}_{14}\text{Si} + \,^0_1 e^+ + \nu$. The change in mass in this process is

$$\Delta m = [m(^{30}_{14}\text{Si}) - m(^{30}_{15}\text{P})] + 2\,(0.00054858 \text{ u})$$

$$\Delta m = 29.97377022 - 29.9783138 + 0.00109716 = -0.0034464 \text{ u}$$

The negative sign means that the process is spontaneous. The change in energy of the system is also negative

$$\Delta E = (-0.0034464 \text{ u}) \times \left(\frac{931.494 \text{ MeV}}{1 \text{ u}}\right) = -3.21032 \text{ MeV}$$

The kinetic energy of the products equals $-\Delta E$ and is distributed among them. The positron has its maximum kinetic energy when the other decay products get none. This maximum is | 3.21032 MeV |.

c) The quick way to get the fraction of ^{30}P atoms left after 450 s is to recognize that 450 s equals three 150 s half-lives. The fraction is then obviously $(1/2)^3$ or | 0.125 |. The rate constant is $\ln 2/t_{1/2}$; it is | 4.62×10^{-3} s^{-1}. |

Tip. The energy equivalent of the mass of the positron and the extra electron in the mass-change equation is 1.0220 MeV. Omitting this term introduces a big error (about 32%).

19.51   The activity of the gallium-67 has decayed to 5.0% of the initial activity when the number of atoms of Ga-67 reaches 5.0% of the original number of atoms: $N = 0.050N_\text{i}$. Insert this relationship into the first-order decay law

$$\ln \frac{N}{N_\text{i}} = -kt \qquad \text{to obtain} \qquad \ln 0.05 = -kt$$

The rate constant k equals $\ln 2$ divided by the half-life, which is given as 77.9 hours. Therefore

$$\ln 0.05 = -\left(\frac{\ln 2}{77.9 \text{ h}}\right) t \qquad \text{from which} \qquad t = \boxed{340 \text{ h}}$$

19.53   Obtain the number of atoms of ^{14}C in the 1.00 g of modern charcoal from the activity of the sample

$$N_{\text{C-14}} = \frac{A_{\text{C-14}}}{k} = A_{\text{C-14}} \left(\frac{t_{1/2}}{\ln 2}\right)$$

where $t_{1/2}$ is the half-life of ^{14}C and $A_{\text{C-14}}$ is its activity. The unit of activity A is the becquerel, which is a reciprocal second (s^{-1}). Convert the given half-life of C-14 from years to seconds so that it may cancel the unit of A. Then

$$N_{\text{C-14}} = 0.255 \text{ s}^{-1} \times \left(\frac{(5730 \text{ yr})(3.156 \times 10^7 \text{ s yr}^{-1})}{\ln 2}\right) = 6.65 \times 10^{10} \text{ atoms}$$

Thus ordinary carbon in the biosphere has $\boxed{6.65 \times 10^{10} \text{ atoms } {}^{14}\text{C (g C)}^{-1}}$.

b) Assume that the charcoal is pure carbon and is representative in the isotopic composition of carbon throughout the whole biosphere. The number of carbon atoms in a 1.00 g sample is

$$N_{\text{C}} = 1.00 \text{ g} \times \left(\frac{1 \text{ mol}}{12.01115 \text{ g}} \right) \left(\frac{6.022 \times 10^{23} \text{ atoms}}{1 \text{ mol}} \right) = 5.014 \times 10^{22} \text{ atoms}$$

From part **a)**, 1.00 g of ordinary carbon in the biosphere contains 6.65×10^{10} atoms of ${}^{14}\text{C}$. The number of atoms of C of all isotopes in 1.00 gram of C is 5.014×10^{22} atoms, which is vastly larger. The required fraction is the first number divided by the second, which is $\boxed{1.32 \times 10^{-12}}$.

19.55 Compute the mass of K-40 that was present in the rock at formation. In the following, the first unit-factor is valid because atoms of K-40 and Ar-40 have essentially the same mass. The second unit-factor deals with the fact that only 10.7%[1] of the original K-40 gives Ar-40.

$$(m_{\text{K}-40})_{\text{i}} = 0.42 \text{ mg } {}^{40}\text{Ar} \times \left(\frac{1 \text{ mg } {}^{40}\text{K by EC}}{1 \text{ mg } {}^{40}\text{Ar}} \right) \left(\frac{100 \text{ mg}}{10.7 \text{ mg } {}^{40}\text{K by EC}} \right) = 3.925 \text{ mg } {}^{40}\text{K}$$

Since K-40 decays by first-order kinetics

$$m_{\text{K}-40} = (m_{\text{K}-40})_{\text{i}} \, e^{-kt}$$

where the use of masses instead of number of atoms is valid because the number of atoms of K-40 is directly proportional to its mass. Take the logarithm of both sides and substitute the two masses and the rate constant. The rate constant k equals $\ln 2$ divided by the half-life of ${}^{40}\text{K}$, which is available in text Table 19.2. The result is

$$\ln \left(\frac{m_{\text{K}-40}}{(m_{\text{K}-40})_{\text{i}}} \right) = -kt = \ln \left(\frac{1.00 \text{ mg}}{3.925 \text{ mg}} \right) = \left(\frac{-\ln 2}{1.28 \times 10^9 \text{ yr}} \right) t \qquad t = \boxed{2.5 \times 10^9 \text{ yr}}$$

19.57 The incomplete equation for the nuclear reaction is ${}^{10}_{5}\text{B} + {}^{1}_{0}n \longrightarrow ? + {}^{4}_{2}\text{He}$. It must be balanced by the insertion of a single symbol for the question mark. The mass number on this symbol must be 7 and the atomic number 3. The element of atomic number 3 is lithium. Hence the other atom that is formed is a $\boxed{{}^{7}_{3}\text{Li}}$ atom.

19.59 Use unit factors to obtain the mass of TNT that must explode to release the same amount of energy as the small A-bomb

$$m_{\text{TNT}} = 1.2 \text{ kg U} \times \left(\frac{1 \text{ mol U}}{0.238 \text{ kg U}} \right) \left(\frac{2 \times 10^{13} \text{ J}}{1 \text{ mol U}} \right) \left(\frac{1 \text{ ton TNT}}{4 \times 10^9 \text{ J}} \right) = \boxed{2.5 \times 10^4 \text{ ton TNT}}$$

19.61 The Earth orbits the Sun at a distance R of 1.50×10^8 km. Imagine that a sphere having this radius surrounds the Sun. The surface area of this immense sphere is $4\pi R^2$. Radiation from the Sun streams out in all directions, cutting through this sphere. Represent the 6371 km radius of the Earth as r. From the Sun, the Earth appears as a disk of area πr^2. This disk, minuscule in comparison to the surface area of the big sphere, intercepts a fraction f of the total radiation in proportion to the area of the big sphere that it covers

$$f = \frac{\pi r^2}{4\pi R^2} = \frac{1}{4} \left(\frac{r}{R} \right)^2 = \frac{1}{4} \left(\frac{6371 \text{ km}}{1.50 \times 10^8 \text{ km}} \right)^2 = 4.5 \times 10^{-10}$$

A radiant flux of 0.135 J s^{-1}cm^{-2} impinges on this Earthly disc. It follows that the Earth intercepts the radiant power

$$P(\text{Earth}) = (0.135 \text{ J s}^{-1}\text{cm}^{-2}) \times \pi (6.371 \times 10^8 \text{ cm})^2 = 1.72 \times 10^{17} \text{ J s}^{-1}$$

---

[1]Only the proportion of K-40 that decays by electron capture (EC) gives Ar-40. See Text Table 19.2.

The total power output of the Sun is $P(\text{Earth})$ divided by the fraction of the Sun's radiant power that hits the Earth

$$P(\text{Sun}) = \frac{P(\text{Earth})}{f} = \frac{1.72 \times 10^{17} \text{ J s}^{-1}}{4.5 \times 10^{-10}} = 3.8 \times 10^{26} \text{ J s}^{-1}$$

The mass equivalent of energy is given by the equation $\Delta E = c^2 \Delta m$. The Sun emits energy, so its ΔE is negative. The rate of change of its mass is

$$\Delta m/t = \frac{\Delta E/t}{c^2} = \frac{-3.8 \times 10^{26} \text{ J s}^{-1}}{(3.0 \times 10^8 \text{ m s}^{-1})^2} = \boxed{-4.2 \times 10^9 \text{ kg s}^{-1}}$$

CUMULATIVE PROBLEMS

19.63   The problem compares aspects of generating one year's energy from three different 500 megawatt power plants. First, calculate the annual energy production. It equals the rated power multiplied by the time of operation

$$E = 500 \text{ MW} \times 1 \text{ yr} \times \left(\frac{10^6 \text{ J s}^{-1}}{\text{MW}}\right)\left(\frac{3.155 \times 10^7 \text{ s}}{\text{yr}}\right) = 1.58 \times 10^{16} \text{ J} = 1.58 \times 10^{13} \text{ kJ}$$

a) Find the mass of coal burned in the coal-fired plant and the mass of ash using unit-factors

$$m_{\text{coal}} = 1.58 \times 10^{13} \text{ kJ} \times \left(\frac{1 \text{ kg coal theory}}{3.2 \times 10^4 \text{ kJ}}\right)\left(\frac{100 \text{ kg coal actual}}{25 \text{ kg coal theory}}\right) = \boxed{2 \times 10^9 \text{ kg}}$$

$$m_{\text{ash}} = 2.0 \times 10^9 \text{ kg coal} \times \left(\frac{0.10 \text{ kg ash}}{1 \text{ kg coal}}\right) = \boxed{2.0 \times 10^8 \text{ kg ash}}$$

b) For the nuclear power plant

$$m_{\text{235U}} = 1.58 \times 10^{13} \text{ kJ} \times \left(\frac{1 \text{ mol }^{235}\text{U theory}}{1.9 \times 10^{10} \text{ kJ}}\right)\left(\frac{100 \text{ mol }^{235}\text{U actual}}{25 \text{ mol }^{235}\text{U theory}}\right)\left(\frac{0.235 \text{ kg }^{235}\text{U}}{\text{mol }^{235}\text{U}}\right)$$

$$= \boxed{7.8 \times 10^2 \text{ kg}}$$

$$m_{\text{fuel}} = 782 \text{ kg }^{235}\text{U} \times \left(\frac{100 \text{ kg fuel}}{4 \text{ kg }^{235}\text{U}}\right) = \boxed{2.0 \times 10^4 \text{ kg}}$$

c) To figure the area of the "solar farm" power plant, use the required power output

$$A = 500 \text{ MW} \times \left(\frac{1000 \text{ kW}}{\text{MW}}\right)\left(\frac{1 \text{ m}^2}{1.5 \text{ kW}}\right)\left(\frac{100 \text{ m}^2 \text{ actual}}{25 \text{ m}^2 \text{ theory}}\right)\left(\frac{24 \text{ h real}}{6 \text{ h useful}}\right) = \boxed{5 \times 10^6 \text{ m}^2}$$

This is nearly 2 square miles.

19.65   **a)** Use Hess's law. The ΔH°_{298} of the reaction $N_2H_4(l) + O_2(g) \longrightarrow N_2(g) + 2 H_2O(g)$ is the sum of the ΔH°_f's at 298 K of the products less the sum of the ΔH°_f's at 298 K of the reactants

$$\Delta H^\circ_{298} = 2\underbrace{(-241.82)}_{H_2O(g)} + 1\underbrace{(0.00)}_{N_2(g)} - 1\underbrace{(0.00)}_{O_2(g)} - 1\underbrace{(50.63)}_{N_2H_4(l)} = \boxed{-534.27 \text{ kJ}}$$

where the numbers come from text Appendix D (recall that elements in standard states at 298 K have ΔH°_f's of zero).

b) The change in the standard internal energy during the constant-pressure reaction is

$$\Delta U^\circ_{298} = \Delta H^\circ_{298} - P\Delta V$$

The $P\Delta V$ term equals $\Delta n_g RT$, if it is assumed that the gases in the reaction are ideal and that the volume of the liquid hydrazine is negligible.[2] As a mole of $N_2H_4(l)$ burns in $O_2(g)$ at 298.15 K, Δn_g of the system equals $+2$ mol, making $\Delta n_g RT$ equal to 4.96 kJ. Hence, for the combustion of one mole of $N_2H_4(l)$:

$$\Delta U^\circ_{298} = \Delta H^\circ_{298} - P\Delta V = -534.27 - 4.96 = \boxed{-539.23 \text{ kJ}}$$

c) Use ΔU°_{298}, the standard change in internal energy associated with the chemical reaction, as ΔE in the Einstein equation

$$\Delta m = \frac{\Delta E}{c^2} = \frac{-539.23 \times 10^3 \text{ J}}{(2.9979 \times 10^8 \text{ m s}^{-1})^2} = \boxed{-5.9998 \times 10^{-12} \text{ kg}}$$

Tip. The chemical reaction of 32 g of hydrazine and 32 g of oxygen occasions a mass loss by the system of about 6 nanograms. This is too small to detect.

19.67   a) To produce ^{228}Ra, the ^{232}Th must emit an alpha particle in the first step of its decay:

$$^{232}_{90}\text{Th} \longrightarrow \,^{228}_{88}\text{Ra} + \,^{4}_{2}\text{He}$$

In the next step, the ^{228}Ra emits a beta particle (an electron) to give ^{228}Ac

$$^{228}_{88}\text{Ra} \longrightarrow \,^{228}_{89}\text{Ac} + \,^{0}_{-1}e^- + \bar{\nu}$$

The decay products over the two steps are an alpha particle, an ^{228}Ac atom, an electron and an anti-neutrino. These products have the same total mass as an ^{228}Ac (228.031015 u) plus a $^{4}_{2}$He atom (4.0026033 u). The sum is 232.033618 u. The mass of the beta particle (electron) is included in the mass tabulated for a neutral actinium atom, and the anti-neutrino is massless. The mass of the reactant is 232.038050 u so Δm for the process is -0.004432 u. This Δm is equivalent to -4.128 MeV. Hence, $\boxed{4.128 \text{ MeV}}$ is the energy lost by the system consisting of one Th atom. The energy appears in the surroundings as kinetic energy of the particles formed by the reaction.

b) The thorium decays to radium in a first-order process, and the radium goes on to decay to actinium in a second first-order process. Let k_1 be the rate constant for the first step and k_2 the rate constant for the second. When the radium is present in a steady-state amount then

$$\frac{dN_{\text{Ra}}}{dt} = 0 = k_1 N_{\text{Th}} - k_2 N_{\text{Ra}} \qquad \text{from which} \qquad N_{\text{Ra}} = \frac{k_1}{k_2} N_{\text{Th}}$$

For each first-order process, the rate constant is $\ln 2$ divided by the half-file $t_{1/2}$. Applying this fact to the previous equation gives

$$N_{\text{Ra}} = \left(\frac{t_{1/2,2}}{t_{1/2,1}}\right) N_{\text{Th}} = \left(\frac{6.7 \text{ yr}}{1.39 \times 10^{10} \text{ yr}}\right) N_{\text{Th}} = (4.8 \times 10^{-10}) N_{\text{Th}}$$

The number of ^{228}Ra nuclei equals $\boxed{4.8 \times 10^{-10}}$ times the number of ^{232}Th nuclei.

---

[2]This is the same kind of assumption made in problem **12.43**.

Chapter 20

Molecular Spectroscopy and Photochemistry

Experimental Methods in Molecular Spectroscopy

20.1 A percent transmittance T of 20.0% means that at this wavelength the intensity of the light detected through the sample cell (symbolized I_S) is 20.0% of the intensity detected through the reference cell (I_R). The absorbance A is defined as the negative logarithm of the ratio of these two intensities

$$A = -\log \frac{I_S}{I_R}$$

In this case then $A = -\log 0.200 = \boxed{0.699}$. Assume that the Beer-Lambert law applies. Use it to obtain ϵ

$$\epsilon = \frac{A}{c\ell} = \frac{0.699}{(5 \times 10^{-4} \text{ mol L}^{-1})(1.0 \text{ cm})} = \boxed{1 \times 10^3 \text{ L mol}^{-1} \text{ cm}^{-1}}$$

20.3 Assume that the Beer-Lambert law (Beer's law) applies and that the absorbances due to compounds A and B are additive at both 400 nm and 500 nm. Thus, A and B do not, for example, react with each other. Compute the absorbances of solutions 1, 2, and 3 at 400 and at 500 nm

$$\text{At 400 nm} \begin{cases} A_1 &= -\log(10/100) = 1.000 \\ A_2 &= -\log(80/100) = 0.0969 \\ A_3 &= -\log(40/100) = 0.3979 \end{cases} \quad \text{At 500 nm} \begin{cases} A_1 &= -\log(60/100) = 0.2218 \\ A_2 &= -\log(20/100) = 0.6990 \\ A_3 &= -\log(50/100) = 0.3010 \end{cases}$$

Solution 1 does not contain compound B. Therefore

$$A_{1,400} = c_A(\epsilon\ell)_{A,400} = (0.0010 \text{ mol L}^{-1})(\epsilon\ell)_{A,400}$$

$$A_{1,500} = c_A(\epsilon\ell)_{A,500} = (0.0010 \text{ mol L}^{-1})(\epsilon\ell)_{A,500}$$

Solve for $(\epsilon\ell)$ in each equation and insert the A's and c's

$$(\epsilon\ell)_{A,400} = \frac{1.000}{0.0010 \text{ L mol}^{-1}} = 1000 \text{ L mol}^{-1}$$

$$(\epsilon\ell)_{A,500} = \frac{0.2218}{0.0010 \text{ L mol}^{-1}} = 221.8 \text{ L mol}^{-1}$$

Solution 2 does not contain compound A, so a similar development is possible

$$(\epsilon\ell)_{B,400} = \frac{0.0969}{0.0050 \text{ L mol}^{-1}} = 19.38 \text{ L mol}^{-1}$$

$$(\epsilon\ell)_{B,500} = \frac{0.6990}{0.0050 \text{ L mol}^{-1}} = 139.8 \text{ L mol}^{-1}$$

283

The absorbances due to A and B are additive in Solution 3. The molar extinction coefficients (the ϵ's) of A and B do not change. Therefore, the Beer-Lambert law for Solution 3 gives

$$A_{3,400} = c_A(\epsilon\ell)_{A,400} + c_B(\epsilon\ell)_{B,400} \qquad A_{3,500} = c_A(\epsilon\ell)_{A,500} + c_B(\epsilon\ell)_{B,500}$$

$$0.3979 = c_A(1000 \text{ L mol}^{-1}) + c_B(19.38 \text{ L mol}^{-1}) \quad 0.3010 = c_A(221.8 \text{ L mol}^{-1}) + c_B(139.8 \text{ L mol}^{-1})$$

where it is also assumed that the length ℓ of the optical path through the sample is the same in all three spectroscopy experiments. Solving the last two equations, which are simultaneous, gives

$$c_A = \boxed{3.7 \times 10^{-4} \text{ mol L}^{-1}} \quad \text{and} \quad c_B = \boxed{16 \times 10^{-4} \text{ mol L}^{-1}}$$

20.5 The Beer-Lambert law states the degree to which a gas reduces the intensity I_0 of radiation that passes through it as a function of the absorption cross-section σ of the molecules of the gas, the number density (the number N of molecules per volume V) of the gas, and the length ℓ of the path along which the radiation traverses the gas. It reads

$$\ln\left(\frac{I}{I_0}\right) = -\left(\frac{N}{V}\right)\sigma\ell$$

where I is the intensity of the radiation after its passage through the gas. The problem states the σ of ozone at a specific wavelength (320 nm), the number density of the ozone molecules, and the degree to which the intensity of the radiation is to be reduced (by 60%, or to 0.40 of its original intensity). Substitute the known quantities and solve the equation for the path length, the only unknown

$$\ln\left(\frac{40}{100}\right) = -\left(\frac{5 \times 10^{12} \text{ molecule}}{\text{cm}^3}\right)\left(\frac{5 \times 10^{-20} \text{ cm}^2}{\text{molecule}}\right)\ell$$

$$\ell = 3.7 \times 10^6 \text{ cm} = \boxed{37 \text{ km}}$$

Tip. The absorption cross-section of a molecule is the effective area that it presents for absorption of photons. It ranges from 0 to about 10^{-16} cm^2 depending on the species and the wavelength of the photons. A value of 10^{-18} to 10^{-17} cm^2 is considered large.

20.7 Use text Figure 20.1b, which deals with HCl, to map the changes in energy in this similar diatomic molecule.[1] At the start, the OH molecule is in its ground electronic state, which is labeled S_0, and also in its lowest vibrational state, which has a vibrational quantum number v'' equal to 0. Its vibrational energy is easily computed

$$E_{\text{vib}} = 3738 \text{ cm}^{-1}\left(v'' + 1/2\right) = 3738 \text{ cm}^{-1}(0 + 1/2) = 1869 \text{ cm}^{-1}$$

The OH molecule gains energy by absorbing a photon from a laser beam. A big portion, labeled E_{el} in Figure 20.1b, of the new energy goes to a redistribution of electrons that lifts it from the ground or S_0 state to the excited electronic state in the figure. This state is often labeled $\tilde{A}$. This big portion equals 32684 cm^{-1}. The rest goes to increase the frequency of the O—H vibration. The OH molecule ends up in the $v' = 1$ vibrational state of $\tilde{A}$ where its vibrational energy is

$$E_{\text{vib}} = 3179 \text{ cm}^{-1}(v' + 1/2) = 3179 \text{ cm}^{-1}(1 + 1/2) = 4768.5 \text{ cm}^{-1}$$

The total change in the energy of the molecule comes from a movement down from $v'' = 0$ to the bottom of the ground electronic state, then up to the bottom of the $\tilde{A}$ electronic state, and then further up to the $v' = 1$ vibrational state of the $\tilde{A}$ electronic state:

$$\Delta E = -1869 + 32684 + 4768.5 = \boxed{35583.5 \text{ cm}^{-1}}$$

---

[1]Energies in Figure 20.1b and in the statement and solution of this problem are given in reciprocal centimeters, a unit much loved by spectroscopists. Conversion from cm^{-1} to J (joules) requires nothing more than multiplication by hc (Planck's constant in J s times the speed of light in cm s^{-1}.)

Tip. The redistribution of electrons associated with the $S_0 \longrightarrow \tilde{A}$ transition weakens the O—H bond, as shown by the fact that its fundamental vibrational energy falls from 3738 cm^{-1} to 3179 cm^{-1}.

20.9 The thermal population of molecular quantum levels follows the Boltzmann distribution. Write text equation 20.4 replacing the subscripts i and j with a 0 and a 1 to mark the 0th and 1st vibrational energy level of H_2

$$\frac{N_1}{N_0} = \frac{g_1}{g_0} \frac{\exp(-E_1/k_BT)}{\exp(-E_0/k_BT)} = \frac{g_1}{g_0} \exp\left(\frac{-(E_1 - E_0)}{k_BT}\right) = \frac{g_1}{g_0} \exp\left(\frac{-\Delta E}{k_BT}\right)$$

Write text equation 20.12 twice, once for the 0th vibrational energy level and once for the 1st vibrational energy level and subtract the first from the second to obtain the difference in energy between the two vibrational levels.

$$E_1 - E_0 = \Delta E = h\nu(1 + \tfrac{1}{2}) - h\nu(0 + \tfrac{1}{2})$$

Compute the difference in energy associated with the observed frequency of the transition. The problem gives this observation as a wave number. Multiplying by hc converts it to an energy

$$\Delta E = 6.62606 \times 10^{-34}\,\text{J s}\,(2.9979 \times 10^{10}\,\text{cm s}^{-1})(4400\,\text{cm}^{-1}) = 8.7403 \times 10^{-20}\,\text{J}$$

Substitute this ΔE into the equation for the ratio N_1/N_0. Also substitute the degeneracies (the g's) of the vibrational states, which both equal 1

$$\frac{N_1}{N_0} = \frac{g_1}{g_0} \exp\left(\frac{-\Delta E}{k_BT}\right) = \frac{1}{1} \exp\left(\frac{-8.7403 \times 10^{-20}\,\text{J}}{(1.3806 \times 10^{-23}\,\text{J K}^{-1})(3300 + 273)\,\text{K}}\right) = \boxed{0.17}$$

Use the same approach for the calculation on the relative population of the rotational energy levels of H_2. The difference in energy between the two is given. Express it in joules

$$\Delta E = 121.7\,\text{cm}^{-1}\,(2.9979 \times 10^{10}\,\text{cm s}^{-1})(6.62606 \times 10^{-34}\,\text{J s}) = 2.417 \times 10^{-21}\,\text{J}$$

Substitute this $\Delta\epsilon$ into the same equation for the ratio of occupancies that was used in the vibrational case. The degeneracies g of the rotational states of diatomic molecules depend on their quantum number according to $g_J = 2J + 1$. Include this important fact as well

$$\frac{N_1}{N_0} = \frac{g_1}{g_0} \exp\left(\frac{-\Delta E}{k_BT}\right) = \frac{3}{1} \exp\left(\frac{-2.417 \times 10^{-21}\,\text{J}}{(1.3806 \times 10^{-23}\,\text{J K}^{-1})(3300 + 273)\,\text{K}}\right) = \boxed{2.86}$$

The hot combustion chamber provides plenty of energy to excite H_2 molecules into the first excited rotational state. This and the fact that the $J = 1$ rotational quantum state is triply degenerate cause the $J = 1$ state to have almost three times the population of the $J = 0$ state. In contrast, the first vibrational excited state of H_2 is populated substantially less than the ground state. It takes much more energy to excite the molecule from its $\nu = 0$ to its $\nu = 1$ state than to excite it from its $J = 0$ to its $J = 1$ state.

Rotational and Vibrational Spectroscopy

20.11 The rotational energy of diatomic molecules such as NO is quantized according to text equation 20.6a

$$E_J = hB\,J(J+1) = h\left(\frac{h}{8\pi^2 I}\right) J(J+1)$$

where I is the moment of inertia of the molecule and J is the rotational quantum number ($J = 0, 1, 2, \ldots$). The rotational energies in the ground state and first excited state are

$$E_{J=0} = \left(\frac{h^2}{8\pi^2 I}\right)(0) = 0 \qquad \text{and} \qquad E_{J=1} = \left(\frac{h^2}{8\pi^2 I}\right)(2)$$

Obtain data on the $^{14}N^{16}O$ molecule[2] and use them to compute its moment of inertia

$$I_{^{14}N^{16}O} = \mu R_e^2 = \frac{m_N m_O}{m_N + m_O} R_e^2 = \left(\frac{(14.00307 \text{ u})(15.99491 \text{ u})}{(14.00307 + 15.9949) \text{ u}} \right) (1.154 \times 10^{-10} \text{ m})^2$$

$$= (7.4664 \text{ u})(1.154 \times 10^{-10} \text{ m})^2 \left(\frac{1 \text{ kg}}{6.022137 \times 10^{26} \text{ u}} \right) = 1.6511 \times 10^{-46} \text{ kg m}^2$$

Then the desired ΔE is

$$E_{J=1} - E_{J=0} = \frac{h^2}{8\pi^2 I_{^{14}N^{16}O}} (2 - 0) = \frac{(6.62608 \times 10^{-34} \text{ J s})^2}{8\pi^2 (1.6511 \times 10^{-46} \text{ kg m}^2)} (2) = \boxed{6.736 \times 10^{-23} \text{ J}}$$

This is equivalent to $\boxed{40.56 \text{ J mol}^{-1}}$.

20.13   **a)** The spacing between the frequencies of adjacent spectroscopic lines in the pure rotational spectrum of a diatomic molecule is *uniform* and equal to $2\tilde{B} = h/4\pi^2 cI$ where I is the moment of inertia of the species.[3] The spacing in terms of frequency (reciprocal time) rather than wave numbers (reciprocal distance) is $2B = h/4\pi^2 I$ because $c\tilde{B} = B$. Average the two spacings that the problem quotes to estimate $2B$ for the case of $^{12}C^{16}O$. The result is $1.155 \times 10^{11} \text{ s}^{-1}$. Finish by solving for $I_{^{12}C^{16}O}$, substituting the constants, and completing the arithmetic

$$1.155 \times 10^{11} \text{ s}^{-1} = \frac{h}{4\pi^2 I_{^{12}C^{16}O}} \qquad I_{^{12}C^{16}O} = \frac{6.626 \times 10^{-34} \text{ J s}}{4\pi^2 (1.155 \times 10^{11} \text{ s}^{-1})} = \boxed{1.45 \times 10^{-46} \text{ kg m}^2}$$

b) The energy of a rotational state of a linear molecule is given by

$$E_{\text{rot}} = \left(\frac{h^2}{8\pi^2 I} \right) J(J+1) \quad \text{which is the same as} \qquad E_{\text{rot}} = \frac{h}{2} \left(\frac{h}{4\pi^2 I} \right) J(J+1)$$

For the rotation of $^{12}C^{16}O$ molecules, the factor $h/4\pi^2 I = 2B = 1.155 \times 10^{11} \text{ s}^{-1}$. Insert it and $h/2$ into the preceding and complete the arithmetic for the $J = 1$ case. Clearly the $J = 2$ and $J = 3$ answers equal the $J = 1$ result multiplied by 3 and and 6 respectively

$$E_{\text{rot}} = \frac{6.626 \times 10^{-34} \text{ J s}}{2} (1.155 \times 10^{11} \text{ s}^{-1}) J(J+1)$$

$$E_{J=1} = \boxed{7.65 \times 10^{-23} \text{ J}} \quad E_{J=2} = \boxed{23.0 \times 10^{-23} \text{ J}} \quad E_{J=3} = \boxed{45.9 \times 10^{-23} \text{ J}}$$

c) The mass of a ^{12}C atom equals 12 u exactly, and the mass of an ^{16}O atom equals 15.994915 u. The reduced mass of this diatomic molecule is

$$\mu = \frac{m_O m_C}{m_O + m_C} = \frac{(12.000000 \text{ u})(15.994915 \text{ u})}{27.994915 \text{ u}} = 6.856208 \text{ u}$$

The moment of inertia depends on the reduced mass and equilibrium bond distance R_e

$$I_{^{12}C^{16}O} = \mu R_e^2$$

Solve for R_e and substitute the reduced mass and moment of inertia. Make sure that the units work out by converting the reduced mass from atomic mass units to kilograms

$$R_e = \sqrt{\frac{I_{^{12}C^{16}O}}{\mu}} = \sqrt{\frac{(1.45 \times 10^{-46} \text{ kg m}^2)}{(6.856 \text{ u})(1.661 \times 10^{-27} \text{ kg u}^{-1})}} = 1.128 \times 10^{-10} \text{ m} = \boxed{1.13 \text{ Å}}$$

---

[2] The masses come from text Table 19.1; the bond length comes from text Table 3.3.
[3] See Text equation 20.7.

20.15 Linear symmetric molecules such as O_2 (and CO_2 and N_2) do not absorb microwave radiation and consequently do not possess pure rotational spectra. They do possess Raman rotational spectra. The selection rule for Raman rotational spectra for such molecules is $\Delta J = 0, \pm 2$ (instead of $\Delta J = \pm 1$, which is the selection rule in pure rotational spectra). Use this rule to figure out the anticipated spacing of the frequencies of allowed transitions in terms of the rotational constant $\tilde{B}$ of the molecule

$$\Delta \tilde{\nu} = \tilde{\nu}_{J+2} - \tilde{\nu}_J$$
$$= \tilde{B}\big[(J+2)(J+2+1) - J(J+1)\big]$$
$$= \tilde{B}\big[4J+6\big]$$

The $\Delta J = +2$ transitions lead to the band of Stokes lines, and the $\Delta J = -2$ transitions lead to the band of anti-Stokes lines shown in Figure 20.14 on text page 857. The spacing between the lines within the side-bands equals $4\tilde{B}$. The wider spacing across the central zero from the beginning of one side-band to the beginning of the other equals $6\tilde{B}$.

Tip. Text Figure 20.14 labels several lines in the Raman rotational spectrum of N_2 with initial and final values of J. Complete the labeling by recognizing and following the pattern established in the figure. In every case the difference between the J's must equal ± 2.

The text states that in the rotational Raman spectrum of $^{16}O_2$ every other line is missing in comparison to the rotational Raman spectrum of N_2. This occurs whenever the end-for-end rotation of a linear molecule exchanges identical nuclei with nuclear spin $I = 0$. The extra symmetry excludes the possibility of rotational energy levels for which J is an even number. Consequently, the Raman rotational spectrum of $^{16}O_2$ looks similar to the spectrum of N_2 in text Figure 20.14 but with the big difference that lines with even numbers in their labels are extinguished. Only lines with odd-odd labels, such as 1-3 and 3-5, persist. Consequently, the spacing of the lines within the side-bands in the Raman rotational spectrum of $^{16}O_2$ equals $8\tilde{B}$ instead of $4\tilde{B}$ and the difference in frequency between the 3-1 and 1-3 lines, which is the new distance across the gap between the starting points of the two side-bands, equals $10\tilde{B}$ instead of $6\tilde{B}$.

Tip. The Raman rotational selection rule ($\Delta J = \pm 2$) is the *same* for $^{16}O_2$ as it is for N_2. Lines are fewer in the $^{16}O_2$ spectrum because it has fewer rotational energy levels.

The rotational constant $\tilde{B} = 1.45$ cm^{-1} for $^{16}O_2$ in the problem comes from an analysis that includes the preceding considerations. Computation of the bond length of the molecule from $\tilde{B}$ can therefore follow the pattern of text Example 20.4. The rotational constant of a rigid rotor, expressed first as B, a frequency, and then as $\tilde{B}$, a wave number, is

$$B = \frac{h}{8\pi^2 I} \qquad \tilde{B} = \frac{h}{8\pi^2 c I}$$

where I is the moment of inertia of the molecule. The given rotational constant is a $\tilde{B}$, in cm^{-1}, so solve the second equation for I and use the result to obtain I_{O_2}

$$I_{O_2} = \frac{h}{8\pi^2 c \tilde{B}_{O_2}} = \frac{6.626 \times 10^{-34}\text{ J s}}{8\pi^2 (3.00 \times 10^{10}\text{ cm s}^{-1})(1.45\text{ cm}^{-1})} = 1.929 \times 10^{-46}\text{ kg m}^2$$

For a diatomic molecule, $I = \mu R_e^2$ where R_e is the equilibrium bond distance. By the definition of the reduced mass, μ_{O_2} is one half of the mass of an atom of ^{16}O, which equals 15.995 u. Solve the preceding equation for R_e and substitute μ_{O2} and I_{O_2}. The reduced mass must be converted from atomic mass units to kilograms as well

$$(R_e)_{O_2} = \sqrt{\frac{1.929 \times 10^{-46}\text{ kg m}^2}{\frac{1}{2}(15.995\text{ u})(1.661 \times 10^{-27}\text{ kg u}^{-1})}} = 1.21 \times 10^{-10}\text{ m} = \boxed{1.21\text{ Å}}$$

Tip. The answer agrees with the accepted O—O distance, which is 1.211 Å. Not taking account of the effect of nuclear spin creates a big error. It multiplies the answer by $\sqrt{1/2}$.

20.17 In vibrational spectra, strong absorption is allowed only for transitions between adjacent vibrational states. Although the infra-red line in the $^7Li_2$ spectrum is "very weak," assume nevertheless that the change in the vibrational quantum number, Δv_r, equals 1. The change in vibrational energy then equals

$$\Delta E_{\text{vib}} = (\Delta v_r)\, h\left(\frac{1}{2\pi}\right)\sqrt{\frac{k}{\mu}} = [1]\, h\left(\frac{1}{2\pi}\right)\sqrt{\frac{k}{\mu}}$$

where Δv_r is the change in the vibrational quantum number,[4] k is the desired force constant and μ is the reduced mass of $^7Li_2$. The change in vibrational energy is related to the wavelength of the absorbed light λ by $\Delta E_{\text{vib}} = hc/\lambda$. Substitute this relation into the preceding

$$\frac{hc}{\lambda} = h\left(\frac{1}{2\pi}\right)\sqrt{\frac{k}{\mu}}$$

The wavelength of the absorption line is quoted in the problem as 2.85×10^{-5} m. From text Table 19.1, the mass of ^7Li is 7.016004 u; the reduced mass of $^7Li_2$ is half of this or 3.508002 u. Converting the reduced mass to kilograms gives 5.8252×10^{-27} kg. Solve the preceding equation for k, insert the numerical quantities, and complete the arithmetic

$$k = \mu\left(\frac{2\pi c}{\lambda}\right)^2 = (5.8252 \times 10^{-27}\text{ kg})\left(\frac{2\pi(2.998 \times 10^8 \text{ m s}^{-1})}{2.85 \times 10^{-5}\text{ m}}\right)^2 = 25.4\text{ kg s}^{-2} = \boxed{25.4\text{ N m}^{-1}}$$

How does the unit kg s^{-2} suddenly turn into N m^{-1} in the last step? The SI unit of force (as in *force* constant) is the newton (N) and 1 N = 1 kg m s^{-2} (text Appendix Table B.2). Multiply (kg s^{-2}) by 1 in the form (m^1 m^{-1}). The result is (kg m s^{-2} m^{-1}), which is equivalent to N m^{-1}.

Tip. A unit of "force per distance" makes sense as the force required to stretch or compress something, such as a spring, a given distance.

20.19 The frequency absorbed by the "signature" C—H stretch is given by $\nu = (1/2\pi)\sqrt{k/\mu}$. Also, $\nu = c/\lambda$. Eliminate ν between these equations and solve for the force constant k

$$k = \mu\left(\frac{2\pi c}{\lambda}\right)^2$$

Next, insert the given values of λ and μ. The reduced mass μ is approximated as the mass of the H atom, which equals the molar mass of H divided by Avogadro's number

$$k = \frac{0.001008\text{ kg mol}^{-1}}{6.022 \times 10^{23}\text{ mol}^{-1}}\left(\frac{2\pi(2.9979 \times 10^8 \text{ m s}^{-1})}{3.4 \times 10^{-6}\text{ m}}\right)^2 = 510\text{ kg s}^{-2} = \boxed{510\text{ N m}^{-1}}$$

Tip. The approximation in this problem is equivalent to imagining that the H atom, which has a small mass and is easily shoved around, does all the vibrating and that the rest of the molecule does not move.

20.21 Abbreviate the initial quantum state (v_i, J_i) of the HCl molecule as $(0,2)$. The selection rule for vibrational spectra is $\Delta v = \pm 1$. The vibrational quantum number v obviously does not remain unchanged. Therefore v increases by 1. The selection rule for the rotational quantum number is $J = \pm 1$. The most likely final quantum states after absorbing a infrared photon are therefore $(v_f, J_f) = \boxed{(1,3) \text{ and } (1,1)}$.

---

[4] The symbol for the vibrational quantum number is the letter v (vee for vibration). It is easily confused with the symbol for frequency, which is the Greek letter ν. The use of v_r is sometimes recommended to avoid this confusion.

Modify text equation 20.14 to calculate the energies of these two transitions in wave-numbers. This saves labor because the problem gives the vibrational frequency of the HCl molecule and its rotational constant in terms of wave-numbers: $\tilde{\nu}_{vib} = 2990$ cm^{-1} and $\tilde{B}_{rot} = 10$ cm^{-1}. The modified version allows direct use of these values. It is

$$\tilde{\nu}_{photon} = \tilde{\nu}_{vib}\big(v_f - v_i\big) + \tilde{B}_{rot}\Big[J_f(J_f + 1) - J_i(J_i + 1)\Big]$$

The transition $(0,2) \longrightarrow (1,3)$ has $v_f = 1$ and $v_i = 0$ and $J_f = 3$ and $J_i = 2$. Substituting the numbers gives

$$\tilde{\nu}_{photon} = 2990 \text{ cm}^{-1}\big(1-0\big) + 10 \text{ cm}^{-1}\big(3(3+1) - 2(2+1)\big) = 2990 \text{ cm}^{-1} + 60 \text{ cm}^{-1} = \boxed{3050 \text{ cm}^{-1}}$$

The transition $(0,2) \longrightarrow (1,1)$ has $v_f = 1$ and $v_i = 0$ and $J_f = 1$ and $J_i = 2$, so

$$\tilde{\nu}_{photon} = 2990 \text{ cm}^{-1}\big(1-0\big) + 10 \text{ cm}^{-1}\big(1(1+1) - 2(2+1)\big) = 2990 \text{ cm}^{-1} - 40 \text{ cm}^{-1} = \boxed{2950 \text{ cm}^{-1}}$$

Both transitions lie in the infra-red region of the spectrum, which is 400 to 4000 cm^{-1}.

The HCl in its original state $\boxed{\text{can undergo an absorptive rotational transition.}}$ This transition would take the molecule from the rotational state $J_i = 2$ to $J_f = 3$ and leave its vibrational state unchanged. A $\boxed{60 \text{ cm}^{-1}}$ photon would supply the required energy. The absorption appears in the microwave region of the spectrum.

20.23 The ratio of the population of two vibrational quantum levels i and j of different energies is

$$\frac{P(E_i)}{P(E_j)} = \exp\Big(-(E_i - E_j)/k_B T\Big)$$

The problem asks for this ratio for the vibrational ground state and first excited state in N_2 at 450 K. Let the j-th state be $v = 0$, the ground state, and let the i-th state be $v = 1$, the first excited vibrational state. The difference between the energies of these states is just h times the natural oscillation frequency (vibrational frequency) of the system

$$E_1 - E_0 = h\nu = (6.626 \times 10^{-34} \text{ J s})(7.07 \times 10^{13} \text{ s}^{-1}) = 4.685 \times 10^{-20} \text{ J}$$

Obtain the desired ratio by substitution into text equation 20.4. Neither state is degenerate, so the g's equal 1

$$\frac{N(E_1)}{N(E_0)} = \exp\left(\frac{-(4.685 \times 10^{-20} \text{ J})}{(1.3808 \times 10^{-23} \text{ J K}^{-1})(450 \text{ K})}\right) = \boxed{0.00053}$$

Tip. At 450 K, which is hot, the lowest excited vibrational level remains sparsely populated compared to the ground state.

20.25 The vibrational frequency of a diatomic molecule is given by

$$\nu = \frac{1}{2\pi}\sqrt{\frac{k}{\mu}}$$

where k is the force constant of the bond connecting the atoms and μ is the reduced mass of the molecule. In this problem vibrational frequencies are available from observations on four diatomic molecules and force constants are desired. Solve the preceding for k

$$k = (2\pi\nu)^2 \mu$$

The reduced masses (the μ's) of the four molecules equal the product of the masses of their two atoms divided by the sum of the masses of the two. Text Table 19.1 provides the necessary mass data. Use the masses of the isotopes that are specified in the formulas

$$\mu_{^1H^{19}F} = \frac{m_{^1H}m_{^19F}}{m_{^1H} + m_{^19F}} = \frac{(1.00783 \text{ u}) (18.9984 \text{ u})}{(1.00783 + 18.9984) \text{ u}} = 0.95706 \text{ u} = 1.5892 \times 10^{-27} \text{ kg}$$

$$\mu_{^1H^{35}Cl} = \frac{m_{^1H}m_{^35Cl}}{m_{^1H} + m_{^35Cl}} = \frac{(1.00783 \text{ u}) (34.9689 \text{ u})}{(1.00783 + 34.9689) \text{ u}} = 0.97960 \text{ u} = 1.6267 \times 10^{-27} \text{ kg}$$

$$\mu_{^1H^{81}Br} = \frac{m_{^1H}m_{^81Br}}{m_{^1H} + m_{^81Br}} = \frac{(1.007823 \text{ u}) (80.9163 \text{ u})}{(1.00783 + 80.9163) \text{ u}} = 0.99543 \text{ u} = 1.6529 \times 10^{-27} \text{ kg}$$

$$\mu_{^1H^{126}I} = \frac{m_{^1H}m_{^126I}}{m_{^1H} + m_{^127I}} = \frac{(1.00783 \text{ u}) (126.9045 \text{ u})}{(1.00783 + 126.9045) \text{ u}} = 0.99989 \text{ u} = 1.6604 \times 10^{-27} \text{ kg}$$

Passing from atomic mass units to kilograms uses the fact that 1 u equals 1.66054×10^{-27} kg.[5] The vibrational frequencies of the four molecules are quoted in the problem in units of reciprocal length (that is, as wave numbers) rather than in units of reciprocal time. Multiply them by the speed of light c in cm s^{-1} to convert from cm^{-1} to s^{-1}. Then substitute into the expression for k. For the first compound

$$\nu_{^1H^{19}F} = (4139 \text{ cm}^{-1})(2.9979 \times 10^{10} \text{ cm s}^{-1}) = 1.2408 \times 10^{14} \text{ s}^{-1}$$

$$k_{^1H^{19}F} = 4\pi^2(\nu_{^1H^{19}F})^2\mu_{^1H^{19}F}$$

$$= 4\pi^2(1.2408 \times 10^{14} \text{ s}^{-1})^2(1.5892 \times 10^{-27} \text{ kg})$$

$$= 966.0 \text{ kg s}^{-2} = \boxed{966.0 \text{ N m}^{-1}}$$

In the following, the other force constants are computed in the same way except that the conversion from cm^{-1} to s^{-1} is combined with the other arithmetic rather than carried out separately:

$$k_{^1H^{35}Cl} = 4\pi^2 c^2 (2991 \text{ cm}^{-1})^2(1.6266 \times 10^{-27} \text{ kg}) = \boxed{516.3 \text{ N m}^{-1}}$$

$$k_{^1H^{81}Br} = 4\pi^2 c^2 (2449 \text{ cm}^{-1})^2(1.6529 \times 10^{-27} \text{ kg}) = \boxed{351.7 \text{ N m}^{-1}}$$

$$k_{^1H^{126}I} = 4\pi^2 c^2 (2308 \text{ cm}^{-1})^2(1.6604 \times 10^{-27} \text{ kg}) = \boxed{313.8 \text{ N m}^{-1}}$$

The force constants range from 314 to 970 N m^{-1} while the reduced masses range only from about 0.96 to 1.0 u. Clearly the $\boxed{\text{force constants dominate}}$ in determining vibrational frequency in this group of molecules.

Tip. Confirm, as in problem **20.17**, that a kg s^{-2} (kilogram per second squared) equals a N m^{-1} (newton per meter).

20.27   In a diatomic molecule, the frequency of the stretching vibration depends on the force constant of the bond and the reduced mass of the molecule according to

$$\nu = \frac{1}{2\pi}\sqrt{\frac{k}{\mu}}$$

Approximate the stretching vibration of the C—H bond in the phenyl ring as a stretch between an H atom and a very heavy second atom. Make the same approximation for the C—D stretch. The reduced masses of these "diatomic" molecules then essentially equal the mass of the H atom (1 u) and the mass of the D atom (2 u) respectively. This approximation amounts to regarding the phenyl ring as an immovable object. Next, take the the ratio of the C—D stretching frequency to the C—H stretching frequency in terms of the preceding equation. The constants cancel out and

$$\frac{\nu_{CD}}{\nu_{CH}} = \sqrt{\frac{k_{CD}\,\mu_H}{k_{CH}\,\mu_D}}$$

---

[5]This equivalency appears in the table on the inside back page of the text.

The force constants k_{CH} and k_{CD} depend almost entirely on the chemical nature of the bond and so should remain nearly unchanged by the substitution of D for H. If so they cancel out and

$$\frac{\nu_{CD}}{\nu_{CH}} \approx \sqrt{\frac{\mu_H}{\mu_D}} = \sqrt{\frac{1 \text{ u}}{2 \text{ u}}}$$

$$\nu_{CD} \approx \sqrt{1/2}\,\nu_{CH}$$

Simply use the frequency in wave numbers since it is directly proportional to frequency expressed in units of reciprocal time

$$\nu_{CD} \approx 0.707\,(3062 \text{ cm}^{-1}) = \boxed{2165 \text{ cm}^{-1}}$$

The answer $\boxed{6.490 \times 10^{13} \text{ s}^{-1}}$ is equivalent.

20.29   The number of vibrational modes equals three times the number of atoms N in the molecule less six (or five in the case of linear molecules).

a) NH_3 has $3N - 6 = 3(4) - 6 = \boxed{6}$ modes.   **b)** C_2H_4 has $3N - 6 = 3(6) - 6 = \boxed{12}$ modes.

c) CCl_2F_2 has $3N - 6 = 3(5) - 6 = \boxed{9}$ modes.   **d)** CH_3CH_2OH has $3(9) - 6 = \boxed{21}$ modes.

20.31   In these infra-red spectra, the absorption of radiation corresponds to valleys, not peaks.

Prominent absorption bands in the first spectrum include aliphatic C—H stretching modes around 2900 cm^{-1} and —CH_2 and —CH_3 bending modes near 1450 and 1375 cm^{-1}. Prominent absorptions in the second spectrum include a broad band around 3400 cm^{-1} assignable to an O—H stretching vibration (a feature that is characteristic of hydrogen-bonded alcohols), —CH_2 and —CH_3 bending modes near 1450 and 1375 cm^1 (similar to those in the first spectrum), and a C—O stretch around 1075 cm^{-1}, another feature characteristic of alcohols.

Based on these features, the IR spectra are from $\boxed{\text{first nonane, second 1-hexanol}}$.

Nuclear Magnetic Resonance Spectroscopy

20.33   The three molecules are butane ($CH_3CH_2CH_2CH_3$), dimethyl ether (H_3C—O—CH_3), and dimethylamine (H_3C—NH—CH_3). All three have two CH_3 groups (methyl groups). "Low resolution" means that the splitting of nuclear magnetic resonance peaks that is caused by spin-spin coupling is not detected.

Butane. Rapid rotation around the C—C bond axes makes the three protons within the methyl groups chemically equivalent; the two-fold symmetry of the molecule makes the two methyl groups equivalent. Consequently, the six methyl protons are chemically equivalent. The four methylene protons (those in the CH_2 groups) are chemically equivalent to each other because of the rapid rotation around the H_2C—CH_2 bond. However they differ chemically from the six in the methyl groups.

Dimethyl ether. The six protons are made chemically equivalent by the rapid rotations about C—O bond axes.

Dimethylamine. The six protons in the methyl groups are equivalent to each other for the reasons discussed in the case of butane. The amine proton (the one bonded directly to the N) differs chemically from other six protons.

The expected number of peaks and relative peak areas are therefore

Molecule	Number of Peaks	Relative Peak Areas
$CH_3CH_2CH_2CH_3$	2	$6:4$
H_3COCH_3	1	—
H_3CNHCH_3	2	$6:1$

Electronic Spectroscopy and Excited State Relaxation Processes

20.35 The lowest unoccupied molecular orbital of et20.13ene is an antibonding orbital (a π^* orbital). The electron that is gained as the $C_2H_4^-$ ion is formed from C_2H_4 goes into this orbital. An additional antibonding electron means that the bond order $\boxed{\text{decreases}}$ in $C_2H_4^-$ relative to C_2H_4.

20.37 The color of a substance is the complement of the color of the light that the substance absorbs. The complement of orange is blue. Expect absorption $\boxed{\text{around 450 nm}}$.

20.39 The benzene molecule contains a cyclic system of three conjugated double bonds; The cyclohexene molecule contains a single double bond. The absorption of UV light in these molecules occurs with excitation of electrons associated with the multiple bonding. Delocalization of the double bonds in benzene should lower the energy for such a transition. Therefore, the absorption occurs at $\boxed{\text{shorter wavelength}}$ in cyclohexene.

20.41 The bond dissociation energy of Cl—F equals 252 kJ mol^{-1}, according to text Table 3.3. Express this energy change in joules per molecule by dividing it by Avogadro's number. Then calculate the corresponding wavelength

$$\lambda = \frac{hc}{\Delta E} = \frac{(6.626 \times 10^{-34} \text{ J s})(2.9979 \times 10^8 \text{ m s}^{-1})}{(2.52 \times 10^5 \text{ J mol}^{-1})/(6.0221 \times 10^{23} \text{ mol}^{-1})} = \boxed{4.75 \times 10^{-7} \text{ m}}$$

Radiation having a wavelength equal to or less than 475 nm can dissociate ClF molecules.

20.43 The finished plot contains two curves, one for the ground state and the other for the excited state. The ground-state curve has a minimum at a H_3C—I distance of 200 pm. The depth of this minimum is 220 kJ mol^{-1}, corresponding to the stated strength of the bond. Absorption of a 260 nm photon raises the energy of the molecule substantially

$$\Delta E = \frac{hc}{\lambda} = \frac{6.626 \times 10^{-34} \text{ J s } (3.00 \times 10^8 \text{ m s}^{-1})}{260 \times 10^{-9} \text{ m}} \left(\frac{6.022 \times 10^{23} \text{ molec.}}{\text{mol}} \right) = 460 \text{ kJ}$$

This input of energy exceeds the H_3C—I bond energy. The resulting state is dissociative; the potential energy curve has no minimum. Dissociation generates a methyl group and a iodine atom. When the two groups are in this excited state they repel each other, and the strength of the repulsion increases if the distance between the two decreases.

Introduction to Atmospheric Chemistry

20.45 Divide 440 kJ, the energy of a mole of bonds, by Avogadro's number to obtain 7.31×10^{-19} J, the energy change in the dissociation of one bond. Calculate the wavelength corresponding to this energy change as

$$\lambda = \frac{hc}{\Delta E} = \frac{(6.626 \times 10^{-34} \text{ J s})(2.9979 \times 10^8 \text{ m s}^{-1})}{7.31 \times 10^{-19} \text{ J}} = \boxed{2.72 \times 10^{-7} \text{ m}}$$

Light of wavelengths shorter than this supplies more than enough energy to break C—F bonds; light of longer wavelengths cannot dissociate these bonds.

Tip. The 272 nm light also suffices to dissociate the C—C bonds and C—Cl bonds in the chlorofluorocarbon because the energies of these bonds are substantially less than the energy of the C—F bond. See text Table 12.5.

20.47 The best Lewis structures are a resonance pair:

$$\left[\ddot{:}\ddot{O}{-}\ddot{O}{=}\ddot{O} \quad \longleftrightarrow \quad \ddot{:}O{=}\ddot{O}{-}\ddot{O}\ddot{:} \right]$$

The central O atom has a lone pair in both structures. The VSEPR model assigns the central O atom $\boxed{SN\,3}$ and thereby predicts $\boxed{sp^2}$ hybridization and an angle of (approximately) 120° at the central O atom: the ozone molecule is $\boxed{\text{bent}}$. There are two electrons in a bonding π orbital formed from the three $2p_z$ orbitals perpendicular to the molecular plane. The non-bonding and antibonding orbitals in this π system are unoccupied. The total bond order is 3, which corresponds to a bond order of $\boxed{3/2}$ for each O-to-O linkage.

Photosynthesis

20.49 Compute the energy of a single photon of the specified wavelength

$$E = h\nu = \frac{hc}{\lambda} = \frac{(6.626 \times 10^{-34}\ \text{J s})(2.9979 \times 10^8\ \text{m s}^{-1})}{4.30 \times 10^{-7}\ \text{m}} = 4.62 \times 10^{-19}\ \text{J}$$

The energy of 1.00 mol of such photons equals this value multiplied by Avogadro's number. It comes out to 278 kJ. The conversion ADP$\longrightarrow$ATP has ΔG° equal to $+34.5$ kJ. The 278 kJ from 1.00 mol of photons is 8.06 times larger than 34.5 kJ. Therefore 8.06 mol of ATP could be produced by 1.00 mol of 430 nm photons. This means that at most $\boxed{\text{8 molecules}}$ of ATP are produced by a single 430 nm photon.

Tip. The value of ΔG° given in the problem is for "biochemical standard conditions." This kind of ΔG° is often distinguished by adding a prime to the symbol: $\Delta G^{\circ\prime}$. No $\Delta G^{\circ\prime}$'s appear in text Appendix D. Under biochemical standard conditions the activity of H_3O^+ is defined as 1 at pH 7 (rather than at pH 0). Biochemical standard conditions provide a reference state for thermodynamic values that is closer to the conditions prevailing in living cells. The ΔG of the ADP $\longrightarrow$ ATP reaction, like that of most biochemical reactions, is highly dependent on pH.

ADDITIONAL PROBLEMS

20.51 The thermal populations of molecular quantum levels follow a Boltzmann distribution. The ratio of the number of molecules in two quantum states depends on their relative energies and on their degeneracies as well, according to text equation 20.4

$$\frac{N_i}{N_j} = \frac{g_i\ \exp(-E_i/k_B T)}{g_j\ \exp(-E_j/k_B T)}$$

where the g's refer the degeneracy of the quantized energy state that is identified by the subscript, and the E's refer to their energies.

The problem concerns the rotational states of the linear molecule NaH. For convenience, change the subscripts in the preceding to use the quantum numbers J of these rotational states

$$\frac{N_{J_1}}{N_{J_2}} = \frac{g_{J_1}\ \exp(-E_{J_1}/k_B T)}{g_{J_2}\ \exp(-E_{J_2}/k_B T)}$$

Text equation 20.6a gives the energies of the rotational states in linear molecules such as NaH assuming that they are rigid rotors

$$E_J = \frac{h^2}{8\pi^2 I} J(J+1)$$

The energy of the Jth rotational quantum state of NaH is

$$E_J = J(J+1)\left(\frac{(6.626 \times 10^{-34}\ \text{J s})^2}{8\pi^2(5.70 \times 10^{-47}\ \text{kg m}^2)}\right)$$

Divide the negative of this energy by $k_B T$ for use in the Boltzmann distribution

$$-E_J/k_B T = J(J+1)\left(\frac{-(6.626 \times 10^{-34}\ \text{J s})^2}{8\pi^2(5.70 \times 10^{-47}\ \text{kg m}^2)}\right)\left(\frac{1}{(1.38 \times 10^{-23}\ \text{J K}^{-1})(298.15\ \text{K})}\right)$$

$$= J(J+1)(-0.0237)$$

The units all cancel out. This is good because the next step requires taking the exponential of this quantity.

According to text equation 20.5b, the degeneracy of the rotational energy levels of molecules like NaH is

$$g_J = 2J + 1$$

Substitute this expression and the preceding into the Boltzmann distribution

$$\frac{N_{J_1}}{N_{J_2}} = \left(\frac{2J_2 + 1}{2J_1 + 1}\right) \frac{\exp\left(-.0237\, J_2(J_2 + 1)\right)}{\exp\left(-.0237\, J_1(J_1 + 1)\right)}$$

Let state 2 be the ground state. This means $J_2 = 0$. The ratios of the populations in the $J_1 = 5$, $J_1 = 15$ and $J_1 = 25$ rotational states to the population in the ground state of NaH are then

a) $\quad \dfrac{N_{J_5}}{N_{J_1}} = \left(\dfrac{2(5) + 1}{2(0) + 1}\right) \dfrac{\exp\left(-.0237\,(5)(5 + 1)\right)}{1} = \boxed{5.40}$

b) $\quad \dfrac{N_{J_{15}}}{N_{J_1}} = \left(\dfrac{2(15) + 1}{2(0) + 1}\right) \dfrac{\exp\left(-.0237\,(15)(15 + 1)\right)}{1} = \boxed{0.105}$

c) $\quad \dfrac{N_{J_{25}}}{N_{J_1}} = \left(\dfrac{2(25) + 1}{2(0) + 1}\right) \dfrac{\exp\left(-.0237\,(25)(25 + 1)\right)}{1} = \boxed{1.04 \times 10^{-5}}$

Tip. The point of the problem is that the first dozen or so rotational excited states of this typical small molecule are abundantly populated at room temperature.

20.53 The moment of inertia of a diatomic molecule equals $I = \mu R_e^2$ where R_e is the equilibrium bond distance and μ is the reduced mass. Substitution of the masses[6] of ^{1}H, ^{19}F and ^{81}Br into the formula for the reduced mass gives

$$\mu_{\text{HF}} = \frac{(1.007825032\ \text{u})(18.9984032\ \text{u})}{(1.007825032 + 18.9984032)\ \text{u}} = 0.957055\ \text{u}$$

$$\mu_{\text{HBr}} = \frac{(1.007825032\ \text{u})(80.916291\ \text{u})}{(1.007825032 + 80.916291)\ \text{u}} = 0.99543\ \text{u}$$

In kilograms, the two reduced masses are

$$\mu_{\text{HF}} = \boxed{1.5893 \times 10^{-27}\ \text{kg}} \quad \text{and} \quad \mu_{\text{HBr}} = \boxed{1.6529 \times 10^{-27}\ \text{kg}}$$

The equilibrium bond distances equal 0.926×10^{-10} m for HF and 1.424×10^{-10} m for HBr.[7] Using $I = \mu R_e^2$ gives these moments of inertia:

$$I_{\text{HF}} = 1.363 \times 10^{-47}\ \text{kg m}^2 \quad \text{and} \quad I_{\text{HBr}} = 3.352 \times 10^{-47}\ \text{kg m}^2$$

The rotational spectra of diatomic molecules consist of lines equally spaced in frequency with the separation between adjacent lines equal to $2\tilde{B} = h/4\pi^2 I$. Therefore

$$\left(\frac{h}{4\pi^2 I}\right)_{\text{HF}} = \frac{6.626 \times 10^{-34}\ \text{J s}}{4\pi^2(1.363 \times 10^{-47}\ \text{kg m}^2)} = \boxed{12.3 \times 10^{11}\ \text{s}^{-1}}$$

$$\left(\frac{h}{4\pi^2 I}\right)_{\text{HBr}} = \frac{6.626 \times 10^{-34}\ \text{J s}}{4\pi^2(3.352 \times 10^{-47}\ \text{kg m}^2)} = \boxed{5.01 \times 10^{11}\ \text{s}^{-1}}$$

---

[6] From Text Table 19.1.
[7] Text Table 3.3.

The large increase in molecular mass from HF and HBr increases the reduced mass of the diatomic molecule only a little. Why? In HF, the center of rotation is already very close to the F atom because F is 19 times heavier than H; the H atom does most of the moving about the center of rotation. Even a big increase in the mass of the heavy atom (replacement of F by Br) moves the center of mass only a little closer to the heavy atom.

20.55   The reduced masses of the three diatomic molecules are

$$\mu_{\text{NaH}} = \frac{(22.9897697\ \text{u})(1.007825032\ \text{u})}{(22.989770 + 1.007825032)\ \text{u}} = \boxed{0.9654995\ \text{u}}$$

$$\mu_{\text{NaCl}} = \frac{(22.9897697\ \text{u})(34.96885271\ \text{u})}{(22.9897697 + 34.96885271)\ \text{u}} = \boxed{13.870686\ \text{u}}$$

$$\mu_{\text{NaI}} = \frac{(22.9897697\ \text{u})(126.904468\ \text{u})}{(22.9897697 + 126.904468)\ \text{u}} = \boxed{19.463754\ \text{u}}$$

The reduced mass of NaD will also be needed

$$\mu_{\text{NaD}} = \frac{(22.9897697\ \text{u})(2.014101778\ \text{u})}{(22.9897697 + 2.014101778)\ \text{u}} = 1.85186266\ \text{u}$$

The force constant of the bond in a diatomic molecule is $k = \mu(2\pi\nu)^2$ where μ is the reduced mass and ν is the vibrational frequency

$$k_{\text{NaH}} = \frac{0.9654995\ \text{u}}{6.022142 \times 10^{26}\ \text{u kg}^{-1}}(4\pi^2)(3.51 \times 10^{13}\ \text{s}^{-1})^2 = \boxed{78.0\ \text{kg s}^{-2}}$$

$$k_{\text{NaCl}} = \frac{13.870686\ \text{u}}{6.022142 \times 10^{26}\ \text{u kg}^{-1}}(4\pi^2)(1.10 \times 10^{13}\ \text{s}^{-1})^2 = \boxed{110\ \text{kg s}^{-2}}$$

$$k_{\text{NaI}} = \frac{19.463754\ \text{u}}{6.022142 \times 10^{26}\ \text{u kg}^{-1}}(4\pi^2)(0.773 \times 10^{13}\ \text{s}^{-1})^2 = \boxed{76.2\ \text{kg s}^{-2}}$$

Insert the reduced mass of NaD and the force constant of NaH into the formula for the vibrational frequency of a diatomic molecule. The reduced mass of NaD is calculated in atomic mass units in the preceding. Converting to kilograms gives $3.0750897 \times 10^{-27}$ kg. The answer is

$$\nu_{\text{NaD}} = \frac{1}{2\pi}\sqrt{\frac{k}{\mu}} = \frac{1}{2\pi}\sqrt{\frac{78.0\ \text{kg s}^{-2}}{3.0750897 \times 10^{-27}\ \text{kg}}} = \boxed{2.53 \times 10^{13}\ \text{s}^{-1}}$$

Tip. The three bond distances were not needed.

20.57   The difference in energy ΔE between the $\nu_{\text{r}} = 1$ and $\nu_{\text{r}} = 0$ vibrational states of HgBr(g) equals $h\nu$ where ν is the given vibrational frequency of the molecule. The ratio of the populations P of the two states depends on $e^{-\Delta E/k_{\text{B}}T}$ and equals 0.127 at some temperature T. Compute T as follows

$$\ln\left(\frac{P_1}{P_0}\right) = \frac{-\Delta E}{k_{\text{B}}T} \qquad \text{or, after substitution:} \qquad \ln(0.127) = \frac{-h\nu}{k_{\text{B}}T}$$

Solve for the temperature and evaluate

$$T = \frac{-h\nu}{k_{\text{B}}\ln(0.127)} = \frac{-(6.626 \times 10^{-34}\ \text{J s})(5.58 \times 10^{12}\ \text{s}^{-1})}{(1.3807 \times 10^{-23}\ \text{J K}^{-1})(-2.0636)} = \boxed{130\ \text{K}}$$

20.59   **a)** There are five C-C double bonds. The isomer to the left of the arrow has four *trans* C-C double bonds in the chain extending to the right from the six-membered ring. The double bond in the six-membered ring is also *trans* when the relative positions of the two largest groups, one of which is the long side-chain, are considered. Hence, this isomer has $\boxed{\text{five}}$ *trans* double bonds. The isomer to the right of the arrow is the same except that the second C-C double bond from the right end of the side chain is *cis*. This isomer has $\boxed{\text{four}}$ *trans* double bonds.

b) The absorption maximum would $\boxed{\text{shift to shorter wavelength}}$. Loss of the ring and the —CHO group would reduce the range of delocalization of electrons in a system of alternating single and double bonds because the ring contains a C=C double bond, and the —CHO group contains a C=O double bond.

20.61 **a)** The carbon atom in formaldehyde is sp^2 hybridized.

b) Formaldehyde has ten valence orbitals: three σ orbitals formed by overlap of sp^2 orbitals on the C atom with $1s$ orbitals on the two H atoms and the $2p$-orbital on the O atom that points toward the carbon; three empty σ^* orbitals with the same parents; two lone-pair $2s$ and $2p$ orbitals on the O atom; one occupied π (bonding) orbital derived from the two remaining $2p$ orbitals, which are directed perpendicular to the plane of the molecule; one empty π^* (antibonding) orbital derived from the same parents.

c) The weaker transition at lower frequency is probably due to excitation of an electron from a lone-pair $2p$ orbital on the oxygen atom to the π^* orbital.

20.63 Nitrogen dioxide is a radical. In the stratosphere, NO_2 would photodissociate to give NO which could catalyze the destruction of O_3. Plausible mechanisms are easy to write:

$$NO + O_3 \longrightarrow NO_2 + O_2 \qquad NO_2 + O \longrightarrow NO + O_2$$

Tip. A radical is a molecule or ion that has one or more unpaired electrons. Nitrogen dioxide has 17 valence electrons, an odd number. It therefore has at least one unpaired electron.

In the troposphere, where the concentration of O_3 is small and the concentration of NO_2 is higher, NO_2 participates in the formation of O_3:

$$NO_2 + h\nu \longrightarrow NO + O \qquad O + O_2 + M \longrightarrow O_3 + M$$

Ozone is bad in the troposphere because of its high toxicity. Unfortunately, O_3 created in the troposphere reacts there with other molecules before it has a chance to diffuse up into the stratosphere, where it might do some good in its role as a filter against actinic ultra-violet radiation from the Sun.

20.65 One would expect the ground-state energy of the special pair of bacteriochlorophyll molecules, the reaction center of a photosynthetic units, to be $\boxed{\text{lower}}$ than the energy of a single bacteriochlorophyll molecule. Molecular orbitals from the single molecules would mix to give new molecular orbitals in the pair. Some of these would have lower energies than their unmixed precursors. Electrons would occupy these lower-energy molecular orbitals in the ground state.

CUMULATIVE PROBLEMS

20.67 The absorbance of a solution of two (or more) species that absorb radiation at a given wavelength equals the sum of the absorbances of the species taken separately if the species do not interact and if the length ℓ of the path that light traverses through the solutions is the same in all measurements (see problem 20.3). In this problem two species absorb radiation but also interact, fortunately, in a simple way. The two are *m*-nitrophenol, an indicator, in its acid form HIn, and in its conjugate-base form In⁻. The concentrations of these two forms are related:

$$K_a = \frac{[H_3O^+][In^-]}{[HIn]} \qquad \text{from which} \qquad pK_a = pH - \log\frac{[In^-]}{[HIn]}$$

The absorbances (the A's) of solutions of the indicator at a concentration of 6.36×10^{-4} mol L^{-1} range from 0.142 at low pH, where essentially all indicator is present as HIn, up to 0.943 at high pH, where essentially all is present as In$^-$. Absorbances of solutions whose pH's have been buffered to intermediate values lie between these extremes. Beer's law (text equation 20.3) states that absorbances are proportional to concentrations. Absorbances of intermediate mixtures are consequently weighted averages of the two extreme absorbances where the weights reflect the relative concentrations of the two. Thus, if [HIn] = [In$^-$], the two species contribute equally to A, which equals their simple average: $A = \frac{1}{2}(0.142) + \frac{1}{2}(0.943) = 0.5425$. If [HIn] > [In$^-$], then the total absorbance is less than 0.5425 because HIn absorbs more weakly than In$^-$. And if [HIn] < [In$^-$], then the total absorbance is more. The $\frac{1}{2}$'s in the preceding need to be replaced with an x and $1 - x$ respectively to cover these situations. Then x fixes the contributions of the two species to the total absorbance according to

$$A = x(0.142) + (1 - x)(0.943)$$

Solving for x in terms of A is easy

$$x = \frac{0.943 - A}{0.801} \qquad \text{which means} \qquad (1 - x) = \frac{A - 0.142}{0.801}$$

The factor x dwindles from 1 (at low pH) toward 0 (at high pH) as [HIn] decreases. Concurrently, $(1 - x)$ and [In$^-$] increase. Thus x and $1 - x$ are proportional to the concentrations of HIn and In$^-$ respectively. Taking the ratio of the second of these two proportionalities to the first gives

$$\frac{[\text{In}^-]}{[\text{HIn}]} = \frac{1 - x}{x}$$

Substituting this result into the equation for pK_a of the indicator then gives

$$pK_a = pH - \log\left(\frac{A - 0.142}{0.943 - A}\right)$$

The problem presents a table of five pH's that were established using buffers of differing ionic strengths I along with the absorbance A at 350 nm of a fixed total concentration of the indicator. Substitute the pH's and the A's from the table into the preceding equation to compute the pK_a's of the indicator in each solution. The answers are

pH	I	A	pK_a
8.321	0.10	0.527	8.355
8.302	0.08	0.518	8.355
8.280	0.06	0.505	8.361
8.251	0.04	0.493	8.359
8.208	0.02	0.470	8.367

Despite a noticeable drop in pH as the ionic strength dropped (top to bottom in the table), the pK_a of the indicator stays nearly constant.

Tip. Knowledge of the total concentration of the indicator is not needed as long as this concentration is the same in all of the experiments.

20.69 Compute the concentrations of the gaseous CO and gaseous CH$_4$ in the air at the given temperature, assuming ideality

$$[\text{CO}] = \frac{n}{V} = \frac{P}{RT} = \frac{1.0 \times 10^{-7} \text{ atm}}{(0.082 \text{ L atm mol}^{-1}\text{K}^{-1})(300 \text{ K})} = 4.07 \times 10^{-9} \text{ mol L}^{-1}$$

$$[\text{CH}_4] = \frac{n}{V} = \frac{P}{RT} = \frac{1.7 \times 10^{-6} \text{ atm}}{(0.082 \text{ L atm mol}^{-1}\text{K}^{-1})(300 \text{ K})} = 6.91 \times 10^{-8} \text{ mol L}^{-1}$$

CO and CH_4 react concurrently with OH in the atmosphere. Both reactions follow second-order kinetics, but it is assumed that the partial pressures (and therefore the concentrations) of the two stay constant. The resulting combined effective rate constant is

$$k = (1.6 \times 10^{11} \text{ L mol}^{-1}\text{s}^{-1})(4.07 \times 10^{-9}) \text{ mol L}^{-1}$$
$$+ (3.8 \times 10^{9} \text{ L mol}^{-1}\text{s}^{-1})(6.91 \times 10^{-8}) \text{ mol L}^{-1} = 651 \text{ s}^{-1} + 262 \text{ s}^{-1} = 913 \text{ s}^{-1}$$

The resulting k is a first-order rate constant. The half-life of OH in the atmosphere under these circumstances is

$$t_{1/2} = \frac{\ln 2}{k} = \frac{0.6931}{913 \text{ s}^{-1}} = \boxed{7.6 \times 10^{-4} \text{ s}}$$

Chapter 21

Structure and Bonding in Solids

Crystal Symmetry and the Unit Cell

21.1   **a)** One side in an isosceles triangle is non-equivalent to the other two; no 3-fold rotation axis exists.

b) A 3-fold axis passes through the center of an equilateral triangle.

c) A 3-fold axis passes through the center of each of the four triangular faces in a tetrahedron and out through the opposite vertex.

d) A 3-fold axis passes along each of the four long diagonals of a cube, that is, from each vertex to the most distant opposite vertex.

Items b), c), and d) have 3-fold rotational symmetry.

21.3   The CCl_2F_2 molecule has two mirror planes. The first is defined by the two Cl atoms and the central C atom, and the second is defined by the two F atoms and the central C atom. The intersection of the two mirror planes coincides with a single 2-fold axis of rotation. This axis passes through the central C atom and bisects the angles Cl1—C—Cl2 and F1—C—F2.

21.5   The Bragg law $n\lambda = 2d\sin\theta$ becomes in this case

$$2(1.660 \text{ Å}) = 2d\sin\left(\frac{54.70°}{2}\right) \qquad \text{from which} \qquad d = \frac{1.660 \text{ Å}}{\sin 27.35°} = \boxed{3.613 \text{ Å}}$$

21.7   The Bragg law $n\lambda = 2d\sin\theta$ becomes

$$4(1.936 \text{ Å}) = 2(4.950 \text{ Å})\sin\theta$$

where $n = 4$ comes from the specification of fourth-order diffraction and 4.950 Å is the interplanar spacing. Solving gives θ equal to 51.46° and 2θ equal to $\boxed{102.9°}$.

Tip. The angle 128.54° (180 − 51.46°) also fulfills the equation. This gives $2\theta = 257.1°$, which is equivalent to −102.9°. This corresponds to "reflection" from the other side of the layers of atoms.

21.9   Solve the Bragg law for θ and substitute the values given for this case of diffraction by crystalline LiCl

$$\theta = \sin^{-1}\left(\frac{n\lambda}{2d}\right) = \sin^{-1}\left(\frac{n\,2.167 \text{ Å}}{2(2.570 \text{ Å})}\right) = \sin^{-1}(0.4216\,n)$$

Inserting integers for n and immediately multiplying the results by 2 gives 2θ equal to $\boxed{\pm 49.87°}$ for $n = 1$ and 2θ equal to $\boxed{\pm 115.0°}$ for $n = 2$. Using higher values of n leads to arguments of $\sin^{-1}$ that exceed 1.00. The inverse sine function is not defined in such cases. Consequently 2θ has only the four possible values.

299

Crystal Structure

21.11 **a)** The cell angles are all $90°$, because the crystal is tetragonal. Then

$$V_{\text{cell}} = abc = (223.5)^2(113.6) = \boxed{5.675 \times 10^6 \text{ Å}^3}$$

b) The volume of the box-shaped crystal is likewise the product of the lengths of the three edges. It equals 3 mm^3—small, but easily visible with the unaided eye. To compare the two volumes divide one by the other and use a suitable unit factor to make sure the units cancel out

$$\frac{V_{\text{crystal}}}{V_{\text{cell}}} = \frac{3 \text{ mm}^3}{5.675 \times 10^6 \text{ Å}^3} \times \left(\frac{1 \text{ Å}^3}{10^{-21} \text{ mm}^3}\right) = \boxed{5 \times 10^{14}}$$

This ratio of volumes equals the number of units cells in the crystal.

21.13 Compute the mass of the contents of the unit cell and divide it by the volume of the cell to obtain the density of the cell. Since the substance consists of many copies of the unit cell side-by-side, this result is the density of the whole crystal. The mass of the contents of the unit cell equals twice the mass of a single formula unit of $Pb_4In_3B_{17}S_{18}$

$$m_{\text{contents}} = 2 \times 1934.235 = 3\,868.47 \text{ u}$$

The volume of the cell is

$$V_c = abc\sqrt{1 - \cos^2\alpha - \cos^2\beta - \cos^2\gamma + 2\cos\alpha\cos\beta\cos\gamma} = abc\sqrt{1 - \cos^2\beta} = abc\sin\beta$$

$$= (21.021 \text{ Å})(4.014 \text{ Å})(18.989 \text{ Å})(0.9924) = 1582.46 \text{ Å}^3$$

The density is then

$$\rho = \frac{m}{V} = \frac{3\,868.47 \text{ u}}{1582.46 \text{ Å}^3} \times \left(\frac{10^{24} \text{ Å}^3}{1 \text{ cm}^3}\right)\left(\frac{1 \text{ g}}{6.022 \times 10^{23} \text{ u}}\right) = \boxed{4.059 \text{ g cm}^{-3}}$$

The two unit-factors convert the answer from an unfamiliar unit of density to a familiar one.

21.15 **a)** The volume of the cubical unit cell in elemental silicon is just the edge of the cell cubed. It equals $(5.431 \text{ Å})^3$, which is 160.19 Å^3. There are 10^8 Å in a centimeter and consequently 10^{24} Å^3 in a cubic centimeter

$$V_{\text{cell}} = 160.19 \text{ Å}^3 \times \left(\frac{1 \text{ cm}^3}{10^{24} \text{ Å}^3}\right) = \boxed{1.602 \times 10^{-22} \text{ cm}^3}$$

b) The mass of the contents of the unit cell of crystalline silicon equals the volume of the unit cell multiplied by its density

$$m_{\text{cell}} = 1.602 \times 10^{-22} \text{ cm}^3 \times \left(\frac{2.328 \text{ g}}{1 \text{ cm}^3}\right) = \boxed{3.729 \times 10^{-22} \text{ g}}$$

c) The unit cell contains eight atoms of silicon for a total mass of 3.729×10^{-22} g. Consequently, a single atom has a mass of $\boxed{4.662 \times 10^{-23} \text{ g}}$.

d) One mole of silicon contains Avogadro's number of atoms of silicon. The molar mass of silicon is 28.0855 g mol^{-1}. Divide this molar mass by the single-atom mass of silicon to obtain Avogadro's number

$$\frac{28.0855 \text{ g mol}^{-1}}{4.662 \times 10^{-23} \text{ g}} = \boxed{6.025 \times 10^{23} \text{ mol}^{-1}} = N_A$$

This is only about 0.05% larger than the accepted value.

21.17   The volume of the unit cell in crystalline sodium sulfate equals the product of its three cell edges a, b, and c. This follows because the term under the radical sign in the equation for the volume of a unit cell

$$V_c = abc\sqrt{1 - \cos^2\alpha - \cos^2\beta - \cos^2\gamma + 2\cos\alpha\cos\beta\cos\gamma}$$

equals 1 when α, β, and γ are 90°, as they are in this orthorhombic crystal. The volume comes out to 708.47 Å^3, which is 7.0847×10^{-22} cm^3. The volume V_m of a mole of unit cells of sodium sulfate is Avogadro's number times the volume of one cell

$$V_m = (7.0847 \times 10^{-22} \text{ cm}^3) \times (6.022 10^{23} \text{ mol}^{-1}) = 426.6 \text{ cm}^3 \text{ mol}^{-1}$$

The density of a unit cell equals the density of the substance itself since a crystal consists of many unit cells stacked side by side. Multiplying the volume of a mole of unit cells by the density of the substance gives the mass of a mole of unit cells

$$m = dV_m = \left(\frac{2.663 \text{ g}}{1 \text{ cm}^3}\right)\left(\frac{426.6 \text{ cm}^3}{1 \text{ mol}}\right) = 1136.1 \text{ g mol}^{-1}$$

The molar mass of the formula unit Na_2SO_4 is 142.04 g mol^{-1}. This is far less than 1136.1 g mol^{-1}. The unit cell must hold several formula units. Because 142.04 is almost exactly 1/8th of 1136.1, it follows that each unit cell contains $\boxed{8}$ Na_2SO_4 formula units.

Tip. The unit cell contains 56 atoms. Obviously, all of these atoms cannot reside at the corners of the unit cell. They are distributed throughout the volume of the cell.

21.19   In this crystalline compound, rhenium atoms lie at the eight corners of the unit cell, and oxygen atoms lie at the mid-points of the 12 edges. Start by figuring out the number of atoms of Re and O per cell: each cell has 1 rhenium atom ($8 \times 1/8$) and 3 oxygen atoms ($12 \times 1/4$). The 1/8 and 1/4 appear because every corner of a unit cell is shared among a total of eight cells and every edge is shared among four cells. The ratio of these numbers gives the empirical formula because the compound is composed of many repeats of the unit cell. The chemical formula therefore is $\boxed{ReO_3}$.

Tip. Another way to explore the locations of the atoms is with fractional coordinates. The eight equivalent Re atoms have these coordinates

$$(0,0,0) \ \ (1,0,0) \ \ (0,1,0) \ \ (0,0,1) \ \ (1,1,1) \ \ (0,1,1) \ \ (1,0,1) \ \ (1,1,0)$$

These are the corners of the cube in text Figure 21.13. Only the first of these locations is distinct. The other seven, those containing 1's in their coordinates, are translations ("one cell over") of the first. The twelve equivalent O atoms at the midpoints of the cell edges have these fractional coordinates

$$\begin{array}{cccc}
\left(\frac{1}{2},0,0\right) & \left(\frac{1}{2},1,0\right) & \left(\frac{1}{2},0,1\right) & \left(\frac{1}{2},1,1\right) \\
\left(0,\frac{1}{2},0\right) & \left(1,\frac{1}{2},0\right) & \left(0,\frac{1}{2},1\right) & \left(1,\frac{1}{2},1\right) \\
\left(0,0,\frac{1}{2}\right) & \left(1,0,\frac{1}{2}\right) & \left(0,1,\frac{1}{2}\right) & \left(1,1,\frac{1}{2}\right)
\end{array}$$

Only the three locations at the beginning of the rows are independent. The nine containing one or two 1's are translations ("one cell over") of these three.

21.21   **a)** A body-centered cubic structure means two Fe atoms per unit cell, one in the center of the cell and one at each of the eight corners of the cell (each corner atom is shared by seven neighboring cells). The two atoms have a total mass of 111.694 u and touch along the body diagonal of the cell, but not along the edges. Compute the volume of the unit cell by multiplying its mass by its density

$$V_{cell} = 111.694 \text{ u Fe} \times \left(\frac{1 \text{ g Fe}}{6.0221 \times 10^{23} \text{ u Fe}}\right)\left(\frac{1 \text{ cm}^3}{7.86 \text{ g Fe}}\right) = 2.36 \times 10^{-23} \text{ cm}^3$$

The edge a of the cubic unit cell is the cube root of the volume. It is 2.87×10^{-8} cm, which is 2.87 Å. The nearest-neighbor distance is one-half the body diagonal b of the unit cell. The body diagonal is related to the edge e as follows

$$b = \sqrt{3}e = \sqrt{3}(2.87 \text{ Å}) = 4.97 \text{ Å}$$

Hence nearest neighbors are $\boxed{2.48 \text{ Å}}$ apart.

b) The lattice parameter equals $\boxed{2.87 \text{ Å}}$, the cubic cell's edge. See above.

c) The atomic radius of Fe equals one quarter of the body diagonal of the unit cell because Fe atoms are "in contact" along this diagonal. It is therefore $\boxed{1.24 \text{ Å}}$.

Tip. This answer equals the value tabulated for iron in text Appendix F.

21.23 **a)** A body-centered cubic lattice has two lattice points per unit cell. In metallic sodium, one Na atom is associated with each lattice point to give $\boxed{\text{two}}$ Na atoms per cell.

b) Let r_{Na} equal the radius of the Na atom. In the crystal, Na atoms touch along the body diagonal b of the cubic cell, which has atoms at its corners and center. This means $4r_{Na} = b$. But b is $\sqrt{3}$ times the edge of the cell. Hence $4r_{Na} = \sqrt{3}\,e$. Cubing this equation gives

$$64\,r_{Na}^3 = 3\sqrt{3}\,e^3$$

The volume of a single Na atom is $4/3\pi(r_{Na})^3$. Two Na atoms have twice this volume

$$V_{2Na} = 2 \times \left(\frac{4\pi(r_{Na})^3}{3}\right)$$

Solve this for $(r_{Na})^3$ and substitute into the equation that precedes it. Also recognize that the volume of the cell equals its edge cubed: $V_{cell} = e^3$.

$$64\left(\frac{3\,V_{2Na}}{8\pi}\right) = 3\sqrt{3}\,V_{cell} \qquad \text{from which} \qquad \left(\frac{V_{2Na}}{V_{cell}}\right) = \frac{(3\sqrt{3})(8\pi)}{3(64)} = \boxed{0.680}$$

Tip. This is not the most efficient packing of equal spheres. Putting the spheres at the lattice points of a face-centered cubic lattice, gives a more efficient packing (cubic close packing).

21.25 The atoms making up the simple cubic array are the host atoms. These atoms touch along the edges of the cubic unit cell. A guest interstitial atom sits in the hole at the center of the host unit cell. The body diagonal b of the host unit cell runs between two diagonally opposite host atoms and passes along the diameter of the guest atom. If the guest is as large as it can be without pushing the host atoms out of contact, then

$$e = 2r_{host} \qquad\qquad b = 2r_{host} + 2r_{guest}$$

But the body diagonal of a cube is longer than the edge by a factor of $\sqrt{3}$. Then

$$2r_{host} + 2r_{guest} = \sqrt{3}\,(2r_{host}) \qquad \text{so that} \qquad \frac{r_{guest}}{r_{host}} = \sqrt{3} - 1 = \boxed{0.732}$$

Cohesion in Solids

21.27 Use electronegativity differences and position in the periodic table.
  a) $BaCl_2$–ionic **b)** SiC–covalent **c)** CO–molecular **d)** Co–metallic.

21.29 The network solid SiC should have the highest melting point, and the metallic solid Co should have the second highest. The melting point of the ionic solid $BaCl_2$ should exceed the melting point of the molecular solid CO. The data on the melting points of Co and $BaCl_2$ links these two pairs. Hence $\boxed{CO < BaCl_2 < Co < SiC}$.

21.31   Consider the various possibilities open to the "typical atom" in a crystal. Suppose that it can form zero chemical bonds. Then the only forces available to maintain the crystal are van der Waals forces, which are weaker than chemical bonds. The absence of strong interatomic attractions leads to mechanically weak (soft) crystals of low melting point. If the typical atom forms one bond, then strongly bonded diatomic molecules form, but again the crystals are soft and low-melting because the formation of pairs of atoms uses up the atoms' bonding capacity. If the typical atom forms two bonds, then strong linear (thread-like) structures are possible. If the typical atom forms three or more bonds, then it is possible for such threads to cross-link to give two dimensional or three dimensional networks.

21.33   In the simple cubic CsCl lattice, the positive ion has $\boxed{8}$ Cl^- ions as nearest neighbors. The second nearest neighbors are a set of $\boxed{6}$ Cs^+ ions, and the third nearest neighbors are a set of $\boxed{12}$ yet more distant Cs^+ ions.

To obtain this answer, imagine a Cs^+ ion at the center of a home cell that has Cl^- ions on its eight corners. These are the 8 nearest-neighbor Cl^- ions. The home cell has 6 faces and 12 edges. The 6 second-nearest neighbors are the Cs^+ ions at the centers of the 6 face-adjoining unit cells. The 12 third-nearest neighbors are the Cs^+ ions at the centers of the 12 edge-adjoining cells.

Tip. Avoid getting bogged down using messy sketches to count neighbors and decide which neighbors are nearer. A better way uses fractional coordinates. Define an origin $(0, 0, 0)$ at the Cs^+ ion at the center of the home cell. In fractional coordinates, the edges of the unit cell are used as units of length. This puts a Cl^- ion at coordinates $(\frac{1}{2}, \frac{1}{2}, \frac{1}{2})$. The cubic symmetry means that the x, y and z coordinates are equivalent and that the plus and minus directions on each coordinate are equivalent, too. Permuting equivalent fractional coordinates and changing the signs of the fractional coordinates therefore generate equivalent locations. For $(\frac{1}{2}, \frac{1}{2}, \frac{1}{2})$ the following eight sets of coordinates result from these operations

$$\begin{pmatrix} +\frac{1}{2}, +\frac{1}{2}, +\frac{1}{2} \\ -\frac{1}{2}, -\frac{1}{2}, -\frac{1}{2} \end{pmatrix} \begin{pmatrix} -\frac{1}{2}, +\frac{1}{2}, +\frac{1}{2} \\ +\frac{1}{2}, -\frac{1}{2}, -\frac{1}{2} \end{pmatrix} \begin{pmatrix} +\frac{1}{2}, -\frac{1}{2}, +\frac{1}{2} \\ -\frac{1}{2}, +\frac{1}{2}, -\frac{1}{2} \end{pmatrix} \begin{pmatrix} +\frac{1}{2}, +\frac{1}{2}, -\frac{1}{2} \\ -\frac{1}{2}, -\frac{1}{2}, +\frac{1}{2} \end{pmatrix}$$

These are the coordinates of the 8 nearest neighbors of the Cs^+ ion. The 6 second-nearest neighbors have these coordinates:

$$(0, 0, 1) \quad (0, 1, 0) \quad (1, 0, 0) \quad (0, 0, -1) \quad (0, -1, 0) \quad (-1, 0, 0)$$

And the 12 third-nearest neighbors have these coordinates:

$$(0, 1, 1) \quad (0, 1, -1) \quad (0, -1, 1) \quad (0, -1, -1) \quad (1, 0, 1) \quad (1, 0, -1)$$
$$(-1, 0, 1) \quad (-1, 0, -1) \quad (1, 1, 0) \quad (1, -1, 0) \quad (-1, 1, 0) \quad (-1, -1, 0)$$

In this description of CsCl, triples that contain any half-integers locate Cl^- ions, and triples that contain all whole numbers locate Cs^+ ions. Fractional coordinates also make it easier to compute interatomic distances in a lattice. For a cubic crystal the distance d between two points (x_1, y_1, z_1) and (x_2, y_2, z_2) specified in fractional coordinates is

$$d = a\sqrt{(x_2 - x_1)^2 + (y_2 - y_1)^2 + (z_2 - z_1)^2}$$

where a equals the edge of the cubic cell.

Defects and Amorphous Solids

21.35   The presence of Frenkel defects will $\boxed{\text{not change}}$ the density of a crystal by a significant amount, because the vacancies at lattice sites are compensated for by interstitial atoms. Frenkel defects in large numbers might cause a small bulging of the crystal and a consequent decrease in its density.

21.37 **a)** A sample of 100 g of this iron(II) oxide contains 76.55 g of Fe and 23.45 g of O. This corresponds to 1.3707 mol of Fe and 1.4657 mol of O. Dividing one by the other gives the formulas $FeO_{1.0693}$ or $\boxed{Fe_{0.9352}O}$. It is improper to round off to the stoichiometric formula FeO. The experimental analysis is precise to four significant figures, and the chemical formula should have the same precision.

b) Let a equal the fraction of sites occupied by Fe^{3+} ions and b equal the fraction of sites occupied by Fe^{2+} ions. The Fe^{3+} ions that occur in Fe^{2+} sites compensate with their extra charge for missing Fe^{2+} ions elsewhere and make the compound as a whole electrically neutral. The average positive charge per site must be 2. Also the sum of a and b is 0.9352, as shown by the empirical formula. In equation form this means

$$3a + 2b = 2 \quad \text{and} \quad a + b = 0.9352$$

Solution of these simultaneous equations gives $a = 0.1296$. This equals the fraction of sites occupied by Fe^{3+} ions. The fraction of the iron in the +3 state is the fraction of sites having +3 iron divided by the fraction having iron of either kind: $0.1296/0.9352 = \boxed{0.1386}$.

A DEEPER LOOK... Lattice Energies of Crystals

21.39 The problem requests computation of the lattice energy of RbCl(s). The electrostatic (Coulomb) lattice energy of a crystal is given by

$$\text{Lattice energy} = \frac{N_A e^2}{4\pi\epsilon_0 R_0} M$$

where M is the Madelung constant, R_0 is the distance between neighboring ions. and e is the charge on the electron (and *not*, in this case, in the edge of a cubic unit cell). Substitute the correct values for RbCl(s). The Madelung constant quoted in the problem is for the rock-salt structure, which is the structure of RbCl(s); R_0 equals the sum of the radii of the Rb^+ and Cl^- ions, which is 3.29 Å, or 3.29×10^{-10} m

$$\text{Lattice energy} = \frac{(6.022 \times 10^{23} \text{ mol}^{-1})(1.602 \times 10^{-19} \text{ C})^2}{4\pi(8.854 \times 10^{-12} \text{ C}^2 \text{ J}^{-1}\text{m}^{-1})(3.29 \times 10^{-10} \text{ m})} 1.7476 = 738 \times 10^3 \text{ J mol}^{-1}$$

Reducing this energy by 10% to account for non-Coulomb effects gives $\boxed{664 \text{ kJ mol}^{-1}}$ for the dissociation energy of RbCl(s). The experimental value is 680 kJ mol^{-1}.

21.41 **a)** The lattice energy of LiF(s) is the change in internal energy associated with total disruption of the crystalline lattice into its component ions

$$\text{LiF}(s) \longrightarrow \text{Li}^+(g) + \text{F}^-(g) \qquad \text{lattice energy} = \Delta U$$

Direct experimental measurements of lattice energies are not possible. The Born-Haber cycle is a series of smaller steps taking place at 25°C that add up to the above change. The ΔU of each step *can* be measured. The steps are

1. Decomposition of LiF(s) to give Li(s) and $F_2(g)$;

2. Vaporization of Li(s) to Li(g) and dissociation of $F_2(g)$ to F(g);

3. Transfer of electrons from Li(g) to F(g) to give $F^-(g)$ ions and $Li^+(g)$ ions.

By the first law of thermodynamics, $\Delta U_{\text{cycle}} = \Delta U_1 + \Delta U_2 + \Delta U_3$.

Evaluate ΔU_3 first. It equals the first ionization energy of Li(g) minus the electron affinity of F(g).[1] Both appear in text Appendix F: $\Delta U_3 = 520.2 - 328.0 = 192.2$ kJ mol^{-1}.

ΔU_2 is the energy change accompanying the vaporization of Li(g) to Li(s) plus the energy change accompanying the dissociation of $F_2(g)$ into atoms. The ΔH_{298}°'s of these reactions equal the

---

[1]Note that ΔU_3 is defined in exactly the same way as the ΔE_∞ that appears in text equation 3.22.

enthalpies of formation of $Li(g)$ and $F(g)$, which appear in text Appendix D as 159.37 kJ mol^{-1} and 78.99 kJ mol^{-1} respectively. ΔH_{298}°'s are not the same as ΔU_{298}°'s, but are related as follows

$$\Delta U_{298}^{\circ} = \Delta H_{298}^{\circ} - RT\Delta n_{\mathrm{g}} = \Delta H_{298}^{\circ} - (2.48 \text{ kJ mol}^{-1})\Delta n_{\mathrm{g}}$$

For the vaporization $Li(s) \longrightarrow Li(g)$, Δn_{g} is +1. It follows that

$$\Delta U_{298}^{\circ} \text{ (vaporization)} = 159.37 - (2.48)(1) = (159.37 - 2.48) \text{ kJ mol}^{-1}$$

For the dissociation $1/2\,F_2(g) \longrightarrow F(g)$, Δn_{g} is +1/2. Hence

$$\Delta U_{298}^{\circ} \text{ (dissociation)} = 78.99 - (2.48)(1/2) = (78.99 - 1.24) \text{ kJ mol}^{-1}$$

Add these two

$$\Delta U_2 \text{ (at 298 K)} = (159.37 - 2.48) + (78.99 - 1.24) = 234.64 \text{ kJ mol}^{-1}$$

The first step in the cycle is the decomposition of $LiF(s)$ to $Li(s)$ and $F_2(g)$

$$LiF(s) \longrightarrow Li(s) + \tfrac{1}{2}F_2(g)$$

This reaction is the "un-formation" of $LiF(s)$. Its ΔH_{298}° is the negative of the standard enthalpy of formation of $LiF(s)$ appearing in Appendix D as -615.97 kJ mol^{-1}. Again, compute ΔU_{298}° from ΔH_{298}° and Δn_{g}

$$\Delta U_{298}^{\circ} = +615.97 - (2.48)(1/2) = 614.73 \text{ kJ mol}^{-1}$$

This is ΔU_3. Finally

$$\Delta U_{\text{cycle}} = \Delta U_3 + \Delta U_2 + \Delta U_1 = 192.2 + 234.64 + 614.73 = \boxed{1041.6 \text{ kJ mol}^{-1}}$$

b) The computation of the Coulomb contribution to the lattice energy of $LiF(s)$ follows the pattern of problem **21.39**:

$$\text{Lattice energy} = \frac{N_A e^2}{4\pi\epsilon_0 R_0}M$$

where M is the Madelung constant and R_0 is the distance between neighboring positive and negative ions. For lithium fluoride R_0 is 2.01 Å, or 2.01×10^{-10} m.[2] Obtain the M for rock salt from text Table 21.5. Then

$$\text{Lattice energy} = \frac{(6.022 \times 10^{23} \text{ mol}^{-1})(1.602 \times 10^{-19} \text{ C})^2}{4\pi(8.854 \times 10^{-12} \text{ C}^2\,\text{J}^{-1}\text{m}^{-1})(2.01 \times 10^{-10} \text{ m})}1.7476 = \boxed{1.21 \times 10^6 \text{ J mol}^{-1}}$$

This (theoretical) Coulomb energy is about 16 percent larger than the (experimental) Born-Haber lattice energy. The discrepancy arises because the Coulomb calculation ignores non-Coulomb interactions.

Tip. In part **a)**, if the $\Delta n_{\mathrm{g}}RT$ corrections in step 2 and 3 are omitted, the answer comes out 1046 kJ mol^{-1}. The difference is less than 1%. In view of the experimental uncertainties in many measurements of enthalpies of reaction taking ΔU to equal ΔH is often defensible. Also, the results used in the Born-Haber cycle were from experiments performed at 298.15 K, but the lattice energy is defined at 0 K.

---

[2]The ionic radii of Li^+ and F^- ions are 0.68 and 1.33 (text Appendix F). The ratio of these radii is 0.51, which confirms that $LiF(s)$ adopts the rock-salt structure.

ADDITIONAL PROBLEMS

21.43   The Bragg law $n\lambda = 2d\sin\theta$ in this case of first-order diffraction of water waves has $n = 1$:

$$1(3.00 \text{ m}) = 2(5.00 \text{ m})\sin\theta$$

Solving for θ gives $17.46°$ so 2θ is $\boxed{35°}$.

21.45   **a)** The unit-cell volume V_{cell} of NaCl is the cell edge cubed or $\boxed{179.43 \text{ Å}^3}$.

b) The volume V_p of the primitive unit cell of NaCl equals one-fourth of the volume of the conventional unit cell or $\boxed{44.856 \text{ Å}^3}$.

c)

$$N_{\text{beams}} = \frac{4}{3}\pi\left(\frac{2}{\lambda}\right)^3 V_p = \frac{4}{3}\pi\left(\frac{2}{2.2896 \text{ Å}}\right)^3 (44.856 \text{ Å}^3) = \boxed{125}$$

d) For this shorter wavelength (0.7093 Å) there are far more diffracted beams:

$$N_{\text{beams}} = \frac{4}{3}\pi\left(\frac{2}{\lambda}\right)^3 V_p = \frac{4}{3}\pi\left(\frac{2}{0.7093 \text{ Å}}\right)^3 (44.856 \text{ Å}^3) = \boxed{4212}$$

21.47   In diamond the C—C bond distance equals the distance between any two nearest-neighbor atoms. Reviewing the list of coordinates given in the problem shows one such pair of atoms is the C at $(0,0,0)$ and the C at $(\frac{1}{4}, \frac{1}{4}, \frac{1}{4})$. This is also clear in text Figure 21.22. Other carbons are equally close to each other but none is closer. These carbons are separated by one-fourth of the body diagonal of the unit cell. The body diagonal is $\sqrt{3}$ times the edge of the cell or $3.57\sqrt{3}$ Å. The bond distance is $1/4$ of this or $\boxed{1.55 \text{ Å}}$.

21.49   **a)** The cell is monoclinic so two of the three cell angles automatically equal $90°$.

b) The volume of the cell is

$$V_c = abc\sqrt{1 - \cos^2\alpha - \cos^2\beta - \cos^2\gamma + 2\cos\alpha\cos\beta\cos\gamma}$$

Because both α and γ are $90°$, this becomes (remembering that $\sin^2\beta + \cos^2\beta = 1$)

$$V_c = abc\sqrt{1 - \cos^2\beta} = abc\sin\beta = (11.04)(10.98)(10.92)\sin 96.73° = 1314.6 \text{ Å}^3$$

The volume equals 1.3146×10^{-21} cm^3. The density is

$$\rho = \frac{n_{\text{(S atoms)}}\mathcal{M}_S}{N_A V_c} = \frac{48(32.066 \text{ g mol}^{-1})}{(6.022 \times 10^{23})(1.3146 \times 10^{-21} \text{ cm}^3)} = \boxed{1.944 \text{ g cm}^{-3}}$$

21.51   Any tetrahedral interstitial site can be viewed as occupying the center of a cube that has every other one of its eight corners occupied by spherical atoms of radius r_1. Let the edge of such a cube have length 1. Then the face diagonal f has length $\sqrt{2}$ and the body diagonal b has length $\sqrt{3}$. The four atoms at the alternate corners surround the center and touch each other along the face diagonals of the cube. Therefore

$$2r_1 = \sqrt{2}$$

Let r_2 be the radius of a spherical atom placed at the interstitial site, the center of the cube. The largest such atom will just touch all four atoms at the corners. The body diagonal in that case equals the sum of the diameters of the two atoms

$$2r_1 + 2r_2 = b = \sqrt{3}$$

Dividing the second equation by the first gives

$$\frac{(r_1 + r_2)}{r_1} = \frac{\sqrt{3}}{\sqrt{2}} \quad \text{hence} \quad 1 + \frac{r_2}{r_1} = 1.225$$

Since r_2/r_1 is 0.225, the largest possible value for r_2 is $\boxed{0.225\,r_1}$.

21.53 Non-metals like sulfur and the halogens (fluorine, chlorine, bromine, iodine) form molecular crystals in the their solid states; metals like the transition metals (iron, nickel, etc.) and the alkali metals (lithium, sodium, potassium, rubidium, cesium) form metallic crystals. Elements at the center of the periodic table (in Group IV) such as carbon and silicon form covalent crystals.

21.55 **a)** According to the equations developed in text Section 21.5 the potential energy and intermolecular distance in a face-centered-cubic molecular crystal depend on the Lennard-Jones parameters for the molecules comprising the crystal as follows

$$R_0 \approx 1.09\sigma \quad \text{and} \quad V_{\text{tot}} \approx -8.61\epsilon N_A$$

where σ and ϵ are the Lennard-Jones parameters, R_0 is the equilibrium spacing (at 0 K), and V_{tot} is the potential energy of the lattice. For N_2, σ is 3.70 Å.[3] Therefore, R_0 is about $\boxed{4.03 \text{ Å}}$. For N_2, ϵN_A is 0.790 kJ mol^{-1}. The potential energy of the lattice is accordingly -6.80 kJ mol^{-1}. This makes $\boxed{+6.80 \text{ kJ mol}^{-1}}$ a reasonable estimate of the lattice energy.

b) The density of a crystal is related to the volume of its unit cell by

$$\rho = \frac{n_c \mathcal{M}}{N_A V_c}$$

For $N_2(s)$, ρ is 1.026 g cm^{-3}. The crystal has four N_2 molecules per unit cell and each molecule has a molar mass of 28.014 g mol^{-1}. Solve the preceding for V_c and substitute:

$$V_c = \frac{n_c \mathcal{M}}{N_A \rho} = \frac{4(28.014 \text{ g mol}^{-1})}{6.022 \times 10^{23} \text{ mol}^{-1}(1.026 \text{ g cm}^{-3})} = 181.36 \times 10^{-24} \text{ cm}^3$$

The edge of the cubic cell is the cube root of the volume of the cell. It equals $\boxed{5.660 \times 10^{-8} \text{ cm}}$ (5.660 Å). In a face-centered cubic cell, a nitrogen molecule lies at the center of every face of the cell and at every corner. The face diagonal is $5.660\sqrt{2}$ Å or 8.005 Å long. One-half of this is the distance from an N_2 at a face center to an N_2 at a face corner. This, the intermolecular distance, is $\boxed{4.002 \text{ Å}}$. This result is only about 0.7 percent less than the distance computed using the Lennard-Jones parameter. The good agreement tends to confirm the analysis in text Section 21.5.

21.57 Sodium chloride is an ionic solid. If there are Schottky defects, a fraction of the Na$^+$ sites is vacant. To maintain electrical neutrality an equal fraction of the Cl$^-$ sites must be vacant. The density of defect-free NaCl is 2.165 g cm^{-3}. Introducing 0.0015 mole fraction of Schottky defects reduces the chemical amount of NaCl per cm^3 to 0.9985 of what it had been. Therefore, the mass of NaCl per cm^3 is 0.9985 of what it had been, or $\boxed{2.162 \text{ g cm}^{-3}}$.

Frenkel defects involve displacement from a regular lattice site to an interstitial site. No mass is removed from the crystal, so the density stays at $\boxed{2.165 \text{ g cm}^{-3}}$ as long as the volume of the crystal is not changed.

21.59 **a)** The binary compound is 28.31 percent O and 71.69 percent Ti by mass. 100 g of the compound contains 1.4977 mol of Ti and 1.7694 mol of O. The formula is $Ti_{0.8464}O$ where the $\boxed{0.8464}$ equals the ratio of 1.4977 to 1.7694.

---

[3]See text Table 9.4.

b) Since only 0.8464 of the stoichiometric quantity of Ti is present, 0.1536 of the Ti sites must be vacant. Let a equal the fraction of Ti sites occupied by a Ti^{3+}, and let b equal the fraction of Ti sites occupied by a Ti^{2+}. Clearly, $a + b = 0.8464$. The net positive charge per oxygen must be +2. Each Ti^{3+} contributes +3 and each Ti^{2+} contributes +2. Electrical neutrality requires $3a + 2b = 2$. Solution of the simultaneous equations gives b equal to $\boxed{0.5392}$ and a equal to $\boxed{0.3072}$. About 31% of the Ti sites contain a Ti^{3+} ion, about 15% are vacant, and about 54% contain a Ti^{2+}.

CUMULATIVE PROBLEMS

21.61 **a)** Use the deBroglie relation to obtain the wavelength of the neutrons

$$\lambda = \frac{h}{m_n v} = \frac{6.626 \times 10^{-34} \text{ J s}}{(1.6750 \times 10^{-27} \text{ kg})(2.639 \times 10^3 \text{ m s}^{-1})} = \boxed{1.499 \times 10^{-10} \text{ m}}$$

b) The edge length of the unit cell is the interplanar spacing of the planes doing the scattering. Compute it by solving the Bragg law for d and substituting

$$a = d = \frac{n\lambda}{2\sin\theta} = \frac{2(1.499 \times 10^{-10} \text{ m})}{2\sin(36.26°/2)} = 4.817 \times 10^{-10} \text{ m} = \boxed{4.817 \text{ Å}}$$

c) Sodium hydride adopts the rock-salt structure. Therefore, Na^+ and H^- ions touch along the edges of the unit cell. The Na^+ ions occupy the corners and center of each face of the cell, forming a pattern like the pattern of five dots on the face of a die. Four H^- ions also lie in each face; they occupy the edges between Na^+ ions. The distance from the center of an Na^+ ion to the center of the adjoining H^- is therefore one-half of the edge of the unit cell. This equals $\boxed{2.409 \text{ Å}}$.

d) As established in slightly different words in the preceding, the edge e of the unit cell is the sum of two Na^+ radii and two H^- radii

$$2r_{H^-} + 2r_{Na^+} = 4.817 \text{ Å}$$

Substitution of 0.98 Å for r_{Na^+} gives r_{H^-} equal to $\boxed{1.43 \text{ Å}}$.[4]

21.63 Applying the rule of thumb assigns each water molecule a volume of 18 $Å^3$. The mass of a water molecule is 18.02 u so the density of water would be 18.02 u/18 $Å^3$. Convert this density to g cm^{-3}

$$\rho = \frac{18.02 \text{ u}}{18 \text{ Å}^3} \times \left(\frac{1 \text{ g}}{6.022 \times 10^{23} \text{ u}}\right)\left(\frac{10^{24} \text{ Å}^3}{1 \text{ cm}^3}\right) = \boxed{1.7 \text{ g cm}^{-3}}$$

The density based on the rule of thumb is much higher than the actual density of solid water (0.90 g cm^{-3}). The rule of thumb fails in this case because the hydrogen bonding in ice maintains an abnormally open structure.

---

[4]Text Appendix F lists 1.46 Å as the radius of the hydride H^- ion.

Chapter 22

Inorganic Materials

Minerals: Naturally Occurring Inorganic Materials

22.1 A Lewis structure for the disilicate ion $Si_2O_7^{6-}$ is

The six O atoms on the perimeter of the structure have three lone pairs and single bonds to an Si atom. All six have formal charges of -1. The O atom linking the Si atoms has f.c. zero. The two Si atoms also have f.c. zero. There are 56 valence electrons in the Lewis structure.

The $P_2O_7^{4-}$ and $S_2O_7^{2-}$ ions are isoelectronic (that is, they also have 56 valence electrons and 34 core electrons). The Lewis structure shown above works for $P_2O_7^{4-}$ by simply replacing the Si's with P's. Both P's then have f.c. $+1$. It also works for $S_2O_7^{2-}$ ion by replacing the Si's with S's. Both S atoms then have f.c. $+2$. The analogous compound of chlorine is Cl_2O_7 (dichlorine heptaoxide). Again, the Si's in the above structure can be replaced, this time with Cl's. The two Cl atoms have f.c. $+3$.

Tip. Numerous additional resonances structures can be drawn if the octet rule is broken for the Si (or P or S or Cl) atoms in these structures. Such structures have one or more double bonds from O atoms to the central atoms. They increase the average bond order and lower the formal charge on the central atoms. See problem **3.91**.

22.3 In each part, determine the Si : O ratio for the network. Ignore O atoms found in other groups, such as OH^- groups. Then use text Table 22.1.
a) Tetrahedra. Ca, $+2$; Fe, $+3$; Si, $+4$; O, -2.
b) Infinite sheets. Na, $+1$; Zr, $+2$; Si, $+4$; O, -2.
c) Pairs of tetrahedra. Ca, $+2$; Zn, $+2$; Si, $+4$; O, -2.
d) Infinite sheets. Mg, $+2$; Si, $+4$; O, -2; H, $+1$.

22.5 The problem is just like problem **22.3** except that Al atoms grouped in the formulas with the Si atoms are counted as Si atoms in determining the Si : O ratio.
a) Infinite network. Li, $+1$; Si, $+4$; Al, $+3$; O, -2.
b) Infinite sheets. K, $+1$; Al, $+3$; Si, $+4$; O, -2; H, $+1$.
c) Closed rings or infinite single chains. Al, $+3$; Mg, $+2$; Si, $+4$; O, -2.

Silicate Ceramics

22.7   Firing steatite (soapstone) drives out water

$$Mg_3Si_4O_{10}(OH)_2(s) \longrightarrow 3\,MgSiO_3(s) + SiO_2(s) + H_2O(g)$$

22.9   The reaction for the preparation of the glass is

$$Na_2CO_3(s) + CaCO_3(s) + 6\,SiO_2(s) \longrightarrow Na_2O \cdot CaO \cdot (SiO_2)_6(s) + 2\,CO_2(g)$$

The molar mass of the glass is 479 g mol^{-1}, which is 0.479 kg mol^{-1}. Hence

$$n_{CO_2} = 2.50 \text{ kg glass} \times \left(\frac{1 \text{ mol glass}}{0.479 \text{ kg glass}} \right) \times \left(\frac{2 \text{ mol } CO_2}{1 \text{ mol glass}} \right) = 10.44 \text{ mol}$$

Use the ideal-gas equation to compute the volume of the CO_2 at 0°C and 1 atm pressure

$$V_{CO_2} = \frac{n_{CO_2} RT}{P} = \frac{(10.44 \text{ mol})(0.08206 \text{ L atm mol}^{-1}K^{-1})(273 \text{ K})}{1 \text{ atm}} = \boxed{234 \text{ L } CO_2}$$

22.11   Assume a sample of exactly 100 g of the soda-lime glass and calculate the chemical amount of each element which is present. This requires use of the molar masses of the several binary oxides. The following table summarizes the results

Mass of Oxide	$\mathcal{M}$(g mol^{-1})	Amount of Non-O / mol	Amount of O / mol
72.4 g SiO_2	60.08	1.205	2.410
18.1 g Na_2O	61.98	0.5841	0.2920
8.10 g CaO	56.08	0.1444	0.1444
1.00 g Al_2O_3	101.96	0.01962	0.02942
0.20 g MgO	40.304	0.004962	0.004962
0.20 g BaO	153.33	0.001304	0.001304

The total chemical amount of oxygen in the sample equals the sum of all the listings for O in the right-most column. It is 2.882 mol. The chemical amounts of the various elements per mole of oxygen are their amounts in the above table divided by 2.882. After rounding to the correct number of significant digits, they are

$$\boxed{0.418 \text{ mol Si}, \ 0.203 \text{ mol Na}, \ 0.050 \text{ mol Ca}, \ 0.0068 \text{ mol Al}, \ 0.002 \text{ mol Mg}, \ 0.0005 \text{ mol Ba}}$$

Tip. The proportions of the oxides in this glass differ from the nominal proportions in soda-lime glass. Note however that the total positive charge (computed by multiplying the number of moles of Si by +4, of Na by +1, of Ca by +2, of Al by +3, of Mg by +2 and of Ba by +2 and adding up the results) equals +2.000. This neatly balances the negative charge on 1.000 mol of oxide ions, which is −2.000.

22.13   The reaction for the production of tricalcium silicate is

$$SiO_2(s, quartz) + 3\,CaO(s) \longrightarrow (CaO)_3 \cdot SiO_2(s)$$

The enthalpy of the reaction is the sum of the enthalpies of formation of the products minus the sum of the enthalpies of formation of the reactants. Text Appendix D gives the ΔH_f°'s for $SiO_2(s, quartz)$ and $CaO(s)$ at 298.15 K. Then

$$\Delta H_{298}^{\circ} = 1 \underbrace{(-2929.2)}_{(CaO)_3SiO_2(s)} - 1 \underbrace{(-910.94)}_{SiO_2(s)} - 3 \underbrace{(-635.09)}_{CaO(s)} = \boxed{-113.0 \text{ kJ}}$$

Nonsilicate Ceramics

22.15   The sum of the oxidations numbers of the atoms in the compound must equal zero. Assign the oxidation numbers -2 to oxygen, $+2$ to Ba, and $+3$ to Y. The copper must then have an oxidation number of $\boxed{7/3}$ to bring the sum to zero.

22.17   **a)** $\boxed{SiO_2(s,\, quartz) + 3\,C(s,\, graphite) \longrightarrow SiC(s) + 2\,CO(g)}$.

b) Find necessary ΔH_f°'s in Appendix D. Then

$$\Delta H_{298}^\circ = 1\underbrace{(-65.3)}_{SiC(s)} + 2\underbrace{(-110.52)}_{CO(g)} - 1\underbrace{(-910.94)}_{SiO_2(s)} - 3\underbrace{(0)}_{C(s)} = \boxed{624.6 \text{ kJ}}$$

c) Silicon carbide should, like diamond, be $\boxed{\text{hard, high melting, a poor conductor of electricity}}$.

22.19   The reaction is $SiC(s) + 2\,O_2(g) \longrightarrow SiO_2(s,\, quartz) + CO_2(g)$. Refer to text Appendix D for the necessary values of ΔG_f° at 298.15 K

$$\Delta G_{298}^\circ = 1\underbrace{(-394.36)}_{CO_2(g)} + 1\underbrace{(-856.67)}_{SiO_2(s)} - 1\underbrace{(-62.8)}_{SiC(s)} - 2\underbrace{(0)}_{O_2(g)} = \boxed{-1188.2 \text{ kJ}}$$

The ΔG_{298}° is less than zero so the reaction is spontaneous at 298.15 K. Thus, $SiC(s)$ is thermodynamically unstable in the presence of 1 atm of oxygen at room temperature. Reducing the partial pressure of oxygen to 0.2 atm (which is its value in air) makes the ΔG_{298} slightly more positive, but the compound is still thermodynamically $\boxed{\text{unstable}}$. Its rate of reaction is however vanishingly slow.

Electrical Conduction in Materials

22.21   The conductivity σ is given in text equation 22.3. The cross-sectional area A of a cylinder is πr^2, and Ohm's law applies ($R = V/I$). Substitute these relationships Then

$$\sigma = \frac{\ell}{RA} = \frac{\ell}{\pi r^2}\frac{I}{V}$$

$$= \frac{(55.0 \times 10^{-3} \text{ m})(150 \times 10^{-3} \text{ A})}{\pi(2.5 \times 10^{-3} \text{ m})^2(17.5 \text{ V})} = 24 \text{ A V}^{-1}\text{ m}^{-1} = \boxed{24\,(\Omega\text{ m})^{-1}} = 24 \text{ S m}^{-1}$$

Tip. Electrical resistance is measured in ohms (Ω). Electrical conductance, the reciprocal of resistance, is measured in siemens (S). The siemens, which is the reciprocal of the ohm, was formerly called the "mho" (ohm spelled backward).

22.23   Compute the number density of the Na^+ ions in the solution from their concentration

$$n_{ion}(Na^+) = 0.10 \text{ mol L}^{-1} \times \left(\frac{10^3 \text{ L}}{\text{m}^3}\right)\left(\frac{6.022 \times 10^{23} \text{ ions}}{\text{mol}}\right) = 6.022 \times 10^{25} \text{ ions m}^{-3}$$

The conductivity σ from the Na^+ ions is the charge density contributed by the ions multiplied by their mobility. The charge density equals the number density multiplied by the charge that each ion carries. The sign of the charge is disregarded, as indicated by the use of absolute magnitude signs in the following

$$\sigma_{Na^+} = |z_{Na^+}|\, e\, n_{Na^+}\, \mu_{Na^+}$$

$$= (1)\left(\frac{1.602 \times 10^{-19} \text{ C}}{\text{ion}}\right)\left(\frac{6.022 \times 10^{25} \text{ ion}}{\text{m}^3}\right)(5.19 \times 10^{-4} \text{ cm}^2 \text{ V}^{-1}\text{ s}^{-1})\left(\frac{10^{-4} \text{ m}^2}{\text{cm}^2}\right)$$

$$= 0.5007 \text{ C s}^{-1}\text{ V}^{-1}\text{ m}^{-1} = 0.5007 \text{ S m}^{-1}$$

The Cl^- ions have the same number density as the Na^+ ions because their concentration in the solution is the same. They contribute to the conductivity in a similar way

$$\sigma_{Cl^-} = |z_{Cl^-}| \, e \, n_{Cl^-} \, \mu_{Cl^-}$$
$$= (1) \left(\frac{1.602 \times 10^{-19} \text{ C}}{\text{ion}} \right) \left(\frac{6.022 \times 10^{25} \text{ ion}}{\text{m}^3} \right) (7.91 \times 10^{-4} \text{ cm}^2 \text{ V}^{-1} \text{ s}^{-1}) \left(\frac{10^{-4} \text{ m}^2}{\text{cm}^2} \right)$$
$$= 0.7631 \text{ C s}^{-1} \text{ V}^{-1} \text{ m}^{-1} = 0.7631 \text{ S m}^{-1}$$

The total conductivity is the sum of the conductivities contributed by the mobile charged species, as shown in text equation 22.8. Therefore

$$\sigma_{tot} = \sigma_{Na^+} + \sigma_{Cl^-} = (0.5007 + 0.7631) \text{ S m}^{-1} = \boxed{1.26 \text{ S m}^{-1}}$$

Tip. See the solution to problem **22.25** for help with the units.

22.25   Use text equation 22.9, which states the Drude model for the conductivity of a metal. Solve it for n_{el}, the number density of the electrons in the metal

$$n_{el} = \frac{\sigma_{el}}{e \, \mu_{el}}$$

Substitute the conductivity of copper from text Table 22.6 for σ_{el} and the room-temperature mobility of electrons in copper, which is given in the problem, for μ_{el}

$$n_{el} = \frac{6.0 \times 10^7 \text{ S m}^{-1}}{(1.602 \times 10^{-19} \text{ C})(3.0 \times 10^{-3} \text{ m}^2 \text{ V}^{-1} \text{ s}^{-1})}$$
$$= (1.25 \times 10^{29}) \frac{\text{S m}^{-1}}{\text{A m}^2 \text{ V}^{-1}} = 1.25 \times 10^{29} \text{ m}^{-3}$$

The cancellation of the units makes use of the fact that an ampere (A) equals a coulomb per second $(C \text{ s}^{-1})$ and that a siemens (S), the SI unit of conductivity, equals an ampere per volt $(A \text{ V}^{-1})$. Next, compute the number density of Cu atoms in metallic copper from the mass density, which is given in the problem as 8.9 g cm^{-3}

$$n_{Cu} = 8.9 \text{ g cm}^{-3} \times \left(\frac{1 \text{ mol}}{63.55 \text{ g}} \right) \left(\frac{6.022 \times 10^{23}}{1 \text{ mol}} \right) \left(\frac{10^6 \text{ cm}^3}{1 \text{ m}^3} \right) = 8.43 \times 10^{28} \text{ m}^{-3}$$

The number density of the mobile electrons divided by the number density of the copper atoms equals the number of mobile electrons per atom

$$\frac{n_{el}}{n_{Cu}} = \frac{1.25 \times 10^{29} \text{ m}^{-3}}{8.43 \times 10^{28} \text{ m}^{-3}} = \boxed{1.5}$$

Band Theory of Conduction

22.27   The band gap is an energy. In this case it equals the energy of light of wavelength 920 nm. Use Planck's relation

$$E_g = h\nu = \frac{hc}{\lambda} = \frac{(6.626 \times 10^{-34} \text{ J s})(2.9979 \times 10^8 \text{ m s}^{-1})}{920 \times 10^{-9} \text{ m}} = \boxed{2.16 \times 10^{-19} \text{ J}}$$

22.29   Substitute in the equation that appears in the problem. The band-gap energy E_g is 8.7×10^{-19} J, which is equivalent to 5.24×10^5 J mol^{-1}. With T at 300 K and R equal to 8.3145 J K^{-1}mol^{-1} the formula becomes

$$n_e = (4.8 \times 10^{15} \text{ cm}^{-3} \text{ K}^{-3/2})(300 \text{ K})^{3/2} \exp \left(\frac{-5.24 \times 10^5 \text{ J mol}^{-1}}{2(8.3145 \text{ J K}^{-1}\text{mol}^{-1})(300 \text{ K})} \right)$$
$$= 6 \times 10^{-27} \text{ cm}^{-3}$$

In a 1.00-cm^3 diamond (pretty big for a diamond) at room temperature, only 6×10^{-27} electrons are excited to the conduction band; there are essentially $\boxed{\text{no electrons}}$ in the conduction band.

Semiconductors

22.31 **a)** Phosphorus-doped silicon is an $\boxed{n\text{-type}}$ semiconductor because substitution of a P (five valence electrons) at an Si (four valence electrons) site populates the conduction band. The carriers of electric current are mobile electrons.

b) Zinc-doped indium antimonide is a $\boxed{p\text{-type}}$ semiconductor. Mobile holes are the charge carriers in this material.

22.33 Use the Planck equation to compute the wavelength that corresponds to a difference in energy of 2.9×10^{-19} J

$$\lambda = \frac{hc}{\Delta E} = \frac{(6.626 \times 10^{-34} \text{ J s})(2.9979 \times 10^8 \text{ m s}^{-1})}{2.9 \times 10^{-19} \text{ J}} = \boxed{6.8 \times 10^{-7} \text{ m}}$$

Light of this wavelength is $\boxed{\text{red}}$.

Pigments and Phosphors: Optical Displays

22.35 At room temperature, zinc white does not absorb in the visible region, although it does absorb ultraviolet light. It appears white. When zinc white is heated, the absorption in the UV is shifted into the blue end of the visible region. Yellow is the color complement of blue, so the absorption of blue light makes the substance appear yellow. The shift into the blue from the UV is a shift to lower frequency and indicates a $\boxed{\text{decrease}}$ in the band gap.

ADDITIONAL PROBLEMS

22.37 Use text Table 22.1. In all of the minerals, Si is in the +4 oxidation state and O is in the −2 oxidation state.

a) Apophyllite contains infinite sheets of silicate units. The oxidation states of the non-Si, non-O elements are F, −1; K, +1; Ca, +2. The water of hydration in the mineral has H in the +1 and O in the −2 oxidation states.

b) Based solely on text Table 22.1 rhodonite contains either closed rings or infinite single chains of SiO_4 units. Either way, the Ca and Mn are both in their +2 oxidation states. Additional information establishes that rhodonite is a pyroxene (containing infinite single chains of silicate units).

c) Margarite is an aluminosilicate. It contains infinite sheets of aluminosilicate groups. The Ca and non-infinite-sheet Al are in the +2 and +3 oxidation states, respectively. The Si, Al, and O in the aluminosilicate framework are in the +4, +3 and −2 states; the hydroxide H and O are in the +1 and −2 states respectively.

22.39 Let x equal the oxidation number of the Fe, and write an equation to express the electrical neutrality of the compound

$$\underbrace{2.36(2)}_{\text{Mg}} + \underbrace{0.48(x)}_{\text{Fe}} + \underbrace{0.16(3)}_{\text{Al}} + \underbrace{2.72(4)}_{\text{Si}} + \underbrace{1.28(3)}_{\text{Al}} + \underbrace{10(-2)}_{\text{O}} + \underbrace{2(-1)}_{\text{OH}} + \underbrace{0.32(2)}_{\text{Mg}} = 0$$

Solving gives $x = 3$; the oxidation state of the iron is $\boxed{3}$.

22.41 For the weathering of anorthite to kaolinite

$$\boxed{CaAl_2Si_2O_8(s) + CO_2(aq) + 2\,H_2O(l) \longrightarrow CaCO_3(s) + Al_2Si_2O_5(OH)_4(s)}$$

Lowering the pH will solubilize the product $CaCO_3$, which dissolves readily in acid (see the following problem). Thus, by LeChâtelier's principle a lower pH $\boxed{\text{increases}}$ the extent of weathering.

22.43   **a)** The dissolution of $CaCO_3(s)$ in water involves the equilibria

$$CaCO_3(s) \rightleftharpoons Ca^{2+}(aq) + CO_3^{2-}(aq) \qquad K_{sp} = 8.7 \times 10^{-9}$$

$$CO_3^{2-}(aq) + H_2O(l) \rightleftharpoons HCO_3^-(aq) + OH^-(aq) \qquad K_{b1} = 2.08 \times 10^{-4}$$

$$HCO_3^-(aq) + H_2O(l) \rightleftharpoons H_2CO_3(aq) + OH^-(aq) \qquad K_{b2} = 2.32 \times 10^{-8}$$

The law of mass action for the two acid-base equilibria gives

$$K_{b1} = \frac{[OH^-][HCO_3^-]}{[CO_3^{2-}]} \quad \text{and} \quad K_{b2} = \frac{[OH^-][H_2CO_3]}{[HCO_3^-]}$$

Note that K_{b1} is K_w divided by the K_{a2} of H_2CO_3 and K_{b2} is K_w divided by K_{a1} of H_2CO_3.[1]

Assume that the temperature is 25°C. Since the solvent has a pH of 7, $[OH^-]$ is 1.0×10^{-7} M. Substitute this value into the two K_b expressions and rearrange

$$[HCO_3^-] = (2080)[CO_3^{2-}] \quad \text{and} \quad [H_2CO_3] = (483)[CO_3^{2-}]$$

Let S equal the solubility of the $CaCO_3(s)$. Then S is equal to $[Ca^{2+}]$ and is also equal to the sum of the concentrations of the three carbon-containing species. This statement is a material-balance condition for the carbonate. Expressed mathematically

$$S = [Ca^{2+}] \quad \text{and} \quad S = [CO_3^{2-}] + [HCO_3^-] + [H_2CO_3]$$

Substituting the independent expressions for the bicarbonate and carbonic acid concentrations into the second equation gives

$$S = [CO_3^{2-}](1 + 2080 + 483) = [CO_3^{2-}](2564)$$

Combine this equation with the K_{sp} expression for $CaCO_3(s)$ to calculate S:

$$8.7 \times 10^{-9} = [Ca^{2+}][CO_3^{2-}] = S\frac{S}{2564} \quad \text{which gives} \quad S = \boxed{0.0047 \text{ mol L}^{-1}}$$

b) Decreasing the pH will $\boxed{\text{increase}}$ the solubility of $CaCO_3(s)$ in water. More carbonate ion is converted to bicarbonate ion or carbonic acid at lower pH.

c) The river's annual flow is 8.8×10^{12} L. This much water would, at equilibrium, dissolve 4.1×10^{10} mol of $CaCO_3$ ($\mathcal{M} = 100$ g mol^{-1}) which is $\boxed{4.1 \times 10^6 \text{ metric tons}}$.

Tip. Do not thoughtlessly assume that $HCO_3^-(aq)$ is the only carbon-containing species present in significant concentration. This assumption corresponds to leaving out the first and third terms in the parentheses in the equation for S and leads to an S of 0.0042 M, which is 10 percent less than the correct answer. A worse error is to ignore the acid-base interaction of the carbonate ion with the water entirely. This corresponds to omitting the second and third terms within the parentheses and gives an S of 9.3×10^{-5} M, about 50 times too low.

22.45   The balanced equation is

$$\boxed{Mg_3Si_4O_{10}(OH)_2(s) + Mg_2SiO_4(s) \longrightarrow 5\,MgSiO_3(s) + H_2O(g)}$$

Trial-and-error balancing gives this answer. Another approach is to note that Mg_2SiO_4 loses O^{2-} ions on a per silicon basis, and $Mg_3Si_4O_{10}(OH)_2$ gains O^{2-} ions on the same basis. This approach

---

[1]These K's and the K_{sp} are from text Tables 15.2 and 16.2 and apply strictly only at 25°C.

puts O^{2-} ion in this reaction in the role played by the electron in the standard method of balancing redox reactions. Arranging the gain of O^{2-} to equal the loss of O^{2-} rapidly gives a balanced equation

$$O^{2-} \text{ loss} \qquad Mg_2SiO_4(s) \longrightarrow MgSiO_3(s) + O^{2-} + Mg^{2+}$$
$$O^{2-} \text{ gain} \qquad O^{2-} + Mg^{2+} + Mg_3Si_4O_{10}(OH)_2(s) \longrightarrow 4\,MgSiO_3(s) + H_2O(g)$$

b) Use LeChâtelier's principle. Increasing the total pressure increases the activity of $H_2O(g)$ and shifts the reaction to the left. The products are $\boxed{\text{disfavored}}$.

c) The slope of the coexistence curve is given by the Clapeyron equation

$$\frac{dP}{dT} = \frac{\Delta H}{T\Delta V} = \frac{\Delta S}{\Delta V}$$

For this reaction ΔV is clearly positive because the products include 1 mol of gas, which has a large volume, but the reactants include no gas. The ΔS of the reaction is positive, for the same reason. The slope of the curve is accordingly $\boxed{\text{positive}}$.

22.47 To impart a red color to the pot, the oxidation state of the iron must be high. The iron in iron oxides will be in a high oxidation state if bound to many oxide anions, so an air-rich (oxygen-rich) atmosphere should be employed. To impart a black color to the pot, the oxidation state of the iron must be low. A smoky fuel-rich atmosphere contains relatively little O_2. In such an atmosphere the iron is not oxidized. $\boxed{\text{For a red pot use an air-rich atmosphere; for a black pot use a fuel-rich atmosphere}}$.

22.49 100.0 g of pure dolomite contains 45.7 g of $MgCO_3$ and 54.3 g of $CaCO_3$. Divide these masses by the molar masses of the two compounds, which are 84.31 g mol^{-1} and 100.08 g mol^{-1}. The chemical amounts of the two compounds equal 0.542 mol. Because the two are present in equal chemical amount the formula is "$(MgCO_3)_1 \cdot (CaCO_3)_1$" which is better written $MgCO_3 \cdot CaCO_3$ or $MgCaC_2O_6$, or $\boxed{MgCa(CO_3)_2}$.

22.51 The equation for the reaction of silicon nitride with hydrofluoric acid is

$$\boxed{Si_3N_4(s) + 12\,HF(aq) \longrightarrow 3\,SiF_4(g) + 4\,NH_3(aq)}$$

In this case, no significant kinetic barrier exists to slow the reaction. Parts fabricated from silicon nitride are rapidly corroded by $HF(aq)$.

22.53 The hybridization of silicon atoms in $Si(s)$ is sp^3, giving rise to a 3-dimensional network of tetrahedral silicon atoms that is just like the diamond structure of carbon. Graphite consists of parallel sheets of hexagonally arrayed σ-bonded carbon atoms. Less directional bonds join the sheets. Graphite is an excellent electrical conductor in a direction parallel to the sheets as a result of extensive electron delocalization in the out-of-plane π system. If silicon were to adopt the graphite structure, one would expect $\boxed{\text{high}}$ electrical conductivity.

22.55 The empty seat will appear to move $\boxed{\text{to the right}}$ at a rate of $\boxed{\text{1 position per five minutes}}$. The analogy with hole motion in p-type semiconductors is this: each seat is a lattice site, the empty seat is the hole, and the people are the electrons.

Chapter 23

Polymeric Materials and Soft Condensed Matter

Polymerization Reactions for Synthetic Polymers

23.1 The balanced equation for the addition polymerization of 1,1,-dichloroethene is

$$n\, Cl_2C=CH_2 \longrightarrow -(CCl_2-CH_2)_{\overline{n}}$$

23.3 Addition polymerization does not split out any small molecules. From the formula of the polymer then, the starting monomer must be $\boxed{\text{formaldehyde}}$ ($H_2C=O$).

23.5 **a)** As glycine (NH_2CH_2COOH) polymerizes to the polypeptide, one molecule of $\boxed{\text{water}}$ is lost ("split out") in the formation of each peptide bond.

b) The repeating structure in the polypeptide is

$$\begin{array}{c} H \quad H \quad O \\ | \quad\;\; | \quad\;\; || \\ -N-C-C- \\ |\quad\quad \\ H \end{array}$$

Applications for Synthetic Polymers

23.7 The repeating unit in the polyamide nylon 66 is $C_{12}H_{22}N_2O_2$. This formula is the sum of the molecular formulas of adipic acid $C_4H_8(COOH)_2$ and hexamethylenediamine $NH_2(CH_2)_6NH_2$ minus twice the formula of water (the diacid and diamine polymerize by condensation with loss of one molecule of water for each acid unit and one for each amine unit added to the chain. The molar mass of the repeating unit is 226.32 g mol^{-1}. Hence

$$n_{\text{monomer unit}} = 1.00 \times 10^3 \text{ kg polymer} \times \left(\frac{1 \text{ kmol of units}}{226.32 \text{ kg polymer}} \right) = 4.419 \text{ kmol}$$

The synthesis requires 4.419 kmol of adipic acid (molar mass 146.1 g mol^{-1}) and 4.419 kmol of hexamethylenediamine (molar mass 116.2 g mol^{-1}). These amounts are $\boxed{646 \text{ kg}}$ of adipic acid and $\boxed{513 \text{ kg}}$ of hexamethylenediamine.

Tip. 1159 kg of reactants give 1000 kg of nylon 66. The other 159 kg is by-product water. 159 kg of water is 8.83 kmol which is 2 × 4.42 kmol, as required by the stoichiometry of the equation.

23.9 Polyethylene, or poly(ethene), is formed by addition polymerization. This means that the mass of the monomer used to make the polymer equals the mass of the polymer that is formed. No mass is

316

split out in the form of water or other by-product, as in condensation polymerization. Use this fact in a train of unit-conversions and then do an ideal-gas law calculation

$$n_{C_2H_4} = 4.37 \times 10^9 \text{ kg LDPE} \times \left(\frac{1 \text{ kg } C_2H_4}{1 \text{ kg LDPE}} \right) \left(\frac{1 \text{ mol } C_2H_4}{0.02805 \text{ kg } C_2H_4} \right) = 1.558 \times 10^{11} \text{ mol } C_2H_4$$

$$V_{C_2H_4} = \frac{n_{C_2H_4}RT}{P} = \frac{(1.558 \times 10^{11} \text{ mol})(0.08206 \text{ L atm mol}^{-1}\text{K}^{-1})(273 \text{ K})}{1.00 \text{ atm}} = \boxed{3.49 \times 10^{12} \text{ L}}$$

Tip. This equals 3.49 km^3 (cubic kilometers!) of gaseous ethene.

Liquid Crystals

23.11 The entropy of the $\boxed{\text{isotropic phase}}$ (the liquid phase) exceeds the entropy in the smectic liquid crystal phase of a substance. Compare the degrees of order apparent in text Figures 23.9a and 23.9c. The enthalpy of the $\boxed{\text{isotropic phase}}$ exceeds that of the smectic phase because heating the smectic phase converts it to the isotropic phase.

23.13 A $\boxed{\text{micelle should form}}$, one containing the hydrocarbon in the interior and water on the outside, with the amphiphile facilitating the separation.

Natural Polymers

23.15 The compound β-D-galactose has $\boxed{\text{five}}$ chiral centers. They are the C atoms numbered 1 though 5 in the structure on the left below. β-D-galactose is an isomer of β-D-glucose. The only difference is an exchange in the positions of the —H and —OH at C-4.

β-D-galactose β-D-glucose

Tip. The preceding structures are Haworth projections. The portion of the ring containing carbons 2 and 3 is supposed to be closer to the viewer, like a tilted platter. They resemble the drawings in text Figure 23.14 except that the puckered six-membered ring is flattened out and rotated a bit. Compare the relative positions of the OH's and H's at each carbon in the Haworth projection of β-D-glucose to their positions in the diagrams in text Figures 23.14 (c) and 23.17 (b).

Tip. The molecules of these two sugars are *not* mirror images of each other. The mirror image of the structure of β-D-galactose at the left would have the locations of the —H and —OH swapped at all five chiral centers. The compound with molecules of that structure is named α-L-galactose.

23.17 A tripeptide is a chain of three monomer units and has distinguishable ends. Any of the three kinds of building block can go in the first position, any of the three can go in the second position, and any of the three can go in the third position. There are accordingly $3^3 = \boxed{27}$ possible tripeptides.

23.19 The pentapeptide is

The pentapeptide has all non-polar side-groups. It should be more soluble in octane than in water.

23.21 The line formula of phenylalanine is $C_6H_5CH_2CH(NH_2)COOH$, equivalent to the empirical formula $C_9H_{11}NO_2$. The polypeptide forms with the removal of an H from the amine end of the molecule and an OH from the carboxylic acid end. Except for the two monomer units at the two extreme ends, which are negligible, each phenylalanine loses one HOH as the polymer forms. The empirical formula of the polymer is therefore C_9H_9NO . The molar mass of a C_9H_9NO unit is 147.2 g mol^{-1}. If the molar mass of the polypeptide is 17500 g mol^{-1} it contains $17500/147.2 =$ 119 monomer units.

ADDITIONAL PROBLEMS

23.23 A catalyst is by definition not consumed in a reaction; it is taken up in one step of a mechanism but regenerated in a subsequent step. In the polymerization of acrylonitrile, the butyl lithium is consumed and the butyl anion is incorporated into the product. Doing this creates a new anion by which chain-building propagates. The butyl anion is not a catalyst.

23.25 Polyvinyl chloride is $-\!(CH_2-CHCl)_{\overline{n}}$. The monomer unit contains 62.50 g mol^{-1}. Polyvinyl chloride is formed by polymerization of ethylene dichloride CH_2ClCH_2Cl ($\mathcal{M} = 98.96$ g mol^{-1}) with the loss of one molecule of HCl per monomer unit added to the chain. The theoretical yield from 950 million pounds of monomer is therefore

$$950 \text{ million lb} \times \frac{62.50}{98.96} = 600 \text{ million lb}$$

If the actual yield of polymer is 500 million pounds, then the percent yield is $500/600\times100\% =$ 83% . If the actual yield gets as high as 550 million pounds, the percent yield is 92% .

23.27 Hair consists of polymeric chains of amino acid cross-linked by —S—S— bridges. To curl hair, treat the hair with a reducing agent, which breaks some of the cross-links. Then arrange the strands of hair in the desired curls, and treat the hair with an oxidizing agent. The disulfide bridges then reform, but in different locations. The new cross-links maintain the curls.

23.29 Two kinds of amino acid can appear at each of the 22 positions in the polypeptide chain. The two ends of the chain are distinguishable. Therefore, there are 2^{22} or about 4.2 million possible isomeric molecules.

23.31 The low-pH form of alanine can donate a hydrogen ion from both its carboxylic acid end

$$^+H_3N\text{---}CH(CH_3)\text{---}COOH(aq) + H_2O(aq) \rightleftharpoons\ ^+H_3N\text{---}CH(CH_3)\text{---}COO^-(aq) + H_3O^+(aq)$$

and from its amino end

$$^+H_3N\text{---}CH(CH_3)\text{---}COOH(aq) + H_2O(aq) \rightleftharpoons\ H_2N\text{---}CH(CH_3)\text{---}COOH(aq) + H_3O^+(aq)$$

Abbreviate the low-pH form of alanine as HA$^+$; abbreviate the product of its first reaction (which has a plus charge on one end and a minus charge on the other, a type of structure called a zwitterion) as Z; abbreviate the product of its second reaction (the ordinary neutral form of alanine) as A. The preceding chemical equations and their mass-action expressions are then

$$HA^+ \rightleftharpoons Z + H_3O^+ \qquad \frac{[H_3O^+][Z]}{[HA^+]} = 10^{-2.3} = 5.0 \times 10^{-3}$$

$$HA^+ \rightleftharpoons A + H_3O^+ \qquad \frac{[H_3O^+][A]}{[HA^+]} = 10^{-9.7} = 2.0 \times 10^{-10}$$

At pH 7, $[H_3O^+]$ equals 1.0×10^{-7} M, and the two mass-action expressions become

$$\frac{(1.0 \times 10^{-7})[Z]}{[HA^+]} = 5.0 \times 10^{-3} \qquad \text{and} \qquad \frac{(1.0 \times 10^{-7})[A]}{[HA^+]} = 2.0 \times 10^{-10}$$

$$\frac{[Z]}{[HA^+]} = 5.0 \times 10^4 \qquad \text{and} \qquad \frac{[A]}{[HA^+]} = 2.0 \times 10^{-3}$$

a) The fraction of the alanine in the Z-form equals the concentration of Z divided by the sum of the concentrations of all three forms

$$f_Z = \frac{[Z]}{[HA^+] + [Z] + [A]} = \frac{5.0 \times 10^4 [HA^+]}{[HA^+] + (5.0 \times 10^4)[HA^+] + (2.0 \times 10^{-3})[HA^+]} = \boxed{0.99998}$$

Essentially $\boxed{\text{all}}$ of the molecules are in the zwitterion-form at pH 7.

b) The fraction of alanine in the A-form is the concentration of that form divided by the sum of the concentrations of all three forms

$$f_A = \frac{[A]}{[HA^+] + [Z] + [A]} = \frac{2.0 \times 10^{-3} [HA^+]}{[HA^+] + (5.0 \times 10^4)[HA^+] + (2.0 \times 10^{-3})[HA^+]} = \boxed{4.0 \times 10^{-8}}$$

23.33   Diesters of phosphoric acid have the general formula $O=P(OH)(OR)_2$. The H in this formula is an acidic hydrogen. The compounds are therefore acids as well is esters.

23.35   Set up a series of unit-factors to compute the length of the DNA molecule

$$l = 2.8 \times 10^9 \text{ g mol}^{-1} \times \left(\frac{1 \text{ base pair}}{650 \text{ g mol}^{-1}} \right) \left(\frac{3.4 \text{ Å}}{\text{base pair}} \right) \left(\frac{10^{-7} \text{ mm}}{\text{Å}} \right) = \boxed{1.5 \text{ mm}}$$

The number of base pairs in this DNA molecule is

$$N = 2.8 \times 10^9 \text{ g mol}^{-1} \times \left(\frac{1 \text{ base pair}}{650 \text{ g mol}^{-1}} \right) = \boxed{4.3 \times 10^6 \text{ base pair}}$$

Appendix A

Scientific Notation

A.1 The trailing zeros in parts **d)** and **e)** must not be omitted when the number is put into scientific notation.
a) 5.82×10^{-5} **b)** 4.02×10^2 **c)** 7.93 **d)** -6.59300×10^3 **e)** 2.530×10^{-3} **f)** 1.47

A.3 **a)** 0.000537 **b)** $9,390,000$ **c)** -0.00247 **d)** 0.006020 **e)** $20,000$

A.5 The number is 746 million kilograms or 746,000,000 kg.

Experimental Error

A.7 **a)** Statistical methods for deciding when to omit an outlier are not developed in the Appendix. Instead the appeal is to use good judgment. The value 135.6 g is grossly out of line with the others.

b) The mean is 111.34 g.

c) The standard deviation is 0.22 g and the 95% confidence limit is

$$\text{confidence limit} = \pm \frac{t\sigma}{\sqrt{N}} = \pm \frac{2.57(0.22 \text{ g})}{\sqrt{6}} = \pm 0.23 \text{ g}$$

where t comes from text Table A.2.

A.9 The measurement of mass in problem A.7 is more precise.

Significant Figures

A.11 **a)** five **b)** three **c)** ambiguous (two or three significant figures) **d)** three **e)** four.

A.13 **a)** 14 L **b)** $-0.0034°$C **c)** 3.4×10^2 lb **d)** 3.4×10^2 miles **e)** 6.2×10^{-27} J

A.15 $2\,997\,215.55$

A.17 **a)** -167.25 **b)** 76 **c)** 3.1693×10^{15} **d)** -7.59×10^{-25}

A.19 **a)** -8.40 **b)** 0.147 **c)** 3.24×10^{-12} **d)** 4.5×10^{13}

A.21 The area of the triangle is 337 cm^2. Three significant figures appear in the answer reflecting the three significant figures in "16.0 cm", a measured quantity. The "1/2" in the formula has an infinite number of significant figures.

Appendix B

SI Units and Unit Conversion

B.1 **a)** 6.52×10^{-11} kg **b)** 8.8×10^{-11} s **c)** 5.4×10^{12} kg m^2 s^{-3} **d)** 1.7×10^4 kg m^2 s^{-3} A^{-1}
Note that the unit kilogram has a prefix yet nevertheless is a base unit.

B.3 **a)** $4983°$C, but it is very hard to measure such a high temperature to $\pm 1°$C.
b) $37.0°$C **c)** $111°$C **d)** $-40°$C.

B.5 **a)** 5256 K **b)** 310.2 K **c)** 384 K **d)** 233 K.

B.7 **a)** 24.6 m s^{-1} **b)** 1.15×10^3 kg m^{-3} **c)** 1.6×10^{-29} A s m
d) 1.5×10^2 mol m^{-3} **e)** 6.7 kg m^2 s^{-3} = 6.7 W.

B.9 One kWh is equal to 3.6×10^6 J. Hence, 15.3 kWh is 5.51×10^7 J.

B.11 The engine displacement is 6620 cm^3 or 6.62 L.

The Concept of Energy: Forms, Measurements, and Conservation

B.13 A mile is 1609.344 m, and an hour is 3600 s. 100 mph is 44.70 m s^{-1}. Use the definition of kinetic energy

$$\mathcal{T} = \tfrac{1}{2}mv^2 = \tfrac{1}{2}(0.270 \text{ kg})(44.70 \text{ m s}^{-1})^2 = 2.7 \times 10^2 \text{ J}$$

The answer could also be rounded off to 3×10^2 J significant figure depending on how the phrase "near to 100 mph" is interpreted.

B.15 The 98 mph tennis ball is moving at 43.81 m s^{-1}. The kinetic energy of the tennis ball is

$$\mathcal{T} = \tfrac{1}{2}mv^2 = \tfrac{1}{2}(2 \text{ oz})(0.0284 \text{ kg oz}^{-1}(43.81 \text{ m s}^{-1})^2 = 55 \text{ J}$$

Appendix C

Using Graphs

C.1 The slope is 50 mi h^{-1}.

C.3 **a)** The equation is in the required form with m (slope) equal to 4 and b (y intercept) equal to -7.
b) The equation is $y = 7/2x - 5/2$. The slope is $\frac{7}{2}$, and the y intercept is $-\frac{5}{2}$.
c) The equation is $y = -2x + \frac{4}{3}$. The slope is -2 and the y intercept is $\frac{4}{3} = 1.333\ldots$.

C.5 The graph of y versus x for the equation $y = 2x^3 - 3x^2 + 6x - 5$ is not linear. The value of y rises from -45 at $x = -2$ and $+11$ at $x = +2$. The graph cuts the x axis at $x = 1$.

Solution of Algebraic Equations

C.7 **a)** $x = -5/7$ **b)** $x = 3/4$ **c)** $x = 2/3$.

C.9 The answers are given to 4 significant figures. **a)** $x = 0.5447, -2.295$ **b)** $x = -0.6340, -2.366$ **c)** $x = +0.6340, +2.366$.

C.11 **a)** Assuming that x is small compared to 2.00 gives $x = 6.5 \times 10^{-7}$. There are also two complex roots.
b) There are three roots because this is a third-degree equation: $x = 4.07 \times 10^{-2}, 0.399, -1.011$.
c) The only real root is $x = -1.3732$. It can be arrived at graphically or using a scientific calculator. The other two roots are complex numbers and are of little interest in chemical applications.

Powers and Logarithms

C.13 **a)** 4.551 (the three significant figures appear in the mantissa)
b) To help understand the use of significant figures, divide the exponent in the number by 2.302585093 to re-express it as a power of 10: $10^{-6.814}$. The "6" plays the role of the characteristic when the antilog is taken and the mantissa has three significant figures. Hence the answer has three significant figures: 1.53×10^{-7}. **c)** 2.6×10^8 **d)** -48.7264

C.15 Take $10^{0.4793}$ using a calculator. The answer is 3.015.

C.17 Some calculators do not accommodate a number with an exponent exceeding 99 in absolute value. To answer this problem using such a calculator, write

$$\log 3.00 \times 10^{121} = \log(3.00) + \log 10^{121} = 121 + 0.477 = 121.477$$

C.19 Simply change the characteristic from 0 to 7 or from 0 to -3 and add it to the same mantissa: $7 + 0.751 = 7.751$ and $-3 + 0.751 = -2.249$.

C.21 The problem is to find x in the equation $\log \ln x = -x$. One way to proceed is to guess an x, put it into the left side of the equation and see on a calculator if the indicated operations gives back the guess. Adjust the guess and repeat until satisfied. The answer is 1.086.

Slopes of Curves and Derivatives

C.23 **a)** $8x$ **b)** $3\cos 3x - 8\sin 2x$ **c)** 3 **d)** $1/x$.

Areas under Curves and Integrals

C.25 **a)** 20 **b)** 78125/7 **c)** 0.0675.